Advances in Coal Science and Technology

Advances in Coal Science and Technology

Edited by Ezekiel Peterson

STATES
ACADEMIC PRESS
www.statesacademicpress.com

Published by States Academic Press,
109 South 5th Street,
Brooklyn, NY 11249, USA
www.statesacademicpress.com

Advances in Coal Science and Technology
Edited by Ezekiel Peterson

International Standard Book Number: 978-1-63989-021-7 (Hardback)

Cataloging-in-Publication Data

Advances in coal science and technology / edited by Ezekiel Peterson.
p. cm.
Includes bibliographical references and index.
ISBN 978-1-63989-021-7
1. Coal. 2. Coal mines and mining. 3. Coal mines and mining--Technological innovations.
I. Peterson, Ezekiel.
TN802 .A28 2022
553.24--dc23

TABLE OF CONTENTS

PREFACE

Coal is a black combustible sedimentary rock. It is formed when plants decay into peat. It transforms into coal due to pressure and heat of the deep burial over millions of years. It is a primary energy source which is used to generate a significant share of world's electricity. It is also used in other industries such as iron and steel. The extraction of coal from ground is known as coal mining. Some of the techniques used for extraction of coal are contour mining, surface mining, underground mining, mountain top removal mining and strip mining. Burning of coal has resulted in some negative health and environmental impacts. New technologies are being developed to mitigate these effects. This book explores all the important aspects of coal science and technology in the present day scenario. It presents researches and studies performed by experts across the globe. This book is appropriate for students seeking detailed information in this area as well as for experts.

Various studies have approached the subject by analyzing it with a single perspective, but the present book provides diverse methodologies and techniques to address this field. This book contains theories and applications needed for understanding the subject from different perspectives. The aim is to keep the readers informed about the progresses in the field; therefore, the contributions were carefully examined to compile novel researches by specialists from across the globe.

Indeed, the job of the editor is the most crucial and challenging in compiling all chapters into a single book. In the end, I would extend my sincere thanks to the chapter authors for their profound work. I am also thankful for the support provided by my family and colleagues during the compilation of this book.

Editor

1

Management of coal processing wastes: studies on an alternate technology for control of sulfate and chloride discharge

Paul T. Behum[1] · Yoginder P. Chugh[2,3] · Liliana Lefticariu[4]

Abstract Management of coal mining and coal processing wastes, particularly of high sulfur coals, can generate excessive amounts of sulfate (SO_4^{2-}) and chloride (Cl^-) in mine drainage that are known to negatively impact quality of both surface and ground water. The U.S. Environmental Protection Agency provides guidance to states on allowable SO_4^{2-} and Cl^- discharges from mine sites. This research evaluates the hypothesis that co-disposal of CCPW and FCPW with appropriate compaction can result in improved geochemical and geotechnical environments that will minimize acid mine drainage formation and SO_4^{2-} and Cl^- discharges. Addition of ground limestone (ag-lime) to the mix was also evaluated as a drying agent and for improvement in overall geochemistry by buffering higher pH values within the coal waste. These objectives were to develop and implement innovative concepts for engineered co-management of CCPW and FCPW at coal mining sites. The authors performed long-term field column leaching studies to analyze improvement in SO_4^{2-} and Cl^- in water quality. Requirements for stricter standards in some states led to the need for development of potentially improved environmental practices. This paper presents the overall encouraging results of the field kinetic studies.

Keywords Kinetic testing · Acid mine drainage · Coal chemistry · Interior Basin · Refuse co-disposal

1 Introduction

A typical underground coal mining complex in the mid-continental USA consists of a combination of mine shafts and slopes to access the coal seam/s along with coal preparation and refuse disposal facilities located on the surface as shown in Fig. 1. The term "coal refuse" in this paper refers to reject rock from crushers prior to run-of-mine (ROM) coal entering the processing plant and other waste rock derived from the coal cleaning processes. The reject rock from the crushers is typically a small portion of the coal refuse. The ROM coal is generally wet processed to remove mineral matter and sulfur and to improve the quality of coal shipped to power plants. The marketable coal recovery typically varies between 50% and 60% with the remaining fraction disposed as waste on-site. Coal refuse typically consists of two size fractions: (1) coarse coal processing waste (CCPW), which is generally larger than 150 microns (100 mesh) in size and in some cases, is larger than 3.2 mm (1/8-inch) in size; and (2) fine coal processing waste (FCPW), which is generally a slurry (± 15% solids content) with solids being less than 150 microns (100 mesh) in size. CCPW refers to reject from heavy media vessels, jigs, cyclones, and spirals, while FCPW refers to reject from flotation columns and cells and effluent from filter presses, screen-bowl centrifuges, and de-sliming cyclones, all of which is typically concentrated in a thickener. The FCPW constitutes about 5%–10% of the ROM coal processed from the Interior Basin coal mines in the USA.

✉ Yoginder P. Chugh
ypchugh1940@gmail.com

[1] Office of Surface Mining and Reclamation Enforcement, Alton, IL 62002, USA

[2] National Academy of Inventors, Tampa, FL 33612, USA

[3] Department of Mining and Mineral Engineering, Southern Illinois University Carbondale, Carbondale, IL 62901, USA

[4] Department of Geology, Southern Illinois University Carbondale, Carbondale, IL 62901, USA

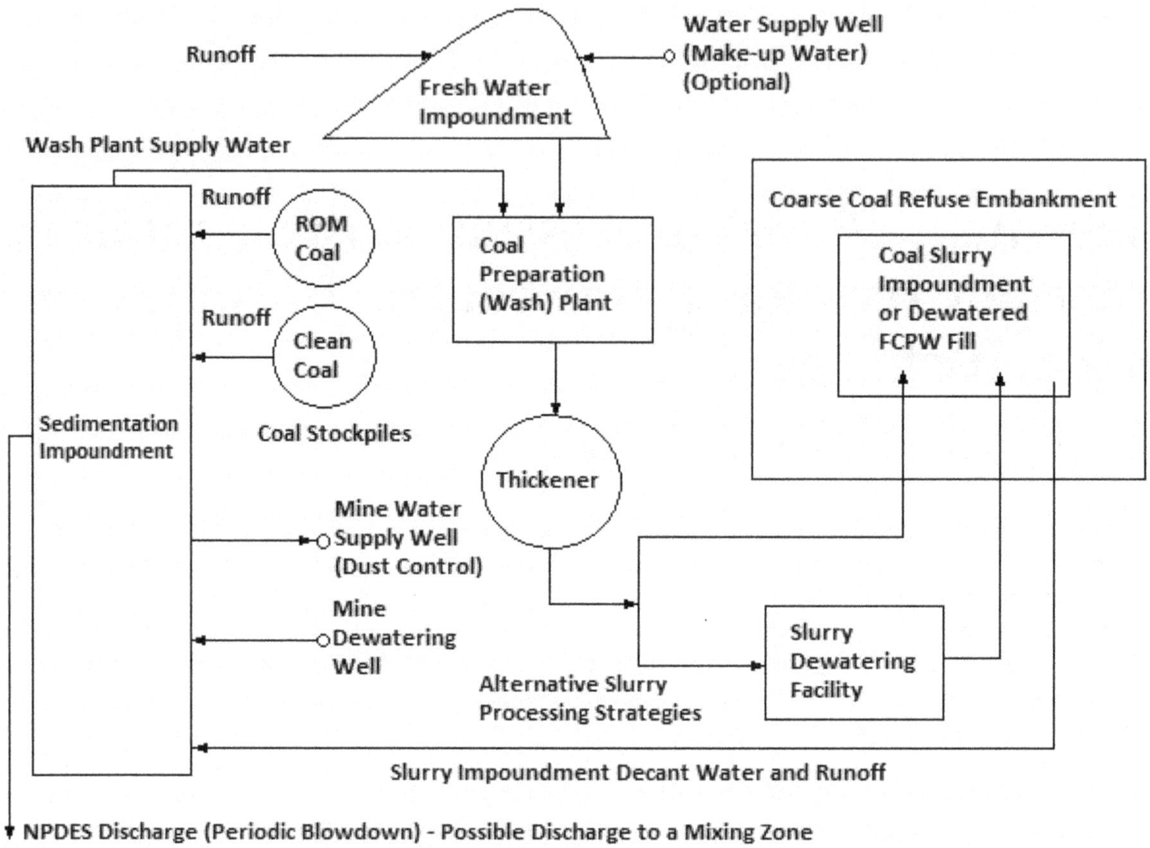

Fig. 1 Schematic of coal processing and water handling operations on the surface at a typical underground coal mine complex

FCPW is typically disposed as: (1) a slurry contained within an embankment constructed of CCPW (conventional practice) or incised ponds, (2) dewatered FCPW (50%–70% solids), contained within CCPW embankments as illustrated in Fig. 1, or potentially (3) a mono-fill constructed of a blend of CCPW and dewatered FCPW. The sizes of coal refuse disposal areas have been increasing in the Interior Basin due to the recent trends of mining thinner coal seams with larger equipment capable of cutting roof and floor rock. This has increased the percentage of the out-of-seam dilution (OSD) material to 50% or more of the ROM coal. Figure 1 also shows that underground coal mine complexes within the region typically operate a closed loop water-handling system where onsite water resources are continuously recycled and any regulated surface water discharges are from sedimentation basins, but typically only during larger precipitation events when either less stringent regulations is applied or there is a considerable amount of dilution water available.

2 Problem statement

Coal waste disposal facilities have been identified as a principal source of elevated SO_4^{2-} discharges by both state and federal environmental regulatory agencies (The Advent Group 2005; Chugh et al. 2007a, b). Infiltration of oxygen- and ferric iron-bearing water into coal waste stockpiles contributes to the oxidation of pyrite and other sulfide minerals with the subsequent release of SO_4^{2-} and metals into mine drainage (Moses et al. 1987; Moses and Herman 1991; Evangelou 1995). Development of good management practices (GMPs) for refuse disposal and water management to reduce pyrite oxidation and SO_4^{2-} and chloride levels in these discharges are necessary to meet compliance levels for regulated constituents in mine drainage. Such practices should: (1) reduce water treatment costs during mining and subsequent reclamation of the facility, (2) reduce liability associated with maintaining and abandoning a large permanent impoundment structure, and (3) allow release of reclamation bonds in a timely fashion.

Weathering of pyrite can be controlled if water rich in oxygen and ferric ions is prevented from contacting sulfide minerals inherent in the waste rock (Evangelou 1995; Nordstrom 1982). Dissolved oxygen concentration, temperature, pH of water, reactive surface area of pyrites, form of pyrite, catalytic agents, flushing frequencies, and time allowed for contact are some of the variables that control the rate of pyrite oxidation and the associated SO_4^{2-} discharge. The inherent pyrite content within the coal and the associated strata is an important factor in determining the amount of sulfate and metals in mine drainage. Although abiotic chemical conversion of Fe (II) to Fe (III) is an extremely slow reaction, the presence of iron-oxidizing bacteria increases reaction rate by an order of magnitude and, as a result, the production of SO_4^{2-} during the weathering process (Singer and Stumm 1970; Kleinmann and Crerar 1979). Within the Interior Basin, several researchers have provided noteworthy contributions to the control of acid mine drainage (AMD), including Nawrot and Gray 2000; Naumann and Wiram 1995; Wiram 1984, which has emphasized the advantages of adding alkalinity producing materials during coal mine reclamation. In China, the largest coal production country in the world, coal waste management and AMD are also very important. More than 350 mt of coal waste is produced each year, and is disposed as coal waste piles. Since most coal waste containing high sulfur content is are located in arid or semi-arid regions, AMD issue has not drawn much attention to date. The focus of research has been on the spontaneous combustion because of the subsequent potential for oxidation of pyrite in coal wastes. The spontaneous combustion of coal waste piles has resulted in serious environmental problems, with several reclaimed coal waste piles re-igniting after establishing vegetation. Chinese scholars have developed new techniques for dealing with these problems (Hu et al. 2009a, b). The key new techniques are monitoring and mapping of the surface temperature by infrared instrument and photogrammetry, putting out fire by deep grouting, controlling oxidation process and acid generation by chemical and biological treatments and soil cover system with compaction (Hu et al. 2008, 2009b; Chen et al. 2014; Xu et al. 2014). However, more attention should be paid to AMD treatment, when spontaneous combustion is controlled.

3 Research hypothesis and overall research

This research has evaluated the hypothesis that co-disposal of CCPW and FCPW with or without addition of alkaline materials, and with appropriate compaction can improve geochemical and geotechnical parameters of coal waste that will minimize AMD formation and SO_4^{2-} discharges. This hypothesis was developed based on data available in the literature and also collected as part of a previous study (Chugh et al. 2007a) that concluded that the different-size CCPW fractions have a higher acid-forming potential (potential acidity or PA) and lower neutralization potential (NP) as compared to FCPW. In current practice, CCPW is typically disposed separately as a structural fill in surface disposal areas, while FCPW is disposed as a dilute slurry (about 15% solids) in impoundments that provide an overall lower disposal cost (NRC 2002). Previous research also indicates that, in coal, the minerals which provide NP (e.g. calcite) weather faster than minerals that are acid-producing such as pyrite (e.g. Nawrot and Gray 2000; Hossner and Doolittle 2003). Therefore, drainage from coal waste, which is initially alkaline, remains alkaline for a certain period of time and then tends to become acidic. This period can be extended by addition of alkaline materials which is a widely accepted mine reclamation practice (Dreher et al. 1994; Nawrot and Gray 2000; Stewart et al. 2001). Chugh et al (2007a, b) and others have suggested that differential AP and NP rates could be used advantageously in waste management operations to allow for improved placement and compaction practices, and reduction of SO_4 discharges. There is a potential improvement in the overall NP and a reduction in the PA in a blend of CCPW and FCPW because the FCPW has a net neutralization potential (NNP; NNP = NP − PA) that in some cases is near neutral or, at least, a level slightly above (−) 5 tons of $CaCO_3$ equivalent per 1000 tons of material. Most permitting agencies consider a NNP less than (−) 5 tons of $CaCO_3$ equivalent per 1000 tons of material (in this case coal waste) potentially acid-forming and requiring management practices to control problematic mine drainage. Co-disposal of CCPW and FCPW offers an interesting alternative concept and is discussed in the next section.

Most Illinois mine operators would consider co-disposal if FCPW could be dewatered inexpensively to about 60%–65% solids and the long-term stability of a co-disposal fills can be ensured. As it is a relatively few number of U. S. mining operations currently practice refuse co-disposal. However, there remains a need to minimize disposal area space requirements and to eliminate the need for slurry disposal areas, which are becoming increasingly difficult and more expensive to permit. None of the operators interviewed in the preceding studies (Chugh et al. 2007a, b), have considered the potential improvements in geochemical and geotechnical environments as the result of co-disposal. A judicious addition of limestone or other alkaline materials to the co-disposal blend can provide additional advantages by buffering drainage pH, reducing high levels of total dissolved solids, and minimizing SO_4^{2-} discharge (Nawrot and Gray 2000). Both hypotheses presented above were evaluated in a field setting by constructing six large-size leaching columns—two each of 100% CCPW (the control sample); 90% (by volume) CCPW and 10% FCPW; and 84% CCPW, 8% FCPW, and 8% limestone. Material in each column was compacted to approximately 50% of the maximum Proctor density. All columns were equipped with a lysimetric and gravity drainage ports for collecting water samples.

This research was motivated by the Illinois Environmental Protection Agency (IEPA) policy to implement lower standards for SO_4^{2-} and Cl^- discharge from coal mining and processing operations. Historically, Illinois allowed compliance level was 3500 mg/L for SO_4^{2-} discharges and 1000 mg/L for Cl^- discharges, whereas a "General Use" standard applicable to other industries had been set at 500 mg/L for both of these anions (Chugh et al. 2007a, b). The IEPA and the USEPA agreed that this difference was not legally defensible and implemented more restrictive SO_4^{2-} levels (USEPA 2009). These IEPA regulations also established a SO_4^{2-} discharge levels in a range of 500–2000 mg/L based on the hardness and Cl^- levels in the receiving stream; Cl^- discharge levels were also set at 500 mg/L.

3.1 Overall research

The overall goal of this study was to investigate the rate of SO_4^{2-} and Cl^- leaching from simulated kinetic tests in the field to demonstrate commercial potential for co-disposal technology. Co-disposal should enhance geotechnical stability through improved cohesion and compaction, while improving the geochemical environment. As a result, the overall refuse facility liabilities should be lowered. Addition of ground limestone (ag-lime) was also evaluated for use as a drying agent and as measure to improve in overall geochemistry by buffering higher pH values (< 8.0) inherent in fresh refuse from the basin. The study would aid mine operators in developing and implementing innovative concepts for engineered waste disposal. Although field testing was conducted at an Illinois mining operation, the concepts involved should be applicable throughout the Interior Basin and, with minor modifications, at other coal mines in the USA.

4 Research studies

These included: (1) physical and chemical characterization of CCPW, FCPW, and limestone materials; (2) characterization of geotechnical properties for identified waste disposal options; and (3) field-scale kinetic tests and data analysis.

4.1 Physical properties of coal waste

Both FCPW and CCPW were collected from a large coal preparation plant cleaning ROM coal from two commonly mined seams: Springfield (No. 5) and Herrin (No. 6). Most of this coal was produced from two underground mines. An approximately 18.9-liter (L) (5-gallon) sample of CCPW was collected and dried at low temperature (< 38 °C or < 100°F), and then screened for particle size distribution. The sampled materials were crushed to less than 10 mm (3/8-inch) in size. Proctor compaction tests were conducted in 15 cm (6-inch) diameter molds to develop moisture-density data for optimizing compaction. The mold samples were also tested for uniaxial compressive strength for an undrained condition (Chugh et al. 2007a, b).

A similar sized sample of FCPW was also collected from the thickener underflow discharge. FCPW was decanted and air-dried before drying in a low temperature furnace. A 3.79-L (1-gallon) sample of agricultural ground limestone was also dried at low temperature and analyzed for particle size distribution.

4.1.1 Results

Table 1 shows average particle size distributions for CCPW and limestone samples. Since FCPW is uniformly less than 200 mesh (0.074 mm or 0.0029 inches), additional particle size analyses were not conducted on this fraction. A majority of CCPW (64.52%) is in the 4.75 to 50.8 mm (0.187–2.0 inch) size range. Agricultural ground limestone was a finer-grained material with a majority (72.97%) in the 1.70–19.05 mm (0.066–0.75 inch) size range. The initial moisture content of CCPW was 11.5%, whereas dewatered FCPW was 31.9% (68.1% solids content). In contrast, the initial moisture content of agricultural limestone was only 0.1%. Therefore, by blending limestone the moisture content could be significantly reduced.

4.2 Chemical properties of coal waste

The geochemical evaluation of CCPW, FCPW, and agricultural limestone included: (1) acid-base accounting, (2) determination of major and key trace elements in high temperature ash (HTA) product, and (3) x-ray fluorescence (XRF) analysis of all materials. Following SIUC laboratory determination of paste pH, the U.S. Geological Survey Energy Resources Program (USGS-ERP), the State of Illinois Office of Mines and Minerals (ILDNR-OMM), and

Table 1 Particle size distribution for CCPW and ground limestone

Sieve opening (mm)	Percent finer	
	CCPW	Ground limestone
50.8	94.31	100
19.05	62.20	100
4.75	29.79	78.09
1.7	15.09	27.03
0.425	4.44	13.17
0.075	0.53	4.58
< 0.075	0.52	3.58

a commercial testing laboratory determined the total and pyritic sulfur content along with an elemental analysis of all materials. Sulfur values were used to calculate potential acidity (PA) for CCPW and FCPW with respective values of 120.37 and 63.91 tons $CaCO_3$ per ton of coal waste.

Neutralization potential (NP) was also determined for all materials to determine the net neutralization potential (NNP). One CCPW and one FCPW coal waste sample was subjected to HTA analysis following heating in a 750 °C air-recirculating chamber furnace using ICP/MS and ICP-OES. Limestone elemental analysis was conducted using a Varian 725 ICP-OES for major elements (except SO_3, which was determined using a LECO sulfur analyzer) and presented in oxide form. A hand-held XRF analyzer was used to determine a large suite of elemental concentrations in whole dried and ground CCPW, FCPW and limestone samples.

4.2.1 Results

Acid-base accounting analyses results are presented in Table 2. The "mine permit" application provided representative data, while the remaining data was compiled from analyses performed in this study. Chugh et al. (2007a) presented additional acid-base accounting data for CCPW and FCPW samples collected from another similar preparation plant and found that: (1) the FCPW represented 10%–15% of the coal waste, and (2) the sulfur content is typically lower for FCPW as compared to CCPW. Elemental data and chemical analyses of HTA are presented in Tables 3 and 4. The second row of limestone data in Table 3 was provided by the quarry operator.

4.3 Geotechnical characteristics of different coal waste disposal options

Three waste disposal options were considered: (1) CCPW disposal in embankments or fills (the control)—Disposal Practice 1 or DP 1; (2) Co-disposal of blended CCPW (90%) and FCPW (10%)—Disposal Practice 2 or DP 2; and (3) Co-disposal of blended CCPW (84%) and FCPW (8%) with limestone addition (8%)—Disposal Practice 3 or DP 3. Moisture-density relationships were developed using standard proctor tests for all disposal options. All tests were performed in a 15-cm (6-inch) mold using ASTM D698-12e1 Method C; ASTM (2013) prior to column construction (Behum et al. 2010). The mold size limited the top size of the material to about 20 mm (¾-inch).

4.3.1 Results

Particle size distribution for the materials used in Proctor tests are summarized in Table 5. These area based on three tests for DP 1 materials, two for DP 2 materials and two for DP 3 materials. The data is based on percent retained rather than percent passing.

The Proctor test results for all disposal options and tests are summarized in Fig. 2. For DP 1, the testing moisture contents varied between 2% and 17%. The field columns had maximum dry unit weight = 18.5 kN/m^3 or 118 pcf. Similar data for DP 2 and DP 3 Proctor tests were 2%–9.5%, and 18.0 kN/m^3 or 115 pcf; 2.5%–11.5% and 18.52 kN/m^3 or 118 pcf. Moisture contents for maximum Proctor density values for DP 1, DP 2, and DP 3 mixes were 6.0%, 5.5%, and 7.5%, respectively. The addition of ground limestone advantageously increased the moisture content at which maximum density was achieved by about 2%.

4.4 Field kinetic testing

Field-scale kinetic tests were conducted at an active coal mining complex. During the first phase of field-scale testing, three sets of duplicate test columns (FC-1 through FC-6) were constructed using 200-L (55-gallon) barrels. These columns were monitored for 19 months. Unfortunately, a tornado damaged some columns. A new set of field-scale kinetic-leaching columns were assembled from relatively

Table 2 Geochemical properties of coal waste and ground limestone samples

Coal waste fraction	Mean sulfur content (%)		Median paste pH	Tons $CaCO_3$ per 1000 Tons of waste		Net Neutralization potential (NNP)
	Total	Pyritic		Maximum potential acidity (PA)	Neutralization potential (NP)	
Mine permit (coarse)	5.70 (n = 2)	3.41 (n = 47)	7.12 (n = 47)	106.4 (n = 47)	23.8 (n = 47)	− 84.5 (n = 47)
Coarse	4.84 (n = 2)	3.90 (n = 1)	5.92 (n = 1)	121.88 (n = 1)	1.51 (n = 1)	− 120.37 (n = 1)
Fine	2.56 (n = 2)	2.13 (n = 1)	6.31 (n = 1)	66.53 (n = 1)	2.65 (n = 1)	− 63.91 (n = 1)
Limestone	NT	0.17 (n = 1)	8.35 (n = 1)	5.31 (n = 1)	58.17 (n = 1)	52.86 (n = 1)

n no. of samples, *NT* not tested

Table 3 Comparison of major elements in coal waste HTA and ground limestone

Sample ID	Al_2O_3 (%)	CaO (%)	Fe_2O_3 (%)	K_2O (%)	MgO (%)	Na_2O (%)	P_2O_5 (%)	SiO_2 (%)	SO_3 (%)	TiO_2 (%)
Coarse	17.9	3.24	10.7	3.59	1.40	0.817	0.249	59.3	2.33	0.699
Fine	18.2	4.30	8.12	3.50	1.35	0.732	0.135	58.0	2.39	0.721
Limestone	0.22	49.67	0.31	0.06	1.63	0.050	0.001	10.1	0.38	0.040
Limestone	0.50	54.52	0.30	0.05	0.64	0.030	0.043	1.07	0.11	0.0005

Table 4 Comparison of trace elements in coal waste HTA and ground limestone

Sample ID	Ba (ppm)	Cr (ppm)	Cu (ppm)	Li (ppm)	Mn (ppm)	Mo (ppm)	Ni (ppm)	Pb (ppm)	Sr (ppm)	V (ppm)	Zn (ppm)
Coarse	480	132	59.5	46.1	264	43.6	80.0	114	113	279	293
Fine	561	125	54.2	61.1	282	21.6	54.3	80.1	109	205	213
Limestone	294	25.8	10.9	NT	171	2.5	11.5	ND	532	ND	34.5

Note ND not detected, *NT* not tested

Table 5 Particle size distribution of materials used in Proctor tests

Sieve opening (mm)	Percent retained		
	DP 1	DP 2	DP 3
4.76	63	58	54
1.68	20	18	20
0.42	11	10	11
0.75	5	4	4
< 0.075	1	10	11

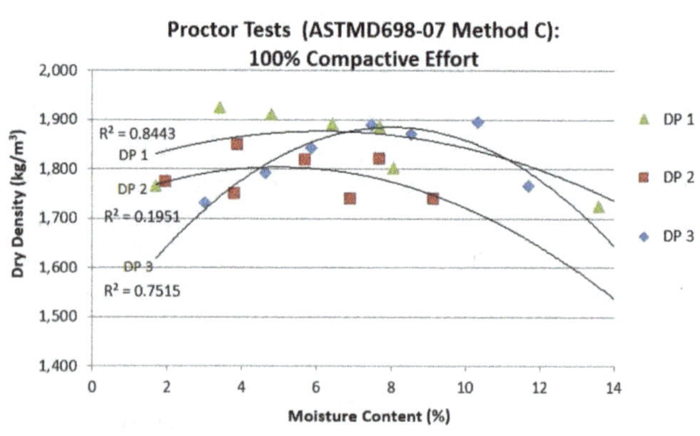

Fig. 2 Composite Proctor test results for DP 1, DP 2 and DP 3 options

inexpensive and readily available components; this second field experiment is the subject of this study. The revised field test columns were enlarged by using about 380–L (100-gallon), polycarbonate oval-tanks intended for livestock activities. These stock tanks measured 113.4 cm (52.5 inches) long by 94 cm (37 inches) wide by 53.3 cm (21 inches) tall (Fig. 3). Two test columns received 100% CCPW; two received a blend of 93.3% CCPW and 6.7% dewatered FCPW by volume; and two received a blend of 86.7% CCPW, 6.7% dewatered FCPW, and 6.7% ground limestone by volume. Assuming a normal porosity of 16%, approximately 300 kg (650 lb.) of CCPW was packed into each column. Sand cone tests were performed on four columns to determine dry unit weight and moisture content achieved during construction. Proctor dry unit weights of 21.55 kN/m^3 (137.3 pcf) and 14.27 kN/m^3 (90.92 pcf) were achieved for DP 1 and DP 3 mix materials, respectively. The kinetic test columns were leached in free-draining mode for 19.3 months (16 leach cycles; Fig. 3).

4.4.1 Leachate sampling and analysis

Pore-water and leachate samples were collected at about 38 days intervals using a 500 mL Nalgene® bottle. Field measurements of pH, temperature, and dissolved oxygen (DO) were conducted on each pore water and leachate sample with a professional-grade multi-parameter meter. In some cases, there was insufficient pore water sample volume to measure DO. Temperature, pH, and conductivity (specific conductance—SC) of pore water and leachate samples were measured immediately upon return to the laboratory. Alkalinity and ferrous iron were determined within 24 h after sample collection. All water sample quality analyses were statistically analyzed using spreadsheet-based models.

4.4.2 Anion analyses

Concentrations of SO_4^{2-}, Cl^-, F^-, and nitrate (NO_3^{2-}) anions were determined using an ion chromatograph (IC). For comparison, SO_4 concentration was also determined

Fig. 3 Field kinetic test columns (left); photograph of field test site (right)

Table 6 Average concentration of selected mine drainage parameters in leachate.*

refuse type	Interval (months)	pH	ORP	SO_4	Cl	TDS	Alkalinity	Acidity	Fe	Mn	Al
CCPW-Only	≤ 7	8.02	0.132	3437	198.0	3865	236.6	9.0	0.76	0.89	1.22
	> 7	2.50	0.769	5449	10.8	10445	0.0	4909	1258	78.8	140.6
CCPW/FCPW blend	≤ 7	8.32	0.077	2490	197.2	2968	266.2	2.7	0.23	0.62	0.66
	> 7	3.56	0.621	4502	12.7	5253	18.2	1008	204.6	23.1	44.00
CCPW/FCPW/limestone blend	≤ 7	7.83	0.133	3093	126.2	3698	203.7	1.5	0.08	0.66	0.01
	> 7	5.82	0.454	3456	11.4	2549	35.9	100.2	12.43	7.10	5.14

* Data in mg/L except pH (median value), ORP (volts), acidity/alkalinity (mg/L CCE)

using colorimetric analysis (Hach Method 8051; Hach Company 2004).

4.4.3 Major and trace metal analyses

Major metals and a suite of trace metals were determined by either an atomic absorption mass spectrometer (AA-MS) or by an ion-coupled plasma optical emission spectrometer (ICP-OES). Additional metals analyses were performed by a commercial laboratory. These were supplemented by colorimetric tests for key mine drainage components: iron (Fe), manganese (Mn), and aluminum (Al) using the Hach FerroVer Method (Method 8008), Periodate Oxidation Method (Method 8034), and the Aluminon Method (Method 8012), respectively (Hach Company 2004).

Geochemical modeling was performed to predict reaction pathways and evaluate reaction kinetics. Models were used to better understand geochemical conditions such as pore water and leachate composition in column materials. Geochemical and modeling studies results are described in detail elsewhere (Behum et al. 2009, 2010; Lefticariu et al. 2014).

4.4.4 Kinetic studies results

4.4.4.1 Leachate chemistry Because this paper is focused more on assessing the overall geochemical advantages of the proposed DP 2 and DP 3 co-disposal practices over DP 1, only broad-based geochemical observations are made here. Table 6 shows the median pH of leachate collected from all columns during the initial leach period ($\leq$ 7 months) ranged from 7.83 to 8.32, which was well above the carbonate buffer pH level of 6.37. However, there was a sharp decline in pH after that time for all DP 1 columns, and for three of the four blended refuse columns (DP 2 and DP 3). The median values of pH for DP 1, DP 2, and DP 3 during months 8–19 were 2.50, 3.56 and 5.82, respectively.

Table 6 also shows results for major anions monitored, including Cl^-, HCO_3^- and SO_4^{2-}. Of these, the HCO_3^- and Cl^- were the most easily leached anions. HCO_3^- is desired because it provides alkalinity needed to buffer the pH, but it declined after 7 months of testing. Cl^- declined from an average of 198.0 mg/L in the initial 7-month period to an average of 10.8 mg/L during the remaining test period for DP 1. Similarly, DP 2 and DP 3 showed declines in Cl^- from 197.2 to 12.7 mg/L, and 126.2 to 11.4 mg/L, respectively. Thus, the discharge of the relatively easily leached and mobile anion Cl^- was relatively unaffected by the alternate disposal practices.

Conversely, the average SO_4^{2-} concentrations were overall greater and rose significantly after 7 months of leaching from an average of 3437–10445 mg/L in DP 12, 490–4502 mg/L in DP 2, and 3093–3456 mg/L in DP 3. Thus, the SO_4 concentrations in leachate from the blended columns were considerably lower especially in DP 3. Similar results were also obtained for major and trace cation (Ca^{2+}, Mg^{2+}, Na^+, K^+, Fe^{2+}, Mn^{2+}, Ni^{2+}, Zn^{2+}, Cd^{2+}, Sr^{2+} and Pb^{2+}) concentrations (Lefticariu et al. 2014). The principle cations during early leaching (< 7 months) were alkali metal ions sodium (Na^+) and potassium (K^+), and alkali earth elements calcium (Ca^{2+}) and magnesium (Mg^{2+}), which is consistent with the earlier laboratory leaching tests (Behum et al. 2009, 2010). Initial Na^+ concentrations were elevated in the initial 7-month period averaging 912.4, 839.1, and 809.1 mg/L for CCPW, CCPW/FCPW, and limestone blend leachate, respectively. Subsequently, Na^+ concentrations declined rapidly to relatively low levels in the final (> 7 months) leach period (203.5, 375.7, and 179.1 mg/L for DP 1, DP 2, and DP 3, respectively). Ca^{2+} and Mg^{2+} concentrations both increased during the study; however, Mg^{2+} increased at a much higher rate (from 30.8 to 253.4 mg/L in DP 1; similar, but less significant, increases in Ca^{2+} and Mg^{2+} concentrations were observed in DP 2 and DP 3 columns. Metals commonly associated with pyrite weathering (Fe, Mn, Al, Ni, Zn, Cu, and Cd) increased in the later leaching period (> 7 months), but were again lower in the DP 2 and DP 3 columns. These results are presented in detail in a research report by Lefticariu et al. (2014).

4.4.4.2 Pore-water chemistry Overall, the volume of pore water collected was much lower than the leachate volume. After 7 months of kinetic testing, all but two of the lysimeter ceramic sample cups ceased producing pore water samples because of Fe-rich precipitates. The remaining two lysimeters ceased producing pore water samples after 9 months of testing. This limited analysis of pore water chemistry during later leaching periods (> 7 months). During the kinetic testing, the average pore water sample collected was only 30.8 mL. The average SO_4^{2-} levels in pore water samples from DP 1, DP 2, and DP 3 columns were 2844, 3022 and 1995 mg/L, respectively, as compared to 3437, 2490, and 3093 mg/L, respectively, in leachate samples. Similar to the leachate sample observations, higher Cl^- concentrations were observed in pore water samples collected during the initial 7-month leach testing period. Again, additional data and discussion is provided in Lefticariu et al. (2014).

Elemental extraction in column leachate Data for most major and trace constituents were converted to a mass (loading) basis by multiplying concentration values and leachate volume. This conversion allowed determination of the cumulative elemental extraction by calculating the percentage of leachate mass in each cycle, then comparing this to the original mass of the element within the column. Mass data was plotted as a function of time, which is represented by the leach cycle. The complete kinetic testing program consisted of 16 leach cycles completed in 568 days (19 months) with an average of 19813 mL of leachate collected for each cycle from all of the refuse columns. Leachate volume was then compared to the estimated pore volume of 54501 mL to yield an initial liquid-to-solids (L/S) ratio of 0.19. As a result, the average rate of pore volume flushing is approximately 0.36 volumes per leach cycle, which yields an approximate of 5.82 pore volumes that were leached over the course of the 19 month/16-cycle study. The cumulative extraction of S and Cl versus leach cycle is shown in Figs. 4 and 5. Although the extraction percentage was greater for Cl than for S during early leach cycles (< 7 months), overall amount of S extracted was higher, especially for CCPW (DP 1) columns. After 7 months of testing, S extraction increases in all leachate. However, only S extraction from DP 1 column leachate exceeded Cl extraction during this period.

Figure 6 plots the percentage extraction of major and selected trace elements during the 19-month kinetic test. S and Cl are the major anions in high total dissolved solids (TDS) discharges and Na and K are the major cations. Ca extraction was somewhat greater than Mg and Sr extraction; Mg extraction was elevated in the CCPW-only columns (DP 1). Mobility of Ca, Mg, and Sr elements commonly associated with calcareous elements were

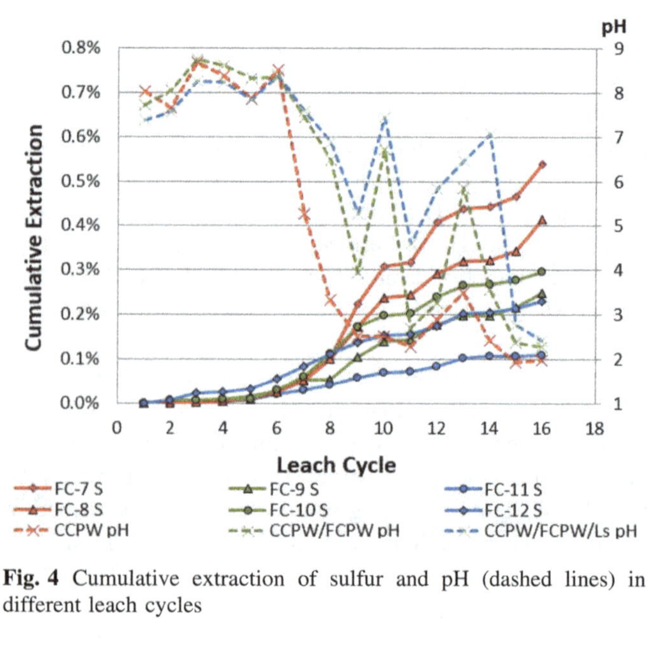

Fig. 4 Cumulative extraction of sulfur and pH (dashed lines) in different leach cycles

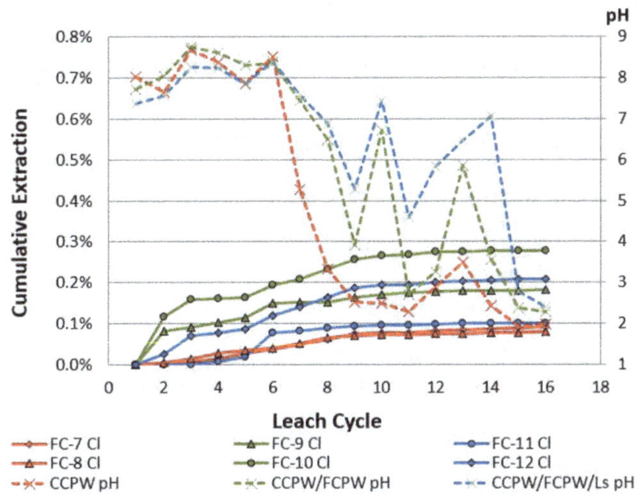

Fig. 5 Cumulative extraction of chloride and pH (dashed lines) in different leach cycles

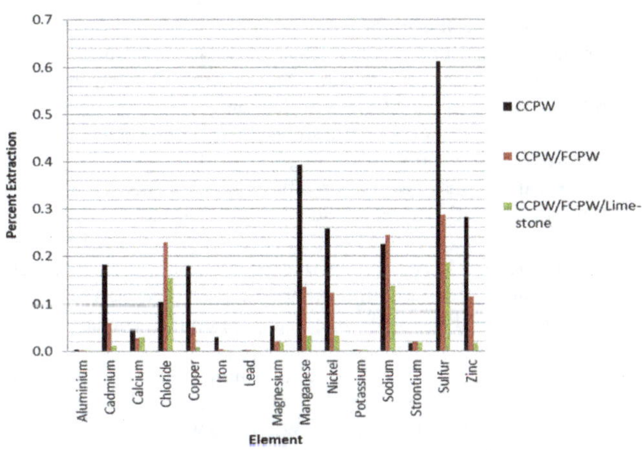

Fig. 6 Elemental extraction during 19-month kinetic test

relatively low during the 19-month kinetic test compared to Mn, Ni, and Zn trace elements that are commonly associated with pyrite (Fig. 6). Fe extraction from the DP 1 columns, and to a much later extent from CCPW/FCPW blend columns (DP 2), increased as pH dropped in later leaching periods (> 7 months). Mn, Ni, and Zn also had higher extraction rates then Fe suggesting considerable precipitation of Fe phases within the test columns. Only a very small amount of Fe extraction occurred in the limestone blend leachate (DP 3), which also maintained an overall higher pH (Lefticariu et al. 2014).

5 Conclusions and recommendations

5.1 Conclusions

(1) The co-disposal practices of CCPW and FCPW with or without limestone addition are a significant improvement over the current practice of disposing CCPW alone. The limestone addition practice further limits SO_4^{2-} release.

(2) The proposed practice had minimal effect on release of chloride (Cl^-). Most of the discharge occurs soon after placement and must consider using water management practices of retention ponds, dilution, and discharge after large precipitation events.

(3) The decline in HCO_3^- concentrations was much greater than sulfate (SO_4^{2-}) concentrations in alternate disposal practices. This suggests that either carbonate weathering rates are faster than pyrite weathering in a mine waste pile environment or that the alkalinity-producing minerals are being coated with mineral precipitates that are limiting dissolution.

(4) Mobility of SO_4^{2-} was significantly lower in alternate disposal technology especially with limestone addition. The higher extraction of S compared to Ca suggests that the formation of calcium sulfate such as gypsum or anhydrite was relatively small.

(5) Elements associated with alkalinity-producing minerals such as calcite and dolomite (Ca, Mg, and Sr) were leached to a greater extent than heavy metals typically associated with pyrite (Mn, N, and Zn) and lithophile elements (Al and K).

(6) For the duration of experiments, the percentage of S extracted was smaller than Cl.

(7) There was at least a 7-month period before additional fresh refuse or the final reclamation cover must be placed in order to avoid excessive release of SO_4^{2-}.

5.2 Recommendations

The encouraging results obtained in this study allow us to disseminate this alternate disposal concept to mine operators globally and seek input on how best to pursue commercialization of technology. Variables that might affect its success in the field need to be identified in cooperation with industry prior to wide-scale implementation of this technology. The authors have identified several areas for further studies that would aid commercialization: (1) a larger scale leaching study should employ commercially dewatered fine coal processing waste; (2) laboratory- and field scale leaching studies on DP 2 and DP 3 (co-disposal) alternatives should consider alternative alkaline amendment materials such as cement kiln dust (CKD), acetylene production waste lime, and steel slag, which may be available at a lower cost than agriculture ground limestone; (3) field demonstration of commercial dewatering technologies to dry coal waste to the necessary solids content of about 65% is needed to develop an economic analysis; and (4) a comprehensive economic analysis of the alternative disposal technologies should be completed. The research team believes that the concepts presented here should be considered for developing other innovative concepts for management of coal waste materials to improve both the economics and environmental impact of disposal.

References

American Society for Testing and Materials (ASTM) (2013) ASTM D698-12e1: standard test methods for laboratory compaction characteristics of soil using standard effort (12400 ft-lbf/ft3 (600 kN-m/m3)). In: Annual book of ASTM standards, vol. 04.08, West Conshohocken PA. website: http://www.astm.org/Standards/D698.htm. Accessed 8 Jan 2015

Behum PT, Chugh YP, Teklehaimanot Y, Lefticariu L (2009) Geochemistry of coal mine drainage in response to improved material handling: preliminary results of kinetic testing. In: Paper no. 28-2, geological society of America, North-Central Section, 43rd annual meeting, Rockford IL, April 2–3

Behum PT, Lefticariu L, Chugh YP (2010) Results of simulated weathering of coal stockpiles and coal refuse disposal areas. Geochimica et Cosmochimica Acta 74(12), Supplement 71

Chen S, Hu Z, Chen S (2014) Construction of isolation layer for preventing spontaneous combustion of coal waste piles and its effects. Trans Chin Soc Agric Eng 30(2):235–243

Chugh YP, Mohanty S, Behum PT, Nawrot J, Bell W, Chen S (2007a) Best management practices to minimize sulfate and chloride discharges. Final technical report on Illinois Clean Coal Institute Project Number DEV05-8

Chugh YP, Behum PT, Nawrot J, Mohanty S (2007b) Identification of good management practices in Illinois mining operations to minimize sulfate and chloride discharges: an overview. In: Singhal RK, Fytas K, Jongsiri S, Ge H (eds) Proceedings, 16th international symposium on mine planning and equipment selection (MPES) and 10th international symposium on environmental issues and waste management in energy and mineral production (SWEMP), Bangkok, Thailand, December 11–13

Dreher GB, Roy WB, Steele JD (1994) Laboratory studies of fluidized bed combustion residues in mixtures with coal slurry solids. In: Proceedings of management of high sulfur coal combustion residues: issues and practices, Southern Illinois University at Carbondale, Springfield, Illinois, 91–101

Evangelou VP (1995) Pyrite oxidation and its control. CRC Press, Boca Raton

Hach Company (2004) DR/890 Colorimeter procedures manual. Hach Company, Loveland, p 616

Hossner LR, Doolittle JJ (2003) Iron sulfide oxidation as influenced by calcium carbonate application. J Environ Qual 32:773–780

Hu Z, Zhang M-L, Ma B-G, Wang P (2008) Selective bactericides for at-source pollution control of acid coal waste piles. Res Environ Sci 21(5):23–26

Hu Z, Zhang M, Chen S, Ma B, Zhao Y, Li J (2009) Pollution generation, migration and in situ control measures on coal refuses. In: proceedings of The 6th international conference on mining science and technology, Elsevier publisher, Xuzhou, p. 1186–1190

Hu Z, Ma B-G, Zhang M-L, Wang P (2009b) Bioremediation of sulfuric contamination from coal wastes by sulfate-reducing bacteria (SRB). J China Coal Soc 34(3):400–404

Kleinmann RLP, Crerar DA (1979) *Thiobacillus Ferroxidans* and the formation of acidity in simulated coal environments. Geomicrobiol J 1(4):373–387

Lefticariu L, Chugh YP, Behum PT (2014) Field demonstration of alternative coal processing waste disposal technology. Final report, Illinois Clean Coal Institute, p 28

Moses CO, Herman JS (1991) Pyrite oxidation at circumneutral pH. Geochim Cosmochim Acta 55(2):471–482

Moses CO, Nordstrom DK, Herman JS, Mills AL (1987) Aqueous pyrite oxidation by dissolved oxygen and by ferric iron. Geochim Cosmochim Acta 51(6):1561–1571

National Research Council (NRC) (2002) Coal waste impoundments, risks, responses and alternatives. National Academy Press, Washington, pp 658–667

Naumann HS, Wiram VP (1995) Alkaline additions to the backfill—a key mining/reclamation component to acid mine drainage prevention. In: Proceedings, 16th West Virginia surface mine drainage task force symposium, Morgantown WV, April 4–5

Nawrot JR, Gray B (2000) Principles and practices of tailings reclamation: coal refuse. In: Barnhisel RI, Darmody RG, Daniels WL (eds) Reclamation of drastically disturbed lands, chapter 18. American Society of Agronomy, Crop Science Society of America, Soil Science Society of America, Madison, pp 451–488

Nordstrom DK (1982) Aqueous pyrite oxidation and the consequent formation of secondary iron minerals. In: Kittrick JA, Fanning DS, Hossner LR (eds) Acid sulfate weathering, chapter 3. Soil Science Society of America, Madison, pp 37–56

Singer PC, Stumm W (1970) Acidic mine drainage: the rate redetermining step. Science 167(3921):1121–1123

Stewart BR, Daniels WL, Zelazny WL, Jackson ML (2001) Evaluation of leachates from coal refuse blended with fly ash at different rates. J Environ Qual 30:1382–1391

The Advent Group, Inc (2005) Changing the water quality standard for sulfate in discharges from coal mines: the economic impact on the coal industry. Final technical report on Illinois Clean Coal Institute Project Number 03-1/US-4

US Environmental Protection Agency (2009) EPA's rationale for approval of amendments to the existing Illinois pollution control board regulation. In: 35 Illinois administrative code 302.102, 302.208, 309.103, 405.109, 406.100, 406.203, and 407. Revised water quality criteria for sulfate and total dissolved solids (WQSTS IL2008-274), March 18

Wiram VP (1984) Challenges to mining coal in the midwest. In: Proceedings, 5th annual West Virginia surface mine drainage task force symposium, Morgantown WV, March 21–22

Xu J-J, Hu Z, Zhao Y-L, Wei BL (2014) The bactericide of acid thiobacillus ferroxidans in acid coal waste piles. China Min Mag 23(01):62–65

2

Effects of rock dusting in preventing and reducing intensity of coal mine explosions

Yi Luo[1] · Deming Wang[2] · Jianwei Cheng[2]

Abstract As an explosion control measure, rock dusting has been used in underground coal mines in many major coal producing countries with different standards. The effectiveness of the rock dust in reducing explosion intensity has been proven by historic events and laboratory experiments. The main functions of rock dust in controlling mine explosions (i.e., isolator, physical heat sink and chemical energy absorber) have been quantitatively studied and results are presented in this paper.

Keywords Coal mine explosions · Rock dusting · Regulatory standards · Isolator · Heat sink · Energy absorber

1 Introduction

Explosions are the most feared hazards in underground mines. A coal mine explosion generally starts with the ignition of methane first with the possibility for the coal dust to participate in and to intensify the explosion. Mining coal generates 40 g, on average, of float coal dust per ton of coal mined. Most of this float coal dust is deposited at or near the working sections. The possibility of methane ignition is also high at the working faces due to methane liberated from newly exposed coal face and possible frictional sparks generated in mining process. The effective control of coal dust on the working section is critical to prevent small methane ignitions at or near the working face from transitioning to a large coal dust explosion. Methane explosion can ignite the suspended coal dust, intensify the explosion event and produce more devastating consequences. Rock dusting has been used as a primary explosion prevention and intensity reduction measure in many major coal producing countries. Even though with different standards, rock dusting is mandated in underground coal mines in the USA, Australia, South Africa, India, Russia, Germany, Poland,... Apparently, the frequency and severity of coal mine explosions are much smaller in these countries after this mine safety measure was in place than those in countries without mandating the rock dusting measure.

This paper briefly reviews the different rock dusting standards in the mine safety regulations established by the major coal producing countries. It shows the difference in explosion intensities between the methane only explosions and methane–coal–dust explosions. The evidences of applying rock dusting to the control of coal mine explosions are also shown. The main part of the paper is placed on gaining a basic qualitative and quantitative understanding about the mechanism of rock dusting in preventing and controlling coal mine explosions. With a better understanding, the strategy and effectiveness of using rock dusting as an important safety measure in underground coal mines can be improved.

✉ Yi Luo
yi.luo@mail.wvu.edu

1 Department of Mining Engineering, West Virginia University, Morgantown, WV, USA

2 College of Safety Engineering, China University of Mining Technology, Xuzhou, Jiangsu, China

2 Histories and regulatory standards for rock dusting

Rock dusting has been widely used in major coal producing countries. However, the quality and quantity standards for the rock dust vary to some degree. The bases for the variations could be related to the coal dust size, methane concentration in mine air, rank of coal, etc.

2.1 Rock dusting standards

Based on the US MSHA (Mine Safety and Health Administration) definition, the rock dust used in underground coal mine is pulverized and light colored limestone, dolomite, gypsum, anhydrite, shale, adobe, or other inert material. All the particles should be smaller than 841 μm with 70% or more by weight being smaller than 74 μm (MSHA 2014). Generally speaking, using such rock dust can greatly remove the amount of heat generated in an explosion event. Thus, the fire flame could be eliminated and the propagation of explosion within the underground ventilation system is prevented.

Due to such thermal inhibitive ability, applying rock dust (or rock dusting) to prevent or mitigate mine explosions is wildly used in US coal mines nowadays. Mine operators routinely cover the floor, rib and roof areas of mine entries with a generous application of inert rock dust (Man and Teacoach 2009).

It is generally agreed that the effectiveness of rock dust lies in its ability to be simultaneously dispersed with coal dust, and, by serving as a heat sink, thus prevent flame propagation. The regulatory requirements on rock dust are various, some are more and some less stringent among the countries (Cashdollar et al. 2010):

After the deadly Farmington mine explosion in US in 1968, the new Federal Coal Mine Health and Safety Act of 1969 mandated the use of rock dusting to prevent mine explosion. This mine safety law made significant improvement in mine safety and health in US underground coal mines, particularly in reducing coal mine explosions. Figure 1 shows the number of fatalities in each of the coal mine explosion disasters (defined as accident with five or more fatalities) and the cumulative mine explosion fatalities since 1900. Apparently, the frequency (measured by the density of the data points) and severity (the deaths per accident) before and after Act 1969 are significantly different. In the 70 years prior to this Act, 403 mine explosion disasters occurred resulting in a total of 10204 fatalities, or 5.76 accidents per year and nearly 25 deaths per accident on average. In the 40 years after the Act, there were 16 explosion disasters resulting in 210 deaths, average 0.4 accidents per year and 13 deaths per accident. There was apparently declining trends both in frequency and severity in the decade after the pass of the Act.

Initially, the rock dusting standard in US coal mines was mandated that 65% incombustible content (IC) in non-return airways and 80% IC in return airways (Harris et al. 2010). When methane is present in any ventilating current, it requires adding an extra 1% of incombustible material

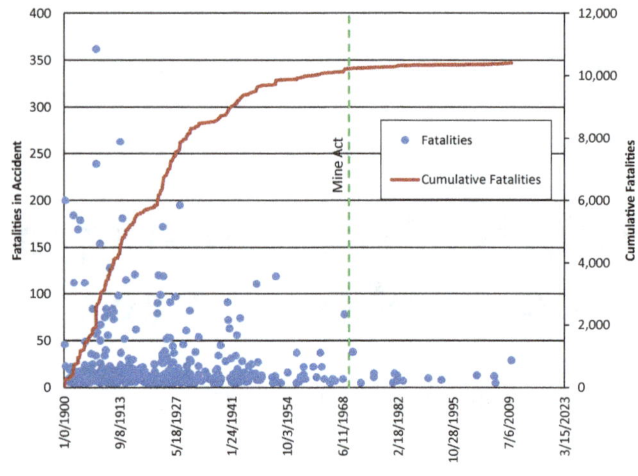

Fig. 1 US coal mine explosion disasters (>5 fatalities/accident), 1910—present

per 0.1% methane for air in intakes and adding 0.4% incombustible per 0.1% methane in the returns (Man and Teacoach 2009). To face the finer coal dust particle size produced by modern coal mining machines, MSHA increased the rock dusting standard on September 23, 2010. The new regulations require the total incombustible content (TIC) of combined coal dust, rock dust, and other dust to be at least 80% in underground bituminous coal mines with an additional requirement to raise the incombustible content by 0.4% for each 0.1% of methane present in the ventilation air.

Canadian coal mines follow different rock dusting criteria according to their locations. In Alberta and British Columbia, the minimum total incombustible content is 65% in intake airways while 75% is required in the province of Nova Scotia (Cain 2003). However, 80% TIC is needed for all return airways.

Regulations in Australia also vary. The required TIC ranges from 70% to 85% in intake airways and 80%–85% in returns in the state of Queensland. However, the maximum TIC is only 80% in intakes in New South Wales (Cashdollar et al. 2010).

Most East European countries, such as Czech Republic, Slovakia, Ukraine, etc., use the same standard requiring TIC to be 80% both in intakes and returns when the methane concentration is lower than 1% otherwise 85% is demanded. The requirements for TIC in Polish and Russian coal mines are less stringent. Differing with mine type, 60% and 70% are required in the intake and return airways for "non-gassy" mines and 75% and 80% are required for "gassy" mines, respectively.

South African mine researchers started the needs assessment of rock dusting in later 1980s. The last major coal mine explosion occurred in 1993 with 53 deaths. Currently, 80% TIC (both in intakes and returns) is required by the government regulations which were established as law in the year of 2002 (Cain 2003).

Although the criteria of TIC are mostly based on the methane concentration, some countries establish their criteria based on the volatile matter (VM) content of coal. In United Kingdom, TIC ranging from 50% to 75% is required as the VM content varies from 20% to 35%. Japanese regulations specify a minimum TIC of 78% when the volatile matter content is over 35%. More specific requirements are also needed depending on ash, moisture, the gassiness of the coal seam, and the fineness of the rock dust used.

2.2 Considerations in rock dusts

As shown early, the TIC in rock dust is the most important quality factor for preventing the propagation of mine explosion. Laboratory experiments show that different rock dusts (i.e., limestone, dolomite and magnesite) have similar inerting effectiveness for coal dust. However, tests with dolomite show that a decrease in rock dust particle size could significantly lower the amount required to inert (Amyotte et al. 1992). Limestone or other rock dusts could easily cohere to form a cake when wetted and then dried making dispersion more difficult in an event of a light blast of air. Hence, the effectiveness of the alternative inerting agents has been experimentally investigated in countries including Germany, US, South Africa, etc. Fly ash and calcium sulphate ($CaSO_4$) may be the good candidates but more testing works are required to prove this.

Rock dust bags are extensively used in some counties to create barriers for stopping explosion propagation. In order to make this explosion mitigation measure effective, the barriers should be properly placed and a certain minimum pressure is required to activate them. Various reports show more failure than successes in using this explosion mitigation measure (Zou and Panawalage 2001). Most failures are attributed to improper location, improper design, poor installation, and inadequate maintenance. If the explosion is too weak to disperse the selves—often as a result of placing the rock dust barriers closer than 60 m from the ignition point, a failure may occur. In addition, if the pressure front or the flame front moves too fast or is too short, the flame front could have easily passed the barrier locations before the dispersion of inert material is completed (Dixon et al. 1994).

The rock dust is considered as nuisance dust that should have no adverse effect to the health of the miners, especially to the workers performing the rock dusting operations. However, it should be noted that research on the effects of rock dust to miners' health is still limited. The most important factor to consider is the content of silica (SiO_2). Any extended accumulation of dust with high silica content in human lungs, together with the tissue reaction to its presence, is usually referred to as pneumoconiosis (Du Plessis 2014). Therefore, for miners' health, MSHA requires that the rock dust not contain more than 5% combustible matter or more than a total of 4% free and combined silica (SiO_2). However, further medical studies should be performed on silica and other mineral components in the rock dust to determine their effects to human health.

3 Current Chinese coal mine explosion prevention measures and results

Watering, water flushing, water mist spray and water barriers (Fig. 2) have been extensively used to prevent and reduce intensity of coal mine explosions in China. Coal mines in the eastern China, especially for the coal fields such as Zhaozhuang, Yanzhou, Huainan, Huaibei, etc., often apply the methods of watering and water flushing to clear float coal dusts in underground. The underground roadways with serious coal dust deposition problems are regularly washed with water. Since most of the Chinese coal mines are designed to use gravity drainage system and drainage ditch is built in each roadway, using water to wash the float coal dust generally will not create drainage problems. The period of washing is determined by the dust deposition density and the minimum dust concentration required by a coal dust explosion. The washing work should be done routinely every day especially in the sections less than 30 m from the dust sources. In addition, due to their shallow mining depths, high productions and water shortage problems, coal mines in the western China often spread rock dust to prevent coal dust explosion.

Chinese coal mines also spread the rock dust in the underground to cover the exposed coal. It is required that the incombustible content of combined coal and rock dusts

Fig. 2 Water barriers in Chinese underground coal mines

must be at least 60% or 90% when the CH_4 concentration is over 0.5%. The quality of rock dust must meet the following quality requirements: (1) the combustible content < 5%, (2) the content of free silica (SiO_2) < 10%, (3) free of any toxic or harmful mixture, such as phosphorus and arsenic. The rock dust, mainly limestone dust, should entirely pass through 50 mesh sieve (<300 μm in sizes), and 70% of it pass through 200 mesh sieve (<75 μm in sizes) (Safety SAOW 2006). The rock dusting requirements, established in 2006, have not been made into the mining law and only serve as the suggested mine safety measure.

In preventing underground explosion, China initially adopted the Polish measures to use the rock dust barriers. However, the "caking" problems of the rock dust limited its effectiveness in explosion control. Since 1980s, water barriers have been extensively used as an explosion mitigation measure in Chinese underground mines. The principles of water barrier are similar as rock dust barrier. In an explosion event, the precursor shock wave would overturn the water barriers or break the water bags to release the stored water into the air. The released water, if fully mixed with the flame front, can greatly reduce the explosion intensity by acting as heat sinker and flame retardant. The effects of water on preventing mine gas/dust explosion under specific experimental conditions are evident. However, when the water is poured from its containers in underground conditions, it is often poorly dispersed in space and can only suspend in air for a short duration. Therefore, the potential effectiveness of the water barriers in suppressing the explosion is greatly reduced. The water barriers' poor performances in preventing underground explosions have been evidenced by a number of serious mine explosion accidents in Chinese coal mines. Hence, the necessity and reliability of water barriers are now questioned by a number of Chinese mine safety experts and engineers.

4 Mechanism of rock dusting in controlling coal mine explosions

In order to effectively use the rock dust in controlling coal mine explosions, the mechanism for the applied rock dust to control explosion should be qualitatively identified and the effectiveness should be quantitatively evaluated. To control an on-going coal mine explosion, the first important step is to lower the air temperature as fast as possible for the purpose of lowering the explosion pressure. The second important step is to prevent coal dust from participating into explosion so that the total thermal energy produced and the intensity of an explosion can be controlled. Properly applied rock dust can achieve both the control steps.

4.1 Main mechanisms

In the underground coal mine settings, the essential conditions for coal dust to participate in an explosion are: (1) fine coal dust suspended in air with a minimum concentration of 100 g/m^3 for biyuminous coal and (2) sufficient ignition energy with a minimum temperature of 440 °C (Stephan). The first condition ensures sufficient reaction surface between the fine coal dust particles and oxygen in the air. The second condition provides sufficient thermal energy to ignite the coal dust and to sustain the chain explosion process. It should be noted that among the combustible contents in the coal (mainly the volatile matters and the fixed carbon), the volatile matters should be the first part and most likely the only part to be involved in the explosion process considering the short contact time between the passing hot air and the coal dust particles. However, the volatile matters have to be devolatilized, stimulated with sufficient thermal energy, into vapor before the start of dust explosion.

Proper rock dusting in underground coal mines can effectively eliminate the two conditions for coal dust to participate in explosion. The inert rock dust can function as isolator, physical heat sink and chemical energy absorber. The heavier rock dust applied over the lighter coal dust can serve as an isolator to prevent the coal dust underneath from becoming easily airborne. The isolator is most likely to function only when the explosion front is propagate at a relatively low velocity. Without suspending the deposited coal dust into the mine air, it is very unlikely for the lower explosive limit of coal dust to be reached.

In a methane explosion event, the explosion propagation can quickly reach supersonic and thus generates shock wave. The shock wave is well capable of suspending the settled coal and rock dusts into mine air. Under this condition, the suspended rock dust along with the unexploded coal dust can act as a physical heat sink to absorb the heat in the passing hot air for creating a temperature equilibrium between the dust and air. Based on the law of energy conservation, a simplified model can be derived for assessing the rock dust's heat sink effect. For example, a coal dust load is assumed to be 1.2 kg per meter length of a 2 m × 6 m mine entry to create dust-air mixture at the minimum explosive concentration of 100 g/m^3 if fully suspended in the air. To demonstrate the heat sink effect of rock dusting, the entry is dusted with limestone powder at 0.00, 2.23 and 4.80 kg per meter of entry to create 0%, 65% and 80% rock dust concentrations, respectively. In an explosion event, the passing hot air at 2000 °C suspends and fully mixes with the coal and rock dusts at a point. The airborne coal and rock dusts (originally at ambient temperature of 16 °C) absorb the heat from the hot air and lower its temperature.

Figure 3 shows the resulting air temperature profiles in the entry for the three simulated cases. Without rock dusting (0% RD) and assuming that the airborne coal dust will not explode due to insufficient concentration, the air temperature in the entry decreases slowly. The air temperature is reduced to about 440 °C, the minimum ignition temperature for coal dust, at a distance of 31 m from the source. On the other hand, it takes about 14 and 8.5 m distances to lower the air temperature to 440 °C when 65% and 80% rock dusts are applied in the entry, respectively. The shorter distance for the mine air to remain above the minimum ignition temperature of coal dust, the less is the possibility for the coal dust to participate in mine explosion. Therefore, the heat sink effect of the proper rock dusting can greatly reduce the possibility of coal dust ignition in a mine explosion event.

The third function for the rock dust to control mine explosion is to chemically absorb the thermal energy in a calcination process. At temperature above 800 °C, the main component of the rock dust, limestone ($CaCO_3$), can be converted into lime (CaO) in an endothermic chemical reaction process as shown in Eq. (1). A complete chemical reaction of one kg of pure limestone dust absorbs about 1786 kJ heat (Zeman 2008). The thermal energy consumed in the limestone-to-lime chemical conversion process can reduce the chance and intensity of the methane and coal dust explosion.

$$CaCO_3 + \text{heat} = CaO + CO_2 \tag{1}$$

The combustion process of fine coal particles at high temperature involves two stages: the ignition and combustion of volatile maters (VM) first and then the fixed carbons (Prins et al. 1989). At 850 °C, the ignition of the fixed carbon normally occurs more than 10 s after the ignition of VM. In a mine explosion event, the contact time between the coal dust particles and hot air in the chemical reaction zone is very short. Therefore, it is most likely that only the VM of the coal or even only a portion of the VM can be vaporized and participated in mine explosion. This is the reason that anthracite (VM less than 10%) dust is impossible to explode while the explosibility of the coal dust increases as the rank of the coal decreases (higher VM content). In an idealized condition where all and only the VM in the coal dust participates in explosion and the calcination process is complete, a balance in the thermal energies produced by coal dust explosion and consumed by the calcination process can be determined using the following equation.

$$x = \frac{1786\alpha}{(1-\alpha)Q} \tag{2}$$

In this equation, x is the VM content, α is the rock dust concentration and Q is the heating value of the coal. If 80% rock dust is applied and $Q = 30{,}000$ kJ/kg for typical coal, the VM content required to reach an energy equilibrium is determined to be 23.8%. In other words, if a coal contains less than 23.8% VM, the complete calcination process can absorb more heat than that the coal dust can generate in the explosion. Consequently the calcination process will consume thermal energy in the hot air and lowers its temperature. Even for the coals having higher VM contents, the energy consumed in the calcination process of the rock dust can still greatly reduce the net thermal energy added by coal dust to the explosion event.

4.2 Anecdote evidences

The anecdote evidence for rock dust to control coal dust explosion is shown in the demonstration tests conducted at the South Africa's CSIR (Council for Scientific and Industrial Research) Kloppersbos facility (Fig. 4). In the

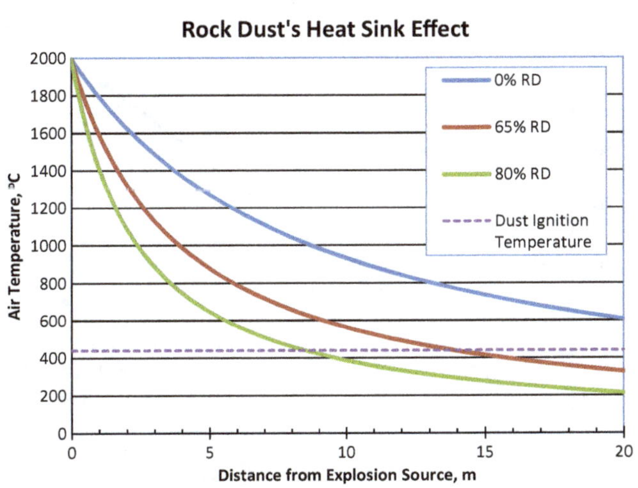

Fig. 3 Rock dust's heat sink effect for lowering the mine air temperature

Fig. 4 Coal dust explosion tests of one teaspoon of coal dust without (*left*) and with (*right*) four spoon of rock dust (80% rock dust)

demonstrations, the explosion of one teaspoon of coal dust in a steel tube results in a large fire ball above the tube as shown in the left of figure and a loud noise. When one teaspoon of coal dust is mixed with 4 teaspoons of rock dust (80% RD), the explosion only resulted in a quiet puff of white dust cloud above the mouth of the tube as shown in the figure on the right. It indicates that the rock dust has considerably reduced the total energy output from the dust explosion.

The recent two explosions in the US coal mines (Sago in 2006 and Upper Big Branch in 2010) show the importance of rock dusting in reducing the intensity of mine explosion. In the Sago mine explosion, the mine was properly rock dusted. The methane explosion was originated from a newly sealed mine area that completely destroyed the mine seals. The back calculated explosion pressures at the mine seals was about 641 kPa (Gates et al. 2007). The immediate impact of the explosion caused one death to a mining crew of 13 located about 123 m from the destroyed mine seals (about 34.5 kPa explosion pressure). The other 11 fatalities in this accident were caused by CO poisoning. Another crew of 13 miners located about 344 m away from the destroyed mine seals (about 14 kPa explosion pressure) escaped from the accident unharmed. The sharp decrease of the explosion pressure within a short distance from the explosion source implies the effectiveness of rock dusting in reducing the explosion intensity.

On the other hand, the explosion in the Upper Big Branch mine almost instantly killed 29 miners located within a radius of more than 2500 m from the source of the explosion (Phillips 2012) indicating that the explosion pressure in this accident is significantly higher than that in Sago accident. Substandard rock dusting was common in this mine and the mine had been cited numerously for coal dust piles and insufficient rock dust by federal and state mine inspectors. The high intensity indicated heavy participation of coal dust in the explosion event.

4.3 NIOSH explosion tests

After the Sago Mine explosion, the US National Institute for Occupational Safety and Health (NIOSH) conducted a series of explosion tests in its Lake Lynn experimental mine (Zipf et al. 2013). In the explosion tests, natural gas was injected into an ignition zone (most time with a volume of 21.5 m^3 and a few with larger volume). The natural gas is composed of 97%–98% methane, 1.5% ethane, and small percentages of other higher hydrocarbons. Most of the tests used a 9% methane–air concentration in the ignition zone. A few of the tests were also conducted with 6%–10% methane–air concentrations. In six of the tests, shelves of pulverized bituminous coal dust (7.3, 14.5 and 120 kg, respectively) were located at a distance of 3–12 m away from the ignition zone as a means to increase the explosion overpressures. A number of important explosion parameters (i.e., pressure, temperature, time of flame front, etc.) were measured during the tests. The author reanalyzed some of the data. One of the findings of the data analysis is the nearly linear relationship between the maximum static explosion pressures and the total heating value possibly released in the explosion as shown in Fig. 5. The maximum static explosion pressures were measured at the location about 4 m away from the ignition zone. The total amount of methane and only the volatile matter (30% VM for bituminous coal) of the coal dust are used to determine the total heating values. Figure 5 shows when only methane is involved in the explosion, the maximum explosion pressure at the experiment sitting reaches only about 234 kPa (34 psi). However, as the coal dust is added in the explosion, the pressure increases considerably. In the last explosion test with 120 kg of coal dust being placed in the testing site, the maximum explosion pressure reached about 620 kPa (90 psi). The good linear relationship also indicates that it is reasonable to assume only the VM content of the coal to participate in the mine explosion as previously shown in the quantification of the rock dust's chemical energy absorber function.

5 Rock-dusting in underground coal mines and regulatory compliance

Rock dusting is currently the primary means of defense against coal dust explosions in US underground mines mining bituminous coals. All areas of a coal mine that can be safely traveled must be adequately rock dusted within 12 m of all working faces. The chance of propagation and risk of widespread explosion disasters in bituminous coal

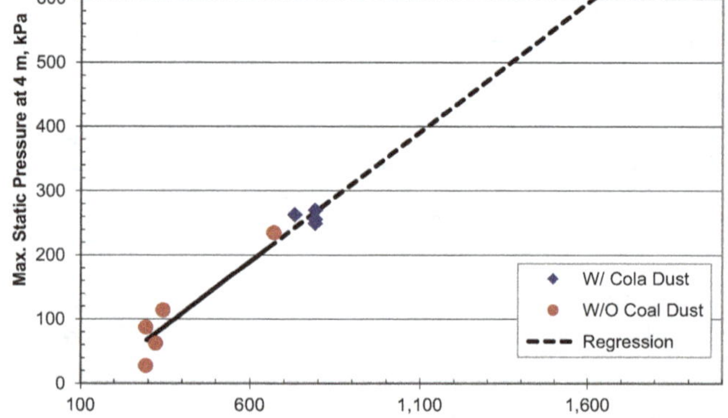

Fig. 5 Total heating value versus maximum explosion pressure (reanalyzed NIOSH test results)

Fig. 6 Manual application of rock dust in a new face area

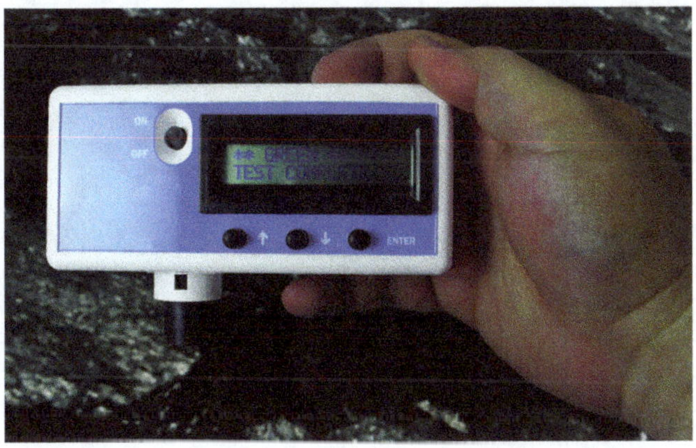

Fig. 7 NIOSH developed coal dust explosibility meter (CDEM)

mines can be nearly eliminated when rock dust is applied liberally and maintained properly. Figure 6 shows a manual rock dusting operation in a new coal working face. Various mechanized rock dusters are commercially available. The mine operators are responsible for maintaining the required minimum incombustible content in the mine.

The incombustible content of rock dust is evaluated during regular and spot inspections. This evaluation includes the collection and analysis of mine dust samples. To assess the compliance, mine dust samples are collected from various locations within the coal mine. The band or perimeter method is used to collect dust samples from the roof, ribs, and floor creating one "band" sample. This band sample includes 25 mm deep material from the floor. The collected sample is thoroughly mixed, coned, and quartered to take a portion for analysis. This sampling essentially assumes a homogeneous mixture of coal, rock, and other dust on all surfaces.

Traditionally, the collected samples are sent to the National Air and Dust Laboratory for moisture and low temperature ash analysis to determine the percent of incombustible content of the minus 20 mesh fraction. A new handheld device, the coal dust explosibility meter (CDEM) shown in Fig. 7, has been developed by the NIOSH. The CDEM can be used at the sampling sites to accurately determine the explosibility of rock and coal dust mixtures nearly in real time.

6 Conclusions

Rock dusting has been used in many major coal producing countries as the primary means of defense against coal dust explosions in underground coal mines. A proper rock dusting should maintain adequate incombustible content in the mine dust mixtures of float coal dust and rock dust. The rock dust applied can function as isolator, physical heat sink and chemical energy absorber that prevent the settled coal dust from participating into explosion and reduce the total thermal energy contained in the explosion event. Through such functions, the rock dust can either stop a coal mine explosion process or reduce its intensity.

References

Amyotte PR, Mintz KJ, Pegg MJ (1992) Effectiveness of various rock dusts as agents of coal dust inerting. J Loss Prev Process Ind 5(3):196–199

Cain P (2003) The use of stone dust to control coal dust explosions: a review of international practice. Report to Natural Resources Canada, p 46

Cashdollar KL, Sapko MJ, Weiss ES, Harris ML, Man CK, Harteis SP, Green GM (2010) Recommendations for a new rock dusting standard to prevent coal dust explosions in intake airways. Centers for Disease Control and Prevention, NIOSH RI 9679, p 49

Dixon DW, Fidler WM, Ediz IG (1994) The preliminary analysis of the pressure wave transmission/reflection characteristics of explosion doors using one-imensional finite isentropic wave theory. In: Proceedings of the 9th Coal Congress of Turkey, Zonguldak, pp 319–330

Du Plessis J (2014) Ventilation and occupational environment engineering in mines. Mine Ventilaiton Scociety of South Africa, Johannesburg

Gates RA, Phillips RL, Urosek JE, Stephan CR, Stoltz RT, Swentosky DJ, Harris GW, O'Donnell JR, Dresch RA (2007) Report of Investigation Fatal Underground Coal Mine Explosion January 2, 2006 Sago Mine, MSHA, p 613

Harris ML, Weiss ES, Man CK, Harteis SP, Goodman GV, Sapko MJ (2010) Rock dusting considerations in underground coal mines. In: Proceedings of the 13th US/North American Mine Ventilation Symposium, Sudbury, Ontario, Canada, MIRARCO-Mining Innovation, pp 267–271

Man CK, Teacoach KA (2009) How does limestone rock dust prevent coal dust explosions in coal mines? Min Eng 61(9):69

MSHA (2014) Rock dusting. arlweb.msha.gov/S&HINFO/RockDusting/RockDusting.asp

Phillips CA (2012) Report of Investigation into Mine Explosion at the Upper Big Branch Mine, WV Office of Miners' Health, Safety and Training, p 230

Prins W, Siemons R, Van Swaaij WPM (1989) Devolatilization and ignition of coal particles in a two-dimensional fluidized bed. Combust Flame 75:57–79

Safety SAOW (2006) Technical specifications of comprehensive dust control measures for underground coal mine (in Chinese)

Stephan CR. Coal dust explosion hazards, p 10. https://arlweb.msha.gov/S&HINFO/TECHRPT/P&T/COALDUST.pdf

Zeman F (2008) Effect of steam hydration on performance of lime sorbent for CO_2 capture. Int J Greenh Gas Control 2(2):203–209

Zipf RK, Gamezob VN, Sapkoa MJ, Marchewkaa WP, Mohameda KM, Oranb ES, Kesslerb DA, Weissa ES, Addisa JD, Karnacka FA, Sellersa DD (2013) Methane–air detonation experiments at NIOSH Lake Lynn Laboratory. J Loss Prev Process Ind 26(2):295–301

Zou DH, Panawalage S (2001) Passive and triggered explosion barriers in underground coal mines—a literature review of recent research. Report to CANMET Natural Resources Canada, p 65

3

Physico-chemical and elemental investigation of aqueous leaching of high sulfur coal and mine overburden from Ledo coalfield of Northeast India

Madhulika Dutta[1,2,3] · **Puja Khare**[4] · **Sanchita Chakravarty**[5] · **Durlov Saikia**[2] · **Binoy K. Saikia**[1]

Abstract Acid mine drainage (AMD) is one of the major problems in high sulfur coal mining areas generating acidic water. The acidic mine water generated contain hazardous elements in varying concentrations. The Northeastern Indian coalfield produces considerable amounts of AMD. The AMD and metal leaching from coal and mine over burden (OB) are the two important naturally occurring processes. In order to know the AMD potential, the aqueous leaching experiment of a few coal and OB samples from the Ledo colliery of the Northeastern coalfield, Mergherita (India), at different time periods (1, 3, 5, and 8 h) and at different temperatures (25, 45, 65, and 90 °C) were performed in the laboratory. The physico-chemical analysis of the aqueous leachates shows the pH, electrical conductivity (EC), and total dissolved solid (TDS) in the range of 1.62–3.52, 106–2006 μs/cm, and 106–1003 ppm for the raw coal samples respectively. The OB samples produced pH, EC, and TDS in the range of 3.68–6.92, 48.6–480 μs/cm, and 69.5–240 ppm respectively. From the study, it was revealed that the concentrations of major (Si, Al, K, Na, Fe, Ca, Mg), minor (Mn) and trace/hazardous elements (As, Ba, Cd, Co, Cr, Cu, Hg, Ni, Pb, Se, Zn) considerably change with leaching time as well as with leaching temperature. Out of these elements As, Cd, Hg, Pb, Cr, and Se are of greater environmental importance. Alteration of the physico-chemical structure of the coal and OB samples resulting from leaching was also studied by field emission scanning electron microscope- energy-dispersive X-ray spectroscopy method. The release of the potentially hazardous elements from the raw coal and OB during leaching time periods to the leachates was detected by inductively coupled plasma-atomic emission spectroscopy and ion-chromatographic analyses. The major minerals found in coal and OB are quartz (SiO_2), pyrite (FeS_2), hematite, marcasite, and kaolinite. The association of different functional groups in minerals and their mode of association were studied by Fourier-transform infrared spectroscopy and X-ray diffraction analytical techniques. The present laboratory study will be useful in relating the characteristics of aqueous leaching from coal and mine OB with the natural weathering condition at the coal mine area.

Keywords AMD · High sulfur coal · Coal mine overburden · Aqueous leaching · Elements in coal and OB · Coal quality assessment

✉ Binoy K. Saikia
bksaikia@neist.res.in; bksaikia@gmail.com

[1] Polymer Petroleum and Coal Chemistry Group, Materials Sciences and Technology Division, CSIR-North East Institute of Science and Technology, Jorhat 785006, India

[2] School of Applied Sciences, Department of Chemistry, University of Science and Technology, Ribhoi 793101, Meghalaya, India

[3] Department of Chemistry, CNB College, Bokakhat 785612, India

[4] Agrotechnology Division, CSIR- Central Institute of Medicinal and Aromatic Plants, Lucknow 226015, India

[5] Analytical Chemistry Division, CSIR- National Metallurgical Laboratory, Jamshedpur 831007, India

1 Introduction

Coal is a heterogeneous mixture of different organic and inorganic materials. Majority of inorganic materials are occurred in mineral forms. The minerals present in coal consist of several metals and nonmetals especially Si, S, Ca, Mg, Fe, Al etc. (Tomeczec and Palugniok 2002; Mukherjee and Borthakur 2003). The common minerals found in coal and associated mine rejects (OB) are quartz (SiO_2), pyrite (FeS_2), hematite, marcasite, and kaolinite. Coal is composed of different elements, metals, and nonmetals as a long biological and physical alteration process taken place during the formation of the coal (Swine 1990; Singh and Singh 1995). Before mining these elements are in an immobile state, but during and after mining they have a tendency to disperse into the surrounding areas (Equeenuddin 2015).

Most of these elements are found in trace amount which are toxic in nature beyond their threshold limit. Trace elements are those whose concentrations are lower than 0.1% in the coal (Swaine 2000; Singh et al. 2012a). The study of these trace elements is very important due to their complex changes, inability to decompose in natural processes like weathering and high toxicity to the environment followed by their release and dispersal to the atmosphere, soil and water through the pathways of leaching and weathering (Zhou et al. 2014; Querol et al. 2008; Dai et al. 2005; Chen et al. 2013). Leaching is one of the primary pathways for trace elements to enter into the ecosystem (Yang et al. 2016, Baruah et al. 2004). Swaine (2000) proposed the possibility of the presence of twenty six trace elements in coal that have potential environmental impacts, including As, Cr, Cd, Hg, Pb, Se, B, Mn, Ni, Cu, V, Zn, Co, Sn, Cl, F, Br, Mo, P, Th, U, Ba, Ra, Sb and Tl. Out of these elements As, Cd, Hg, Pb, Cr and Se are of greater environmental importance (Swaine 2000; Prachiti et al. 2011; Singh et al. 2015a, b). Most of the Se in coal is associated with organic constituents; however, some of the Se is also associated with the sulfide minerals of coal, mainly pyrite (Swaine 1990). In coal, Cd is predominantly associated with Sphalerite (ZnS) (Alper et al. 2008). Lead (Pb) occurs in coal mainly as the sulfide mineral galena (PbS) which is the most common form of lead in coal (Finkelman 1994; Swaine 1990). These trace elements present in coal and its mine rejects/overburden enter into the environment through different processes including leaching of coal and oxidation of pyrite (Equeenuddin 2015). Coal dumps are also an important factor in the environmental degradation because the potentially toxic elements are directly released to the environment through the leaching and oxidation of pyrite. To understand the release of potential hazardous elements (PHE's) from coal and coal mine rejects in potential environmental and ecological conditions, the leaching behavior of coal and mine rejects in open cast coal mining is important. During the coal mining activities, the pyrite present in coal is exposed and comes into contact with the atmospheric oxygen, moisture or rain water. This weathering process leads to the generation of AMD. AMD generation and metal leaching processes from coal and mine OB are natural processes. These processes are accelerated by intensification of mining operations (Banerjee 2014). Such mining activities have a detrimental effect on the local environment as well as on the ecology (Kimmel 1983; Johnson 1987; Kaeser and Sharpe 2001).

Due to the formation of sulfuric acid by exposing pyrite to atmospheric oxygen and water, acid mine drainage is generated in coal mining areas. The metals in coal and coal wastes, which are minerals associated with the inorganic part of the coal structure are easily dissolved in highly acidic AMD. Thus, the AMD (pH 2.0–3.0) (Baruah et al. 2006; Dowarah et al. 2009) has a strong leaching potential for releasing a large number of potentially PHE's into the aquatic environment. In addition to the AMD formation, under different atmospheric conditions, pyrite (FeS_2) in coal is often found in association with the hazardous elements such as As, Cd, Cu, Co, Pb, Hg, and Zn (Monterroso and Macias 1998; Singh et al. 2016). The formation of elemental sulfur takes place during depyritization of coal by leaching of coal wastes (Schippers et al. 1996). The leachates generated from OB dumps are rich in heavy metals, especially Fe, Cu, Mn and Ni (Tiwari 2001). The leaching property of different elements depends on several factors like specific element solubility and availability or release potential (Moyo et al. 2012). Also the other factors that enhance the metal leaching include swiftly weathering metal containing minerals and the mine drainage conditions that can increase the solubility and flow rates through the contaminated minerals (Banerjee 2014).

The sulfur present in Northeastern (NER) Indian coal is mainly in three forms-sulfate sulfur, pyritic sulfur, and organic sulfur (Baruah and Gogoi 1998). The major part of the inorganic sulfur present in coal is in pyrite (FeS_2). The Northeastern Indian coals are rich in sulfur (2%–11%) (Chabukdhara and Singh 2016; Singh et al. 2012b, 2013). The sulfate formed from pyritic sulfur in coal by atmospheric oxygen is not only responsible factor for AMD generation but the oxidation of Fe(II) ion to Fe(III) ion during the process of weathering of pyrite can produce a noticeable amount of AMD (Baruah et al. 2006). Equeenuddin (2015) reported that coal dumps cause environmental degradation due to the release of potentially hazardous elements through the process of leaching of pyrite. The amount of impact of waste and coal containing contaminants in a form of toxic elements on the environment depend not only on the type and concentrations of these elements, but mainly on their mobility. The mobility of pollutants contained in the waste is expressed as the potentiality of them to leach in water (Kalembkiewicz and Palczak 2015). Leaching methods are classified into two types depending upon whether the leaching liquid is added at a single instance (static) or it is renewed (dynamic) (Moyo et al. 2011). The potential to produce acidity of drainage from mine water and mine wastes can be predicted by both geochemical methods in static and kinetic leaching processes (Ferguson and Erickson 1998). The static methods are further classified into two classes, batch leaching where a specific amount of sample is placed in a given amount of leachant solution (US DOE 2005; Hesbach et al. 2005; Kim et al. 2003) and column leaching.

Thus, the quality of surface and ground water in the streams, drains, well, contaminated with AMD water and leachates of coal and overburden was also analyzed to study the possible impact of coal mining on the nearby water resources. During leaching, the elements present in coal and OB dissolve or undergo some chemical and physical reactions and as a result they transfer into the leaching solution (Fotoupoulou et al. 2010; Wang et al. 1999). In the present work, the static aqueous leaching experiments are mainly carried out to evaluate the potential of release of selected metals. The aqueous leaching experiments of different coal and overburden samples, from the Ledo colliery are carried out and the results are also compared with those of field data.

1.1 Study area

The Makum coalfield is the largest coalfield among all the coalfields of the Northeastern region (NER) of India. The coalfield is located in the Tinsukia district of Assam and includes five working collieries namely Ledo, Baragolai, Tirap, Tikak, and Tipong. Among them, Ledo is the most recent and active colliery where open cast mining activities are operational. The mining area in Ledo-Margherita range is located within 27°15′–27°25′N latitudes and 95°40′–95°50′E longitudes which includes the probable mine affected area. In this study, the nearby areas influenced by the Ledo open cast mining activities have also been chosen. The chosen area extends from Ledo mining area to the Tirap mining area including the Tirap River, Ledopani River, and Kachanalla River. The areas near the Kachanalla and Tirap rivers are large agricultural areas with a tea garden, rural habitation area, and a colony of mining workers. A number of streams and ponds are also found which come across the mine area. The major sources for drinking water for the people of these areas are tube-wells and dig-wells. A location map of the study areas is shown in the Fig. 1.

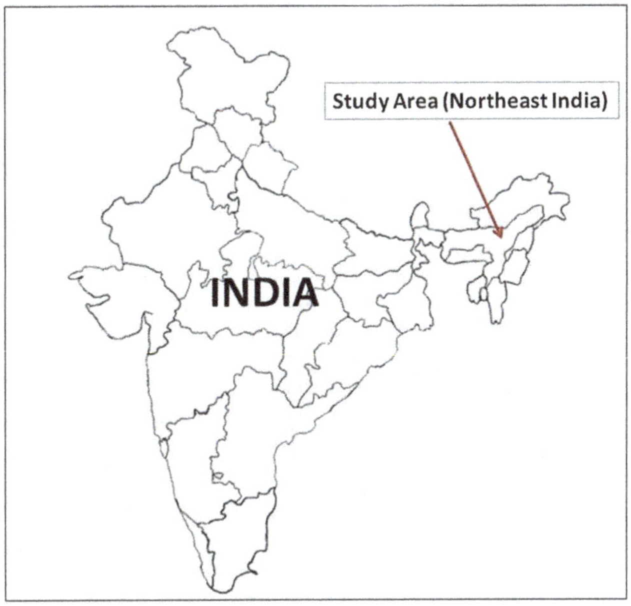

Fig. 1 Map showing the location of Ledo colliery (study area) in Northeast India region (not to scale)

2 Experimental sections

2.1 Sampling and aqueous leaching

Freshly mined representative coal samples (LC-20A, LC-60A) and overburden samples (LOB-15A, LOB-15B) (dump sites) were collected from the Ledo colliery (India). The coal and overburden samples were ground into-72 BS (0.211 mm) size, preserved in a desiccator for subsequent experiments. Mine water samples (LW-15A, LW-15B, LW-15C, LW-15D, LW-15E, and LW-15F) were also collected from the AMD source and from different points within the 5 km radius from the mining source of the Ledo colliery as shown in Fig. 2.

100 g of each finely ground sample was mixed with distilled water (300 ml) for 1, 3, 5, and 8 h at 25, 45, 65, and 90 °C respectively. Then the mixture was mechanically stirred in an electronic shaker (VELP, Scintifica JLT6) for the time periods 1, 3, 5, and 8 h at a stirring rate of 250 rpm. The supernatant liquid was filtered from the solid phase with a Whatmann-1 filter paper. The filtrate was collected and used for different physical, chemical, and elemental analyses. The residues of the leached samples were also subjected to selected chemical analyses as mentioned in later part.

2.2 Physico-chemical analysis of aqueous leachates

The aqueous leachates of the coal and overburden (OB) samples obtained were subjected to pH, TDS (total

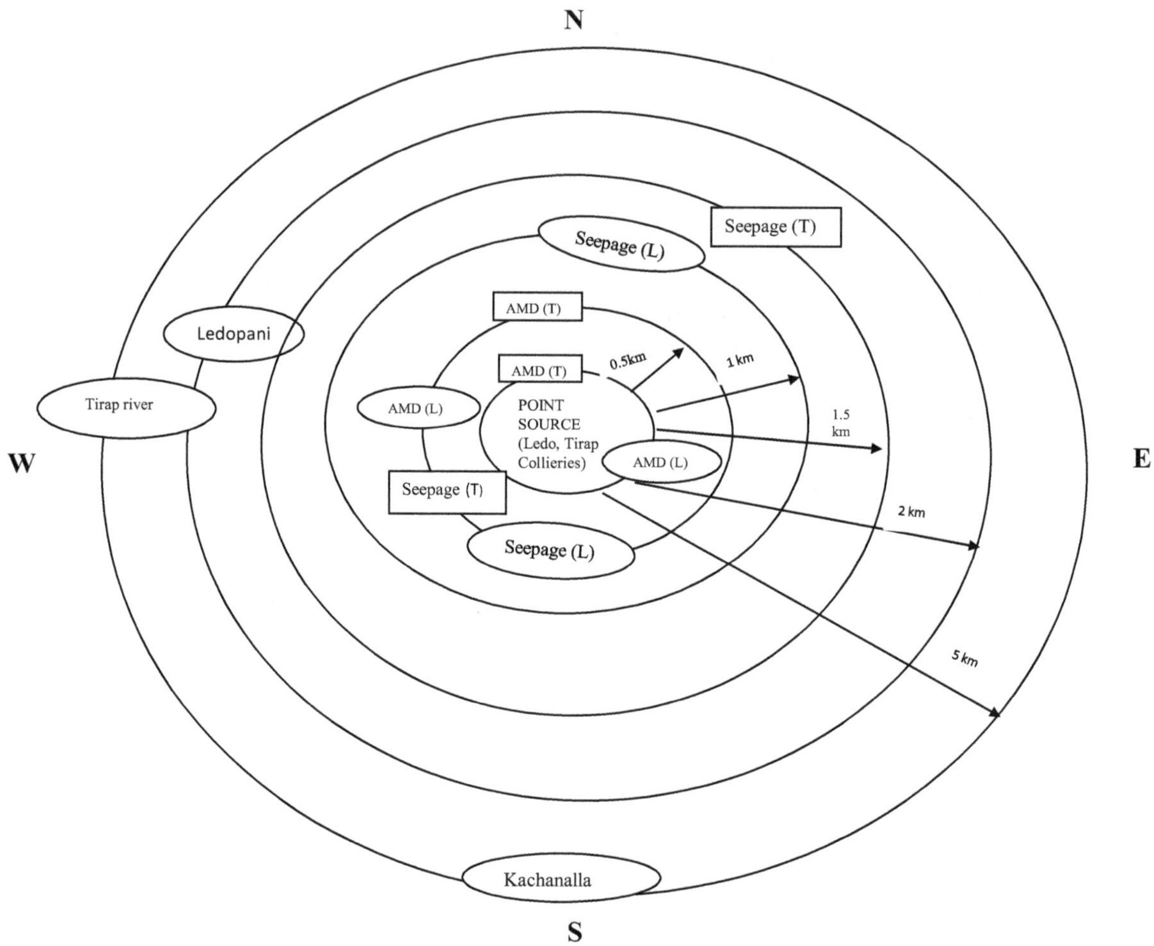

Fig. 2 Schematic diagram showing the actual sampling locations in around Ledo colliery (Open cast mining)

dissolved solid), and EC (electrical conductivity) analysis by using a EUTECH PC700 pH/EC/TDS meter.

2.3 Chemical analysis of aqueous leachates

2.3.1 Ion-chromatographic analysis

Ion-chromatographic (IC) analysis of the aqueous leachates of coal and overburden samples at different time and temperatures were subjected to IC analysis to determine the concentrations of different cations and anions (in ppm) by using an ion chromatography system-Metrohm 882 Compact IC Plus using standards purchased from Merck and Sigma Aldrich. Before analysis, the liquid samples were filtered through 0.22 μm Millipore filter paper. The anions were analyzed using 4 mm × 250 mm Metrosep A Supp 5 analytical column with the chemical suppression sulfuric acid and Metrosep RP2 3.5 guard column. Similarly, the cations were analyzed using a 4 mm × 150 mm Metrosep C-4 analytical column and Metrosep RP2 3.5 guard column.

2.3.2 Inductively coupled plasma optical emission spectrometry (ICP-OES) and atomic absorption spectrometry (AAS)

The trace elements were determined by using an inductively coupled plasma optical emission spectrometry (ICP-OES). In this process 0.2 g from reference material for digestion was used. Wet digestion of samples was performed using 5 ml mixtures of acids: HNO_3:$HClO_4$ (3:1). Each sample was heated up to 180 °C for 3.0 h (sample should be transparent) on the heating digestion block. Then the acid digest sample was allowed to cool and filtered into a 25 mL volumetric flask, using Whatmann filter paper and made up to mark with de-ionised water. The blank digests sample was similarly processed. The digested samples were followed by ICP-OES, (Perkin Elmer, Optima 5300 V).Calibration curves for elements were created from mixtures of high-purity element standard solutions (Merck, Germany) in a 4% (v/v) HNO_3 matrix.

Atomic absorption spectrometry (AAS), flame atomic absorption spectrometry (FAAS), and graphite furnace atomic absorption spectrometry (GFAAS) analytical methods were used to detect different elements. GAAS is based on the same principles as FAAS with the main difference in the atomization of the samples. Instead of the flame in FAAS, the GFAAS uses an electro thermal atomizer (a small graphite tube). The samples were prepared for both analyses by digesting the samples with HNO_3–H_2O_2 in a ratio of 4:1 using microwave assisted acid digestion process. For FAAS analytical method, different standard solutions were used like Fe [(p/n 8003-0906), Zn (p/n 8003-0942)], Cu (p/n 5190-8280). In GFAAS analytical method, stock standard solutions are prepared from analytical reagent grade high purity metals, oxides, or nonhygroscopic salts using reagent water and redistilled HNO_3 or HCl.

2.3.3 X-ray diffraction analysis

X-ray diffraction analysis (XRD) of the samples were carried out in an X-ray diffractometer, model JDXII PZA, JEOL, Japan, using Cu Kα Fe-filtered X-ray radiation (λ = 1.5426 Å) to study the presence of different minerals in the leached samples. Operating parameters were as follows: start angle: 5.0, target: Cu (Fe-filtered), stop angle: 80.0, measuring time: 0.5 and step angle: 0.03, data processing conditions included: smoothing points, Goniometer radius (R): 240 mm, equatorial angle subtended at the specimen by the detector slit (β): 1°.

2.3.4 Fourier transform infrared spectroscopy analysis

The Fourier transform infrared spectroscopy (FTIR) analyses of the leached samples were done by using Perkin-Elmer system 2000, model 640B with KBr pellet. The detector used was deuterated triglycine sulphate (DTGS). The total number of scans was 50 with the spectral resolution of 4 cm^{-1} during the recording of the spectra. The samples were subjected to FTIR analyses to determine the presence of different functional groups, including sulfur containing groups.

2.3.5 Field emission-scanning electron microscopic analysis

Field emission-scanning electron microscopy (FESEM) provides elemental information at magnifications of 10×–300,000× with virtually unlimited depth of field. Compared with convention scanning electron microscopy (SEM), field emission SEM provides clearer, less electrostatically distorted images with spatial resolution down to 11/2 nm—three to six times better. Field emission scanning electron microscopy (FESEM) images were obtained from a Carl Zeiss Sigma FESEM instrument. Energy dispersive X-ray spectroscopy (EDS) pattern and elemental mapping analysis was recorded on Oxford X Max 20 equipment.

3 Results

3.1 Physico-chemical characteristics of aqueous leachates

The physico-chemical parameters including pH, EC, and TDS were determined for the aqueous leachates obtained at increasing time intervals and temperatures. Table 1 shows that the pH values of the leachates of the samples decreased gradually from 1 h of leaching to 5 h of leaching time at room temperature. The pH of the leachates however increased slightly at 8 h of leaching time. The leaching at 45 °C for 1, 3, 5 and 8 h respectively demonstrate a gradual increase in pH values. The electrical conductivity (EC) of the aqueous leachates of the samples at room temperature (25 °C) gradually increased from 1 to 8 h of leaching. Similarly, the TDS values were found to increase gradually with increase in time period of leaching and with increase in leaching temperature (from room temperature to 45 °C). Various physico-chemical properties of the aqueous leachates of coal and OB samples are given subsequently. The pH of the leachates (leached in different time periods at different temperature) from the Ledo coal is highly acidic (pH 2.5) while other OB samples release a nearly neutral discharge (4–7.4). In the cases of leaching at a temperature higher than room temperature (45 °C), the pH increased up to 9.40.

3.2 Elemental characteristics of aqueous leachates

3.2.1 Ions present in the aqueous leachates of coals and overburdens

The Ion-Chromatographic analyses were carried out for aqueous leachates of coal and overburden samples after leaching for 1, 3, 5, and 8 h at 25, 45 and 65 °C respectively. The concentration of cations like Na^+, Mg^{2+}, Ca^{2+}, NH_4^+ and Li^+ in the leachates of coal and overburden decreased with an increase in leaching temperature (Figs. 3, 4, 5, 6, 7, 8, 9, 10). The leaching of Li^+ ion in aqueous solution is found to be decreased with leaching time from 1 to 8 h as shown in Table S5 (see in supplementary information).

From Figs. 11, 12, 13 and 14, it can be observed that the concentrations of F^-, Cl^- and SO_4^{2-} ions from the leachates of coal and overburden samples of Ledo colliery increased with an increase in the leaching time at room temperature. With the increase in leaching temperature

Table 1 Physical analysis of aqueous leachates at room temperature at different leaching times

Sample type	Sample code	Leaching time (h)	pH	TDS (ppm)	EC (μs/cm)
Coal	LC-20A	1	1.99	989	1912
	LC-60A		2.65	106	209
	LC-20A	3	1.80	983	1970
	LC-60A		2.27	119	238
	LC-20A	5	1.62	1003	2006
	LC-60A		3.52	106.9	210
	LC-20A	8	2.10	790	313
	LC-60A		3.50	213	106
Overburden	LOB-15A	1	6.72	120	242
	LOB-15B		5.85	24.2	48.6
	LOB-15A	3	4.76	129	129
	LOB-15B		6.92	89.2	89.2
	LOB-15A	5	3.68	240	480
	LOB-15B		3.71	186	373
	LOB-15A	8	5.36	139	278
	LOB-15B		6.30	69.5	129.5

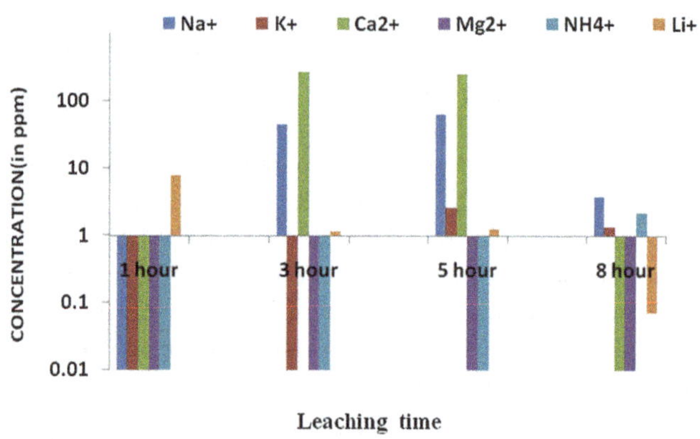

Fig. 3 Change in concentration (in ppm) of cations in aqueous leachates of Ledo coal (LC-20A) with leaching time (in hour) at 25 °C

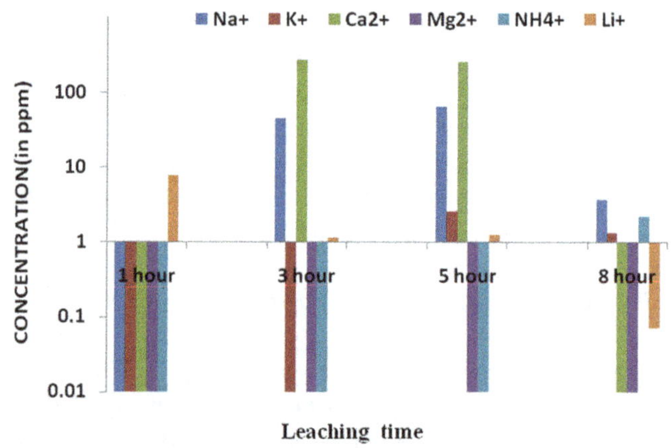

Fig. 5 Change in concentration (in ppm) of cations in aqueous leachates with leaching time (in hour) of Ledo overburden (LOB-15A) at 25 °C

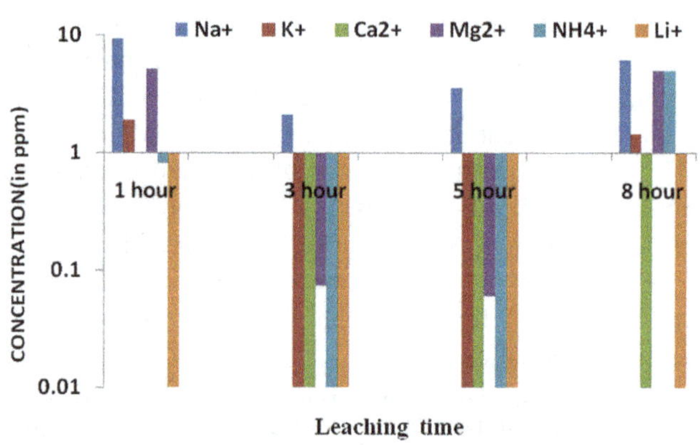

Fig. 4 Change in concentration (in ppm) of cations in aqueous leachates of Ledo coal (LC-60A) with leaching time (in hour) at 25 °C

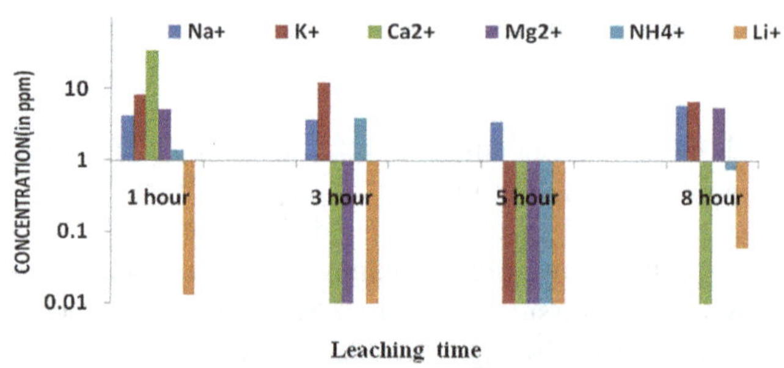

Fig. 6 Change in concentration of cations (in ppm) in aqueous leachates of Ledo overburden (LOB-15B) with leaching time (in hour) at 25 °C

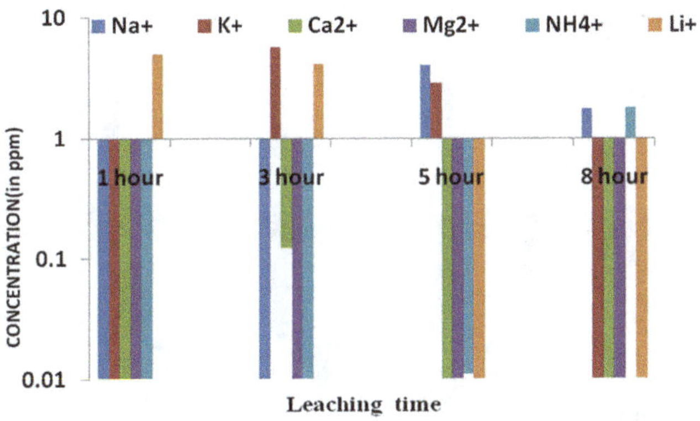

Fig. 7 Change in concentration of cations (in ppm) in aqueous leachates of Ledo coal (LC-20A) with leaching time (in hour) at 45 °C

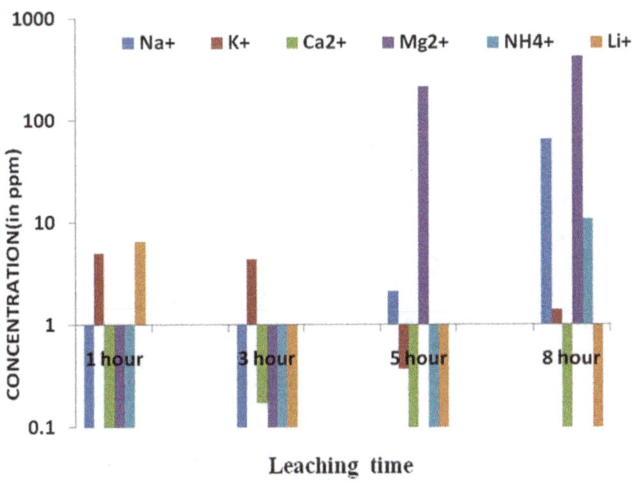

Fig. 8 Change in concentration of cations (in ppm) in aqueous leachates of Ledo coal (LC-60A) with leaching time (in hour) at 45 °C

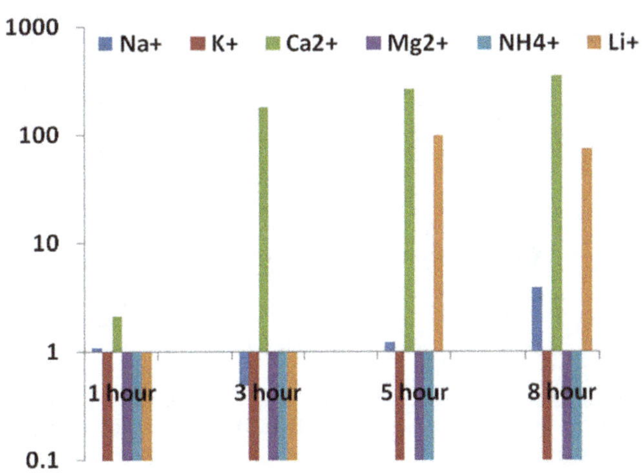

Fig. 9 Change in concentration of cations (in ppm) in aqueous leachates of Ledo coal (LC-20A) with leaching time (in hour) at 65 °C

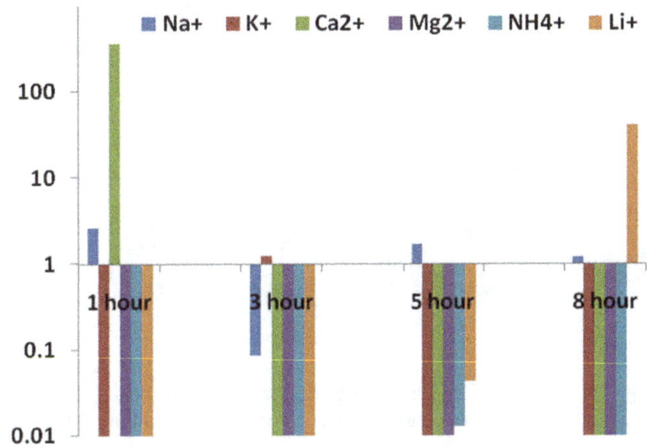

Fig. 10 Change in concentration of cations (in ppm) in aqueous leachates of Ledo coal (LC-60A) with leaching time at 65 °C

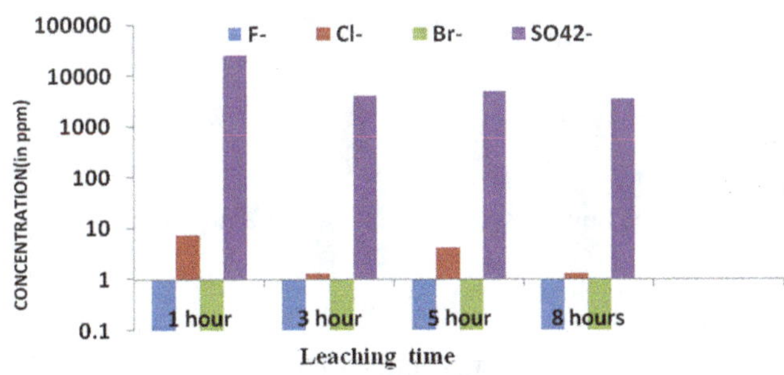

Fig. 11 Change in concentration (in ppm) of anions in aqueous leachates of Ledo coal (LC-20A) with leaching time (in hour) at 25 °C

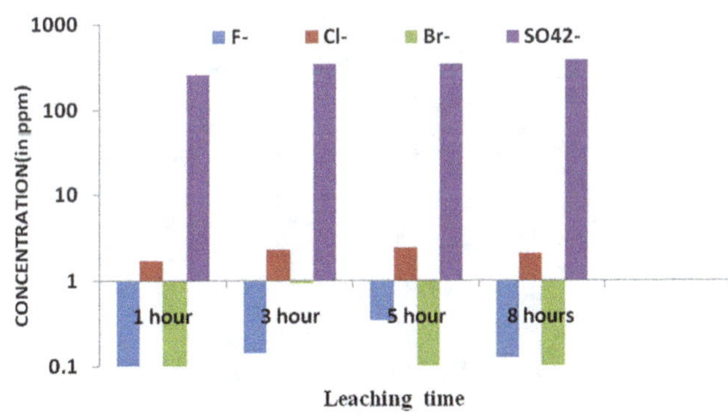

Fig. 12 Change in concentration (in ppm) of anions in aqueous leachates of Ledo coal (LC-60A) with leaching time (in hour) at 25 °C

from 25 to 45 °C however, the concentration of F^-, Cl^- became lower (Figs. 15, 16, 17, 18). The leaching of SO_4^{2-} ion from coal and overburden was increased with leaching temperature from 25 to 65 °C and time (see Figs. 15, 16, 17, 18, 19, 20, 21, 22).

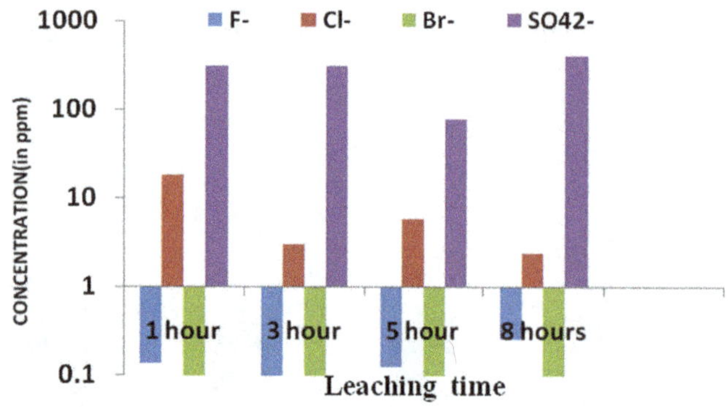

Fig. 13 Change in concentration (in ppm) of anions in aqueous leachates of Ledo overburden (LOB-15A) with leaching time (in hour) at 25 °C

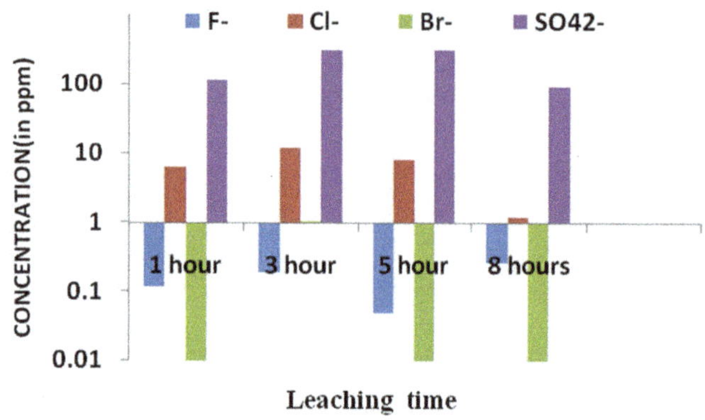

Fig. 14 Change in concentration (in ppm) of anions in aqueous leachates of Ledo overburden (LOB-15B) with leaching time (in hour) at 25 °C

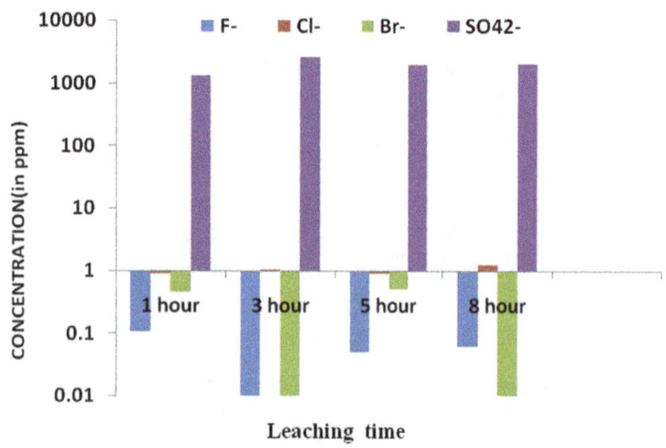

Fig. 15 Change in concentration (in ppm) of anions in aqueous leachates of Ledo coal (LC-20A) with leaching time (in hour) at 45 °C

3.2.2 *Major, minor, and trace/hazardous elemental analysis*

The mode of occurrence of an element plays an important role for prediction of its nature during the process of

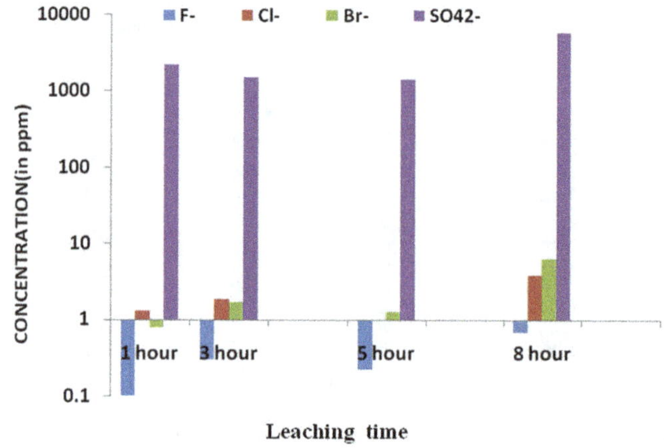

Fig. 16 Change in concentration (in ppm) of anions in aqueous leachates of Ledo coal (LC-60A) with leaching time (in hour) at 45 °C

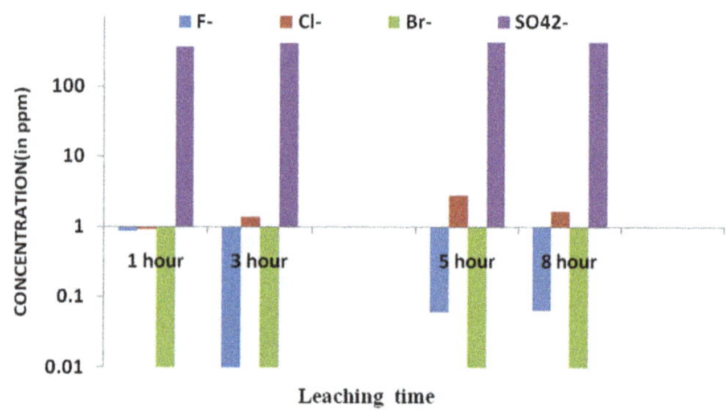

Fig. 17 Change in concentration (in ppm) of anions in aqueous leachates of Ledo overburden (LOB-15A) with leaching time (in hour) at 45 °C

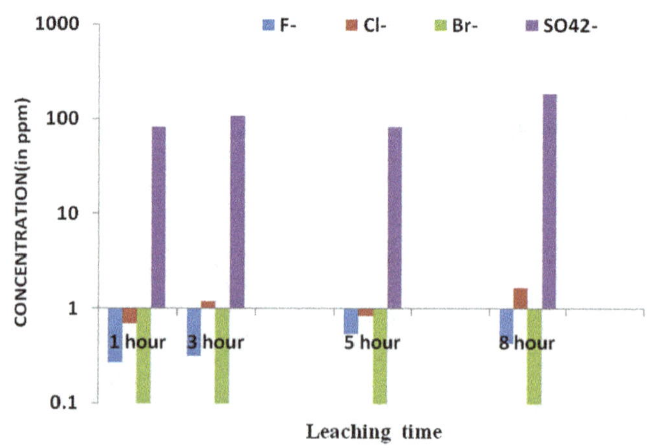

Fig. 18 Change in concentration (in ppm) of anions in aqueous leachates of Ledo overburden (LOB-15B) with leaching time (in hour) at 45 °C

leaching. The modes of occurrences of major elements like Al, Ca, Fe, Mg, Na, K, etc., minor element Mn and the trace/hazardous elements As, Ba, Cd, Cr, Co, Ni, Cu, Pb,

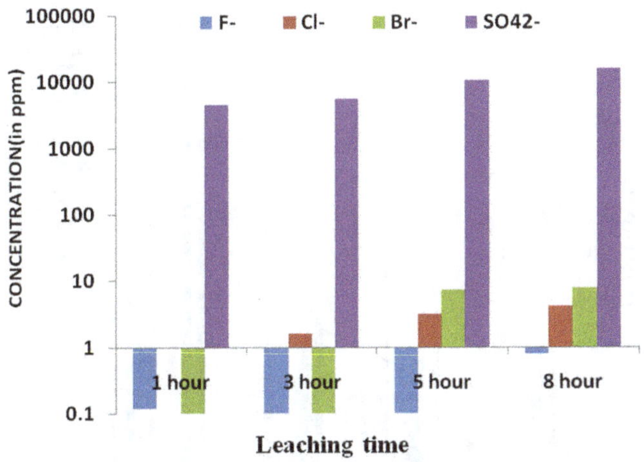

Fig. 19 Change in concentration (in ppm) of anions in aqueous leachates of Ledo coal (LC-20A) with leaching time (in hour) at 65 °C

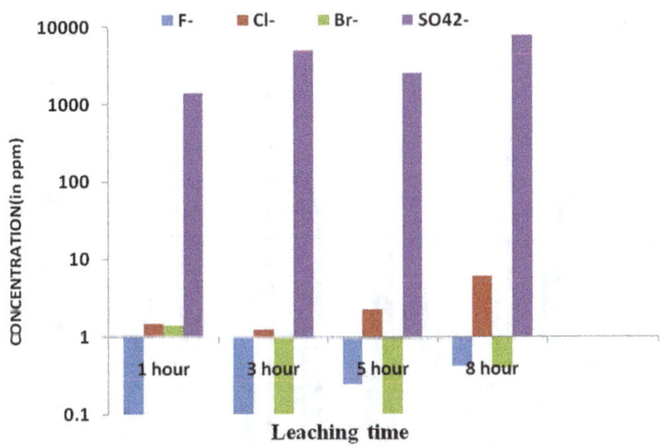

Fig. 20 Change in concentration (in ppm) of anions in aqueous leachates of Ledo coal (LC-60A) with leaching time (in hour) at 65 °C

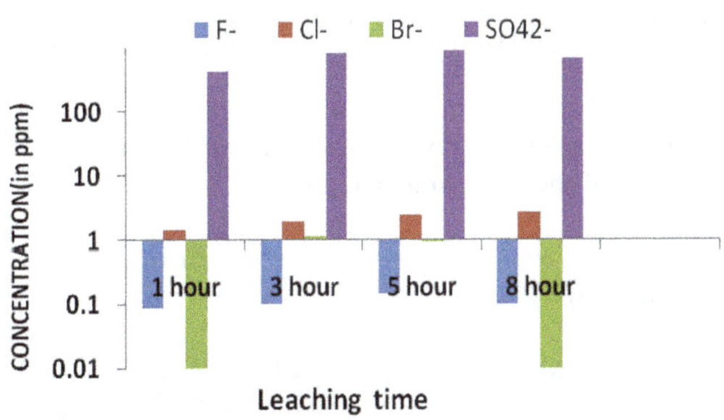

Fig. 21 Change in concentration (in ppm) of anions in aqueous leachates of Ledo OB (LOB-15A) with leaching time (in hour) at 65 °C

Hg, Ni and Zn as well as the minerals in coal can vary from coal to coal and according to their association in coal, the concentration of various elements in aqueous leachates can

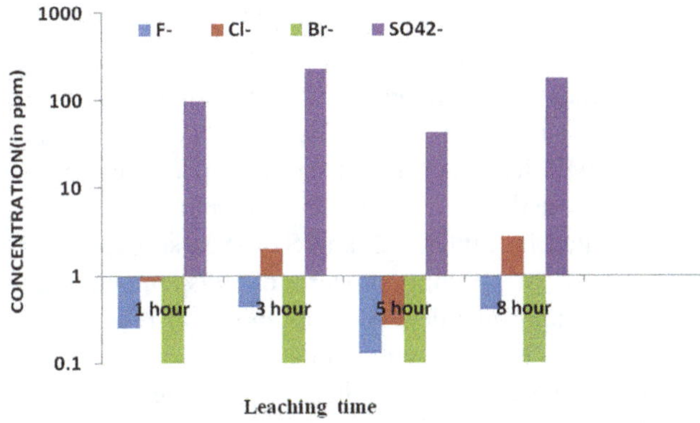

Fig. 22 Change in concentration (in ppm) of anions in aqueous leachates of Ledo OB (LOB-15B) with leaching time (in hour) at 65 °C

be detected. In most of the coals, the metals occur as mineral matters, or as discrete minerals, or as free or embedded in organic matter (Swain and Goodarzi 1995). From nano-mineralogical analysis of Northeast Indian coal and overburden samples, it was found that the Si minerals like kaolinite [$Al_2Si_2O_5(OH)_4$], sulphate minerals such as barite [$BaSO_4$], jarosite [$KFe_3(SO_4)_2(OH)_6$)], pickeringite[$MgAl_2(SO_4)_4{\cdot}22(H_2O)$], sulphide mineral galena [PbS], pyrite [FeS_2], and oxide mineral hematite[Fe_2O_3], are most common minerals which contains elements in their matrix. The presence of minerals like illite (K, H_2O)(Al, Mg,Fe)$_2$(Si,Al)$_4O_{10}$[$(OH)_2{\cdot}(H_2O)$)] and marcasite (FeS_2) in Northeast Indian coals are also reported by Mossbauer spectroscopic analysis elsewhere (Dutta et al. 2017).

The ICP-OES analysis of the aqueous leachates of the coal reveals that the concentration of different elements like Al, Ba, Co, Ni, Fe, Mn, Cu, and Zn increased with

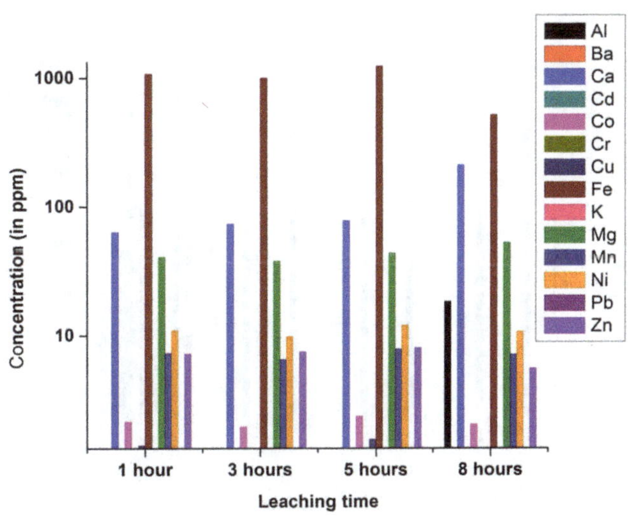

Fig. 23 Change in elemental concentration (in ppm) in aqueous leachates of Ledo coal (LC-20A) with leaching time (in hour) at 25 °C

leaching time at room temperature (25 °C) (Figs. 23, 24). From Figs. 23, and 24, it is found that the concentrations of major elements like Al, Fe, etc., minor element Mn and trace/hazardous elements like Ba, Co, Ni, Cu, Zn, etc. increased with leaching time at room temperature (25 °C). The rates of leaching of each element from raw coals are also indicated in Tables S23 and S24 (See supplementary information). The leaching property of Ni is temperature as well as leaching time dependent as shown in Figs. 23, 24, 25, 26, 27 and 28. The increase rates of leaching of Ni with time and temperature are indicated in Tables S23, S24, S25, S26, S27 and S28 (details are in supplementary information). Here the concentration of Ni in the aqueous leachates of coal is maximum at leaching time of 8 h. In the case of Fe, Mn, Cu, Zn, Ca and Mg the concentration was found to increase with leaching time and temperature (Figs. 23, 24, 25). But when the temperature is low (25 and 45 °C) the concentration of elements increase at maximum time of leaching of 8 h (Figs. 23, 24, 25). The release of Pb is maximum at 45 °C which is independent of leaching (Fig. 25). The release of Cu, Co and Zn from the coal and overburden is temperature as well as the leaching time dependent (Figs. 25, 26, 27, 28) Leaching rates of these elements increases gradually with time and temperature as indicated by Tables S25, S26, S27 and S28 (in supplementary information). The leaching of Al is fully leaching temperature dependent (Figs. 25, 26, 27, 28). From the Figs. 29 and 30 it is found that at high temperature most of the elements present in coal are leached when the leaching time is 1 h (Figs. 31, 32, 33, 34). The temperature dependence leaching of elements at a constant leaching time is shown in Tables S29 and S30 (see in supplementary information). In the case of Fe, the concentration is higher in leachates than in raw samples (the maximum Fe concentration in raw coal was 43.1 and 2320 ppm in the leachates of the coal samples).

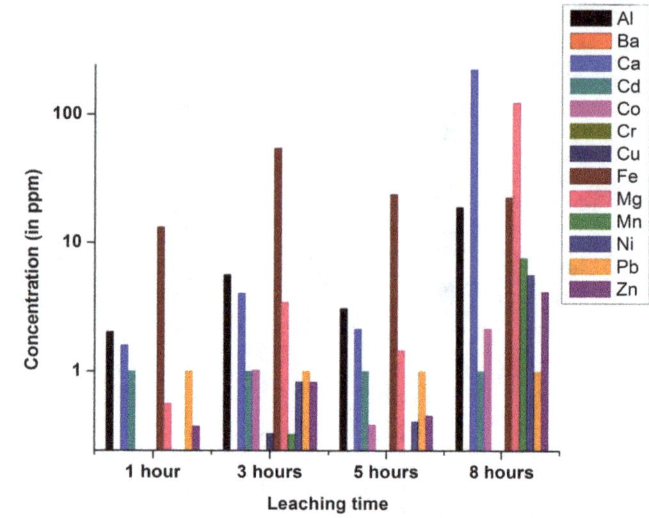

Fig. 25 Change in elemental concentration (in ppm) in aqueous leachates of Ledo coal (LC-20A) with leaching time (in hour) at 45 °C

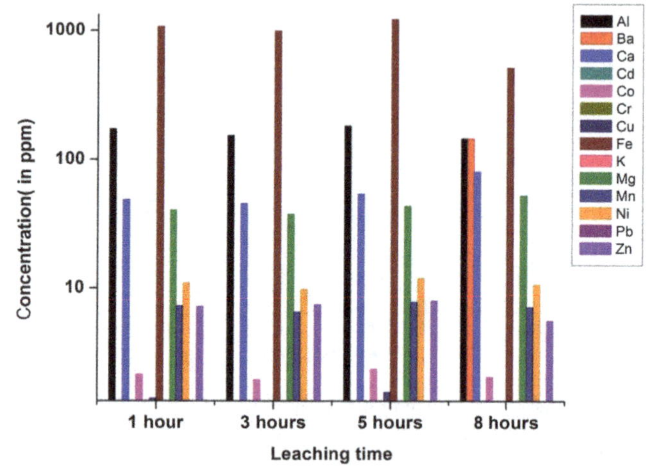

Fig. 26 Change in elemental concentration (in ppm) in aqueous leachates of Ledo coal (LC-60A) with leaching time (in hour) at 45 °C

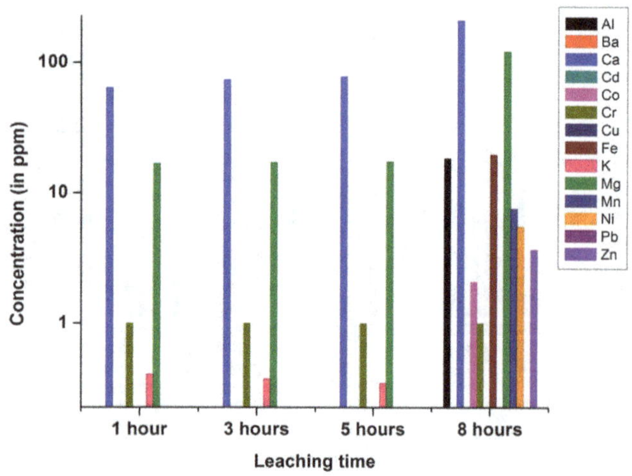

Fig. 24 Change in elemental concentration (in ppm) in aqueous leachates of Ledo coal (LC-60A) with leaching time (in hour) at 25 °C

3.2.3 Field-emission-scanning-electron microscopic analysis

In the FE-SEM analysis of raw coal samples of the Ledo coal (Fig. 35) it was found that barite ($BaSO_4$) and galena (PbS) were associated with pyrite (FeS_2) which is not shown by EDS analysis of leached coal (Fig. 36). From the FE-SEM/EDS analysis of both the raw and leached samples, it was found that the atomic percentage of Fe in raw coal and overburden sample (LC-20A, LOB-15A) were 18.33% and 8.43% whereas in leached coal it was 7.84%

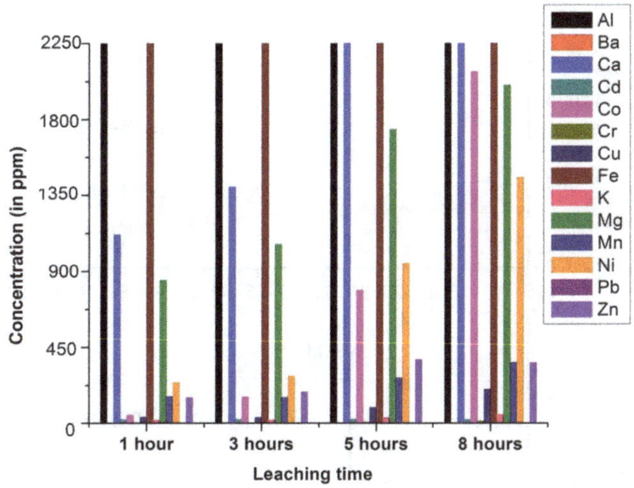

Fig. 27 Change in elemental concentration (in ppm) in aqueous leachates of Ledo coal (LC-20A) with leaching time (in hour) at 65 °C

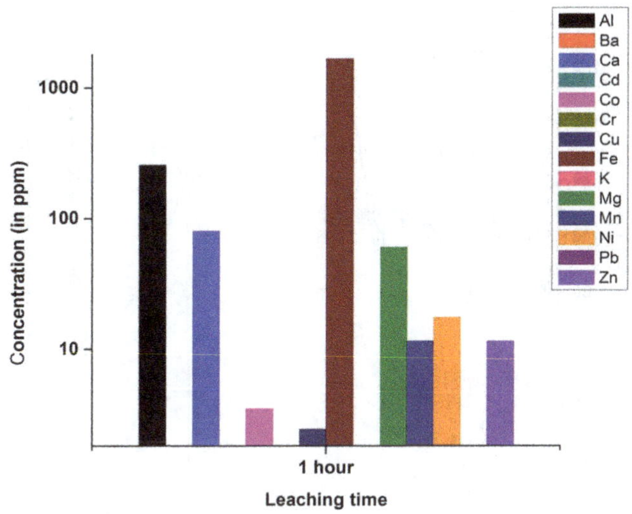

Fig. 29 Change in elemental concentration (in ppm) in aqueous leachates of Ledo coal (LC-20A) leaching for 1 h at 90 °C

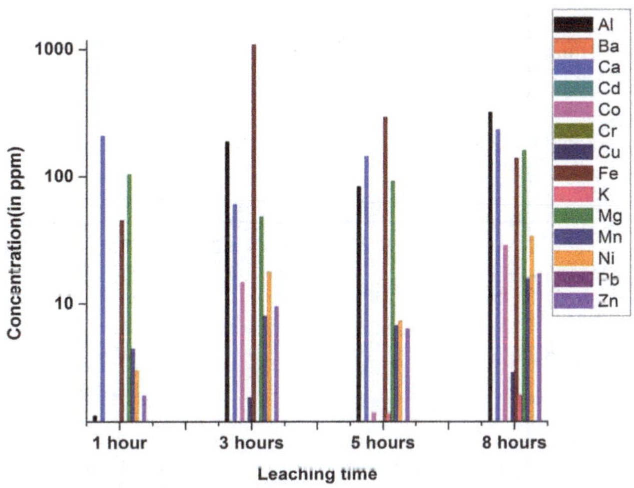

Fig. 28 Change in elemental concentration (in ppm) in aqueous leachates of Ledo coal (LC-60A) with leaching time (in hour) at 65 °C

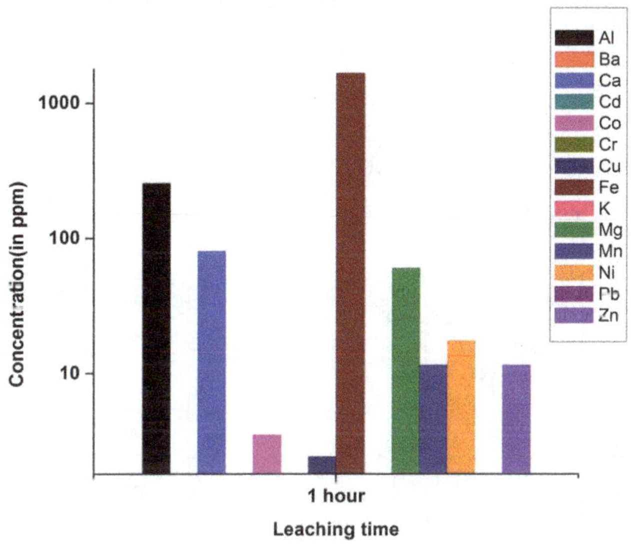

Fig. 30 Change in elemental concentration (in ppm) in aqueous leachates of Ledo coal (LC-60A) leaching for 1 h at 90 °C

and 4.80% respectively. The FESEM–EDS analysis of leached coal and OB of the Ledo coal mine were carried out to find the approximate concentration of metals in the residues of leached samples.

4 Discussions

4.1 Physico-chemical characteristics of aqueous leachates

From the Table 1 it is found that the leaching at 45 °C for 1, 3, 5 and 8 h respectively demonstrate a gradual increase in pH values The gradual increase in pH (decrease in acidity) values implies that the continuous rainfall at normal temperature for 5 h can generate AMD at the coal mine site. Thus, AMD from coal and mine OB are highly acidic at the coalfield site during the rainy season. The gradual increase in pH values of leachates at 45 °C from 1 to 5 h indicates the low acidity of leachates with rise in leaching temperature. A high pH value of leachates may be attributed to the formation of $Ca(OH)_2$ from $CaCO_3$ present in overburden due to long time leaching (up to 8 h), since the pH of a solution is an important determinant for counting aquatic toxicity. The amount of toxicity tends to be higher at low pH or high pH situations than at an approximately neutral pH (near pH 7), since the dissolution of many metals can be considered as amphoteric, with a high tendency to dissolve and can form cations at low pH

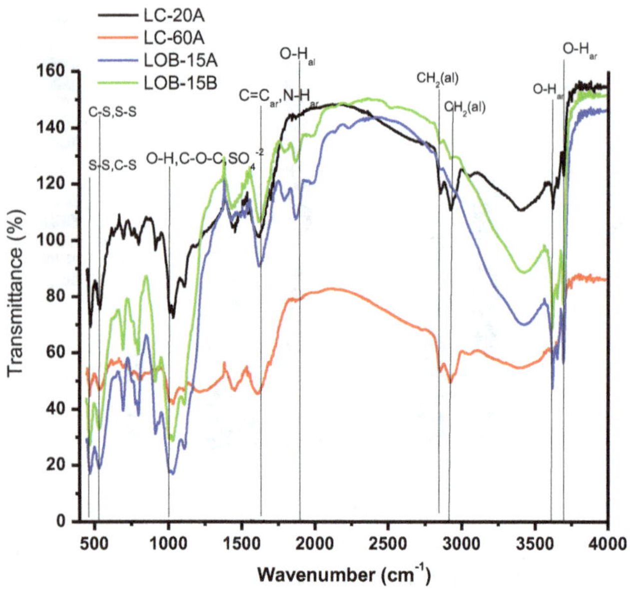

Fig. 31 FT-IR spectra for the leached coal and OB of the Ledo colliery

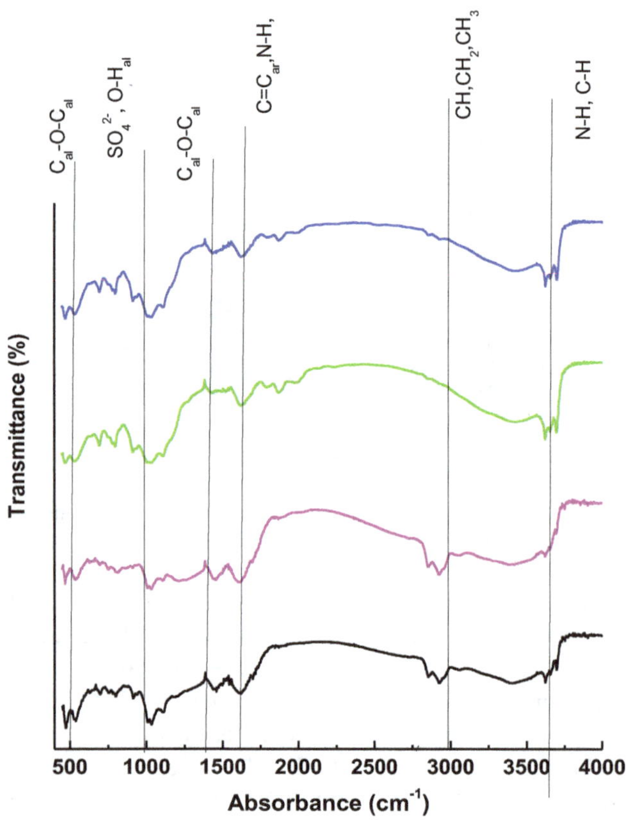

Fig. 32 FT-IR spectra for raw coal and OB of Ledo colliery

or anions at high pH values (Langmuir 1997). For instance, $Al(OH)_3$ and different aluminosilicate minerals like kaolinite $[Al_2Si_2O_5(OH)_4]$ have their minimum solubility at pH values 6–7 (Nordstrom and Ball 1986; Bigham and

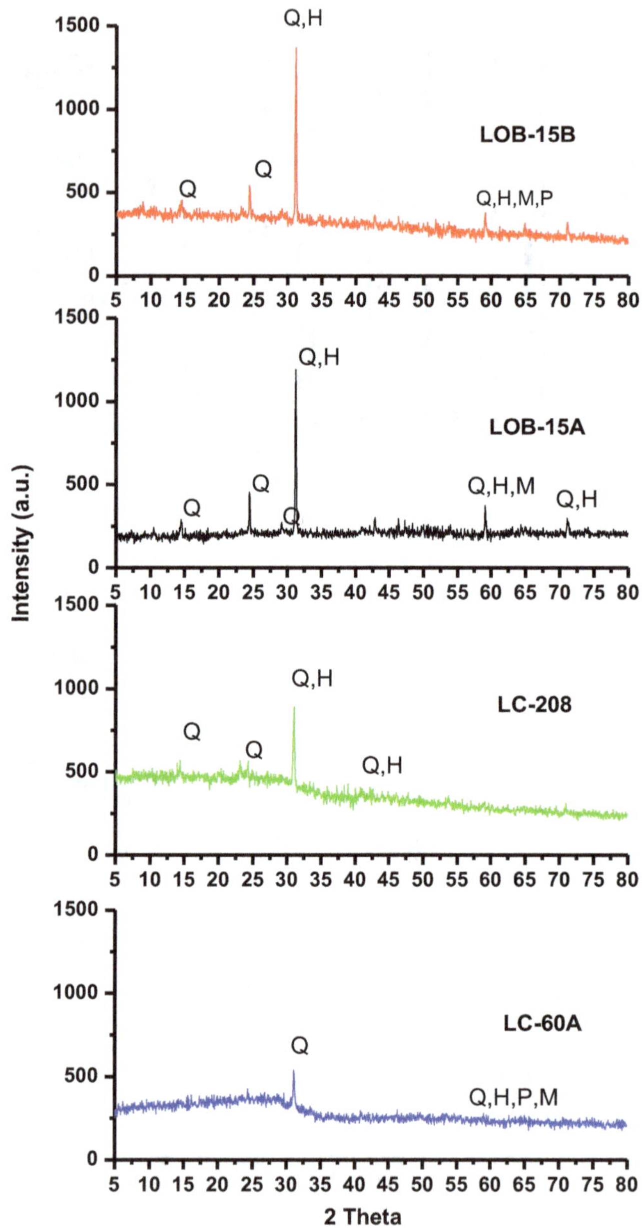

Fig. 33 XRD pattern of leached coal and OB of Ledo colliery

Nordstrom 2000) and even relatively low concentrations of dissolved Al can exert toxic effect to fish and other aquatic organisms (Baker and Schofield 1982; Elder 1998). The Environmental Protection Agency (EPA 2000, 2002a, b) recommends a pH 6.5–9.0 for protection of freshwater aquatic life and pH 6.5–8.5 for public drinking purposes. In the present study the maximum pH values for leachates of overburdens were found to have a pH 9.4. This indicates a lower contamination of water resources from overburden than for coal. It should however be noted that pH is not the only determining factor of metals dissolution. Comparing the physical parameters of aqueous leachates of coal and overburden with the mine waste water it was found that the elements from coal and overburden can leach easily under normal weathering processes.

Again, from the Table 1, it was revealed that at room temperature the value of EC and TDS of leachates increase with leaching time. The high value of EC and TDS at a temperature higher than room temperature indicates the high mobility of ions present in the samples which were leached out of the sample as a result of leaching.

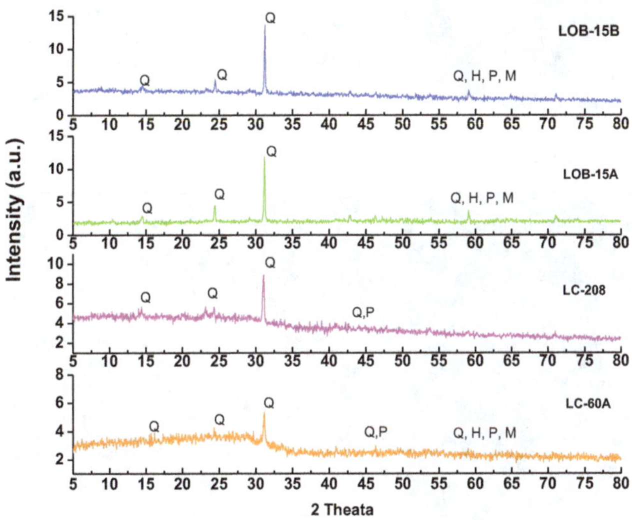

Fig. 34 XRD graph for raw coal and OB of Ledo colliery

4.2 Elemental characteristics of aqueous leachates

4.2.1 Ions present in the aqueous leachates of coals and overburdens

Figures 9 and 10 show that the leaching of Ca^{2+} ion from Ledo coal is maximum at high temperature 65 °C which is an indication of maximum leaching of Ca^{2+} ion in monsoon season. The low concentration of Na^+, Mg^{2+}, NH_4^+ and Li^+ ions in the leachates of coal and overburden at high temperature indicates minimum leaching of these ions in monsoon season. Table S5 shows the decrease in leaching of Li^+ ion with leaching duration. Since lithium is generally associated in coal and overburden as lithium silicate and the decrease in concentration of Li^+ ion in leachates with leaching time is attributed to the production of orthosilicic acid (Li et al. 2017). Figures 11, 12, 13 and 14 show an increase in F^-, Cl^- and SO_4^{2-} ion concentrations in aqueous leachates of Ledo coal and OB with leaching time which is due to the fact that leaching time is

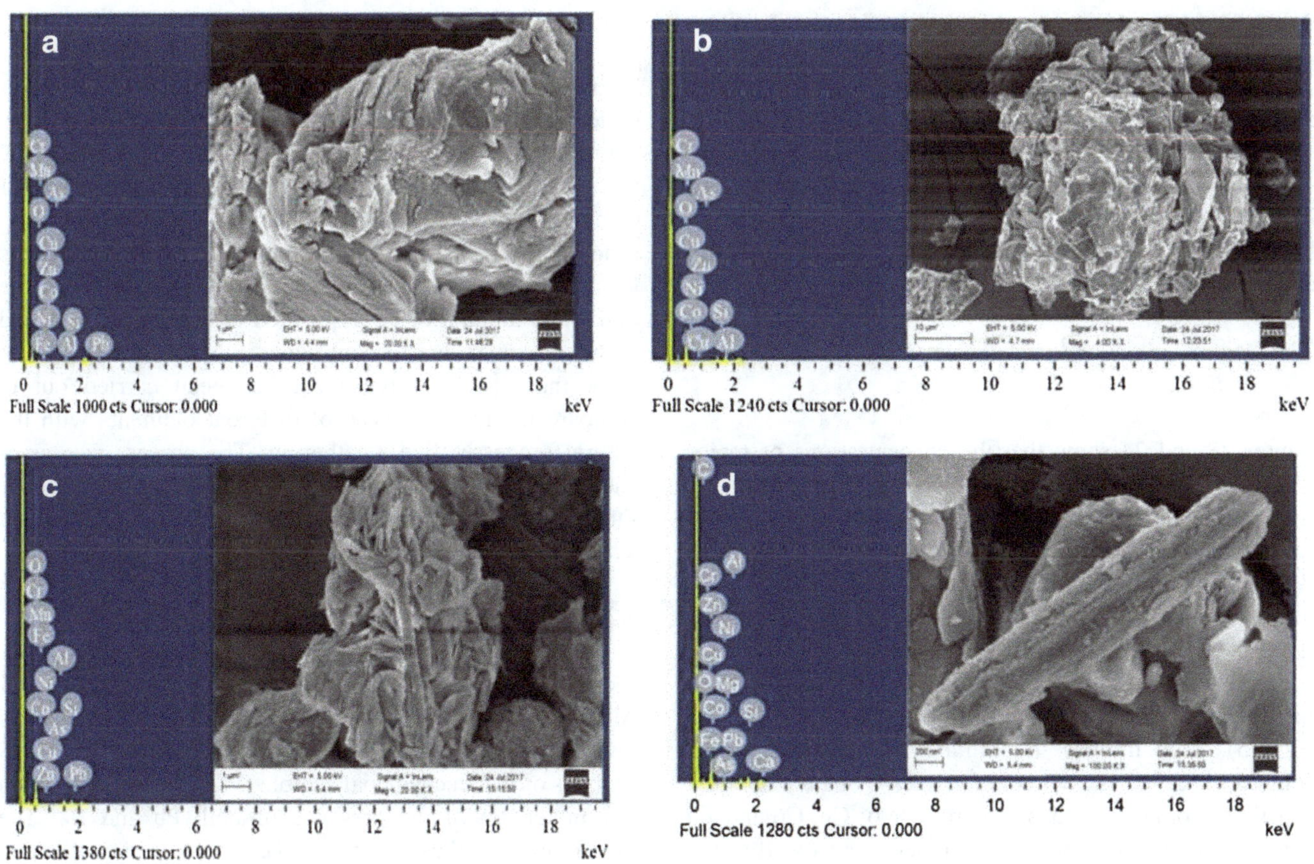

Fig. 35 FESEM-EDS micrographs of raw coal samples LC-20A (**a**), LC-60A (**b**) and OB samples LOB-15A (**c**), LOB-15B (**d**)

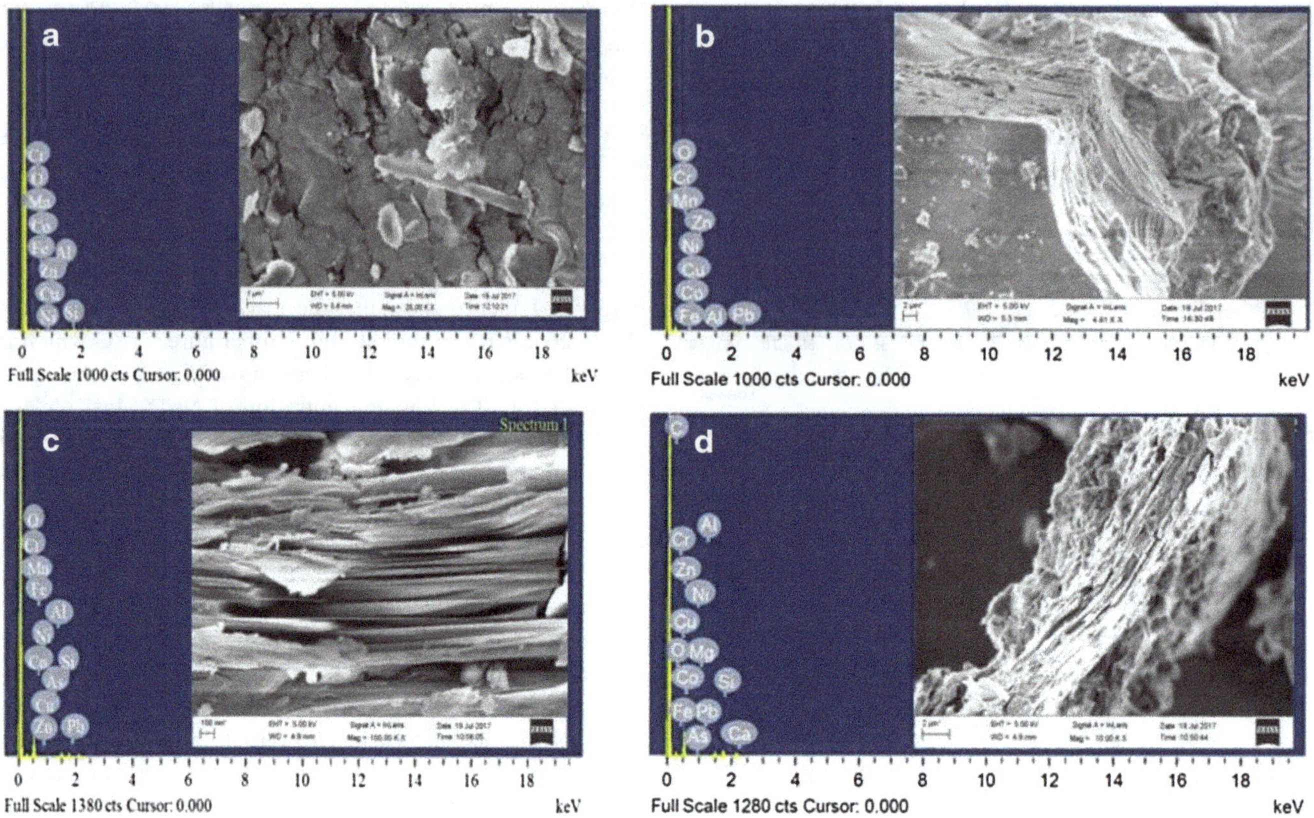

Fig. 36 SEM-EDS of leached coal LC-20A (**e**), LC-60A (**f**) and OB samples LOB-15A (**g**), LOB-15B (**h**)

an important factor which determine the amount of elements leached from the raw coal (Moyo et al. 2011). The leaching of SO_4^{2-} ion from coal and overburden was increased with leaching temperature from 25 to 65 °C and time (in Figs. 15, 16, 17, 18, 19, 20, 21, 22) which reveals the dependence of leaching of sulfate ion on leaching temperature and time.

4.2.2 Major, minor and trace/hazardous elemental analysis

Figures 23 and 24 show that the concentrations of major elements like Al. Fe, etc., minor element Mn and trace/hazardous elements like Al, Ba, Co, Ni, Fe, Mn, Cu, Zn, etc. increased with leaching time at room temperature (25 °C). But after 5 h of leaching the concentration of these elements gradually decreased which implies that at 25 °C most of the elements are released from the coal due to continuous raining of up to 5 h. The release of Pb is maximum at 45 °C which is independent of leaching (Fig. 25). This implies the mining temperature (ranges from 25 to 55 °C) is favorable for release of Pb from coal as well as from overburden. The release of Cu, Co and Zn from the coal and overburden is temperature as well as the leaching time dependent (Figs. 25, 26, 27, 28) which can be interpreted from the ICP-OES results of the mine water collected in different season and the leachates of coal and overburden at different temperatures. From the Figs. 29 and 30 it can be concluded that at high temperature most of elements are leached from coal even at a short leaching time. Since the concentration of Fe is more in aqueous leachates than in raw samples. This indicates that the maximum amount of Fe associated in coal can be leached with rain water due to continuous raining. In the present work the leaching experiment has been carried out to compare the leaching type of different elements with the natural processes like weathering. The process was associated with continuous stirring of the water mixture of the samples for increasing time interval (1, 3, 5 and 8 h) at temperatures 25, 45, 65 and 90 °C.

4.2.3 Field-emission-scanning-electron microscopic (FE-SEM) analysis

Although association of barite and galena with pyrite was revealed by FE-SEM analysis (Fig. 35), but the EDS analysis of the leached coal sample (Fig. 36) has not shown any indication of the presence of both Pb and Ba. The association of these sulfur minerals is responsible for highly acidic leachates due to weathering under natural condition. Ba and Pb present in coal can easily be released from coal at a lower pH (Giere and Stille 2004). Under the natural weathering condition, the coal associated with Pb and Ba undergoes leaching to release them and dissolve in AMD water, which was also confirmed by ICP-OES analysis of aqueous leachates of the Ledo coal at various time intervals. From the FE-SEM/EDS analysis of raw and leached samples (LC-20A, LOB-15A), it was found that the atomic percentage of Fe is different in both samples (raw and leached). It is happened because during the leaching process, the coal samples lost their structural behavior due to increased porosity owing to release of inorganic minerals from their surface (Manoj et al. 2009). From the FE-SEM/EDS analysis it was found that in Ledo coals, (LC-20A and LC-60A) the concentration of Cd, Cr and Ba were found in very low concentrations which resembles with the ICP-OES analysis of aqueous leachates of coal. On the other hand, both analyses reported the excess concentrations of the elements Ni, Co, Zn, Cu, Fe (more than regulatory level). The FE-SEM/EDS elemental data of leached samples also can be co-related with the ICP-OES analytical data of aqueous leachates. The aqueous leachates of coal and OB at different leaching times and temperatures revealed maximum concentrations of Fe (1670 ppm), Cr (0.62 ppm), Cd (0.002 ppm), Co (92.5 ppm), Ni (64.7 ppm), Pb (0.181 ppm), Ba (0.213 ppm), Cu (9.02 ppm), and Zn (17.1 ppm) in ICP-OES analysis. A high concentration of these elements in the leachates indicates their transfer from coal and OB under different chemical processes to the aqueous medium. It is to be noted that the intensity of the peaks found in the EDS graph of a sample cannot give a quantitative measure of elemental concentration; however, the EDS analysis could be used for qualitative identification of elements in the respective sample (Kutchko and Kim 2006). The composition of the elements is expressed in atomic and weight percentage (wt%) and the detection limit of EDS is about 0.1 wt% (Kursula 2000).

5 Mineralogy and functional groups in leached coals

From the XRD analyses, it is reported that the minerals leached from coal and overburden (Fig. 33) are absent after leaching. In raw samples of coal (Fig. 34) the minerals quartz and hematite (d-values 3.3404 and 1.5789 respectively) were present, but in leached samples hematite was not found. The minerals quartz, hematite, pyrite, marcasite (d-values 3.3404, 1.5789, 1.8252 respectively) were found in the overburden (OB) samples of both collieries. On the contrary to this, the leached samples of the overburden contain only quartz. This may be an indication of leaching of different minerals which are associated with different elements that can be leached in natural processes, like weathering. These minerals are mainly associated with Fe. The absence of these minerals in the leached samples of coal and overburden indicates the easy leaching with rain water to form acid mine drainage (AMD).

FTIR analysis of raw and leached coal and overburden of Ledo colliery (see Figs. 31, 32) revealed the absence of some of the leached components from the samples of coal and overburden. The absorption bands for CH, CH_2, CH_3 (al) are absent in the leached OB and the bands for both $-OH_{ar}$, $-OH_{al}$ are absent in both coal and OB. Although the absorption bands for C_{al}–O–C_{al}, SO_4^{2-} etc. are present in both raw and leached samples of coal and OB but the intensity of these peaks in the FTIR spectra of the leached samples are weak. This is due to the leaching of these samples which causes a low carbon and sulfur contents in the leached samples.

The major minerals present in coal are quartz, kaolinite, illite, calcite, carbonate, dolomite and sulfide mineral like pyrite (Ward 2002). From the XRD and FTIR analyses it was revealed that quartz, kaolinite and pyrite are the major minerals in the Ledo coal. Other Indian coals like Bokaro coal samples contain mostly kaolinite and quartz, but other minerals such as siderite and analcine are found as intermediate phases. In Neyeveli coal, dolomite and kaolinite are the major constituents with trace amounts of hematite and bassanite (Equeenuddin 2015). From analytical data it was found that the aqueous leachates of the Ledo coal are highly acidic (pH range 1.99–3.09). But the leachates of other Indian coals have the normal pH range (6.8–7.4) (Equeenuddin 2015). The strong acidic nature of the leachates of Ledo coal is due to the leaching of pyrite which is present in high amounts in the Ledo coal. The high concentration of iron (1670 ppm) in the leachates is attributed to maximum leaching of pyrite. Low pH of aqueous leachates of Ledo coal leads to leach out most of the heavy metal from the coal.

6 Comparison of aqueous leachates with coal mine water characteristic

The aqueous leachates generated from coal and OB are compared with the AMD water collected from the coal mine. The parameters (pH, EC, TDS) are found to be similar as given in the Tables 1 and 2 respectively. The increase of pH values in mine water samples in comparison with aqueous leachates may be due to dilution during raining as well as ground water contamination. The comparable data reveals the simulation experiment of AMD on a large scale.

Table 2 Geochemical analysis of mine water samples

Sample type	Distance from AMD source (km)	pH	EC (μs/cm)	TDS (ppm)
LW-15A	Sample point	4.0	1226	927
LW-15B	0.5 km from sample point	4.1	1234	915
LW-15C	0.5 km from sample point	3.50	2290	1140
LW-15D	1 km from sample point	3.30	2490	1180
LW-15E (Ledopani river)	2 km from sample point	4.4	936	0.0468
LW-15F (Kachanala stream)	5 km from sample point	5.0	659	0.0328

7 Conclusions

Leaching is found to be one of the primary pathways for trace elements entering into the ecosystem. Dumping of coal and overburden causes large environmental degradation due to direct release of potentially hazardous elements through leaching and oxidation of pyrite minerals. The leaching property of different elements has been found to be dependent to several factors like specific element solubility and availability or release potential. Also the other factors that enhance the metal leaching include swiftly weathering metal containing minerals and the mine drainage conditions that can increase the solubility and flow rates through the contaminated mineral. The potential to produce acidity of drainage from mine water and mine wastes can be predicted by both geochemical methods static and kinetic leaching processes. The leachates generated from overburden dump are found to be rich in heavy metals specially Fe, Cu, Mn and Ni. The concentration of some of these elements increases with increase in leaching time and temperature. High concentration of elements in AMD water and other mine affected water resources is attributed to the continuous raining. Thus Ledo coal mine can cause more threat to the quality of water resources, ground as well as the surface water.

Some measures can be taken to minimize the natural leaching including re-mining, reclamation, and vegetation. These steps can reduce acid loads by covering acid producing materials, decreasing infiltration rates, removing the residual coal which at large area is the source of most of the pyrite. Vegetation can alter the soil cover systems by way of holes through roots and can uptake and transport contaminants present below the soil cover. Also, vegetation growth may increase the organic matter of the soil cover, and the decay of the plants consumes oxygen thereby decreasing the oxygen availability.

Acknowledgements Authors are thankful to Director, CSIR-NEIST (Jorhat) for his keen interest to this work. The funding from CSIR, New Delhi (OLP-2003) is duly acknowledged. Special thanks go to Prof. Frans Wandeers for editing the draft.

References

Alper B, Gulbin G, Fatma S, Ozgur O (2008) Effect of leaching temperature and pH on leachability of metals from fly ash. A case study: can Thermal power plant, Province of Canakkale, Turkey. Environ Monit Assess 139:287–298

Baker JP, Schofield CL (1982) Aluminum toxicity to fish in acid waters. Water Air Soil Pollut 18:289–309

Banerjee D (2014) Acid drainage potential from coal mine wastes: environmental assessment through static and kinetic tests. Int J Environ Sci Technol 11:1365–1378

Baruah MK, Gogoi PC (1998) A new form of sulphur in coal: the discovery of an iron-sulphur coordination compound. Fuel 77:979

Baruah BP, Saikia BK, Gogoi P, Kotoky P (2004) Acid mine drainage in Indian coal industry. International seminar on mineral processing technology

Baruah BP, Saikia BK, Kotoky P, Rao PG (2006) Aqueous leaching of high sulfur sub-bituminous coals in Assam, India. Energy Fuels 20:1550–1555

Bigham JM, Nordstrom DK (2000) Iron and aluminum hydroxysulfate minerals from acid sulfate waters. In: Jambor JL, Alpers CN, Nordstrom DK (eds) Sulfate minerals, crystallography, geochemistry and environmental significance, mineralogical society of America reviews in mineralogy and geochemistry, vol 40, pp 351–403

Chabukdhara M, Singh OP (2016) Coal mining in northeastern India: an overview of environmental issues and treatment approaches. Int J Coal Sci Technol 3:87–96

Chen J, Liu G, Kang Y, Wu B, Sun R, Zhou C, Wu D (2013) Atmospheric emissions of F, As, Se, Hg, and Sb from coal-fired power and heat generation in China. Chemosphere 90:1925–1932

Dai S, Ren D, Tang Y, Yue M, Hao L (2005) Concentration and distribution of elements in late Permian coals from western Guizhou Province, China. Int J Coal Geol 61:119–137

Dowarah J, DekaBoruah HP, GogoiJ Pathak N, Saikia N, Handique AK (2009) Eco-restoration of a high-sulfur coal mine overburden dumping site in northeastern India: a case study. J Earth Syst Sci 118:579–608

Dutta M, Saikia J, Taffarel SR, Waanders FB, Medeiros D, Cutruneo CMNL, Silva LFO, Saikia BK (2017) Environmental assessment and nano-mineralogical characterization of coal, overburden, and sediment from Indian coal mining acid drainage. Geosci Front 8:1285–1297

Elder JF (1998) Metal biogeochemistry in surface-water systems—a review of principles and concepts. US Geol Surv Circ 1013:32

EPA (2000) United States Environmental Protection Agency: methodology for deriving ambient water quality criteria for protection of human health.EPA/822/B-00/004, October 2000

EPA (2002a) US Environmental Protection Agency: guidance on choosing a sampling design for environmental data collection (QA/G-5S). EPA/240/R-02/005, December 2002

EPA (2002b) US Environmental Protection Agency: guidance on environmental data verification and data validation (QA/G-8). EPA/240/R-02/004, Office of the Environmental Information, Washington, DC

Equeenuddin SK SKMd (2015) Leaching of trace elements from Indian coal. J Geol Soc India 86:102–106

Ferguson KD, Erickson PM (1998) Pre-mine prediction of acid mine drainage. In: Environmental management of solid waste-dredge material and mine tailings. Springer, New York, pp 24–43

Finkelman RB (1994) Methods of occurrence of potentially hazardous elements in coal: levels of confidence. Fuel Process Technol 39:1817–1829

Fotoupoulou M, Siavalas G, Inaner H, Katsanou K, Lambrakis N, Christainis K (2010) Combustion and leaching behaviour of trace elements in lignite and combustion by-products from the Mugla basin, SW Turkey. In: Proceedings of the 12th international congress bulletin of the geological society of Greece

Giere R, Stille P (2004) Energy, waste, and the environment: a geochemical perspective. Geological Society, London, Special Publications, vol 236

Hesbach P, Beck M, Eick M, Lee DW, Burgers C, Greiner A (2005) Inter laboratory comparison of leaching methods. World Coal Ash (WOCA), Lexington

Johnson DW (1987) In situ toxicity tests of fishes in acid waters. Biogeochemistry 3(1–3):181–208

Kaeser AJ, Sharpe WE (2001) The influence of acidic runoff episodes on slimy sculp in reproduction in stone run. Trans Am Fish Soc 130(6):1106–1115

Kalembkiewicz J, Sitarz-Palczak E (2015) Efficiency of leaching tests in the context of the influence of the fly ash on the environment. J Ecol Eng 16(1):67–80

Kim AG, Kazonich G, Dahlberg M (2003) Solubility of cations in class F fly ash. Environ Sci Technol 37:4507–4511

Kimmel WG (1983) The impact of acid mine drainage on the steam ecosystem, Resources, Technology and Utilization, Pennsylvania, Pennsylvania Coal

Kursula PK (2000) Accuracy, precision and detection limits of SEM–WDS, SEM–EDS and PIXE in the multi-elemental analysis of medieval glass. X-Ray Spectrom 29.111–118

Kutchko BG, Kim AG (2006) Fly ash characterization by SEM-EDS. Fuel 85:2537–2544

Langmuir D (1997) Aqueous environmental geochemistry. Colorado School of Mines. Prentice Hall, Upper Saddle River

Li S, Qin S, Kang L, Liu J, Wang J, Li Y (2017) An efficient approach for lithium and aluminum recovery from coal fly ash by pre-desilication and intensified acid leaching processes. Metals 7(7):272

Manoj B, Kunjomana AG, Chandrasekharan KA (2009) Chemical leaching of low-rank coal and its characterization using SEM/EDAX and FTIR. J Miner Mater Charact Eng 8(10):821

Monterroso C, Macias F (1998) Drainage waters affected by pyrite oxidation in a coal mine in Galicia (NW Spain): composition and mineral stabilities. Sci Total Environ 126:121–132

Moyo S, Mujuru M, McCrindle RI, Matlala NM (2011) Environmental implications of material leached from coal. J Environ Monit 12:3047–3310

Mukherjee S, Borthakur PC (2003) Effect of leaching high sulfur sub-bituminous coal by potassium hydroxide and acid on removal of mineral matter and sulfur. Fuel 82:783–788

Nordstrom DK, Ball JW (1986) The geochemical behavior of aluminum of acidified surface. Science 232:40–41

Prachiti PK, Manikyamba C, Singh PK, Balaram V, Lakshminarayana G, Raju K, Singh MP, Kalpana MS, Arora M (2011) Geochemical systematics and precious metal content of the sedimentary horizons of lower Gondwanas from the Sattupalli coal field, Godavari Valley, India. Int J Coal Geol 88:83–100

Querol X, Izquierdo M, Monfort E, Alvarez E, Font O, Moreno T, Alastuey A, Zhuang X, Lu W, Wang Y (2008) Environmental characterization of burnt coal gangue banks at Yangquan, Shanxi Province, China. Int J Coal Geol 75:93–104

Schippers A, Jozsa PG, Sand W (1996) Sulfur chemistry in bacterial leaching of pyrite. Appl Environ Microbiol 62:3424–3431

Singh MP, Singh PK (1995) Mineral matter in the Rajmahal coals: study through incident light microscopy and scanning electron micrography. J Geol Soc India 46:557–564

Singh PK, Singh AL, Kumar A, Singh MP (2012a) Mixed bacterial consortium as an emerging tool to remove hazardous trace metals from coal. Fuel 102:227–230

Singh PK, Singh MP, Singh AK, Naik AS (2012b) Petrographic and geochemical characterization of coals from Tiru valley, Nagaland, NE India. Energy Explor Exploit 30(2):171–192

Singh AK, Singh MP, Singh PK (2013) Petrological investigations of oligocene coals from foreland basin of northeast India. Energy Explor Exploit 31(6):909–936

Singh AL, Singh PK, Singh MP, Kumar A (2015a) Environmentally sensitive major and trace elements in Indonesian coal and their geochemical significance energy sources part a: recovery, utilization, and environmental effects, vol 37. Taylor & Francis, New York, pp 1836–1845

Singh PK, Rajak PK, Singh MP, Naik AS, Singh VK, Raju SV, Ojha S (2015b) Environmental geochemistry of selected elements in lignite from Barsingsar and Gurha Mines of Rajasthan, Western India. J Geol Soc India 86:23–32

Singh PK, Rajak PK, Singh MP, Singh VK, Naik AS (2016) Geochemistry of Kasnau-Matasukh lignites, Nagaur basin, Rajasthan (India). Int J Coal Sci Technol 3(2):104–122

Swain DJ, Goodarzi F (1995) Environmental aspects of trace elements in coal. Kluwer, Norwell, pp 111–127

Swaine DJ (2000) Why trace elements are important. Fuel Process Technol 65–66:21–23

Swine DJ (1990) Trace elements in coal. Butterworths, London, p 294

Tiwari RK (2001) Environmental impact of coal mines on water regime and its management. Int J Water Air Soil Pollut 132(1–2):185–199

Tomeczec J, Palugniok H (2002) Kinetics of mineral matter transformation during coal combustion. Fuel 81:1251–1258

US DOE (2005) International energy outlook. US DOE, Washington

Wang Y, Ren D, Zhao F (1999) Comparative leaching experiments for trace elements in raw coal, laboratory ash, fly ash and bottom ash. Int J Coal Geol 40:103–108

Ward CR (2002) Analysis, and significance of the mineral matter in coal mines. Int J Coal Geol 50:135–168

Yang L, Bai JSX, Song B, Wang R, Zhou T, Jia J, Pu H (2016) Leaching behavior and potential environmental effects of trace elements in coal gangue of an open-cast coal mine area, inner Mongolia. China Miner 6(2):50

Zhou C, Liu G, Wu D, Fang T, Wang R, Fan X (2014) Mobility behavior and environmental implications of trace elements associated with coal gangue: a case study at the Huainan Coalfield in China. Chemosphere 95:193–199

4

Surface coal mine permit application for successful reclamation, semi-arid shortgrass prairie (Wyoming, USA)

Anna Krzyszowska Waitkus[1]

Abstract Approximately 40% of USA coal originates in an ecologically sensitive area of semi-arid shortgrass prairie in Wyoming. Before a surface coal mine can begin operation in the USA, it must secure a mining permit and comply with regulations and performance standards under the USA Surface Mining Control and Reclamation Act (SMCRA), other federal environmental acts, and state programs. The Wyoming Department of Environmental Quality (WDEQ)/Land Quality Division (LQD) administers Wyoming's coal regulatory program. The permit application and bonding process for the largest surface coal mine permit in the USA, North Antelope Rochelle Mine (NARM) located in short grassland prairie in the northeast Wyoming, is discussed. The permit application process begins with the collection of baseline environmental data that characterizes premining conditions of the permit area. The permit application includes adjudication information, baseline information, mine and operation plans, and reclamation plans. Fulfillment of permit commitments and requirements of rules and regulations are inspected monthly by the LQD's representative in the field. Before a mine permit is issued, the mine operator must submit a reclamation bond to secure the performance of reclamation obligations that is later revised annually. In Wyoming, four reclamation bond release phases indicate the completion of various stages of the reclamatikon process. NARM's specific bond release verification cirteria, performance standards, and field verificatoins of bond release phases are discussed. The Bond Release Geodatabase (a GIS/GPS approach) was developed for this mine to monitor progress in meeting criteria and performance standards for incremental bond release. The Bond Release Geodatabase significantly reduces the time needed to track bond release progress, reach agreement between operator and regulator, and improve the state inspector's ability to assess reclamation adequacy and progress.

Keywords Surface coal mine permitting · Bonding · Reclamation · GIS/GPS geodatabase · USA rules and regulations · Inspections

1 Introduction

The State of Wyoming produces approximately 40% of all mined coal in the USA (Carroll 2011). The Land Quality Division (LQD) of the Wyoming Department of Environmental Quality (WDEQ) is in charge of Wyoming's coal regulatory program (State of Wyoming 2016). The state's regulations and environmental protection performance standards are important in achieving successful reclamation results. According to WDEQ/LQD Coal Rules and Regulations (Wyoming State Rules and Regulations 2017), the goal of surface coal mine reclamation is to restore the land to its pre-disturbance ecological function. To accomplish this goal, a thorough baseline inventory of environmental conditions must be performed prior to mining (Norton et al. 2010). It is especially important in the semi-arid environment of Wyoming, where a native vegetation of the shortgrass prairie is sensitive to any

The 2nd International Symposium on Land Reclamation and Ecological Restoration, October 20–23, 2017, Xi'an, China.

✉ Anna Krzyszowska Waitkus
enviro@wyoming.com

[1] Environmental Consulting, 80 Eagle Nest, Laramie, WY 82070, USA

disturbance due to the lack of the rainfall (Knight et al. 2014). The baseline inventory includes surveys and assessments of: cultural and historical resources; soils; vegetation; wildlife, surface and groundwater hydrology; climatology; wetlands, and geological data (soil and rock structure, coal seam thickness). The company then develops mining and reclamation plans incorporating all of the environmental components of the baseline inventory.

Before a company can begin mining in the USA, it must secure a mining permit and comply with regulations and performance standards of the USA Federal Surface Mining Control and Reclamation Act (SMCRA) of 1977 (Public Law 95-87, referred to as SMCRA), state programs and other federal environmental acts (USA Government Publishing Office 2016). Verification of state rules and regulations, permit commitments, and performance standards are conducted through field inspections, annual reports, and permit revisions.

Few references discuss surface coal mine permit content and its role in achieving reclamation goals (USA Congress, Office of Technology Assessment 1986). Specific studies conducted in the western USA states discuss the effects of mining and reclamation approaches on geomorphic processes (Toy and Black 2000), soil properties, and management (Pinchak et al. 1985; Schroeder and Vinning 1993; Buchanan et al. 2005; Reynolds and Reddy 2012), and re-vegetation strategies and methods (Williams et al. 2002; Schuman et al. 2005; Strom et al. 2010).

Before a mine permit is issued, the mine operator must submit a reclamation bond to secure the performance of reclamation obligations (Bonogofsky et al. 2015). The operator might reduce a reclamation bond by reclaiming disturbed areas simultaneously with mining coal. Reclamation progress in Wyoming occurs in four phases of bond release together with the verification of bond release criteria and performance standards (Krzyszowska Waitkus and Blake 2011).

This paper discusses the permit content information and the bond release processes of the largest surface coal mine permit in the USA, the North Antelope Rochelle Mine (NARM), located in the semi-arid (less than 300 mm rain/year) shortgrass prairie of northeast Wyoming. Information presented in this article were obtained from the Permit 569-T8 Peabody Powder River Mining LLC North Antelope Rochelle Mine stored in the Cheyenne office of the WDEQ/LQD (Permit 569-T8 2014).

NARM is an active open pit mine that produced over 111 million tons of coal in 2015. The permit area covered 18590 ha in 2015. Since the initial operation began in 1982, approximately 7680 ha have been disturbed. Coal seam thickness averages approximately 20 m with an average overburden thickness approaching 90 m. The mine contains three major pits: East Pit (6 km long), North Pit (8 km long), and West Pit (6 km long). The mine uses three operating draglines and 11 shovels supported by numerous trucks and bulldozers.

2 Rules and regulations

Since 1969 the state of Wyoming has created numerous environmental acts to help regulate the mining industry. They were revised after the primary federal law SMCRA was enacted in 1977, creating a basis for environmental laws, rules and regulations in various states (Squillace 1990). SMCRA established minimum federal standards for the regulation of mining, introduced an idea of bonding, and provided penalties for violations of the laws, regulations, or permit conditions. A federal agency, the Office of Surface Mining, Reclamation and Enforcement (OSMRE), was created to oversee SMCRA and revise it on an ongoing basis. The Code of Federal Regulations lists the rules and maintains an up-to-date electronic version (USA Government Publishing Office 2016).

In addition to SMCRA, mining operations must abide by regulations put forth by a number of other federal and state environmental agencies. In Wyoming these include: USA Office of Surface Mining, Reclamation and Enforcement; USA Army Corps of Engineers; USA Environmental Protection Agency; USA Fish and Wildlife Service; Wyoming Game and Fish Department; and WDEQ.

SMCRA's major components were incorporated into the Wyoming Environmental Quality Act and Industrial Development Information and Siting Act (State of Wyoming 2016) giving WDEQ/LQD the authority to establish standards for reclamation. These standards allowed for the creation of rules and regulations for the coal industry in Wyoming (Wyoming State Rules and Regulations 2017), which became an official part of Wyoming's coal regulatory program after approval by the USA Secretary of the Interior. Rules quoted in this article are from the Wyoming Environmental Quality Act and Industrial Development Information and Siting Act (State of Wyoming 2016) and from WDEQ/LQD Coal Rules and Regulations (Wyoming State Rules and Regulations 2017) that are constantly being updated to reflect changes in SMCRA. All these rules, regulations, and guidelines can be found on the website of the WDEQ/LQD.

3 Content of the permit

Information presented in the permit follows federal and state rules and regulations and contains specific permit commitments. Surface coal mining permits are issued for a term not to exceed 5 years, after which the operator must request and submit a revised permit for a term renewal. A surface coal mining permit must be terminated if the permittee has not commenced surface coal mining operations covered by the permit within 3 years of permit issuance.

The NARM's coal mine permit application consists of four major portions: adjudication Section (2 volumes), supporting data-baseline information (18 volumes), mine plan including mining method, schedule, and mining hydrology (7 volumes), and reclamation plan describing processes of overburden and topsoil handling with a reclamation schedule in the form of numerous maps (2 volumes). More than 60% of the information concerns premining baseline data.

The application, once approved, constitutes the enforcement contractual document (permit) with the WDEQ/LQD that is used during compliance inspections of mining operations. Any discrepancy from the permit commitment found during an inspection or through a revision of an annual report could result in a violation with associated penalties. The permit must contain a time schedule to address the following: mining progression (topsoil removal, overburden removal, coal removal), backfilling, grading, contouring, and seeding. The time schedule uses yearly blocks for 5 years followed by blocks of 5 year periods until the end of mining. Maps and cross-sections included in the permit must follow specific, consistent requirements related to scale, contours, legend, etc. All engineering maps, plans, and designs or reports including cross sections must be certified by a professional engineer registered under the laws of the State of Wyoming (2016). All mining and reclamation maps are presented in the same scale of 1:1500.

3.1 Adjudication information

Adjudication files include forms showing the surface land to be affected by the mine, the license to mine coal, a certificate of public liability insurance, a list of surface and coal owners, documentation of surface owner consent and right of entry, the permit area boundary, and lands to be affected over the life of the mine. New permit applications must contain the reclamation performance bond estimate, covering all disturbances and construction anticipated during the first 12 months of the permitted operation (State of Wyoming 2016). This bond amount is updated annually by the operator through annual reports and approved by the WDEQ/LQD according to WDEQ/LQD Guideline No. 12 (Wyoming Department of Environmental Quality 2017).

A certificate of public liability insurance must be submitted with the permit according to requirements of the State of Wyoming (2016). An appendix of lands within and adjacent (within 0.8 km) to the permit area contains a list of surface owners and coal rights, and records of water rights owners. Surface ownership includes private, federal (e.g., USA Department of Agriculture), and state of Wyoming lands.

3.2 Baseline information

Pre-disturbance inventory is a critical part of reclamation planning and establishes a framework and schedule for successful restoration of severely disturbed land, monitoring, and evaluation (Norton et al. 2010). According to rules and regulations the following information is required in the baseline information of the permit: land use, brief history of the area including description of historic places, archeological and paleontological resources, climatology, topography, geology and overburden assessment, hydrology (groundwater, surface water, water rights), soil assessment, vegetation inventory, and wildlife use (State of Wyoming 2016; Wyoming State Rules and Regulations 2017).

The premining land use summary is delineated on a map, statistically calculated, and presented in a table. For example, the categories of land use within the permit area include: surface coal mining, grazing, oil and gas development, wildlife habitat, and recreational uses. The permit contains the hectares of each land use category including a vegetation type.

Historical overview, cultural resource sites, and paleontological inventories are conducted on all permit areas according to requirements included in the Wyoming Environmental Quality Act. The individual inventory, evaluation, and mitigation reports are compiled in proprietary volumes to protect the confidentiality of cultural resources under the Archeological Resources Protection Act of 1979 (State of Wyoming 2016). The characteristics of climatology presented in the permit include temperature, growing season, precipitation, humidity and evapotranspiration, and winds. Due to the potential for the air pollution from dust, the operator must present air quality monitoring plans including the locations of air quality monitoring stations.

The topography, geology, and overburden assessments describe the premining topography, geology, stratigraphy, quality and quantity of premining overburden (State of Wyoming 2016). Geologic cross sections provide information on the lithology across the permit area including overburden, interburden, and coal thickness. Wyoming mines must follow the requirements of an Approximate Original Contour (AOC) type topography where surface configuration achieved by backfilling and grading of mined areas must closely resemble the general surface configuration of the land prior to mining and complement the drainage patterns of the surrounding terrain (Wyoming State Rules and Regulations 2017). The State of Wyoming and other states (Montana, Colorado, New Mexico, and North Dakota) require the establishment of AOC as per federal regulations unless specifically exempted due to excess or lack of overburden material.

During the mining and reclamation process two sampling programs are employed at NARM to define unsuitable overburden material ahead-of-mining and backfill material. In order to define unsuitable overburden material ahead-of-mining, overburden sampling procedures are established for material moved by a truck/shovel operation only. In NARM, at least one hole per 64 ha within the permit area is sampled for chemical analysis. Sample cuttings are collected on intervals at a maximum of 3 m in specific lifts of 16.5 m to a depth of 148 m. Samples are analyzed for quality by comparing with parameters and concentrations described in WDEQ/LQD Guideline 1 (Wyoming Department of Environmental Quality 2017). The location of unsuitable material collected from the baseline study is delineated on a map titled "Overburden Suitability" included in the permit and also is shown on the Geographic Position System (GPS) unit located on a shovel removing overburden. The map on the GPS of the shovel delineates places where these unsuitable materials could be stored (at least under 1.2 m of suitable material) and cannot be stored (such as in close proximity to a drainage channel or its flood plain).

Postmining backfill quality procedures include a description of the sampling methods, suitability classification, and selective placement. Composite samples of backfill material are collected at depths of 0–0.6 m and 0.6–1.2 m and analyzed for quality by comparing with parameters and concentrations described in WDEQ/LQD Guideline No. 1. Each sample site is located on 150 m centers (2 ha). If unsuitable material is found in the top 1.2 m of regraded spoil material, a mitigation plan is implemented to verify the quality of the initial backfill sample or to more accurately define the area of unsuitability within the initial sample area. The unsuitable backfill material is capped or removed and relocated deeper under at least 1.2 m of suitable material (excluding areas under floodplain channel bottom of reclaim drainages). Chemical properties of backfill material with a map showing the collection sites are submitted with the annual reports. Results are analyzed by the operator and reviewed by the LQD specialist. Best management practices concerning successful revegetation and chemical properties of surface material show a direct connection between the chemical results of the overburden baseline study and the practical implementation of burying unsuitable overburden and backfill material.

Premining hydrology and hydraulics for both surface and ground waters within the permit area and 0.8 km adjacent area around the permit are discussed in the baseline study (State of Wyoming 2016; Wyoming State Rules and Regulations 2017). Surface water includes rivers, creeks, lakes, and reservoirs within and adjacent to the permit area. Descriptions of premining stream characteristics (drainage density, channel sinuosity, stream channel morphology) are discussed, along with drainage basins and surface water runoff quantity and quality. At NARM, baseline water quality data for major streams are collected regularly and analyzed for specific parameters (e.g. ammonia, nitrogen, nitrate, bicarbonate, boron, carbonate, fluoride, sulfate, total dissolved solids, and dissolved arsenic, calcium, chloride, chromium, iron, molybdenum, potassium, sodium, and zinc). Information is presented on aquifers that may be affected, their areal extent and hydraulic properties, groundwater quality, and groundwater recharge and discharge areas. In 2016, NARM 93 groundwater monitoring sites and seven surface water stations were sampled regularly according to the permit commitments.

Soil at the mine is ranked as suitable or unsuitable for salvage according to chemical or physical criteria that affect revegetation success. These criteria are established by the state regulatory authority and reflect characteristics of climate and vegetation (USA Congress, Office of Technology Assessment 1986). All suitable soils must be salvaged and replaced on the reclaimed area. Most prairie soils in Wyoming are thin with low organic carbon content. It was estimated that minimum concentration of 0.52% organic content or 0.89% organic matter is necessary to sustain an adequate level of nutrient cycling in Wyoming reclaimed soils (Ingram et al. 2005).

WDEQ/LQD Guideline No. 1A (Wyoming Department of Environmental Quality 2017) includes criteria for suitability for soils or soil substitutes. Soil assessment requirements include a site specific soil inventory and suitability maps with tabulations of soil salvage depths for affected lands (State of Wyoming 2016). At least one sample for each soil series in the affected area is collected and analyzed for specific elements (e.g. pH, conductivity, particle size, texture, sodium absorption ratio, selenium, boron, and organic matter). Results are compared with the criteria to determine soil suitability. The results of quantitative soil analyses are used to create a soil stripping depth map for affected lands. A table in NARM's permit lists the following for each soil mapping unit: area units where soil removal is planned, average depth of soil to be removed, and volumes of soil to be salvaged. NARM's permit includes a map showing various soil unit descriptions with soil salvage depths. This information assists in determining soil depth application during the reclamation process.

Vegetation inventory requirements include delineation, mapping and tabulation of major vegetation communities, cover sampling of vegetation communities, species lists, shrub density determinations, and tree counts. The vegetation inventory also includes the selection of a reference area located outside of a disturbed area. Data from a reference area is used to compare data from the reclaimed and revegetated areas. NARM's permit includes a map titled "Native Vegetation Types" delineating eight major vegetation communities.

The baseline study includes a description of potential and actual faunal distribution within the permit and areas adjacent to the permit with a list of indigenous vertebrate wildlife species. Wildlife studies are done in accordance with state and federal wildlife agency requirements. NARM's permit includes a map titled "Wildlife Features", showing prairie dog colonies, intact raptor nests, former raptor nests, nesting platforms, and active and inactive sage grouse leks.

3.3 Mine plan

According to State of Wyoming requirements, the mine plan must include a description of type and method of mining, life of the mine in years, affected area units, estimated annual production, mine facilities, blasting plan, equipment list, and mining hydrology. The permit application must contain a map showing the estimated order and progression of mining and reclamation on all proposed affected lands (Wyoming State Rules and Regulations 2017). Mine sequence maps for soil, coal removal, and rough backfilling include identification of the area to be mined during the first 5 years in 1-year blocks during the specific term and in 5-year blocks for the remaining life of the mine. Soil salvage and handling techniques as well as erosion control plans are presented in the mine plan section. Tables of soil volumes that will be salvaged, material excavated and backfilled, and hectares of areas stripped of soil and graded with soil replaced and seeded are presented for each of the first 5 years, and in 5-year blocks thereafter, until the end of mining. This information is included in NARM's permit table titled "Material Handling Summary." The information is verified during annual report reviews and inspections to ensure the operator is following permit commitments and reclamation is occurring according to the schedule. The operator demonstrates the ability to achieve AOC through the calculation of bulking factors (swell factors) and estimation of overburden material volume available for backfilling.

The mine plan must ensure the protection of the quantity and quality of, and rights to, surface water and groundwater both within and adjacent to the permit area (Wyoming State Rules and Regulations 2017). Mining hydrology sections include a plan and timetable for control and treatment of surface and groundwater, discussion, and determination of probable hydrologic consequences within and outside the permit area for the duration of mining. The plan must estimate groundwater drawdown and provide a monitoring plan to include the collection, recording, and reporting of data on water levels and quality.

A hydrologic control plan for surface water includes a hydrologic control map, structure designs, and a maintenance plan. Six of the 29 volumes of NARM's permit application contain engineering plans of sedimentation reservoirs, highwall sumps, and permanent impoundments. Design requirements for each constructed feature of the surface water control plan follow WDEQ/LQD Guideline No. 8 (Wyoming Department of Environmental Quality 2017) and Coal Rules and Regulations (Wyoming State Rules and Regulations 2017).

3.4 Reclamation plan

The reclamation plan describes procedures and methods used to minimize adverse impacts during mining and to restore disturbed land to premining land use (Norton et al. 2009). By definition, reclamation "shall restore the land to a condition equal to or greater than the highest previous use." The land, after reclamation, must be suitable for the previous use or must have a use which is of greater economic or social value than previous uses (Wyoming State Rules and Regulations 2017). At NARM, reclamation must restore the land to livestock grazing and wildlife use functions. The reclamation plan coordinates with the mining plan and contains a time schedule to complete reclamation as quickly as possible.

NARM's reclamation plan includes information on soil removal, storage, protection, and replacement, along with a plan for backfilling, grading, and contouring of affected lands. Also included are methods of monitoring hydrologic restoration of disturbed land, revegetation plans with a means of measuring of success, and information on post-mining land use.

Reclamation begins with the salvage and protection of soil from the disturbed land. Salvaged soil is either stockpiled and protected from erosion and degradation or placed directly on regraded areas (direct haul). According to NARM's experience, direct haul produced the best revegetation results. Direct haul can result in higher mycorrhizae levels, better physical characteristics, and a slightly greater potential for seed bank benefit as compared to stockpiled soil (Schuman 2002; Anderson et al. 2008). However, caution should be taken if soil contains infestations of weeds or other undesirable species. The soil replacement management plan in NARM's permit discusses erosion control and water conservation practices, soil replacement depth range, and soil characteristics for the restoration of critical and important habitat such as wetlands, floodplains etc. Soil is usually applied at a uniform depth on regraded areas in Wyoming surface coal mines. According to the Wyoming State Rules and Regulations (2017), variable soil depths are also allowable: "Soil thickness may also be varied to the extent such variations help meet the specific revegetation goals identified in the permit." Some studies performed in the western USA showed that revegetation results can be affected by varying soil depths (Buchanan et al. 2005; Bowen et al. 2005; Williams et al. 2002). To enhance the growth of hard-to-establish shrubs such as big sagebrush (Artemisia tridentata), NARM has developed two soil replacement depth ranges: typical (15–40 cm) and deep (40–120 cm).

Soil is applied on backfilled and graded areas as soon as possible following mining, according to performance standards (Chapter 4, Section 2(b)) of the Wyoming State Rules and Regulations (2017):

- "Rough backfilling and grading shall follow coal removal as contemporaneously as possible based upon the mining conditions. The operator shall include within the application for a permit to mine a proposed schedule for backfilling and grading with supporting analysis."
- "Backfilled materials shall be replaced in a manner which minimizes water pollution on and off the site and supports the approved postmining land use. Preparation of final graded surfaces shall be conducted in a manner that minimizes erosion and provides a surface for replacement of soil that will minimize slippage."
- "All affected lands shall be returned to their approximate original contour…"

The recontoured surface design has critical effects on postmining vegetation, land stability, and postmining land use (USA Congress, Office of Technology Assessment 1986). In Wyoming's semi-arid environment, a complex (convex on top, concave at bottom) slope of less than 25% has been found to be least prone to erosion. To avoid long uniform slopes, constructed slope lengths do not exceed premining lengths and are shorter than 150 m long. No highwalls may be left unless they are approved by the Administrator of the WDEQ/LQD. During reclamation, pre- and postmining topographies are compared to determine if reclamation achieved requirements of AOC topography (Steward et al. 2006). In NARM's permit, pre- and postmining slope frequency and distribution are illustrated graphically (showing e.g., average slope, median slope, and slope interval). Slope distribution analysis is performed using state-of-the-art terrain modeling and analysis software. Analysis of histograms and cumulative frequency curves indicate a strong correlation between pre- and postmining slopes at NARM.

A rebuilt hydrologic system must be erosionally stable to drain water and blend with surrounding drainage systems. For NARM, surface water quality is predicted to be similar to premining and meet premining class of use standards. Hydrologic restoration plans for surface water discuss postmining watershed characteristics, practices to be used for drainage reconstruction (including channel geometry), locations of permanent impoundments, and a postmining monitoring plan. The function of Porcupine Creek has been restored by rebuilding the creek structure, sinuous channel, and hydrology. The restoration established a maximum amount of wetland and riparian habitat and stabilized streambanks (Hansen and Murphree 2003).

Structures (facilities and flood control reservoirs, sediment ponds, diversions, sediment traps, alternate sediment control measures) constructed in NARM's permit area control runoff and provide sediment control during mining. According to Wyoming State Rules and Regulations (2017), sediment can't leave the permit boundary and cause an off-site impact. Sediment ponds are depicted on the Mine and Reclamation Plan Map titled "Hydrologic Control Plan." Sixty-one permanent impoundments as of 2014 in NARM's permit area will supply water for livestock and wildlife, and will be used as wetland mitigation. Permanent impoundments on reclaimed surfaces are delineated on the Mine and Reclamation Plan Map titled "Post-mining Contours with Hydrologic Restoration."

Hydrologic restoration plans for groundwater describe the source of recharge and provide a postmine monitoring plan with a map of permanent backfill well locations. A discussion of probable hydrologic consequences estimates final ground and surface water quantity and quality. The assessment of ground water quantity includes rate of recovery predictions for the affected and backfill aquifers. A map showing final anticipated potentiometric surfaces of the affected aquifer is provided in NARM's permit.

The major goal of the reclamation plan is to establish stable vegetation communities on chemically suitable soil and backfill material. According to the revegetation performance standard "The operator shall establish on all affected lands a diverse, permanent vegetative cover of the same seasonal variety native to the area or a mixture of species that will support the approved postmining land use in a manner consistent with the approved reclamation plan. This cover shall be self-renewing, and capable of stabilizing the soil" (Wyoming State Rules and Regulations 2017). The revegetated area must support the approved postmining land use. NARM's revegetation plan provides a discussion on vegetation establishment methods (seed mixes, seeding techniques, mulching materials and rates, husbandry practices). Also included are interim vegetation monitoring and final revegetation success plans. Various seed mixes are developed based on baseline studies and onsite evaluation. Native plant species are preferred, with introduced species used only to provide utility in the mixes. The seed mixes correspond to the following vegetation types: grassland, riparian grassland, sagebrush grassland, playa grassland, shrub mosaic and sagebrush mosaics. All of these communities are native to the premining environment. High plant production on reclaimed soils in Wyoming, despite lack of fertilization, suggests that total nitrogen is not limiting production (Ingram et al. 2007). It was found that reclaimed soils in Powder River Basin in Wyoming had recovered structurally towards a native soil condition after a period of 10–15 years (Wick et al. 2009). Reclamation success is determined by a quantitative comparison of data from the reference area with data from the reclaimed area that include: cover (vegetative cover, total ground cover), production (total production, specific species production), density and distribution (shrubs, trees), species diversity, and species composition (Bilbrough and Howlin 2012). This information is presented and discussed in the NARM's annual report. Also specific areas planted in the various seed mixes, along with seeding rates, are included in tables and maps.

Disturbed areas seeded with native shrub and grass species also provide habitat for wildlife. Rock piles are constructed to simulate natural features. To reduce the effects of mining on birds of prey, several mitigation measures have been incorporated into the reclamation plan. Wildlife mitigation and monitoring plans include information on all habitat restoration procedures.

The NARM permit includes following maps (scale 1:1500) and exhibits showing major steps in the mining/reclamation processes together with a time schedule: "Topsoil removal (disturbance) schedule"; "Hydrologic Control Plan"; "Mining schedule"; "Reclamation schedule (topsoil replacement)"; "Postmining contours with hydrologic restoration"; "Postmining contours with surface slope distribution"; "Postmining vegetation"; and "Postmining environmental monitoring locations." These maps are reviewed in the field during inspections. Also reviews of annual reports are important tools in determining if the operator is on time with the planned schedule of mining/reclamation processes. In case of delay or change in the direction of mining and reclamation, the operator must submit a revision to the permit.

4 Reclamation bond performance

Before a mine permit is issued the mine operator must submit a reclamation bond to secure the performance of reclamation obligations. The cost of reclaiming the disturbed areas of a coal mine is revised and updated annually through annual reports submitted to WDEQ/LQD. A reclamation performance bond or financial assurance must be large enough to cover the full costs of reclamation to ensure that the regulatory authority will have sufficient funds to reclaim the site if the permittee fails to complete the reclamation plan or hire a third party to complete the reclamation plan in the event of a bankruptcy of the coal mine company. The bond release can begin 5 years from the date of permanent seeding in the eastern USA and 10 years after permanent seeding in the western USA.

In Wyoming, there are four bond release phases (Area Bond, Phase 1, 2, and 3) for surface coal mines indicating the completion of various reclamation phases (Krzyszowska Waitkus and Blake 2011) (Table 1). In general, performance standards related to the release of performance bonds for Area Bond and Phase 1 include the area being backfilled and regraded with a functional drainage system and applied soil. For Phase 2, vegetation cover must be established in stable areas without erosion. For Phase 3, the revegetation standards and goals must be completed (Wood and Buchanan 2000). Revegetation as a portion of the assessment of reclamation success created a lot of discussion between coal operators, regulators, and environmental specialists for long time in the state of Wyoming (Krzyszowska Waitkus et al. 2000; Schuman 2002).

The amount of the bond is revised annually through annual reports according to WDEQ/LQD Guideline 12 (Wyoming Department of Environmental Quality 2017). Bond release submittals are not part of the permit, but the bond amount and bonding instruments (e.g. letter of credit, certificates of deposit, and treasury bills) are submitted with the permit.

As a result of cooperation between the federal agency, private industry, environmental consultants, and WDEQ/LQD representatives a set of criteria (Table 1) and performance standards was established that have to be fulfilled and verified before the reclamation bond can be released.

Table 1 Verification of bond release criteria

Bond release phase	Verification of bond release criteria
Area bond	Backfilling, regrading completed according to approved plan
Phase 1	Stream channel reconstructed and soil applied
Phase 2	Vegetation established (ocular evaluation)
	Surficial stability established
	Permanent ponds construction design approved
Phase 3	Revegetation success standards established (statistical analysis)
	Mitigation of wetlands (approved by the US Army Corps of Engineers)
	Tree established

Up-to-date information (WDEQ/LQD Guidelines 20, 21, 22, 23, and 25) regarding bond release verification and compliance standards is posted on the Wyoming DEQ website (Wyoming Department of Environmental Quality 2017).

As reclamation phases are completed, the permitte may apply for incremental phased bond release. Phased bonding encourages companies to begin reclamation as soon as possible (called contemporaneous reclamation). Specific bond release requests are approved in the field after joint inspection by the land owners, permit operator, LQD, and OSMRE representatives. Additionally, the operator compares the data gathered during the baseline inventory with data gathered from the reclaimed land. After reviewing the submitted information and performing a field inspection, the LQD representative makes a decision to approve or deny the bond release request. If an Area Bond and Phase 1 request is approved, 60% of the bond can be released, for Phase 2, 15%, and for Phase 3, 25% of the bond can be released.

Various phases of bond release require verification of criteria and performance standards in the field which represents a challenge for the regulatory agency, especially for large mines. These challenges require a system to streamline field data collection which can be utilized in evaluating bond release and compliance requirements. As a result of these challenges, a Bond Release Geodatabase was developed for NARM in Wyoming. The system used a Geographic Information System (GIS) (ESRI Personal Geodatabase, ArcInfo 9.2) to organize bond release data. This geodatabase was the first spatially supported database developed in Wyoming and in the USA to process reclamation bond releases. The geodatabase was designed to support the tracking of areas which have achieved specific bond release criteria and have satisfied performance standards in support of incremental bond release. Compliance features (e.g., topsoil depth, erosion, or wildlife habitat features) were monitored and entered into the geodatabase by LQD personnel during inspections using a Global Position System (GPS) enabled data collector (Trimble GeoExplorer Series GeoXM and Juno handheld GPS unit using ESRI ArcPad 7.1) (Krzyszowska Waitkus and Calle 2008). Verified field data together with information submitted by the operator were compiled into one Bond Release Geodatabase. This geodatabase is a highly effective method which sufficiently reduces the time needed to track the bond release progress and reach agreement between the operator and the regulator. It also improves the state inspector's ability to assess reclamation adequacy and progress (Krzyszowska Waitkus and Blake 2011).

5 Summary

The primary governmental responsibility for the regulation of surface coal mining rests with the rules and regulations of states that follow the requirements of SMCRA, 1977. The coal mine permit is an agreement between the mine operator, state, and federal agencies. The permittee is required to meet performance standards and fulfill reclamation goals to restore disturbed land to premining land use.

Coal mine permit content follows the rules and regulations of state regulatory agencies. The permit presents data requirements and collection methods to assess the adequacy of baseline and monitoring data, predicts the impacts of mining, provides information on reclamation techniques, and presents methods evaluating the success of reclamation. Because mining and reclamation schedules included in the permit are constantly changing due to market demand, the permit must be constantly updated. The schedule, permit commitments, and compliance with rules and regulations are inspected monthly by the WDEQ/LQD's representative in the field. The permit is instrumental for the permittee and for the regulatory agencies to gather all information and present how all disturbed land can be restored to the premining land use for the western states of the USA. The presented case of the permit shows how the regulations together with the operator's experience and their best management practices are advantageous in the reclamation of severely disturbed land.

Reclamation of large surface coal mines can be successful in a disturbance-sensitive, semi-arid environment. Data from the largest surface coal mine in Wyoming shows that a disturbed area can be sufficiently revegetated to fulfill SMCRA requirements over a period of 10 years. The set of criteria and performance standards established for the state of Wyoming accelerate the bond release application submittals and more reclaimed area should be released to the land owners in timely manner. A Bond Release Geodatabase developed by the WDEQ/LQD field inspector has streamlined the bond release process and helped with compliance inspections. The GIS/GPS approach was proven to be successful for the large mining/reclamation activities of severely disturbed areas.

References

Anderson JD, Ingram LJ, Stahl PD (2008) Influence of reclamation management practices on microbial biomass carbon and soil organic carbon accumulation in semiarid mined lands of Wyoming. Appl Soil Ecol 40:387–397

Bilbrough C, Howlin S (2012) Handbook of approved sampling and statistical methods for evaluation of revegetation success on Wyoming coal mines. Wyoming Department of Environmental Quality, Land Quality Division. http://deq.wyoming.gov/lqd/coal/resources/handbook/. Accessed 19th Feb 2017

Bonogofsky A, Jahshan A, Yu H, Cohn D, MacDonald M (2015) Undermined Promise II. National Wildlife Federation, Natural Resource Defense Council, & Western Organization of Resource Councils. http://www.underminedpromise.org/UnderminedPromiseII.pdf. Accessed 19th Feb 2017

Bowen CK, Schuman GE, Olson RA (2005) Influence of topsoil depth on plant and soil attributes of 24-year old reclaimed mined lands. Arid Land Res Manag 19:267–284. https://doi.org/10.1080/15324980590951441

Buchanan B, Schuman GE, Olson RA (2005) Long-term plant community development in response to topsoil replacement depth on mined land in Wyoming. In: Barnhisel RI (ed) Raising reclamation to new heights: Proceedings of a Joint Conference of American Society of Mining and Reclamation, 22nd Annual National Conference, June 18–25, 2005, Breckenridge, Colorado, USA. ASMR, 3134 Montavesta Rd. Lexington, KY 40502, pp 108–114

Carroll C (2011) Wyoming's Coal Resources. Wyoming State Geological Survey. Summary report

Hansen BW, Murphree P (2003) Reclamation of Porcupine Creek riparian and wetland habitats. In: Barnhisel RI (ed) Working together for innovative reclamation: Proceedings of the Joint Conference of the 9th Billings Land Reclamation Symposium and the 20th Annual Meeting of the American Society of Mining and Reclamation and the 9th Billings Land Reclamation Symposium, June 3–6, 2003, Billings, Montana, USA. ASMR, 3134 Montavesta Rd. Lexington, KY 40502, pp 86–89

Ingram LJ, Schuman GE, Stahl PD, Lowell LK (2005) Microbial respiration and organic carbon indicate nutrient cycling recovery in reclaimed soils. Soil Sci Soc Am J 69:1737–1745. https://doi.org/10.2136/sssaj2004.0371

Ingram LJ, Stahl PD, Anderson JD (2007) The influence of management practices on microbial and total soil nitrogen. In: Barnishel RI (ed) 30 years of SMCRA and beyond. Proceedings of the 24th National Meeting of the American Society of Mining and Reclamation 2007: 2–6 June, Gillette, Wyoming, USA. ASMR, 3134 Montavesta Rd. Lexington, KY 40502, pp 346–360

Knight DH, Jones GP, Reiners WA, Romme WH (2014) Mountains and plains: the ecology of Wyoming. Yale University Press, New Haven

Krzyszowska Waitkus A, Blake C (2011) Tracking bond release at a large Wyoming coal mining operation. In: Barnishel RI (ed) Reclamation: science leading to success: Proceedings of the National Meeting of the American Society of Mining and Reclamation, June 11–16, 2011, Bismarck, ND, USA. ASMR, 3134 Montavesta Rd. Lexington, KY 40502, pp 338–349

Krzyszowska Waitkus A, Calle M (2008) Mobile computing and geodatabase application for coal mine inspection. In: Paper presented at the 2nd National Meeting of SMCRA Geospatial Data Stewards, March 26, 2008, OSM Geospatial Conference "Incorporating Geospatial Technologies into SMCRA Business Processes", March 25–27, Atlanta, GA

Krzyszowska Waitkus A, Smith J, Cash G, Moxley M (2000) Successful reclamation techniques and bond release for coal mine in Wyoming. In: Striving for Restoration, Fostering Technology and Policy for Reestablishing Ecological Function. Proceedings of the Billings Land Reclamation Symposium, March 20–24, 2000, Billings, Montana, USA. Reclamation Research Unit Publication No. 00-01 Montana State University-Bozeman, pp 33–44

Norton J, Krzyszowska Waitkus A, Loubsky T (2009) Successful restoration of severely disturbed lands: overview of critical components. University of Wyoming. Cooperative Extension Service. Bulletin No. B-1202

Norton J, Krzyszowska Waitkus A, Strom C (2010) Successful restoration of severely disturbed lands: predisturbance/baseline inventory. University of Wyoming. Cooperative Extension Service. Bulletin No. B-1212

Permit 569-T8. (2014) Peabody Powder River Mining LLC, North Antelope Rochelle Mine. Wyoming Department of Environmental Quality, Land Quality Division, Cheyenne, Wyoming, USA

Pinchak BA, Schuman GE, Depuit EJ (1985) Topsoil and mulch effects on plant species and community responses of revegetated mined land. J Range Manag 38:262–265

Reynolds B, Reddy KJ (2012) Infiltration rates in reclaimed surface coal mines. Water Air Soil Pollut 223(9):5491–5958

Schroeder SA, Vinning KC (1993) Relation of compaction and soil physical parameters to productivity of reclaimed soils. Final Report. North Dakota State University Land Reclamation Research Center

Schuman GE (2002) Mined land reclamation in the Northern Great Plains: have we been successful? In: Barnishel RI, Collins M (eds) Reclamation with a purpose. Proceedings of the 19th National Meeting of the American Society of Mining and Reclamation and the 6th International Affiliation of Land Reclamationists. June 9013, 2002, Lexington, Kentucky, USA. 3134 Montavesta Rd. Lexington, KY 40502, pp 842–864

Schuman GE, Vicklund LE, Belden SE (2005) Establishing *Artemisia tridentata* ssp. wyomingensis on mined lands: science and economic. Arid Land Res Manag 19:353–362

Squillace M (1990) The strip mining handbook. Environmental Policy Institute. https://sites.google.com/site/stripmininghandbook/. Accessed 20th Jan 2017

State of Wyoming (2016) Wyoming Environmental Quality Act and Industrial Development Information and Siting Act. 2016 Edition, LexisNexis

Steward DGM, Burget W, Carlson R, Stowe RR, Vicklund LE (2006) Handbook of western reclamation techniques, second edition. University of Wyoming. https://www.osmre.gov/resources/library/ghm/WestReclHandbook2ndEd.pdf. Accessed 19th Feb 2017

Strom C, Norton J, Loubsky T (2010) Successful restoration of severely disturbed lands: seeding essentials for reclaiming disturbed lands. University of Wyoming. Cooperative Extension Service. Bulletin No.B-1204

Toy TJ, Black JP (2000) Topographic reconstruction: the theory and practice. In: Barnishel RI, Darmody RG, Daniels WL (eds) Reclamation of drastically disturbed lands. Agronomy Monography 41. ASA, CSSA, and SSSA, Madison, WI, pp 41–77

USA Congress, Office of Technology Assessment (1986) Western surface mine permitting and reclamation. Washington, DC: USA Government Printing Office, June 1986, OTA-E-279. Library of Congress Catalog Card Number 86-600506

USA Government Publishing Office (2016) Chapter VII-Office of Surface Mining Reclamation and Enforcement, Department of the Interior. Code of Federal Regulations. Title 30-Mineral Resources. https://www.gpo.gov/fdsys/pkg/CFR-2016-title30-vol3/pdf/CFR-2016-title30-vol3-chapVII.pdf. Accessed 18th Feb 2017

Wick AF, Ingram LJ, Stahl PD (2009) Aggregate and organic matter dynamics in reclaimed soils as indicated by stable carbon isotopes. Soil Biol Biochem 41:201–209

Williams M, Shuman GE, Hild A, Vicklund L (2002) Wyoming big sagebrush density: effects of seeding rates and grass competition. Restor Ecol 10:385–391

Wood MK, Buchanan BA (2000) Reclamation considerations for arid regions of the southwest receiving less than twenty-five centimeters annual precipitation.. In: Barnishel RI, Darmody RG, Daniels WL (eds) Reclamation of drastically disturbed lands. Agronomy Monography 41. ASA, CSSA, and SSSA, Madison, WI, pp 303–322

Wyoming Department of Environmental Quality (2017) Land quality division: guidelines. http://deq.wyoming.gov/lqd/resources/guidelines/. Accessed 24th Jan 2017

Wyoming State Rules and Regulations (2017) Environmental Quality, Department of Land Quality—Coal. http://deq.wyoming.gov/lqd/resources/guidelines/. Accessed 24th Jan 2017

5

Orthogonal experiment design of EMI of security monitoring system in coal mines

Xiang Liu[1] · Chaoqun Jiao[1] · Aifen Yao[1]

Abstract Security monitoring system of coal mines is indispensable to ensure the safe and efficient production of colliery. Due to the special and narrow underground field of the coal mine, the electromagnetic interference can cause a series of misstatements and false positives on the monitoring system, which will severely hamper the safe production of coal industry. In this paper, first, the frequency characteristics of the interference source on the power line are extracted when equipment runs normally. Then the finite difference time domain method is introduced to analyze the effects of the electromagnetic interference parameters on the security monitoring signal line. And the interference voltage of the two terminal sides on the single line is taken as evaluating indexes. Finally, the electromagnetic interference parameters are optimized by orthogonal experimental design based on the MATLAB simulation on the normal operation of equipment.

Keywords Security monitoring system · Finite difference time domain method (FDTD) · Electromagnetic interference (EMI) · Orthogonal experimental design

1 Introduction

With the development of coal mine automation, more and more high power electrical instruments have been used in coal mines. The underground electromagnetic interference has become particularly severe because the on-the-spot space is relatively narrow. At times EMI results in errors of the security monitoring system. Currently, there are hardly any research has been carried out in the field of coal mine EMC. Delogne (1991) generated a regulation of electromagnetic wave transmissions and indicated the difficulty of studying practical transmissions in tunnels. Hill (1989) studied the traveling wave in tunnels and radiation impedance of dipoles. Rappaport (1989) conducted a prediction using a ray tracing method. The references (Sun 1999; Sun et al. 2006) about the investigation into the wireless communication and channel studied the measurement for improving EMC system. These studies largely focused on wireless communication, which did not involve in the EMC studies in coal mines. Sun carried out a study of harsh classification on electromagnetic compatible measures of mining monitoring and communication equipment (Sun et al. 2008), as well as an investigation in coal mining EMC (Liang 2007; Zhang et al. 2007). References (Gao 2004; Liu et al. 2009, 2010) researched the region electromagnetic environment while the immunity of the monitoring was not involved. However, there has been no report about the effect of the electromagnetic interference parameters on the security monitoring system so far.

Security monitoring signal from sensors is often relatively weak, and its anti-interference ability is very poor in the process of transmission, which resulted in vulnerability to external electromagnetic interference. Therefore, it is very significant and important to reduce the interference voltage of security monitoring single line. In this paper, for the normal operation of equipment, the electromagnetic interference parameters of security monitoring signal line are optimized by using orthogonal experiment design

✉ Xiang Liu
13121434@bjtu.edu.cn; chqjiao@bjtu.edu.cn

[1] School of Electrical Engineering, Beijing Jiaotong University, Beijing 100044, China

method based on the finite difference time domain method (FDTD) analyses of multiple conductor transmission line (MTL) model, then the simulation experiments are carried out, and the electromagnetic interference voltage are decreased.

2 The basic electromagnetic interference theory of security monitoring system

2.1 Measurement and extraction of interference sources

The data of interferential voltage on device's power line while working normally can be obtained for many times by using the Amway 2711 spectrum analyzer. For all spectrum data measured, frequency-domain characteristic parameters are mainly analyzed, including the frequency band range of interference and the frequency distribution, etc. In order to investigate the average interference level on device's power line, the amplitude of each frequency point in the frequency domain is measured by using statistical analysis methods. Statistical analysis methods used are as follows: the interference data of the same type measured is analyzed; the average value at each frequency point is calculated; then connect all the points and draw the mean envelope. With analyzing all interference voltage spectrums on device's power line, the results of a statistical test are shown in Table 1. Figure 1 shows that the average value of envelope spectrum of interferential voltage on device's power line while working normally. With the increase of the frequency, the interferential voltage decreases rapidly when the frequency grows from 0.05 to 0.1 MHz; the interferential voltage increases briefly, then it decreases sharply when the frequency is between 0.1 and 0.2 MHz; the interferential voltage begins to flatten when the frequency is in the range of 0.1–0.7 MHz, and then to decline when the frequency is among 0.7–0.8 MHz; finally, when the frequency is higher than 0.8 MHz, the interferential voltage begins to flatten out again gradually, and it is the minimum value at this time.

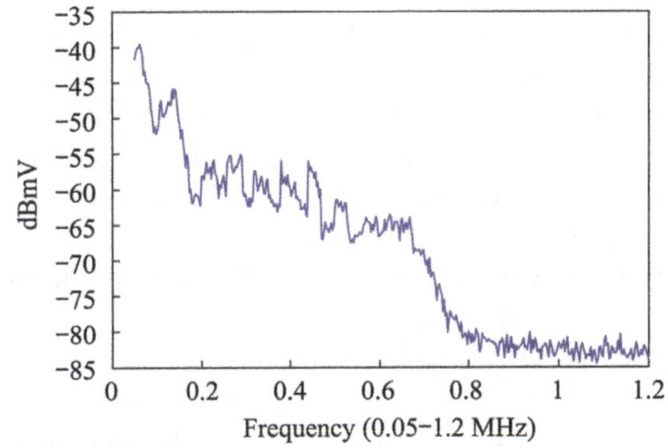

Fig. 1 The average value envelope spectrum of interferential voltage on device's power line while working normally

As shown in Fig. 1, the maximum interferential voltage value is 8.2327 V when the frequency is 0.05 MHz on device's power line while working normally. Therefore, interference sources can be taken as follows:

$$Vs = 8.2327 \times 10^{-6} \times \cos(2\pi \times 0.05 \times 10^{6} t)\ (\mathrm{V})$$

2.2 Numerical prediction model

The multi-conductor transmission line (MTL) structure can be described by the voltage and current iteration equations as follows (Erdin et al. 1998; Paul 2008; Jiao et al. 2012):

$$\frac{\partial V(z,t)}{\partial z} + L\frac{\partial I(z,t)}{\partial t} + RI(z,t) = V_F(z,t) \quad (1)$$

$$\frac{\partial I(z,t)}{\partial z} + C\frac{\partial V(z,t)}{\partial t} + GV(z,t) = I_F(z,t) \quad (2)$$

Table 1 The analysis of frequency-domain characteristic parameters of interference voltage on device's power line while working normally

No.	Frequency band range (MHz)	Frequency distribution (MHz)	0.05–0.4 MHz Mean voltage (dBmV)	0.4–0.8 MHz Mean voltage (dBmV)	0.8–1.2 MHz Mean voltage (dBmV)
1	0.05–0.806	0.067, 0.133, 0.283, 0.453	−53	−66	−83
2	0.05–0.773	0.053, 0.122, 0.378, 0.427	−54	−67	−83
3	0.05–0.788	0.064, 0.125, 0.228, 0.455	−59	−68	−82
4	0.05–0.785	0.056, 0.133, 0.263, 0.447	−55	−67	−83
5	0.05–0.78	0.062, 0.133, 0.26, 0.461	−55	−67	−82
6	0.05–0.772	0.064, 0.142, 0.292, 0.455	−55	−68	−83
7	0.05–0.76	0.053, 0.139, 0.378, 0.432	−55	−67	−83
8	0.05–0.766	0.064, 0.142, 0.323, 0.441	−56	−67	−83
9	0.05–0.749	0.062, 0.142, 0.28, 0.398	−54	−67	−83
10	0.05–0.743	0.073, 0.142, 0.22, 0.461	−55	−68	−83

where $V(z, t) = [V_1(Z, T), V_2(z, t),\ldots, V_n(z, t)]^T$ and $I(z, t) = [I_1(Z,T), I_2(z,t),\ldots, I_n(z,t)]^T$ represent the voltage and current vectors on the z point of line at moment t respectively; R, G, L, and C are resistive, inductive, conductive, and capacitive per-unit-length (p.u.l.) parameters of the line respectively; $V_F(z, t)$ and $I_F(z, t)$, respectively, denotes the excitation voltage source and motivation current source vectors on the z point at time t.

In this paper, the three-conductor transmission line is used as an example based on the coal mine security monitoring system, one of these is power line, the other two are signal lines as shown in Fig. 2. S-terminal and L-terminal represent two terminal sides of the transmission line. Infinite ground plane can be seen as reference, and the three lines are parallel.

There are two kinds of crosstalk of MTL: capacitive crosstalk and inductive crosstalk. The segment of the isometric lines is shown in Fig. 3, where V_S is the excitation source, V_1 and I_1 are the starting point of voltage and current, dz is the space step.

Because R and G in the lossless uniformed MTL are zero, the spatial discretization for MTL is along the transmission line based on the method of FDTD. The following results are obtained as Formula (3) and (4):

$$V_k^{n+1} = V_k^n - \frac{\Delta t}{\Delta z}C^{-1}\left(I_k^{n+1/2} - I_{k-1}^{n+1/2}\right)k = 2, 3, \ldots, NDZ \tag{3}$$

$$I_k^{n+3/2} = I_k^{n+1/2} - \frac{\Delta t}{\Delta z}L^{-1}\left(V_{k+1}^{n+1} - V_k^{n+1}\right)k = 1, 2, \ldots, NDZ \tag{4}$$

where subscript k denotes the spatial discretization sequence, superscript n is the time step point. Use 0 as the voltage superscript initial value, 1/2 as the current superscript initial value. We assume that the initial value of the voltage and current of transmission line is zero. It can be seen in the formula, the current on the conductor could be solved by iteration of the current moment voltage and before a moment voltage. Otherwise the voltage can be solved by iteration of current.

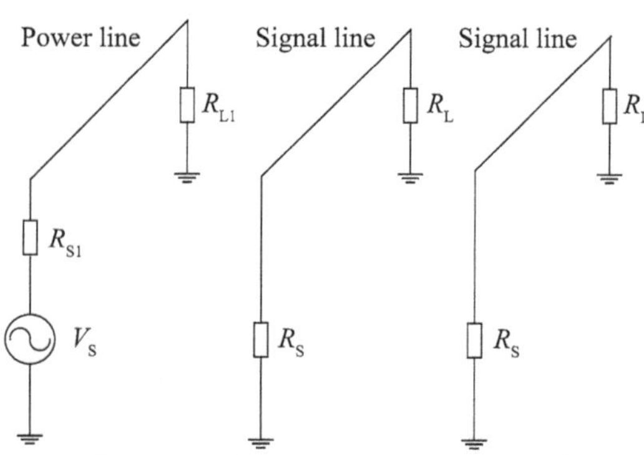

Fig. 2 Model of three parallel conductor transmission lines

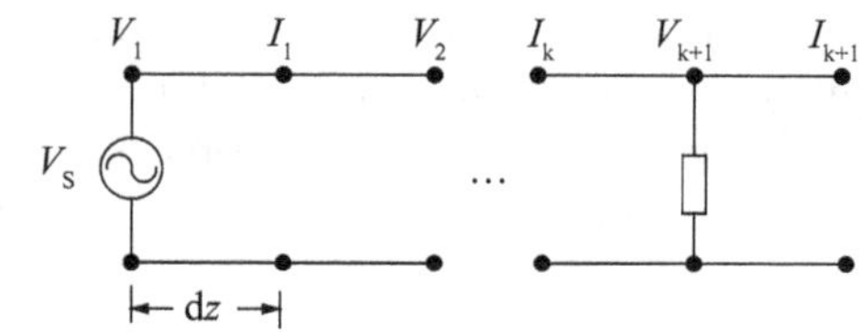

Fig. 3 Spatial discrete of parallel conductors

Combine with the terminal conditions of the transmission line, we can get the difference equation of both S-terminal and L-terminal side as Formula (5) and (6):

$$\begin{aligned} V_1^{n+1} &= \left(R_S\frac{C}{2}\frac{\Delta z}{\Delta t} + \frac{E}{2}\right)^{-1} \times \left[\left(R_S\frac{C}{2}\frac{\Delta z}{\Delta t} - \frac{E}{2}\right) \times V_1^n - R_S\left(I_1^{n+1/2} - I_0^{n+1/2}\right) + \frac{V_S^{n+1} + V_S^n}{2}\right] \\ &= \left(R_S\frac{C}{2}\frac{\Delta z}{\Delta t} + \frac{E}{2}\right)^{-1} \times \left[\left(R_S\frac{C}{2}\frac{\Delta z}{\Delta t} - \frac{E}{2}\right) \times V_1^n - R_S I_1^{n+1/2} + \frac{V_S^{n+1} + V_S^n}{2}\right] \end{aligned} \tag{5}$$

$$\begin{aligned} V_{NDZ+1}^{n+1} &= \left(R_L\frac{C}{2}\frac{\Delta z}{\Delta t} + \frac{E}{2}\right)^{-1} \times \left[\left(R_L\frac{C}{2}\frac{\Delta z}{\Delta t} - \frac{E}{2}\right)V_{NDZ+1}^n - R_L\left(I_{NDZ+1}^{n+1/2} - I_{NDZ}^{n+1/2}\right) + \frac{V_L^{n+1} + V_L^n}{2}\right] \\ &= \left(R_L\frac{C}{2}\frac{\Delta z}{\Delta t} + \frac{E}{2}\right)^{-1} \times \left[\left(R_L\frac{C}{2}\frac{\Delta z}{\Delta t} - \frac{E}{2}\right)V_{NDZ+1}^n + R_L I_{NDZ}^{n+1/2} + \frac{V_L^{n+1} + V_L^n}{2}\right] \end{aligned} \tag{6}$$

The Formula (3), (4), (5) and (6) constitute the basic iterative formula of the FDTD method to analysis of MTL.

3 Electromagnetic interference parameters by using orthogonal experiment design

The electromagnetic interference parameters for security monitoring system in coal mines during devices' normal working are optimized by orthogonal experimental design methods based on the MATLAB simulation on the interference processes.

3.1 Orthogonal experimental design

Among all the interference parameters, length of transmission line (φ), distance between the power line and signal line (d_1), distance between two signal lines (d_2),

Table 2 Interference parameters with their ranges and values at four levels

Interference parameters	Range	Level 1	Level 2	Level 3	Level 4
Length of transmission line φ (m)	90–900	90	210	540	900
Distance between the power line and signal line d_1 (cm)	10–100	10	20	50	100
Distance between two signal lines d_2 (mm)	2–5	2	3	4	5
Distance between the transmission line and ground d_0 (cm)	10–100	10	20	50	100
Load of terminal side of the signal line R (Ω)	10–10000	10	100	1000	10000

Table 3 Orthogonal table and results

No.	Factor					Evaluation value	
	φ	d_1	d_2	d_0	R	Voltage of S-terminal (V)	Voltage of L-terminal (V)
1	1	1	1	1	1	6.27×10^{-9}	6.05×10^{-10}
2	1	2	2	2	2	3.99×10^{-8}	1.31×10^{-8}
3	1	3	3	3	3	1.39×10^{-7}	1.57×10^{-7}
4	1	4	4	4	4	5.71×10^{-7}	6.33×10^{-7}
5	2	1	2	3	4	3.22×10^{-6}	3.48×10^{-6}
6	2	2	1	4	3	7.81×10^{-7}	8.28×10^{-7}
7	2	3	4	1	2	1.14×10^{-8}	4.12×10^{-9}
8	2	4	3	2	1	9.42×10^{-10}	4.24×10^{-11}
9	3	1	3	4	2	6.25×10^{-7}	1.19×10^{-7}
10	3	2	4	3	1	4.57×10^{-8}	1.10×10^{-9}
11	3	3	1	2	4	4.86×10^{-7}	7.01×10^{-7}
12	3	4	2	1	3	3.07×10^{-8}	4.11×10^{-8}
13	4	1	4	2	3	3.13×10^{-6}	4.36×10^{-6}
14	4	2	3	1	4	8.38×10^{-7}	2.34×10^{-6}
15	4	3	2	4	1	6.03×10^{-8}	9.76×10^{-10}
16	4	4	1	3	2	1.70×10^{-7}	2.58×10^{-8}
K_{1i}	K_{11}	K_{11}	K_{13}	K_{14}	K_{15}		
K_{21}	K_{21}	K_{22}	K_{23}	K_{24}	K_{25}		
K_{3i}	K_{31}	K_{32}	K_{33}	K_{34}	K_{35}		
K_{4i}	K_{41}	K_{42}	K_{43}	K_{44}	K_{45}		
R_i	R_1	R_2	R_3	R_4	R_5		

distance between the transmission line and ground (d_0), load of terminal side of the transmission line (R) are the five most important factors for security monitoring system in coal mines. And the interference results were evaluated by interference voltage on two terminal sides (S-terminal and L-terminal) of the transmission line. Finally, an orthogonal experiment of five factors was designed, and four-level orthogonal array $L_{16}(4^5)$ was employed; this array specifies 16 experimental runs and has 5 columns.

The selected interference parameters, along with their ranges and levels, are given in Table 2. Considering the structural characteristics of the underground power lines, the impedance of the S-terminal of the power line is set to 2 Ω and the impedance of the L-terminal of the power line is set to 100 MΩ. The detailed scheme of orthogonal experiment and the numerical simulation results of interference voltage on the two terminal sides of the transmission line at different levels are listed in Table 3.

In the Table 3, K_{ji} represents the experimental indicator when each parameter is at levels 1–4, R_i is the difference between the lowest value and the highest value for each parameter. The deference equation of R_i is:

$$R_i = \max\{K_{1i}, K_{2i}, K_{3i}, K_{4i}\} - \min\{K_{1i}, K_{2i}, K_{3i}, K_{4i}\} \quad (7)$$

According to the above analysis, the minimum interference voltage can be concluded.

Table 4 Interference voltage of S-terminal at different levels (unit: V)

Interference parameters	Length of transmission line	Distance between the power line and signal line	Distance between two signal lines	Distance between the transmission line and ground	Load of terminal side of the transmission line
Level 1	1.89×10^{-7}	8.62×10^{-7}	3.61×10^{-7}	2.22×10^{-7}	2.83×10^{-8}
Level 2	1.00×10^{-6}	4.26×10^{-7}	8.37×10^{-7}	9.15×10^{-7}	2.12×10^{-7}
Level 3	2.97×10^{-7}	1.74×10^{-7}	4.01×10^{-7}	8.93×10^{-7}	1.02×10^{-6}
Level 4	1.05×10^{-6}	1.93×10^{-7}	9.41×10^{-7}	5.09×10^{-7}	1.28×10^{-6}
Maximal difference	8.62×10^{-7}	1.57×10^{-6}	5.80×10^{-7}	6.94×10^{-7}	1.25×10^{-6}

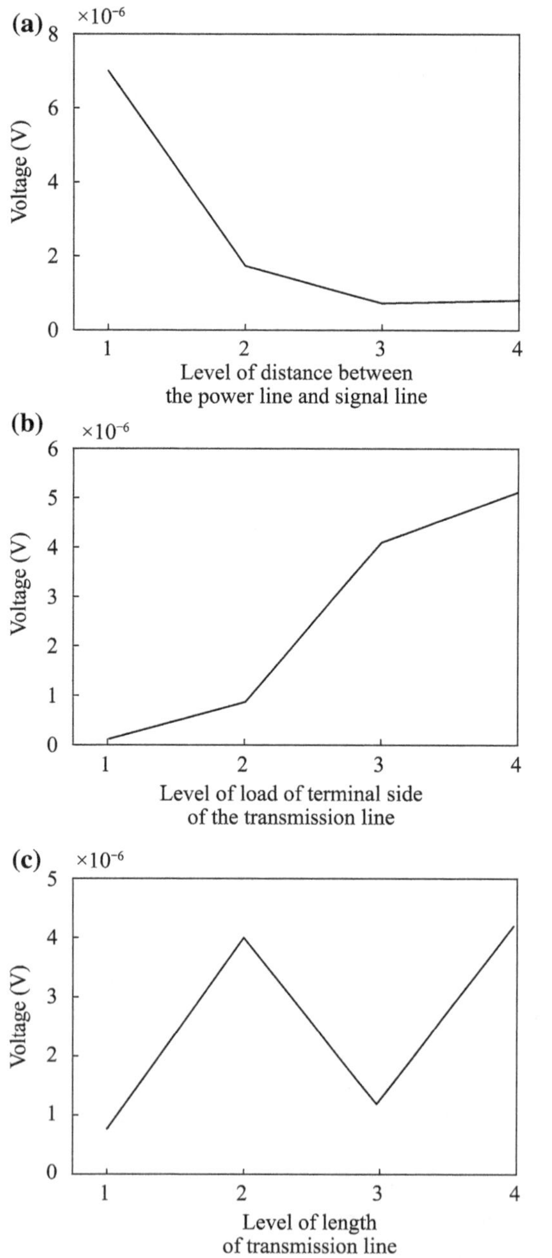

Fig. 4 Interference voltage of S-terminal for parameters at four levels. **a** Level of distance between the power line and signal line; **b** level of load of terminal side of the transmission line; **c** level of length of transmission line

3.2 Analysis of orthogonal experiment result

Considering the equipment to work normally, the interference voltage of S-terminal for each parameter at levels 1–4 and the difference between the lowest value and the highest value are listed in Table 4. By comparing the value of each parameter, it can be found that the three most significant parameters which influence the interference voltage of S-terminal are as follows: distance between the power line and signal line, load of terminal side of the transmission line and length of transmission line. The interference voltage of S-terminal for the three parameters at four levels is plotted in Fig. 4.

Figure 4a shows the variation of the interference voltage of S-terminal with the distance between the power line and signal line. With the increasing of distance between the power line and signal line, the interference voltage of S-terminal decreases at levels 1–4. Figure 4b shows the effect of the load of terminal side of the transmission line on the interference voltage of S-terminal. With the increasing of the load of terminal side of the transmission line, the interference voltage of S-terminal increases at levels 1–4. Figure 4c shows that the variation of the interference voltage of S-terminal with length of transmission line. It is observed that the length of transmission line at level 2 and level 3 are inflection points of the variation of the interference voltage of S-terminal, with the increasing of the length of transmission line; the interference voltage of S-terminal increases when the length of transmission line is lower than 100 Ω, then decreases when the length of transmission line is higher than 100 Ω and lower than 1000 Ω; finally, it increases when the length of transmission line is higher than 100 Ω.

Considering the equipment working in normal, the interference voltage of L-terminal for each parameter at levels 1–4 and the difference between the lowest value and the highest value are listed in Table 5. By comparing the value of each parameter, it can be found that the three most significant parameters which influences the interference voltage of L-terminal are as follows: the distance between the power line and signal line, load of terminal side of the transmission line and length of transmission line. The

Table 5 Interference voltage of L-terminal at different levels (unit: V)

Interference parameters	Length of transmission line	Distance between the power line and signal line	Distance between two signal lines	Distance between the transmission line and ground	Load of terminal side of the transmission line
Level 1	2.01×10^{-7}	1.99×10^{-6}	4.06×10^{-7}	5.97×10^{-7}	6.81×10^{-10}
Level 2	1.08×10^{-6}	7.96×10^{-7}	8.85×10^{-7}	1.29×10^{-6}	4.05×10^{-8}
Level 3	2.33×10^{-7}	2.33×10^{-7}	6.54×10^{-7}	9.17×10^{-7}	1.34×10^{-6}
Level 4	1.68×10^{-6}	1.75×10^{-7}	1.25×10^{-6}	6.95×10^{-7}	1.81×10^{-6}
Maximal difference	1.48×10^{-6}	1.82×10^{-6}	8.44×10^{-7}	3.20×10^{-7}	1.81×10^{-6}

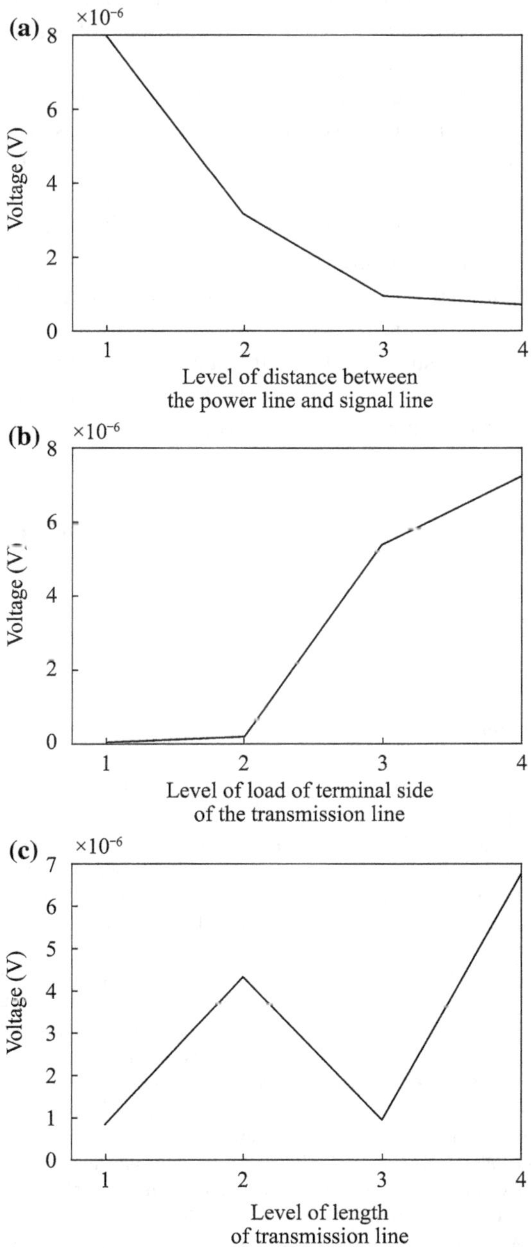

Fig. 5 Interference voltage of L-terminal for parameters at four levels. **a** level of distance between the power line and signal line; **b** level of load of terminal side of the transmission line; **c** level of length of transmission line

interference voltage of L-terminal for the three parameters at four levels is plotted in Fig. 5.

Figure 5 shows the variation of the interference voltage of L-terminal with the distance between the power line and signal line, the load of terminal side of the transmission line and the length of transmission line. It is observed that the variation of the interference voltage of L-terminal is similar to the variation of the interference voltage of S-terminal.

From Table 4 and Fig. 4, it is clear that the interference voltage of S-terminal is a minimal value at the third level of distance between the power line and signal line, at the

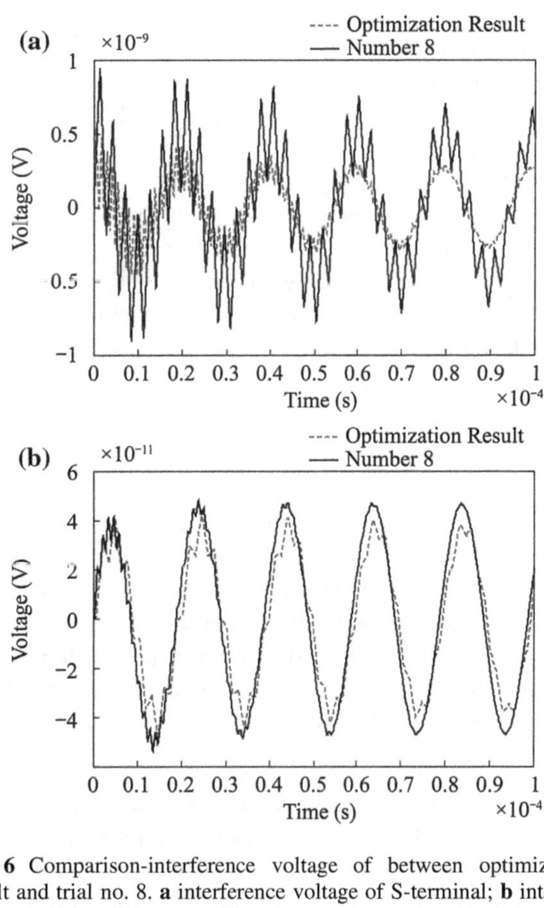

Fig. 6 Comparison-interference voltage of between optimization result and trial no. 8. **a** interference voltage of S-terminal; **b** interference voltage of L-terminal

Table 6 Comparison between the optimized and no. 8 simulation results

Interference voltage	The optimized simulation value (V)	simulation value of No. 8 (V)	Improvement effect (%)
Interference voltage of S-terminal	5.35×10^{-10}	9.42×10^{-10}	−43.21
Interference voltage of L-terminal	2.79×10^{-11}	4.24×10^{-11}	−34.20

first level of the load of terminal side of the transmission line, and at the first level of the length of transmission line. From Table 5 and Fig. 5, it is clear that the interference voltage of S-terminal is a minimal value at the forth level of distance between the power line and signal line, at the first level of the load of terminal side of the transmission line, and at the first level of the length of transmission line.

To sum up, the optimal parameters for the interference voltage of S-terminal and L-terminal are as follows: $\varphi = 90$ m, $d_1 = 50$ cm, $d_2 = 2$ mm, $d_0 = 10$ cm, $R = 10$ Ω.

4 Simulation experiment

From Table 3, it is observed that the minimal value of interference voltage of S-terminal and L-terminal among all the trials occur during the No.8 trial. To verify the optimization results of orthogonal experimental design, the comparison-interference voltage of S-terminal and L-terminal between the optimization result and No.8 trial as shown in Fig. 6.

Based on the orthogonal experimental design, the comparison between the optimized and No. 8 simulation results were performed and the results are listed in Table 6. As observed from the table, after the optimization, the interference voltage of the security supervision signal line was reduced at least 34.20 % before optimization.

5 Conclusions

In this paper, the frequency characteristics of interference source of the power line are extracted when equipment runs normally, the interference parameters are optimized by using orthogonal experiment design method based on the FDTD analyses of MTL on device's power line while working normally, and the results of experimental design are confirmed by MATLAB simulation. Conclusions are as follows:

(1) The interference source on device's the power line while working normally can be obtained by using the Amway 2711 spectrum analyzer; the optimal interference parameters obtained using orthogonal experiment design are as follows: length of transmission line is 90 m, distance between the power line and signal line is 50 cm, distance between two signal lines is 2 mm, distance between the transmission line and ground is 10 cm, load of signal line port is 10 Ω.

(2) Among the interference parameters, the distance between the power line and signal line is the most important one to affect the interference voltage of S-terminal and L-terminal; with the increase of the distance, the interference voltage of S-terminal and L-terminal decreases.

(3) The experiments on the electromagnetic interference of coal mine power system to security monitoring system show that electromagnetic interference can be decreased at least 34.20 %, which providing certain theoretical guidance for engineering practice.

Acknowledgments This study is supported by the Fundamental Research Funds for the Central Universities (No. 2014JBM116) and (No. 2014JBZ017).

References

Delogne P (1991) EM Propagation in tunnels. IEEE Trans Antennas Propag 39(3):401–405

Erdin I, Khazaka R, Nakhla M (1998) Simulation of high-speed interconnects in a multilayered medium in the presence of incident field. IEEE Trans on Microw Theory Tech 46(12):2251–2257

Gao YG (2004) Shielding and earthing. Beijing University of Posts and Telecommunications Press, Beijing, pp 61–65

Hill DA (1989) Reflection coefficient of a waveguide with slightly uneven walls. IEEE Trans Microw Theory Tech 37(1):244–252

Jiao CQ, Xia ZJ, Fu WN (2012) A generalized multiconductor transmission line model and optimized method for the solution of the MTL equations. Int J Antennas Propag 7

Liang ZG (2007) Theory, technique and application of electromagnetic compatibility. China Machine Press, Beijing, pp 35–37

Liu XQ, Wei GH, Pan XD (2009) Simulation and analysis of field performance in reverberation chamber with materials of different permeability. High Volt Eng 35(8):1986–1989

Liu XQ, Wei GH, Pan XD (2010) Numerical analysis on field performance optimization of reverberation chamber using dual-feeding. High Volt Eng 36(5):1234–1239

Paul CR (2008) Analysis of multiconductor transmission line. John Wiley & Sons, New York

Rappaport TS (1989) Characterization of UHF multipath radio channels in factory building. IEEE Trans Antenna Propag 37(8):1058–1069

Sun JP (1999) Harsh classification on electromagnetic compatible measure of mining monitoring and communication equipment. Coal Sci Technol 6:23–25

Sun JP, Pan T, Tian ZJ (2006) Study on electromagnetic compatibility in coal mining. J China Coal Soc 31(3):377–379

Sun JP, Ma FY, Li C (2008) Improved complex filter applied in enhancing EFT/B immunity at a coal mining monitoring substation. J China Univ Min Technol 18(1):22–26

Zhang SX, Xu ZH, Qian ZP (2007) Analysis of shielding effect of metallic box with apertures using FDTD. J PLA Univ Sci Technol (Nat Sci Edn) 8(4):324–327

6

Construction of green infrastructure in coal-resource based city

Jiang Chang[1] · Tinghao Hu[2] · Xiangxu Liu[1] · Xiaogeng Ren[3]

Abstract As an important coal-resource based city in eastern China, coal mining activities have greatly promoted economic development to Xuzhou while resulted in kinds of blocks for sustainable development. Subsidence areas caused by coal mining activities and rapid urbanization process led to the losing of ecological function and decreasing of landscape connectivity in the city. As the important life-support system, urban green infrastructure (GI as short) has great significance in improving human well-beings. Taking Xuzhou urban area as the study object, this paper proposed a green infrastructure construction method based on the principle of ecological priority. Firstly, key ecological patches were identified by GIS and RS methods. Secondly, we established corridors by minimum cost path model. Then, we built GI network and evaluated the ecological importance of different patches. Finally, by comparing the status of patches and coal mining subsidence areas, we proposed a framework for constructing GI network in Xuzhou urban area.

Keywords Coal-resource based city · Green infrastructure · Green space · Xuzhou urban area

1 Introduction

Green infrastructure (GI) is an important life-support system, vital to improving the urban ecology and maintaining sustainable development (Benedict and McMahon 2006; Weber et al. 2006). However, green infrastructure in coal-mining cities is threatened not only by mining activities but also by the ongoing processes of urbanization. In China, 95% of coal is exploited from underground deposits. The extraction causes subsidence, as well as the destruction of cultivated land, residential areas and transportation infrastructure. As a result, There exist a large number of post-mining landscapes in China (Chang et al. 2011; Chang 2013). Clearly, coal-mining cities urgently require a theory for the reconstruction of mining areas in order to ensure their sustainable development (Feng et al. 2016).

Hundred years of coal mining activities have brought considerable economic benefits to Xuzhou while resulted in kinds of tumbling blocks for sustainable development. Subsidence areas caused by coal mining activities not only directly affected various production activities, but also had adverse effects on the urban ecological environment. Nevertheless, whether the damaged post-mining lands have ecological potentiality? The answer is definitely yes. Owning to the long term abandon and less human disturbance, the ecosystem in subsidence areas is in a state of self-recovery, which has formed new important areas with high ecological potential and provides vivo sphere for threatened species (Chang et al. 2011).

✉ Jiang Chang
changjiang102@163.com

[1] School of Architecture and Design, China University of Mining and Technology, Xuzhou, China

[2] School of Mechanics and Civil Engineering, China University of Mining and Technology, Xuzhou, China

[3] School of Architecture, Southeast University, Nanjing, China

2 Theoretical fundamental of green infrastructure

2.1 GI definition

Green infrastructure is an open, systematic ecological function network which makes great significance in improving urban ecology and maintaining sustainable development (Weber et al. 2006). GI was firstly proposed by the United States in 1990s and was implemented into the state, regional and local policies. GI theory encompasses all spatial levels. It focuses on connecting urban green, aiming at multi-functionality, forging a link to the spatial provision of ecosystem services and highlighting the need for close collaboration between scientific, political and administrative actors (Pauleit et al. 2011; Benedict and McMahon 2006; Haase et al. 2014). Green infrastructure and 'grey' technical infrastructure are considered as essential functional systems in cities.

2.2 Functions and components of GI

Green infrastructure is a natural and man-made green network. It consists of core areas, corridors and stepping stones (Li 2009) (see in Fig. 1).

Core area is the core region of green infrastructure network. It plays the role of a variety of natural processes. It appears with various forms and scales, and different scales covers different contents, including nature reserves, forest, parks, woodlands, green space in communities, etc.

Corridor plays a role in connecting elements in green infrastructure network. It maintains key biological processes and provides recreation and leisure spaces for people as well. Corridors link up the whole green infrastructure network and keep the entire system circulating. Ecological linear spaces, rivers, and greenbelts of railways and roads are normal corridors.

The definition of stepping stone is unified and relatively blurry. In some certain scales, it can be regarded as patches that smaller than core areas, or corridors for species migration. Stepping stones can provide ecological nodes for species migration and recreation activities for human beings. It covers small habitats, small green spaces in communities and gardens. An example is shallow water areas in Xuzhou coal mining areas. In Xuzhou, the phreatic water level is 1–1.5 m below the surface, and a large area of water bodies thus appears because of coal mining activities. In coal mining subsidence areas, many of these water bodies still remained in un-reclamation status. After years of self-recovery process, those places have been able to act as habitats of fishes and birds as well as provide some ecological functions. Although these "stepping stones" are mostly in small area and discrete distributed, they can acts as important complements of core areas and corridors in the entire GI network (see in Fig. 2).

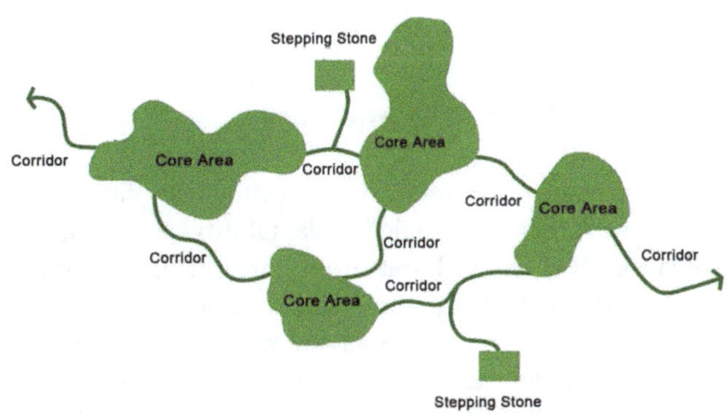

Fig. 1 Structure and components of green infrastructure

3 Study materials

3.1 Study area

Xuzhou (117°20′E, 34°26′N) is located in the northwest of Jiangsu province, China, southeast of the north China plain. The city is rich in mineral resources and is one of the important coal production bases in the eastern plain of China. At present, there are 29 coal mines in total within study area (see in Fig. 3). In this paper, the research focused on the Xuzhou urban area which covers 3126 km^2.

According to the data in 2016, more than 160 km^2 coal mining subsidence areas were found in Xuzhou urban area. They formed a pattern of semi-surrounding the main urban area. According to distribution and pattern, coal mining subsidence areas in Xuzhou urban area could be divided into east and west parts, and nine districts in total. In the west, Pangzhuang and Jiahe District as the core, scattered around Chacheng District, Zhangji District, Yi'an District and Xinhe-Woniu District. Jiawang District, Dahuangshan District and Dongzhuang District formed east part of subsidence areas in Xuzhou urban area (see in Fig. 3).

3.2 Data sources

Using Landsat—TM images (the spatial resolution is 30 m × 30 m) of Xuzhou in July 2016 and reference of land use classification system (Li et al. 2013), we divided land use and land cover into 6 categories: woodland, grassland, cultivated land, water body, roads, urban construction land (see in Table 1 and Fig. 4). We identified Green Infrastructure (GI as short) as ecological sources, nature reserves, scenic spots, forest parks etc., which covers an area of 3.18×10^4 ha and 65.25% of the research area.

Fig. 2 Stepping stones in coal mining subsidence area in Xuzhou urban area

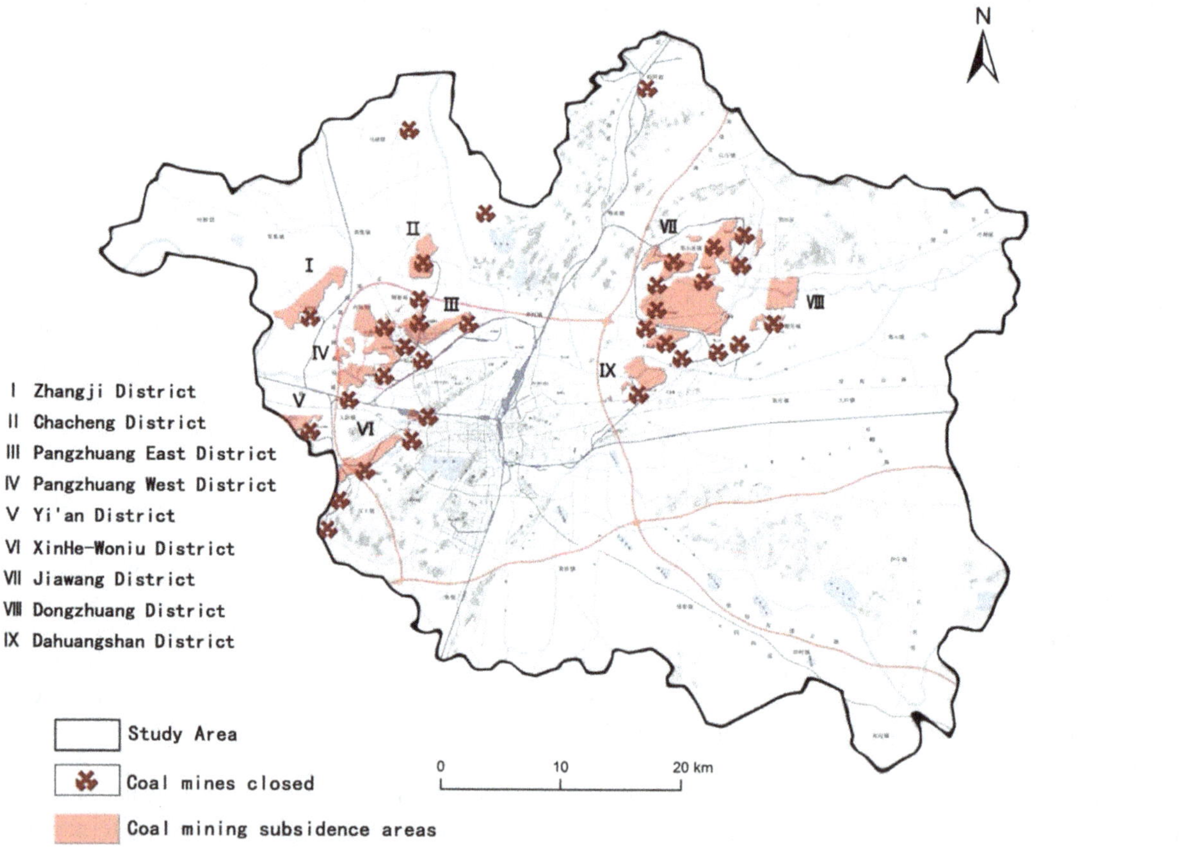

Fig. 3 Distributions of coal mines and subsidence areas in Xuzhou urban area

Table 1 Land use types in Xuzhou urban area

Land use and land cover	Area (km^2)	Proportion (%)
Woodland	374.29	11.94
Grassland	187.07	5.97
Cultivated land	1282.5	40.92
Water body	201.27	6.42
Roads	15.46	0.49
Urban construction land	1073.71	34.26

4 Process

4.1 Ecological sensitivity evaluation

Using classification method of ecological factors in McHager vertical ecological adaptability analysis, the evaluation was carried out from three aspects: human factor, ecological background and natural factor. Based on the features of the present situation of Xuzhou urban area, 10 of most fundamental factors were selected. They were: Land use type, Cultural heritage, Transportation, Pollution source, Key infrastructure, NDVI, Natural reserve area,

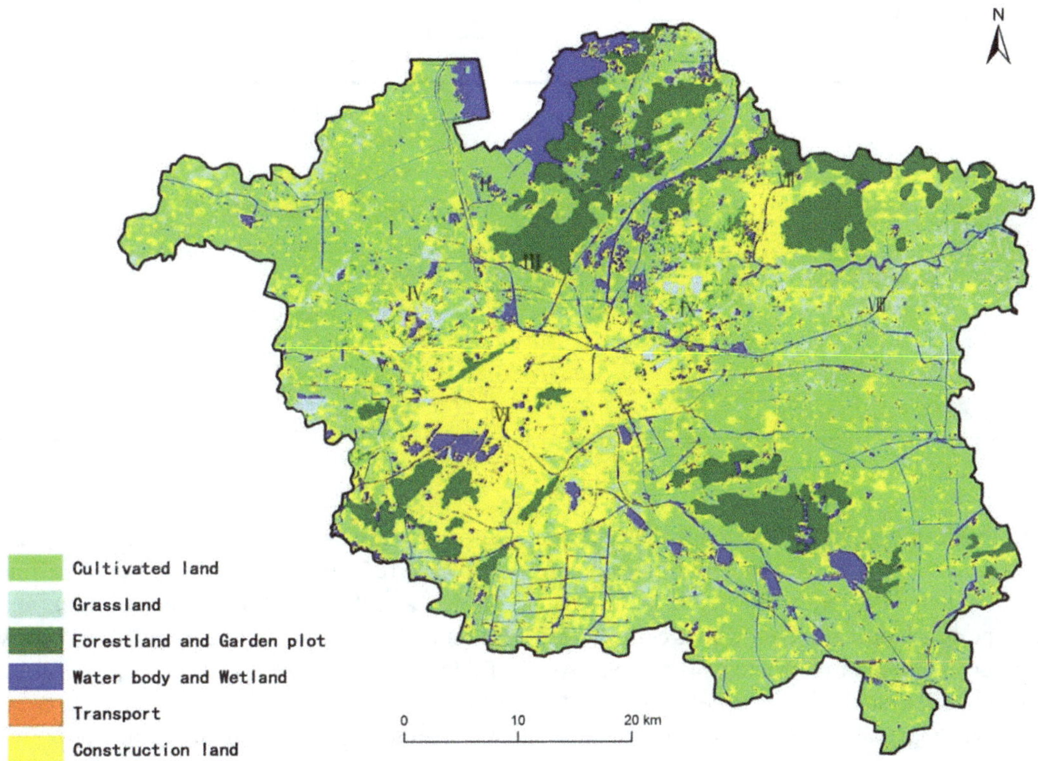

Fig. 4 Land cover and land use types in Xuzhou urban area

Water body, Gradient, and Altitude. These factors had significant impacts on ecological sensitivity in Xuzhou urban area (Shen et al. 2009; He et al. 2011; Polasky et al. 2011; Theobalda et al. 2011). Meanwhile, according to the characteristics of each index layer, the grading standard was established, and the study area was assigned by 30 m × 30 m grid.

In our study, Analytic Hierarchy Process was used to determine the weight of each index. In order to measure the evaluation variables and weights of factors, we consulted 10 specialists who had knowledge about the study area and about different research fields (Urban planning, Landscape planning, Physical geography, and Phytology, etc.) to address the necessary information. The experts discussed and rated the factors and compared the indicators in same layer to obtain the comparison matrix. Then we used MATLAB software to calculate the weight (see in Tables 2, 3).

After determining the factors and weight, the superposition of multiple factors should be considered in ecological sensitivity evaluation process. Therefore, we overlapped the data from every-single factor's partition and put the weighted superposition into the ecological suitability evaluation model (Formula 1) using grid computing function in ArcGIS10.2.

$$S = \sum_{i=1}^{n} w_i x_i \tag{1}$$

where S is the ecological suitability score; x_i is the score of each influencing factor; w_i is the weight of factor.

Then, we used natural break classification method to evaluate the ecological environment sensitivities. They can be divided in 5 groups: high sensitive areas, relatively high sensitive areas, middle sensitive areas, low ecological sensitive areas, and very low sensitive areas. The calculated results were shown in Fig. 5.

4.2 GI element identification

4.2.1 Patches selection process

After the sensitivity evaluation process, High sensitive areas, relatively high sensitive areas, and middle sensitive areas were chosen as patches in GI network. Meanwhile, each patch was determined by using the land use type data. Based on the geological interpretation of remote sensing, we removed the patches that deeply developed or changed by human activities. At the same time, the area less than 100 ha was removed by using the area threshold and 49 ecological patches were selected. The specific result is shown in Fig. 6.

Table 2 Weights of evaluation variables

Factor	Factor weight	Variable	Variable weight
Human factor	0.2599	Land use type	0.0821
		Cultural heritage	0.0410
		Transportation	0.0273
		Pollution source	0.0821
		Key infrastructure	0.0274
Ecological background	0.4126	NDVI	0.2063
		Natural reserve area	0.2063
Natural factor	0.3275	Water body	0.1310
		Gradient	0.0665
		Altitude	0.1310

In order to further identify the core patches that play a key role in the GI network, we introduce the concept of "source". According to the definition of Chen et al. (2006), "Source" landscape is a landscape type which contributes positively to the development of the ecological process. As the center of GI network system, the "source" is generally referred to the large core patches that can provide integrated ecosystem services and maintain ecosystem stability. Thus we further selected 4 source patches according to "Xuzhou metropolitan area planning (2010–2030)". They were: Weishan wetland patch, Dadong mountain scenic patch, Lvliang mountain scenic patch, Yunlong mountain senic patch. These 4 patches were all selected from natural reserve areas and core area of scenic area (see in Fig. 4).

4.2.2 Corridors selection process

"Mining path method" was used in this study to simulate the potential corridors of animal migration. Using the identified source patches, the calculation and simulation of corridors was carried out by the assignment of landscape resistance. The minimum cost model is mainly taken into account of the three aspects: ecological sources, distance and resistance factor. The equation is:

$$C_i = \sum (D_i \times F_j), \quad (i = 1, 2, \ldots, m;\ j = 1, 2, \ldots, n) \tag{2}$$

where, the D_i refers to the actual interval of a landscape unit i to the source; F_j refers to the impedance value of a certain landscape element j in landscape space; C_i refers to the calculation result, that is, the cumulative consumption distance value of the landscape unit i to the source; m and n are the sum of the basic landscape units.

Based on TM remote sensing image data and topographic map data, we determined the different landscape resistance values by investigating the vegetation of different land use types and human disturbance degree. We also referred others previous research results by Pascual-Hortal and Saura. (2011) and Liu, et al. (2010). The resistance values of different land use types were given in Table 4.

4.3 Patch importance evaluation

Using Conefor Sensinode 2.6 and ArcGIS 10.2, we calculated patches importance values by IIC (Integral Index of Connectivity). IIC was regard as basis for evaluating patches importance. Wherein the greater the value of IIC was, the more important the patch will be. The IIC can be indicated as:

$$I_{\mathrm{IIC}} = \frac{\sum_{i=1} \sum_{j=1} \frac{a_i \cdot a_j}{1 + nl_{ij}}}{A_{\mathrm{L}}^2} \tag{3}$$

where, n represents the total number of patches in the landscape; a_i and a_j respectively represent the area of patch i and patch j; nl_{ij} represents the number of connections between patch i and patch j; A_L is the area of the study area. When $0 \leq I_{\mathrm{IIC}} \leq 1$, the value of IIIC is 0, which means there is no connection between the different patches. When I_{IIC} is equal to 1, it means the whole landscape is a habitat patch.

According to the research results of Pascual-Hortal and Saura. (2007) and Fei (2011), the ecological patch connected distance threshold is set to 10,000 m and the probability of directly connected is set to 0.01 (i.e. when the distance between the two patches is less than 10000 m or the connectivity probability is greater than 0.01, there is connectivity between two patches and the connecting value is 1. Otherwise it is 0).

By analyzing the IIC, the importance of each patch was classified. Using natural break classification, IIC value interval was set as 0–0.75, 0.75–2.50, 2.50–49.00 according to differentiate importance rating. There were 17

Table 3 Indices of ecological sensitivity factors

Factor	Variable	First-class object	Second-class object	Value
Human factor	Land use type	Land use type	Forest	5
			River	5
			Grassland	4
			Farmland	3
			Construction land	2
			Roads	1
	Cultural heritage	Cultural heritage	30 m buffer	5
			30–50 m buffer	3
			50–100 m buffer	2
			Out of buffer range	1
	Transportation	Railway	40 m buffer	5
			40–100 m buffer	3
			100–200 m buffer	2
		Highway	30 m buffer	5
			30–80 m buffer	3
			80–150 m buffer	2
		Other transportation factors		1
	Pollution source	Pollution source	300 m buffer	4
			300–500 m buffer	3
			500–800 m buffer	2
			Out of buffer range	1
	Key infrastructure	Water supply and drainage facilities; Power supply facilities	40 m buffer	5
		High pressure corridor, substation	0–80 m buffer	3
		Other key infrastructure factors		1
Ecological background	NDVI	NDVI	$NDVI > 0.3$	5
			$0.2 < NDVI < 0.3$	4
			$0.1 < NDVI < 0.2$	3
			$0 < NDVI < 0.1$	2
			$NDVI < 0$	1
	Natural reserve area	Natural reserve area	Within reserve area	5
			500 m buffer	4
			500–800 m buffer	3
			800–1000 m buffer	2
			Out of buffer range	1
Natural factor	Water body	First and second class river; reservoir; lake	50 m buffer	5
			50–120 m buffer	3
		Brooklet; creek; pond	20 m buffer	5
			20–60 m buffer	3
		Other water body factors		1
	Gradient	Gradient	> 20°	5
			15°–20°	4
			10°–15°	3
			5°–10°	2
			0°–5°	1
	Altitude	Altitude	> 100 m	5
			50–100 m	3
			0–50 m	1

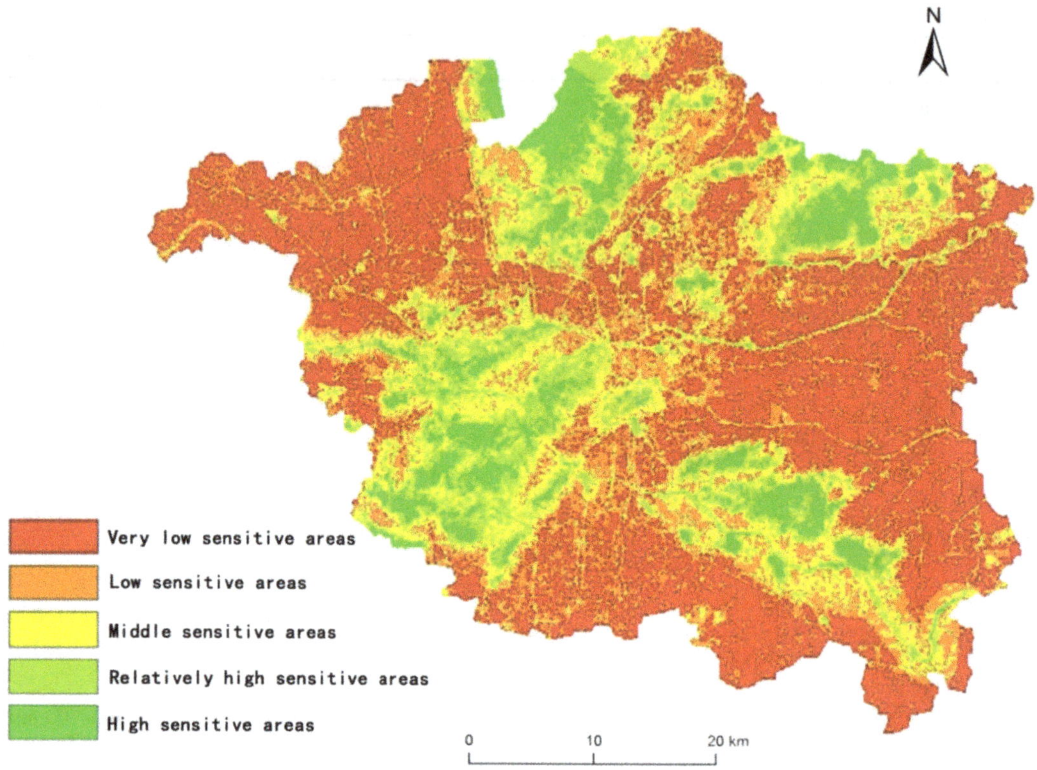

Fig. 5 Comprehensive evaluation of ecological sensitivity in study area

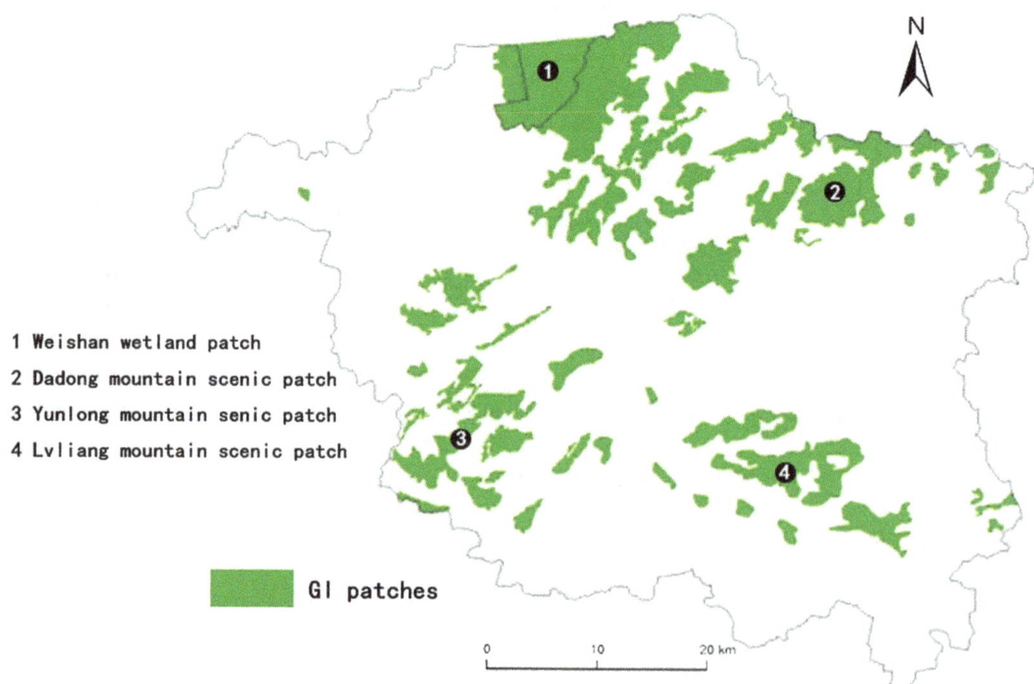

Fig. 6 GI patches in Xuzhou urban area

patches of IIC value were between 0.75 and 2.50. Values between 0.75 and 2.50 occupied 16. The left 16 patches' IIC value were remaining between 2.50 and 49.00 (see in Fig. 7).

4.4 Construction of GI network

The construction of GI network in Xuzhou urban area followed principles of urban functional arrangements, and strictly protected the ecological resources such as mountains, wetlands waters bodies and forests. In our study, the evaluating importance and status quo of patches were both taken into consideration. By carding and linking greenbelt corridors, river corridors, traffic corridors and potential corridors through key patches, we finally drew GI network pattern in Xuzhou urban area.

5 Results

In Xuzhou urban area, we identified 49 patches (Less than 100 ha) and 32 corridors by spatial methods. To obtain the key patches in GI network, we conceptualized patches and merged them within the same ecological radiation scope.17 patches and 9 corridors were selected as the key components of GI network in Xuzhou urban area.

Table 4 Cost value of different land use types

Land use type	Cost value
Forest	1
Grassland	3
Farmland	30
Construction land	1000
Transportation	300
Water body	500

5.1 Patches grading results

According to the importance and area of each patch, we classified patches in 3 levels and finally attained 17 patches in total (see in Table 5). There were 4 patches in level 1. These patches were mainly scenic areas and natural reserve areas and played key role in maintaining ecosystem services and ecological stability. There were 7 patches in level 2. These patches were mostly forest parks, mountains and coal mining subsidence areas. The last 6 patches were comparatively small mountains, urban parks and other coal mining subsidence areas (see in Fig. 8).

5.2 Corridors grading results

Based on the layouts, locations and ecological status of different corridors, we classified 9 corridors in three levels (see in Table 6): regional corridor, sub-regional corridor, and local corridor. There were 4 corridors in regional level. They were responsible for connecting key patches and

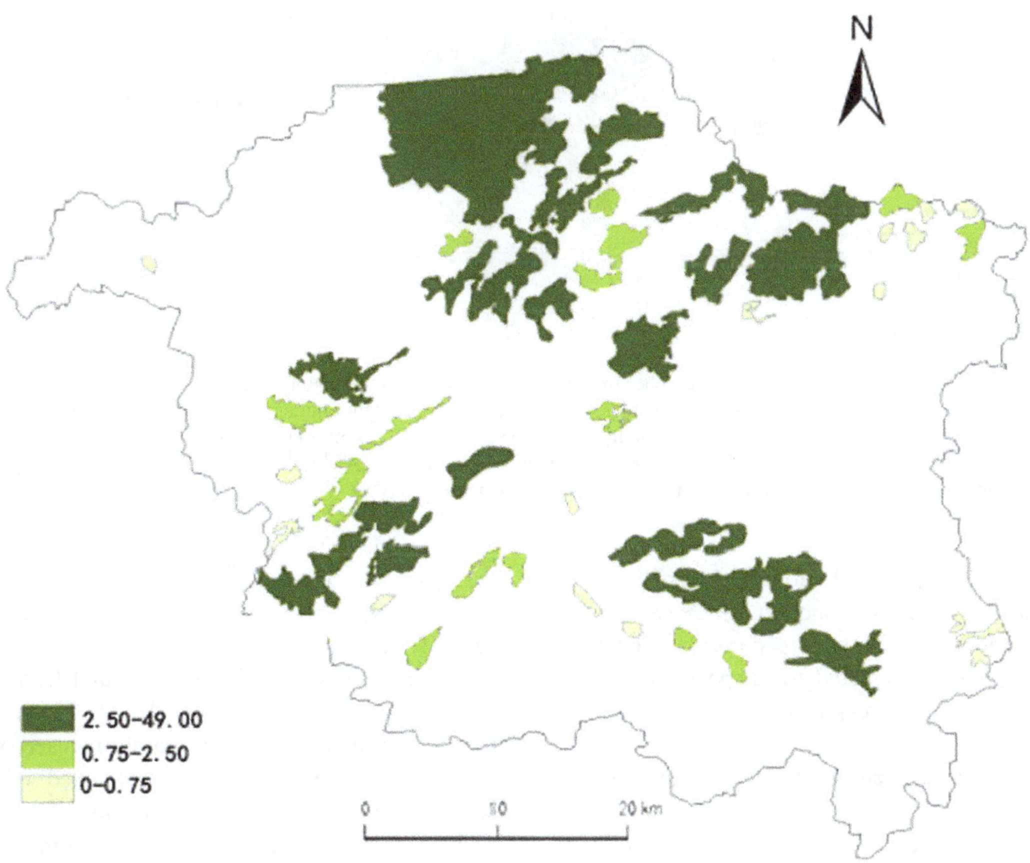

Fig. 7 GI patches importance ranking in Xuzhou urban area

Table 5 Classification of GI patches in Xuzhou urban area

Level	Quantity	Content
Patch level 1	4	Weishan wetland patch, Lvliang mountain scenic patch, Dadong mountain scenic patch, Yunlong mountain senic patch
Patch level 2	7	Northern mountains forest patch, Pan'an wetland patch, Jiuli subsidence patch, Cuihe reservoir patch, Jiawang subsidence patch, Tuan mountain patch, Pangzhuang subsidence patch
Patch level 3	6	Tuolong mountain forest patch, Jiuli mountain patch, Northern Woniu subsidence patch, Sorthern Woniu subsidence patch, Dahuang mountain country park patch, Dongzhuang subsidence patch

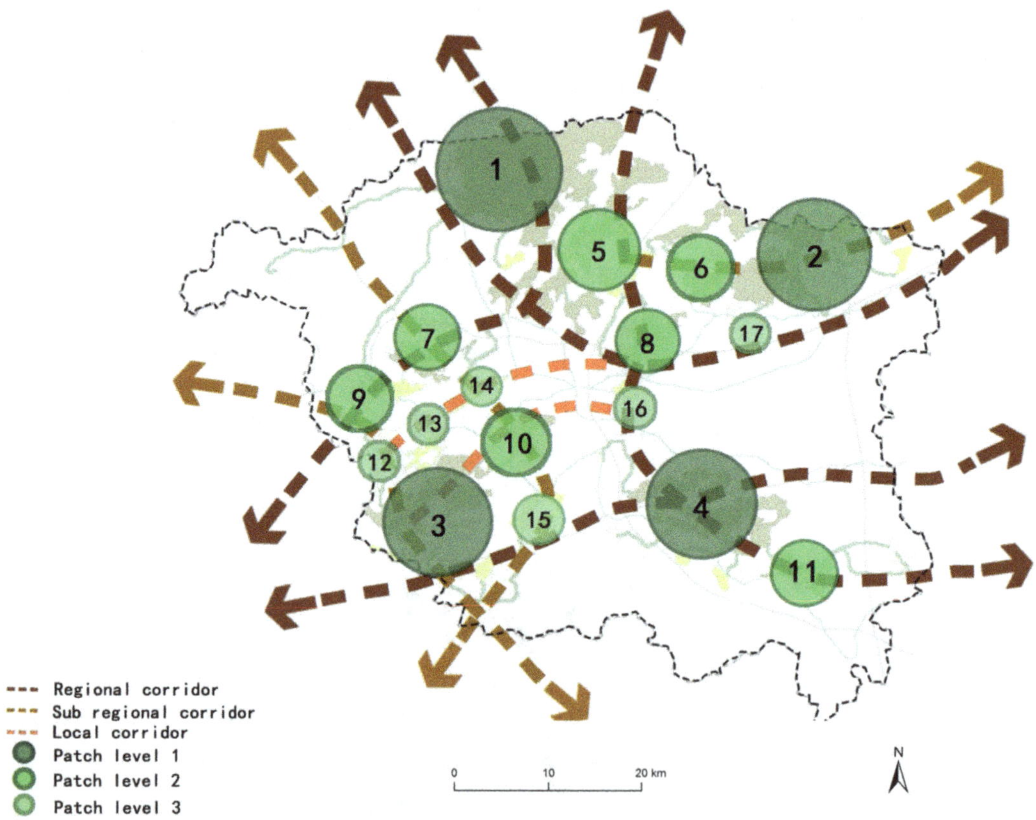

Fig. 8 GI pattern and patches in Xuzhou urban area. Patches in level 1: 1. Weishan wetland patch, 2. Dadong mountain scenic patch, 3. Yunlong mountain senic patch 4. Lvliang mountain scenic patch. Patches in level 2: 5. Northern mountains forest patch, 6. Jiawang subsidence patch, 7. Jiuli subsidence patch, 8. Pan'an wetland patch, 9. Pangzhuang subsidence patch, 10. Tuan mountain patch, 11. Cuihe reservoir patch. Patches in level 3: 12. Sorthern Woniu subsidence patch, 13. Northern Woniu subsidence patch, 14. Jiuli mountain patch, 15. Tuolong mountain forest patch, 16. Dahuang mountain country park patch, 17 Dongzhuang subsidence patch

forming the main skeleton of GI network. 3 corridors were in sub-regional level. Their functions were connecting patches and corridors and extending outward. The left 2 were local corridors. They acted to connect patches and corridors within the GI network (see in Fig. 9).

5.3 Coal mining subsidence patches in GI network

In our study, 10 of 17 patches have been already preserved. While, other 7 patches were not. It has to be emphasized that 5 of unrestored patches were coal mining subsidence areas. They were Jiawang subsidence patch, Pangzhuang subsidence patch, Northen woniu subsidence patch, Southern Woniu subsidence patch and Dahuang mountain forest park patch (see in Fig. 10). In Xuzhou urban area, the current planning system mainly focused on the original ecological places that mostly remained in good natural status, however, ignored the ecological values of coal mining subsidence areas to a great extent.

Table 6 Classification of GI corridors in Xuzhou urban area

Level	Quantity	Content
Regional corridor	4	Beijing-hangzhou canal corridor, Northern mountains—Pan'an wetland—Cuihe reservoir corridor, Yunlong mountain—Tuolong mountain—Lvliang mountain corridor, Weishan lake—Jiuli lake—Pangzhuang subsidence area corridor
Sub regional corridor	3	Jiuli lake—Tuan mountain—Tuolong mountain corridor, Pangzhuang subsidence area—Woniu subsidence area—Yunlong mountain corridor, Weishan lake—Jianwang subsidence area—Dadong mountain corridor
Local corridor	2	Dahuang mountain—Tuan mountain—Yunlong mountain corridor, Pan'an lake, Jiuli lake, Woniu Subsidence area corridor

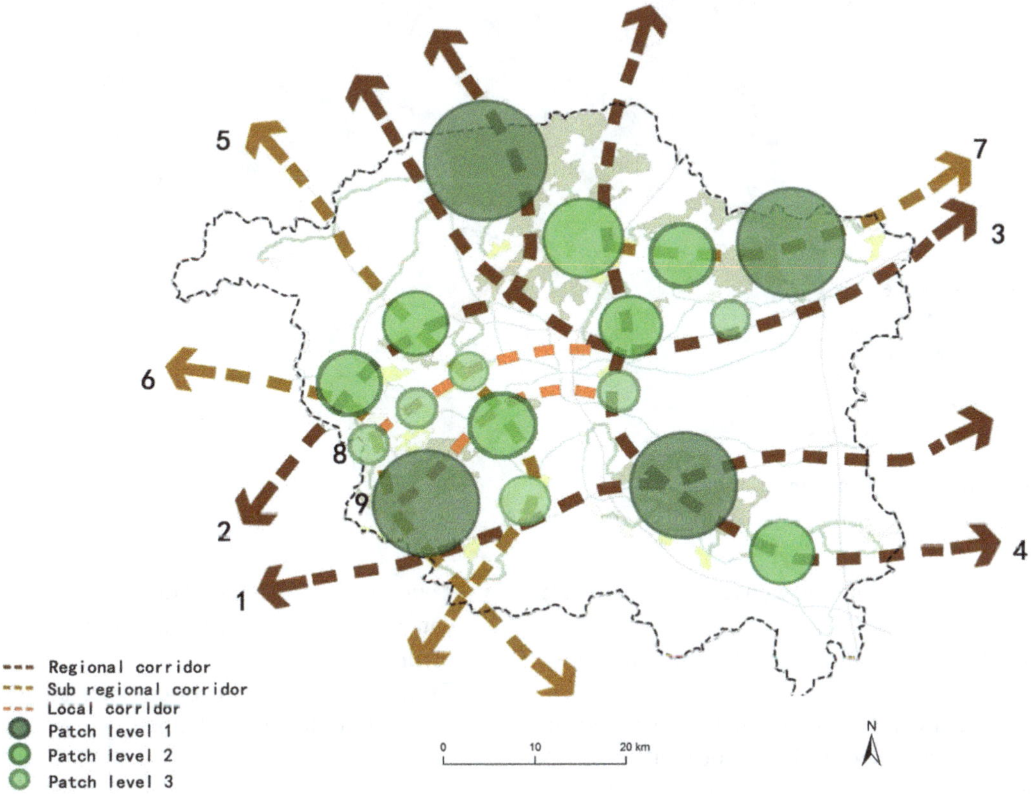

Fig. 9 GI pattern and corridors in Xuzhou urban area. Regional corridor: 1. Yunlong mountain—Tuolong mountain—Lvliang mountain corridor, 2. Weishan lake—Jiuli lake—Pangzhuang subsidence area corridor, 3. Northern mountains—Pan'an wetland—Cuihe reservoir corridor, 4. Beijing-hangzhou canal corridor. Sub regional corridor: 5. Jiuli lake—Tuan mountain—Tuolong mountain corridor, 6. Pangzhuang subsidence area—Woniu subsidence area—Yunlong mountain corridor, 7. Weishan lake—Jianwang subsidence area—Dadong mountain corridor. Local corridor: 8. Dahuang mountain—Tuan mountain—Yunlong mountain corridor, 9. Pan'an lake, Jiuli lake, Woniu Subsidence area corridor

6 Conclusion

6.1 Patches regulation

In order to realize the regulation and management of GI in Xuzhou urban area, we tried to put forward regulation suggestions for the elements in GI network. The patches regulations were proposed due to patches' level and existing ecological background of each one.

6.1.1 Level 1 patches

Level 1 patches contained Weishan wetland patch, Lvliang mountain scenic patch, Dadong mountain scenic patch, Yunlong mountain scenic patch. These patches covered natural reserve areas, scenic areas, forest parks, water conservation areas and other ecological sensitive areas. The ecological spaces in these areas were the core of the whole ecosystem and should be strictly regulated and protected. The specific regulation suggestions were as follows:

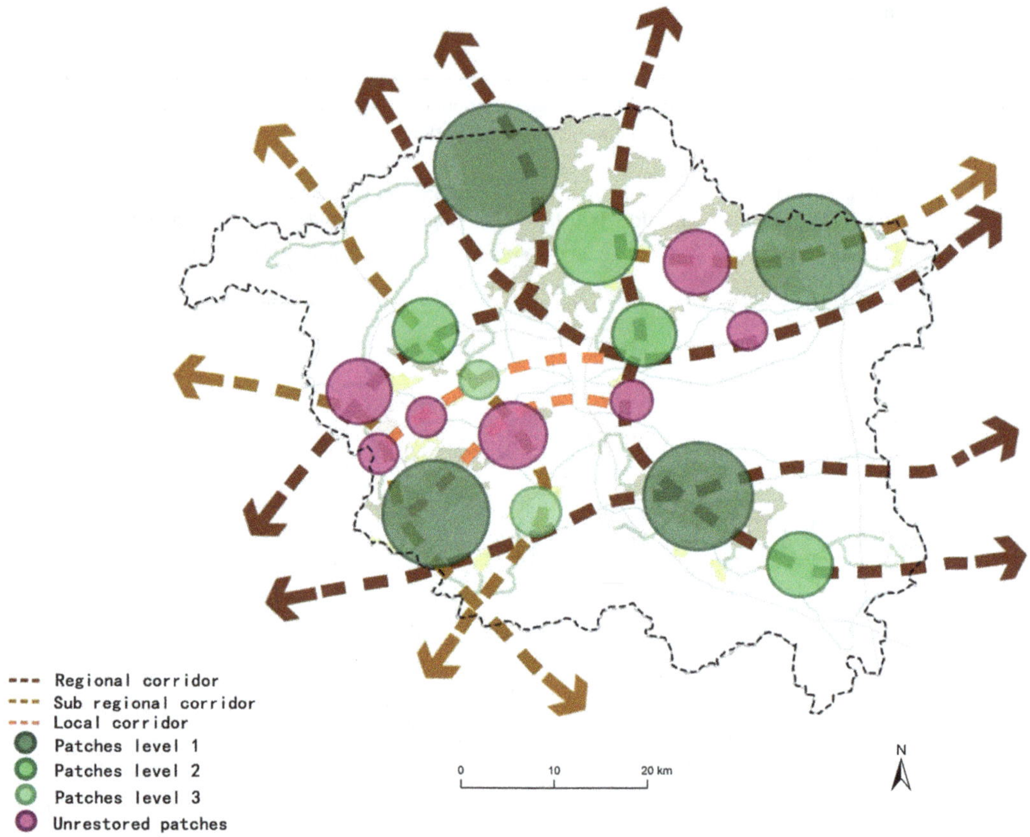

Fig. 10 Unrestored subsidence patches in GI network

- High-strength development and construction work are forbidden in these patches.
- Forest reservation and patches ecological integrality should be guaranteed.
- Ecological tourism can be appropriately carried out in areas where ecological sensitivity is not high.
- Cultural relics and historical sites should be protected and the buffer should be defined.

6.1.2 Level 2 patches

Level 2 patches contained Northern mountains forest patch, Pan'an wetland patch, Jiuli subsidence patch, Cuihe reservoir patch, Jiawang subsidence patch, Tuan mountain patch, Pangzhuang subsidence patch. These patches covered mountains, forest parks, reservoirs and wetlands. The specific regulation suggestions were as follows:

- Projects or constructions that may cause damage to wetlands, water sources and regional ecology should be forbidden.
- Enhance water source protection and recovery water quality and functions.
- Tourism development should be carried out in accordance with the latest master planning and special planning.
- Cultural relics and historical sites should be protected and Buffer should be defined.

6.1.3 Level 3 patches

Level 3 patches contained Tuolong mountain forest patch, Jiuli mountain patch, Northern Woniu subsidence patch, Sorthern Woniu subsidence patch, Dahuang mountain country park patch, Dongzhuang subsidence patch. These patches covered mountains, forest parks, and wetlands. The specific regulation suggestions were as follows:

- Limit the high-strength construction and construction activities in patches, and regulate the existing construction projects in the patches.
- Construction and tourism development can be carried out in accordance with the latest master planning and local standards.
- Cultural relics and historical sites should be protected and the buffer should be defined.

6.2 Coal mining subsidence patches regulation

In regard to coal mining subsidence areas, these places largely distributed in Xuzhou urban area and most of them remained unrestored. According to the patches importance evaluation, we divided coal mining subsidence areas into 3 levels to better integrate them into GI network. Priority Class 1 (high restoration priority), including: Jiuli subsidence patch, Jiawang subsidence patch, Pan'an wetland patch; Priority Class 2 (medium restoration priority), including: Dahuang mountain patch, Pangzhuang subsidence patch and northern Woniu subsidence patch; Priority Class 3 (low restoration priority), including: southern Woniu subsidence patch and Dongzhuang subsidence patch.

"Priority Class 1". These patches were crucial in improving urban ecosystem services and habitat quality and strongly recommended to be integrated into urban GI network. These areas were suggested to be well preserved and restored into natural conservation. Large-scale human construction activities should be forbidden in these areas.

"Priority Class 2". These areas had positive effects in improving urban ecosystem services and habitat quality and recommended to be integrated into urban GI network. These areas were suggested to avoid significant artificial trace and commercial activities, and be developed into parks and open spaces with appropriate human disturbance.

"Priority Class 3". These areas had no obvious positive effects in improving urban ecosystem services and habitat quality and can be integrated into urban GI network when allowed by the objective conditions. These areas can be developed into either urban open spaces or used as construction lands.

6.3 Corridors regulation

6.3.1 Regional corridor

Regional corridor contained Beijing-hangzhou canal corridor, Northern mountains—Pan'an wetland—Cuihe reservoir corridor, Yunlong mountain—Tuolong mountain—Lvliang mountain corridor, Weishan lake—Jiuli lake—Pangzhuang subsidence area corridor. These corridors covered rivers, transportation, greenbelts, etc. These corridors connected key patches and made great significance for ecological resources circulation and urban ecological environment improvement. We suggested that restoration priority should be given to corridors of vital habitat and damaged areas. The human construction work should be strictly controlled on both sides of the corridors. The width of the corridor was suggested as 300–500 m in this level.

6.3.2 Sub regional corridor

Sub regional corridor contained Jiuli lake—Tuan mountain—Tuolong mountain corridor, Pangzhuang subsidence area—Woniu subsidence area—Yunlong mountain corridor, Weishan lake—Jianwang subsidence area—Dadong mountain corridor, which covered rivers, transportation, greenbelts. These corridors mainly connected the regional corridor and promote the internal energy flow. The width of the corridor was suggested as 100–300 m in this level. The development and construction work on both sides of the corridor were restricted.

6.3.3 Local corridor

Local corridor contained Dahuang mountain—Tuan mountain—Yunlong mountain corridor, Pan'an lake, Jiuli lake, Woniu Subsidence area corridor. Types of local corridors were rivers mainly and connected regional corridors and sub regional corridors. The width of the corridor was suggested as 100 m in this level. The development and construction work on both sides of the corridor were restricted. Attention should be paid to purifying water quality and strengthening the ecological restoration on both sides of the corridor.

7 Discussion

In this study, we used a comprehensive approach to determine the patches, corridors and built up the entire GI network in Xuzhou urban area. The study was embodied the principle of "ecological priority", demonstrating how a holistic perspective can be applied to local operation. It emphasizes that the target of ecological restoration should be started from a traditional "available" status to the overall urban ecology recovery and reconstruction. Therefore, this is an expansion in the research that focused on coal-resources based city, compared with the qualitative description and the overall evaluation of conventional methods. The approach and case study can be used as a reference for other coal-resource based cities, especially for those mining industrial cities that have ecological potentials.

However, some indicators can only reflect quantitative variances rather than quality aspects because of the mathematical characteristics of landscape patterns. Also, remote sensing interpretation is the basic of our study. When concerning about the ecology, we have not further discussed the mechanism of ecosystem services and biodiversity in patches and between each other because of the data and methods. Additionally, expert scoring method requires researchers should validate the models to ensure their efficacy. All methods used and proposed in the study will depend on data availability for the region of interest. The formation of GI network is a long-term process with dynamic change. The research framework of this study was not sufficient to consider the dynamic connotation. Therefore, long-phase and multiscale studies are needed directions for future research. Time-scale ecological restoration scenario simulation can be conducted with the support of subsidence prediction and ecological monitoring.

Acknowledgements This work was supported by National Natural Science Foundation of China for funding the project "Research on Space–Time Evolution Laws and Optimization Model of Green Infrastructure in Coal Resource Based Cities" (No. 41671524). We would also like to thank the other members of the projects team who have supported this publication with comments and ideas.

Compliance with ethical standards

Conflict of interest None of the authors have any competing interests in the manuscript.

References

Benedict MA, McMahon ET (2006) Green infrastructure linking landscapes and communities. Island Press, Washington, DC

Chang J (2013) From post-mining landscape to urban space. Flächenmanagement und Bodenordnung 34(75):73–78

Chang J, Wende W, Luo PJ, Deng YY (2011) Re-use of the mining wasteland. Tongji University Press, Shanghai

Chen LD, Fu BJ, Zhao WW (2006) Source-sink landscape theory and its ecological significance. Acta Ecol Sin 26(5):1444–1449

Fei YL (2011) Study on bird species diversity and geographical distribution patterns in Jiangsu Province. Doctoral dissertation, Nanjing: Nanjing Forestry University

Feng S, Chang J, Hou W (2016) A framework for setting restoration priorities for coal subsidence areas based on green infrastructure (GI). Acta Ecol Sin 36(9):2724–2731

Haase D, Larondelle N, Andersson E et al (2014) A quantitative review of urban ecosystem service assessments: concepts, models, and interpretation. Ambio 43(4):413–433

He D, Jin FJ, Zhou J (2011) Urban construction land suitability evaluation in resource-based cities: taking the Grand Canal Ecologic and Economic Area as an example. Geogr Res 30:655–666

Li K (2009) Green infrastructure: concept, theory and practice. Chin Landsc Archit 25(10):88–90

Li JX, Li C, Zhu FG, Song CH, Wu JG (2013) Spatiotemporal pattern of urbanization in Shanghai, China between 1989 and 2005. Landsc Ecol 28(08):1545–1565

Liu CF, Zhou B, He XY, Chen W (2010) Selection of distance thresholds of urban forest landscape connectivity in Shenyang City. Chin J Appl Ecol 21:2508–2516

Pascual-Hortal L, Saura S (2007) Impact of spatial scale on the identification of critical habitat patches for the maintenance of landscape connectivity. Landsc Urban Plan 83:176–186

Pascual-Hortal L, Saura S (2011) Comparison and development of new graph-based connectivity indices: towards the prioritization of habitat patches and corridors for conservation. Handb Internet Stud 2011:251–269

Pauleit S, Liu L, Ahern J, Kazmierczak A (2011) Multifunctional green infrastructure planning to promote ecological services in the city. In: Niemelä J (ed) Urban ecology. Patterns, processes, and applications. Oxford University Press, New York, pp 272–285

Polasky S, Nelson E, Pennington D, Johnson KA (2011) The impact of land-use change on ecosystem services, biodiversity and returns to landowners: a case study in the State of Minnesota. Environ Resource Econ 2011:219–242

Shen SG, Wang H, Jia DP, Kang Gu (2009) Land suitability evaluation methods for urban green space system Planning based on GIS in Changzhou city. J Nanjing For Univ 33(4):72–76

Theobalda DM, Crools KR, Norman JB (2011) Assessing effects of land use on landscape connectivity: loss and fragmentation of western U.S. forests. Ecol Appl 21:2445–2458

Weber T, Sloan A, Wolf J (2006) Maryland's green infrastructure assessment: development of a comprehensive approach to land conservation. Landsc Urban Plan 77(1–2):94–110

7

Modelling spontaneous combustion liability of carbonaceous materials

M. Onifade[1] · B. Genc[1]

Abstract This paper presents predictive models to determine spontaneous combustion liability of carbonaceous materials (coals and coal-shales) using statistical analysis. The intrinsic properties and spontaneous combustion liability index were determined by testing 14 coals and 14 coal-shales from Witbank coalfields, South Africa. The relationship between these intrinsic properties (obtained from proximate, ultimate and petrographic analysis) and spontaneous combustion liability indices (the Wits-Ehac Index and Wits-CT Index) were established. The influence of the intrinsic properties of coal-shales in relation to coal properties affecting spontaneous combustion has been established using a statistical method. The linear regression analysis indicates better linear relationships between some of the selected intrinsic properties and spontaneous combustion liability index and thus, identifies the major intrinsic factors affecting their liability toward spontaneous combustion. It was found that a definite positive or negative correlation coefficient exists between the intrinsic factors and spontaneous combustion liability. A set of models to predict the spontaneous combustion liability was derived. The best significant correlation along with the most appropriate model as indicated by R-squared values, the coefficient of correlations and standard error was used to predict the incident of spontaneous combustion.

Keywords Coals · Coal-shales · Statistical analysis · Wits Ehac Index · Wits-CT Index · Spontaneous combustion

1 Introduction

The event of spontaneous combustion is the reason for several cases of fires in the coal mining industry. It is well-known that carbonaceous materials such as coal and coal-shales undergo self-heating when exposed to atmospheric conditions. Self-heating occurs naturally due to the chemical reactions between oxygen and coal and other carbonaceous materials. The spontaneous combustion characteristics of coal may be influenced by the natural convection within a coal seam, waste dumps and other mining piles (Moghtaderi et al. 2000).

Self-heating of coal-shales has been reported in the South African coal mines to be capable of starting spontaneous combustion but not the coals alone (Onifade and Genc 2018b, c, d). This incident has been observed in areas such as overburden shales, selected bands of a coal seam, spoil heaps and highwall (Figs. 1 , 2). It is known that coal, roof shales and spoil heaps may become self-heated and liberate heat naturally under favourable weather conditions. Sedimentary rocks such as coal and coal-shale, consist of certain amounts of organic matter (macerals) and inorganic materials (mainly crystalline) causing them to likely undergo self-heating leading to spontaneous combustion (Onifade and Genc 2018d). Reactive sedimentary materials have pore spaces embedded in the solid together with the occurrence of a carbon-rich element (Dullien 1979). This renders the rock porous to different fluids like water and

✉ B. Genc
Bekir.Genc@wits.ac.za

M. Onifade
1519496@students.wits.ac.za

[1] The School of Mining Engineering, University of the Witwatersrand, Johannesburg 2050, South Africa

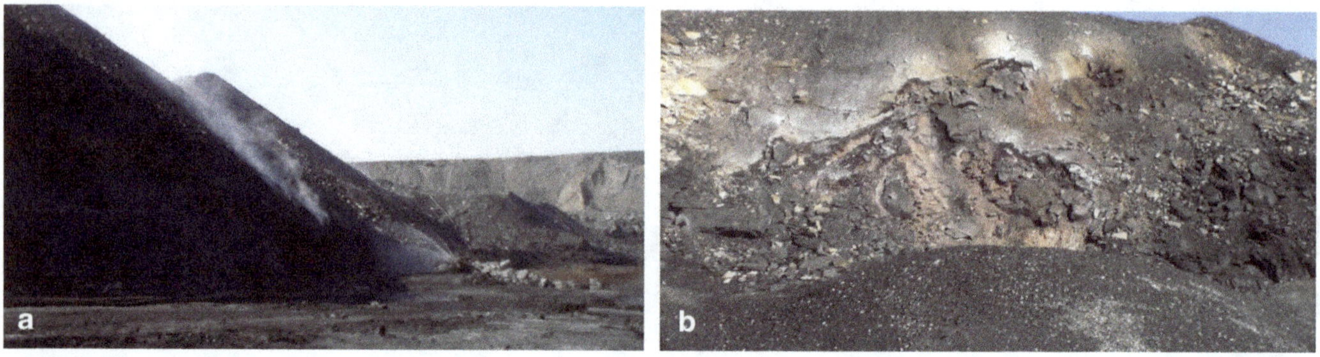

Fig. 1 **a** Self-heating of run-of-mine and **b** spoil heaps at Tweefontein Mine, Witbank, South Africa

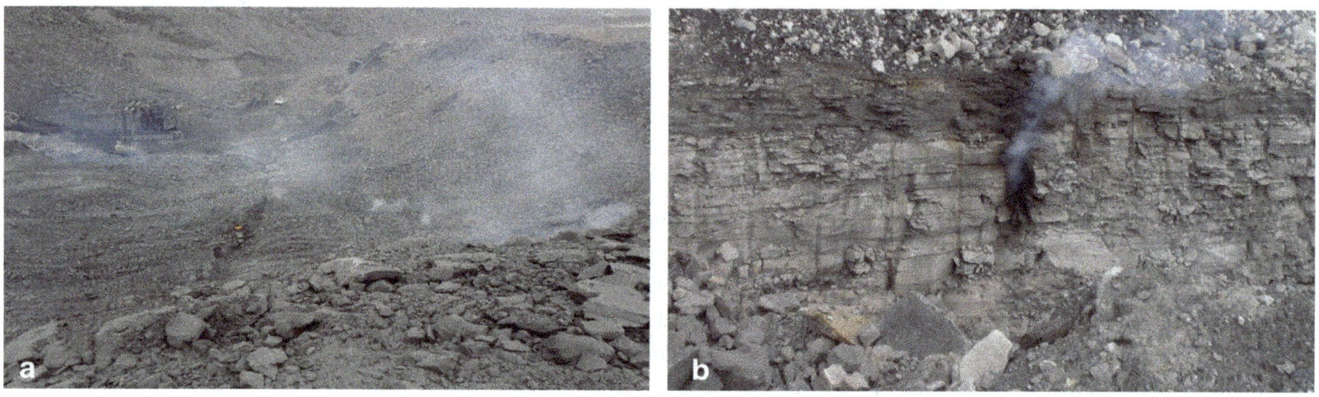

Fig. 2 **a** Signs of self-heating in coal seam at Khwezela Mine (Bokgoni Pit) and **b** self-heating of in-seam shale at Goedgevonden Colliery, Witbank, South Africa

air, increases its surface area, thus making the organic particles reactive as it permits oxidation to occur (Dullien 1979). Such reactive materials might experience spontaneous combustion. Spontaneous combustion of coals and coal-shales is an important phenomenon that requires in-depth investigation. Studies of self-heating of coal-shales in selected bands of coal seams and highwalls are limited (Onifade et al. 2018; Onifade and Genc 2018d). Limited information is available to investigate and predict the spontaneous combustion liability of coal-shales in relation to coals. However, these are materials that comprise a number of varying microscopic and inorganic constituents for spontaneous combustion in old workings, spoil heaps and abandoned mines.

Many investigations have been carried out on the self-heating characteristics of coal, both experimentally and computationally (Gouws and Wade 1989a, b; Genc and Cook 2015; Stracher and Taylor 2004; Kucuk et al. 2003; Carras and Young 1994; Onifade et al. 2018; Onifade and Genc 2018b, c,d), but only limited studies exist on other reactive materials. The occurrence of various intrinsic and extrinsic parameters influencing the start and development of spontaneous combustion is the reason for the difficulties in understanding the mechanism of spontaneous combustion. Most of these parameters have been reported by many researchers (Kaymakci and Didari 2002; Banerjee 1985; Kim 1977; Beamish and Blazak 2005; Falcon 2004; Panigrahi and Sahu 2004; Panigrahi and Saxena 2001; Ren et al. 1999) to predict the self-heating characteristics of coal.

This study evaluated the intrinsic properties of some South African bituminous coals and coal-shales using characterisation tests (proximate, ultimate and petrographic analyses), following standard experimental procedures. Furthermore, a statistical interpretation of coals and coal-shales analysis data and the combined effects of the major intrinsic factors affecting their liability towards spontaneous combustion for predictive purposes were examined. An understanding of these factors may be important in understanding the spontaneous combustion liability of coals and coal-shales in coal mines.

2 Materials and methods

2.1 Sample collection and preparation

Fourteen coals and fourteen coal-shales used for this study were collected (using the ply sampling technique) from four coal mines (iMpunzi Mine, Goedgevonden Colliery, Khwezela Mine, and Tweefontenin Mine) in Witbank, South Africa. The samples were sealed in airtight bags (plastic), immediately after collection to avoid moisture loss and oxidation. The airtight bag is produced using aluminium-coated polyester which allows maximum reduction of vapour and oxygen permeability. Each ply sample bag was clearly labelled with a chosen number. Samples of bituminous coal and coal-shales were collected from a selected band of the coal seams (Witbank coalfields) and highwalls as shown in Figs. 1 and 2. The samples were selected in respect of their high propensity to oxidation from the areas known to be prone to spontaneous combustion. The samples were stored and prepared under laboratory conditions (room temperature). The sample lumps were reduced using a crusher (Rocklabs MK III) to suitable sizes (> 10 mm). The samples were milled to 250 µm size fraction for proximate, ultimate, total sulphur and forms of sulphur, 212 µm for spontaneous combustion tests, while for petrographic analysis the samples were milled to ≤ 1000 µm according to the South African National Standard (SANS, 7404 Parts 1–5), blended and divided using a centrifugal mill (Retsch ZM 200) to generate different sub-samples. Proximate, elemental and petrographic analysis, and spontaneous combustion tests (Wits-Ehac Index and Wits-CT Index) were carried out.

2.2 Characterisation tests

The experimental tests involving the determination of the moisture, ash, volatile matter and fixed carbon contents were carried out according to the American Society for Testing and Material (ASTM) standards (ASTM, D-3173-17a, ASTM, D-3174-11, ASTM, D-3175-17). Fixed carbon was obtained by subtracting the sum of the percentage of ash, moisture and volatile matter from 100 (ASTM, D3176-15). The carbon, hydrogen, nitrogen and total sulphur content were determined using a LECO TruSpec CHN and C and S analyzer according to the International Standard Organisation Standards (ISO 12902: 2001). Oxygen content was calculated by subtracting the sum of the percentage of moisture, ash, carbon, hydrogen, nitrogen and sulphur from 100. The results are reported in weight percent of air-dried (wt%, ad). Pyritic sulphur was detected in selective digestion with hydrogen nitrate (HNO_3) and analysed by inductively coupled plasma optical emission

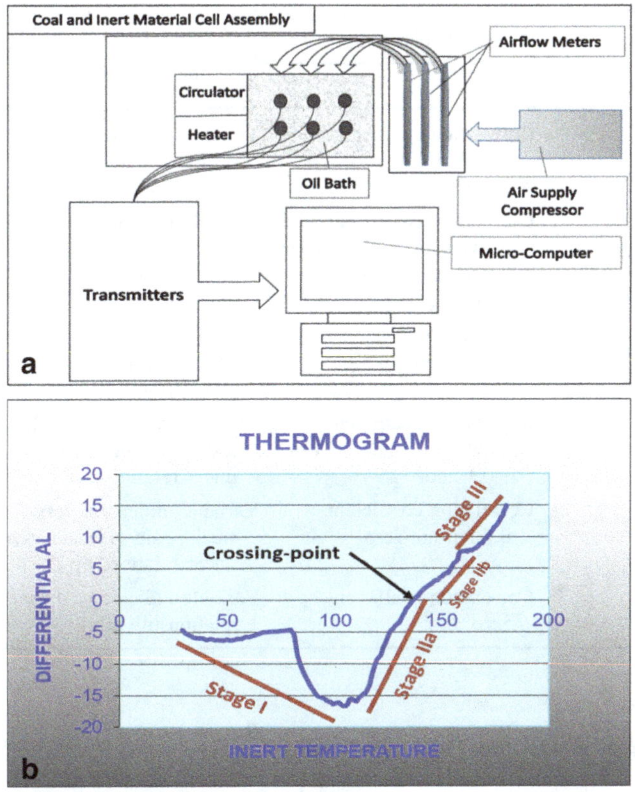

Fig. 3 **a** Wits-Ehac test apparatus set-up (Wade et al. 1987); **b** A typical differential analysis thermogram

(ICP-OES), while sulphate sulphur was detected in selective digestion with hydrogen chloride (HCl) and analysed by photometric method as barium sulphate (BaSO4). Organic sulphur is taken as the difference between the total sulphur and the sum of the sulphate and pyritic sulphur. Maceral analysis including mineral matter was conducted to evaluate the organic and inorganic content of the samples. A Zeiss Axio Imager M2 M reflected light petrographic microscope fitted with oil immersion lens at a total magnification of 500× was used. The petrographic pellets and/or blocks used for the petrographic analyses were produced by mixing 2–3 g of the grain size specified (≤ 1000 µm) according to the South African National Standard (SANS, 7404 Parts 1–5) with epoxy resin in a 3.2 cm diameter phenolic ring form moulds. The moulds were allowed to cure, and ground with the use of 320, 400 and 1000 grit papers. Once ground, they were polished using 1.0 and 0.3 µm alumina slurries on various cloth. Final polishes were achieved with the use of colloidal silica, which provides relief free and scratch-free surfaces for analysis by reflectance microscopy. The macerals results including petrographically observable mineral matter are reported as volume percent (vol%), inclusive of mineral matter or mineral matter free basis (vol%, mmf).

Table 1 Criterion set for factors affecting spontaneous combustion liability of coals and coal-shales

Category	Criterion	Remarks
1	Correlation coefficient between 0.95 and 1 or −0.95 and −1	Variable indicate a perfect positive or negative linear relationship
2	Correlation coefficient between 0.51 and 0.94 or −0.51 and −0.94	Variable indicate a strong positive or negative linear relationship
3	Correlation coefficient between 0.25 and 0.50 or −0.25 and −0.50	Variable indicate a moderate positive or negative linear relationship
4	Correlation coefficient between 0.1 and 0.24 or −0.1 and −2.24	Variable indicate a weak positive or negative linear relationship
5	Correlation coefficient > 0.1 but not zero	Variable indicate a very weak positive or negative linear relationship
6	Correlation coefficient of zero	Variable indicate no linear relationship at all

2.3 Wits-Ehac tests

The Wits-Ehac Index was developed in South Africa in the late 1980s to test the spontaneous combustion liability of coal (Eroglu 1992; Genc et al. 2018; Genc and Cook 2015; Gouws and Wade 1989a, b; Onifade et al. 2018; Onifade and Genc 2018a, b, d; Wade 1989). The testing apparatus consists of an oil bath, six coal and inert material cell assemblies, an oil circular, a heater, a flowmeter, an air supply compressor and a computer (Wade et al. 1987) as shown in Fig. 3a. Freshly pulverized (< 212 μm) dried coal samples (under room temperature) of 20–25 g in weight is used. The test apparatus is used to test coals under predefined conditions and a spontaneous combustion liability index is obtained. The testing apparatus incorporates the determined crossing point temperatures (XPT) and differential thermal analysis (DTA) from selected coal samples with periodic measurement of temperatures to obtain a consistent self-heating liability index termed the Wits-Ehac Index. The coal temperatures are recorded every 30 s by the computer over an average of 4 h for the oil in the bath to be heated to 200 °C. When the temperature difference between the inert material (calcined aluminium) and the coal sample is plotted against the inert temperature, the part of the graph where the coal is heating faster than the inert sample (where an exothermic reaction occurs) is termed Stage II. The index makes use of the fact that coal highly prone to self-heat have a steeper stage II slope and lower XPT than coal not highly prone to self-heat. When using DTA, the difference in temperatures between the coal sample and an inert material sample is measured via a data logger, stored in a computer and plotted against the temperature of the inert material sample. It is important to understand that during DTA, three stages are obtainable. At first, the temperature of an inert material sample is higher than the temperature of the coal sample (Stage l), which is dependent on the cooling effect of the evaporation of the coal sample moisture content. Secondly, evaporation of the moisture content; the coal sample begins to heat up at a higher rate than the heating rate of the inert material (Stage II) and this is based on the liability of coal to self–heating and attempting to reach the temperature of the surrounding temperature (oil bath temperature). Lastly, the high exothermicity is reached at a point where the line crosses the zero base line and is referred to as the XPT. Uludag et al. (2001) indicated the three stages in a study and described that stage I starts with minimal differential and progressively increases towards the XPT where the differential is zero. Stage II continues from the XPT to the point referred to as kick–point. In addition, Stage II is one of the best indicators of spontaneous combustion. Stage III is when the coal starts to burn beyond the kick–point. According to Wade et al. (1987), coals with a spontaneous combustion liability index below 3 are considered to be low risk, 3–5 medium risk, and coals with values > 5 are considered to be highly liable to spontaneous combustion. The index is calculated from the formula in Eq. 1. Using this formula, MS Excel calculates the stages and generates the thermogram (Fig. 3b).

$$\text{Wits-Ehac Index} = (\text{Stage II slope}/\text{XPT}) \times 500 \qquad (1)$$

2.4 Wits-CT tests

The Wits-Ehac Index was developed in South Africa to test the spontaneous combustion liability of coal. However, in some cases, the Wits-Ehac Index fails to produce tangible results when testing coal-shales. To overcome this problem, a new apparatus has been developed to test coals and coal-shales under chemical reactions with oxygen and an index has been obtained (Onifade et al. 2018). A device for predicting the spontaneous combustion liability of coal, coal-shale and other carbonaceous materials under the influence of oxygen without any heating system was recently developed in the School of Mining Engineering, University of the Witwatersrand, and is referred to as the Wits-CT test. The liability of various samples to spontaneous combustion were evaluated for 24 h under the influence of oxygen. This experiment was designed to evaluate the temperature differences within the shortest period of time in a coal mass under the influence of oxygen.

The equipment has a capacity to accept a 15 kg sample, depending on the packing density. In order to determine

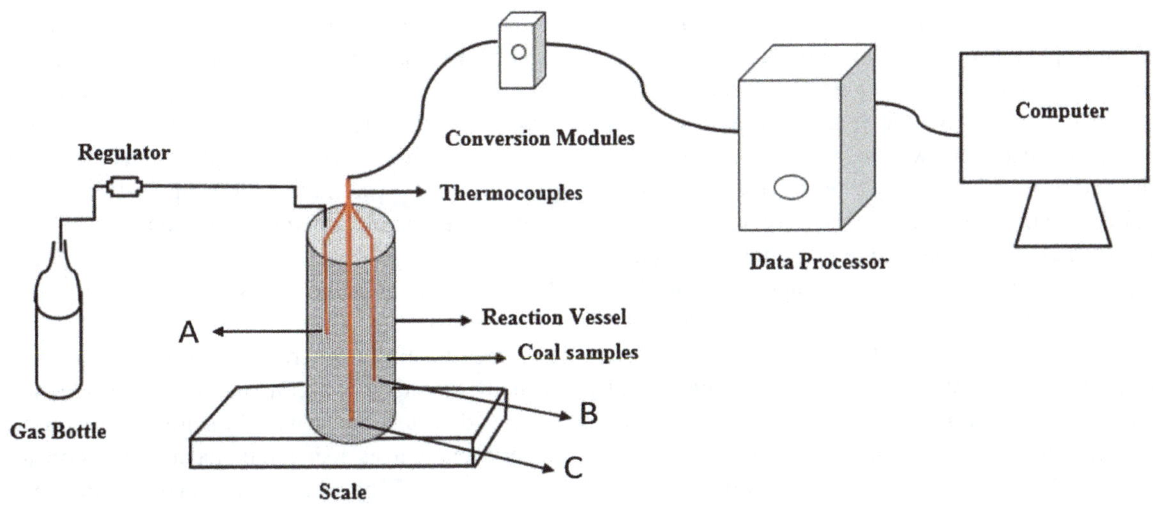

Fig. 4 Wits-CT apparatus (Onifade et al. 2018)

spontaneous combustion liability via self-heating, a representative sample of particle size (< 6.39 mm) is weighed and loaded into an autoclave. The process is repeated until the autoclave is filled to the marked point and the lid is fixed and fastened as soon as the autoclave is full. This technique of preparation gives a closely sized sample that is preferable to measure the rate of oxidation. Three thermocouples each with two temperature probes are inserted into the column as the correct sample level is reached. Three uniformly separated temperature sensors placed along the length of the autoclave are used to check and record the temperature measurements. The sensor arrangement is typically located at different points within the autoclave (locations A, B, and C within the interior of the container). Each of the three thermocouples has two temperature sensors which measure temperatures at different levels at the locations A, B, and C in order to keep a recording of the temperature distribution during the reaction of a carbonaceous material and oxygen. At location A, sensors are located at levels 13 and 23 cm within the container, location B, 53 and 63 cm, and location C, 33 and 43 cm.

Oxygen is supplied and controlled at a fixed flow rate (20 mL/min) by means of a flowmeter before being fed into a manifold attached below the lid of the autoclave Table 1. The changes in temperature distributions are recorded by the data logger every minute over 24 h. The Wits-CT Index uses the total carbon content and the temperature differences due to the reaction with oxygen to predict the spontaneous combustion liability. According to Onifade et al. (2018), coals and coal-shales with a spontaneous combustion liability index below 2.5 are considered to be less reactive, 2.5–5 moderately reactive, 5–7.5 reactive, and > 7.5 are considered to be highly liable to spontaneous combustion. An illustration of the experimental setup is indicated in Fig. 4. The index is calculated from the formula in Eq. 2.

$$\text{Wits-CT index} = (T_M/24 + T_R) \times \%\, C_{ad} \quad (2)$$

where, T_M is the difference between the sum of maximum temperatures of each thermocouple inside the equipment and the room temperature (22 °C), T_R is the difference between the maximum temperature and the initial temperature during oxidation reaction in degree Celsius, $\%C_{ad}$ is the air-dried percentage of carbon content of the sample, × is the multiplication sign and 24 is the duration of the test (24 h) and it is constant.

3 Results and discussion

3.1 Statistical analysis

The results for characterisation tests (proximate, elemental and petrographic analysis) and spontaneous combustion tests (the Wits-Ehac Index and the Wits-CT Index) carried out on the samples are presented in Tables 2, 3, 4, 5 and 6 respectively. The evaluation of coal and coal-shales intrinsic factors affecting spontaneous combustion was at first based on data obtained from the spontaneous combustion liability tests and the intrinsic properties as determined in the laboratory. It is appropriate to group the data into dependent (the Wits-Ehac Index and the Wits-CT Index) and independent (intrinsic properties) variables in order to make the analyses meaningful. The observed temperatures values of spontaneous combustion tests are dependent on the coal and coal-shale properties, coal-forming elements and coal structure. Hence, the statistical analysis was carried out by correlating the samples intrinsic properties and spontaneous combustion liability indices. The range value (Table 1) is the value of the independent variables from the lowest to the highest. The R-squared value is a statistical tool which can either carry out hypothesis testing or predict future result on the basis of related information. It provides a measure of how well observed outcomes are reproduced by a model on the basis of the proportion of total variation of outcomes described by the model (Draper and Smith 1998; Glant 1990). The higher the R-squared value, the better the fit is. In statistics, the correlation coefficient measures the strength and direction of a linear relationship between two variables (dependent and independent variables) on a scatter plot (Deborah 2016; Stigler 1989). The range of values for the correlation coefficient is -1.0 to 1.0 and it is a representation of the linear interdependence of two variables or sets of data. The correlation coefficients were obtained using the data analysis in MS Excel. A correlation coefficient of -1.0 indicates a perfect negative correlation, while a correlation of 1.0 indicates a perfect positive correlation. A correlation coefficient of 1.0 indicates a perfect positive relationship between two variables. i.e. for a positive increase in one variable, there is also a positive increase in the second. A correlation coefficient of -1.0 indicates a perfect negative relationship between the two variables. i.e. the variables move in opposite directions, for a positive increase in one variable, there is a decrease in the second variable. A zero correlation means that for every increase, there is no positive or negative increase (i.e. the two are not related). The R-squared values and the correlation coefficients were used to measure the trends and any significant relationships between the intrinsic properties and the Wits-Ehac Index and Wits-CT Index. The overall database involved the spontaneous combustion test results and intrinsic properties of 14 coals and 14 coal-shales in order to determine the factors affecting spontaneous combustion. The study interpreted the linear relationship between the intrinsic properties and spontaneous combustion liability index based on the set criterion (Table 1).

The criterion is established based on the strength and directions of a linear relationship between the intrinsic properties and spontaneous combustion liability index on a scatter plot. Statistically correlations of $+1$ or -1 are expected, however, this is not possible due to the use of real dataset from experimental tests, and real data are never perfect to get a perfect correlation coefficient (Deborah 2016). Since the results of the real data is expected to determine how close is close enough to indicate a strong, moderate or weak linear relationship, this necessitated the need to interpret the correlation coefficient values, classify and find how close they are using the set criterion (Table 1). Given that the correlation coefficient measures the degree to which how two variables vary together, establishing a criterion means whether each intrinsic property and spontaneous combustion liability index vary strongly, moderately or weakly and if it is a positive or negative relationship. The higher the correlation coefficient, the stronger the relationship is (higher correlation coefficients means a better prediction).

3.2 Linear regression analysis

The results of the characterisation tests were correlated with the spontaneous combustion liability indices (Wits-Ehac Index and the Wits-CT) statistically. Data from the experimental tests using the linear regression analysis is given in the Tables 2, 3, 4, 5 and 6 and the values in brackets are in negative sign.

As a preliminary step, the experimental data was analysed using statistical methods to ascertain whether any of the parameters were linearly related to the oxidation potential. The dataset was divided into dependent and independent variables in order to establish trends and relationships. The correlation coefficients between the independent and dependent variables are seen in Tables 2, 3, 4, 5 and 6. The linear regression analysis was used to identify if there are better linear relationships between dependent and independent variables. It was found that each factor has an influence towards spontaneous combustion based on their relationships.

The statistical analysis demonstrated the better linear relationship between the dependent and independent variables in most cases based on the R-squared values (Figs. 5, 6, 7, 8, 9, 10, 11, 12, 13, 14, 15, 16) and correlation coefficients obtained. The higher correlation coefficients obtained from the data sets suggest that this may be the best way to minimize the overriding effect of the various coals and coal-shales on analysis of the general oxidation trends. Thus, obtaining a result which may be an indication of spontaneous combustion liability of certain carbonaceous materials. It is well-known that carbonaceous materials behaves differently when subjected to similar atmospheric conditions due to differences in their physical and chemical properties. The analysis of variable pairs (moisture/Wits-Ehac Index; moisture/Wits-CT Index; volatile matter/Wits-Ehac Index; volatile matter/Wits-CT Index; ash/Wits-Ehac Index; ash/Wits-CT Index and etc.) demonstrated uniform trends, i.e. the oxidation trend increases with increasing volatile matter, carbon, hydrogen, etc. and decreases with increasing ash and mineral matter, etc. for both coal and coal-shale samples. Working with carbonaceous materials having similar rank make it difficult to establish significant relationships between the petrographic composition and spontaneous combustion liability as was expected before regression. This is in-line with the study reported by Eroglu

(1992). Few significant relationships were found between the spontaneous combustion liability indices and petrographic composition (total vitrinite, total inertinite and total liptinite) from the statistical analysis. However, it was decided to investigate the relationship between the various

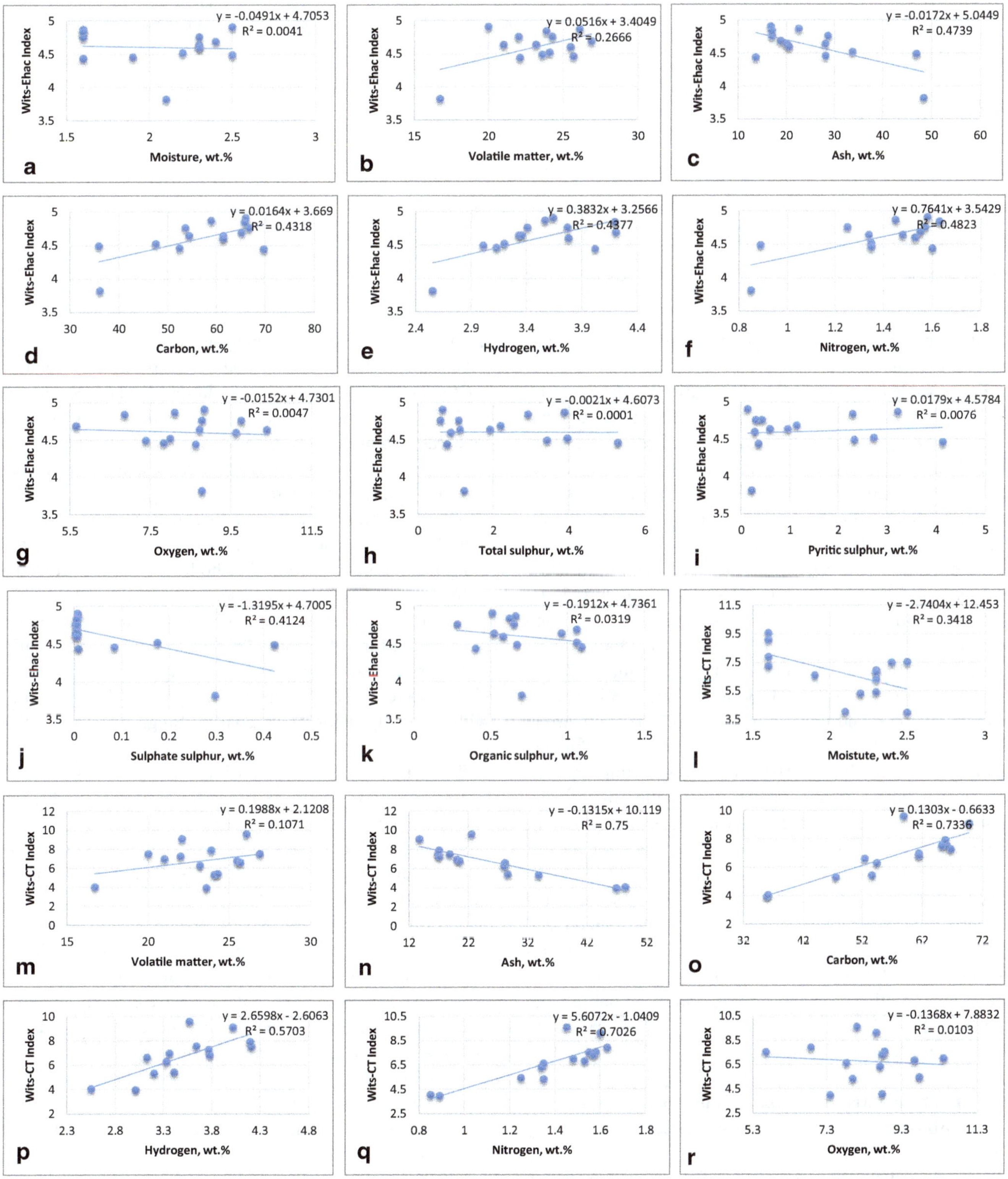

Fig. 5 Influence of proximate and ultimate analysis on spontaneous combustion of coals

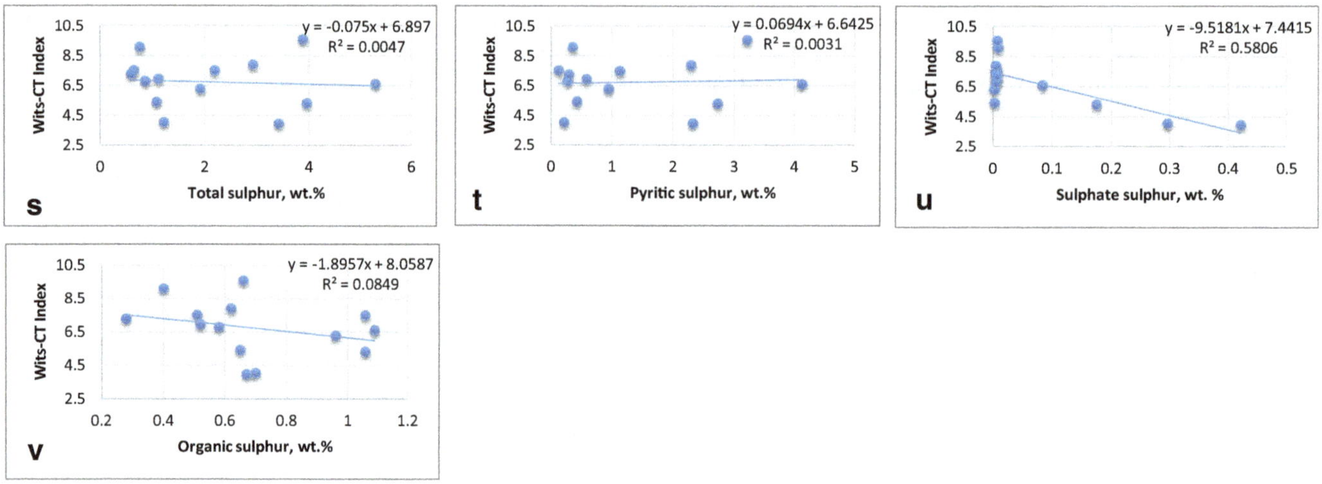

Fig. 5 continued

constituents of each maceral group on spontaneous combustion liability results.

From Tables 2, 3, 4, 5, 6 and 7, based on the criterion set, it was found that contents of volatile matter, 0.5164; ash, −0.6884; carbon, 0.6572; hydrogen, 0.6616; nitrogen, 0.6945; sulphate sulphur, 0.6422; and inertinite macerals [fusinite, 0.5663] with strong effect size are parameters affecting spontaneous combustion liability of coals. Furthermore, it was found that contents of moisture, 0.7715; volatile matter, 0.6389; ash, −0.8352; carbon, 0.7962; hydrogen, 0.5795; nitrogen, 0.6446; total sulphur, 0.5791; and its forms [pyritic, 0.5704; sulphate, 0.5365; and organic sulphur, 0.5933], total inertinite, 0.7360; and its constituents [fusinite, 0.5898; total semifusinite, 0.7776; secretinite, 0.6003; and total inertodetrinite, 0.7749] total maceral, 0.7653; and mineral matter, −0.7653 with strong effect size are factors affecting spontaneous combustion liability of coal-shales. The linear regression analysis shows that both liability indices show better linear relationships with some of the intrinsic properties and thus, identifies the major intrinsic factors affecting spontaneous combustion liability. Previous studies by Falcon (2004); Gouws and Wade (1989a); Sahu et al. 2005; Kaymacki and Didari (2002); and Nimaje and Tripathy (2016) indicated that intrinsic properties are a measure of spontaneous combustion liability. The effects of coals and coal-shales intrinsic properties on spontaneous combustion liability are shown in Figs. 5, 6, 7, 8, 9, 10, 11, 12, 13, 14, 15 and 16 using linear regression analysis.

3.3 Effects of intrinsic properties on spontaneous combustion of coal and coal-shale samples

Figures 5l and 6a show the effect of moisture content on the spontaneous combustion liability of the coals obtained from the linear regression analysis. The results indicated a negative correlation for the coals and a positive correlation for the coal-shales. As the correlation displays R-squared of 0.3418 for coals, it shows that the linear model has a moderate strength. It is different for the coal-shales. Here the R-squared is higher, 0.5952, than the coals. This means the coal-shales have a strong effect size.

The R-squared value for volatile matter is considered as 0.2666 for coals and 0.4763 for coal-shales (Figs. 5b and 6m). This is an indication that both coals and coal-shales have a moderate effect size. There is a positive relationship between volatile matter and spontaneous combustion liability index for both cases. This means that as the volatile matter increases, the oxidation capacity of coals and coals-shales is more likely to increase in general. This is in-line with the study reported on coal by Singh and Demirbilek (1987).

There is a negative relationship between the ash content and spontaneous combustion liability index for both coals and coal-shales. This means that as ash content increases, the spontaneous combustion liability decreases. This is in-line with the study reported on coal by Beamish and Blazak (2005), Singh and Demirbilek (1987). The relationship between the ash content and spontaneous combustion liability index seems to be strong. The R-squared value is 0.75 for coals and 0.8926 for coal-shales (Figs. 5n and 6n). The value for coals and coal-shales have a strong effect size. The decreasing ash content affects the oxidation potential by causing a relatively high spontaneous combustion liability for both coals and coal-shales analysed. This is in-line with the studies reported on coal by Beamish and Arisoy (2008), Humphreys et al. (1981), Panigrahi and Sahu (2004); Smith et al. (1988).

The R-squared value for the carbon content is 0.7336 for coals and 0.9273 for coal-shales (Figs. 5o and 6o). The

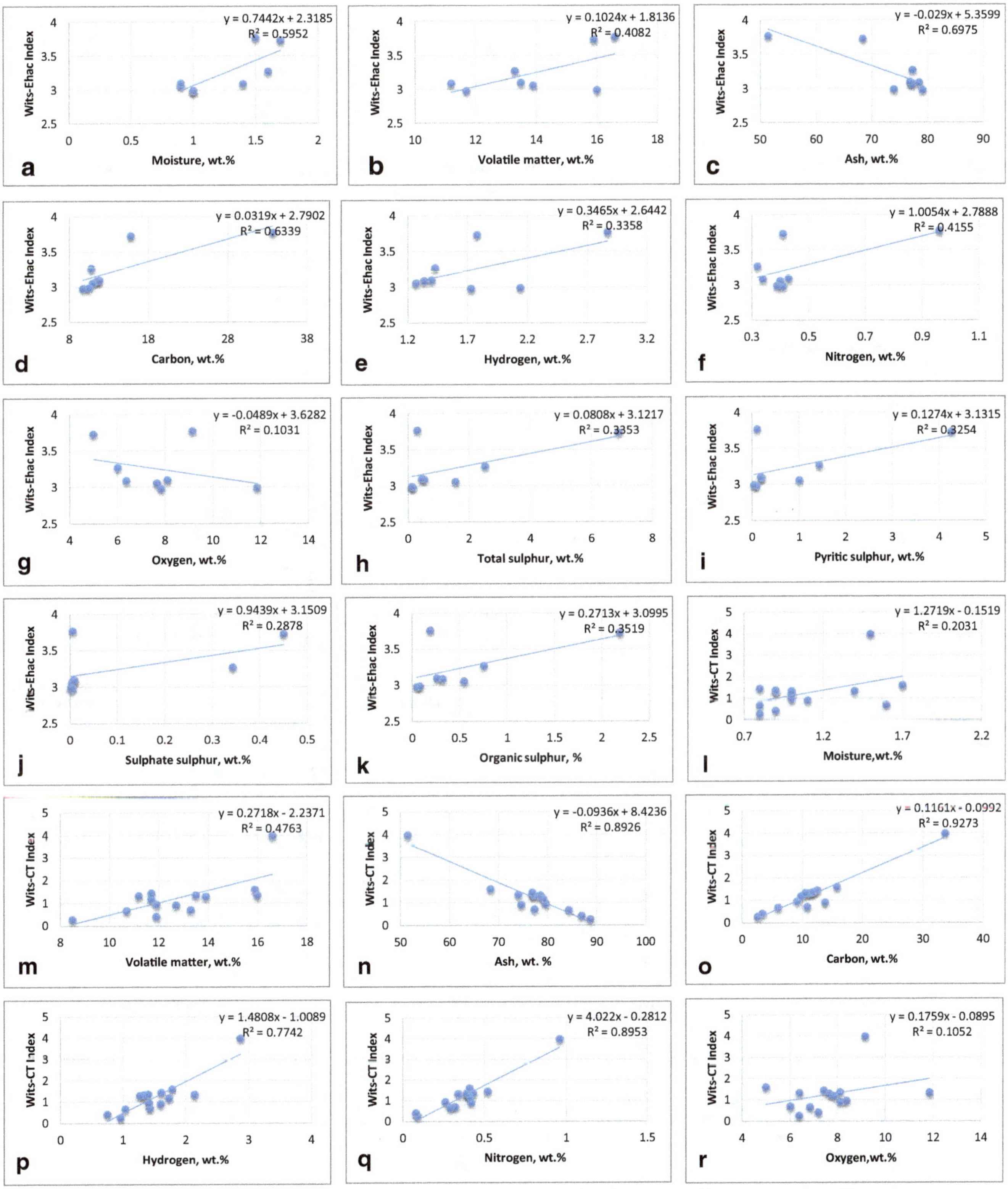

Fig. 6 Influence of proximate and ultimate analysis on sponatneous combustion of coal-shales

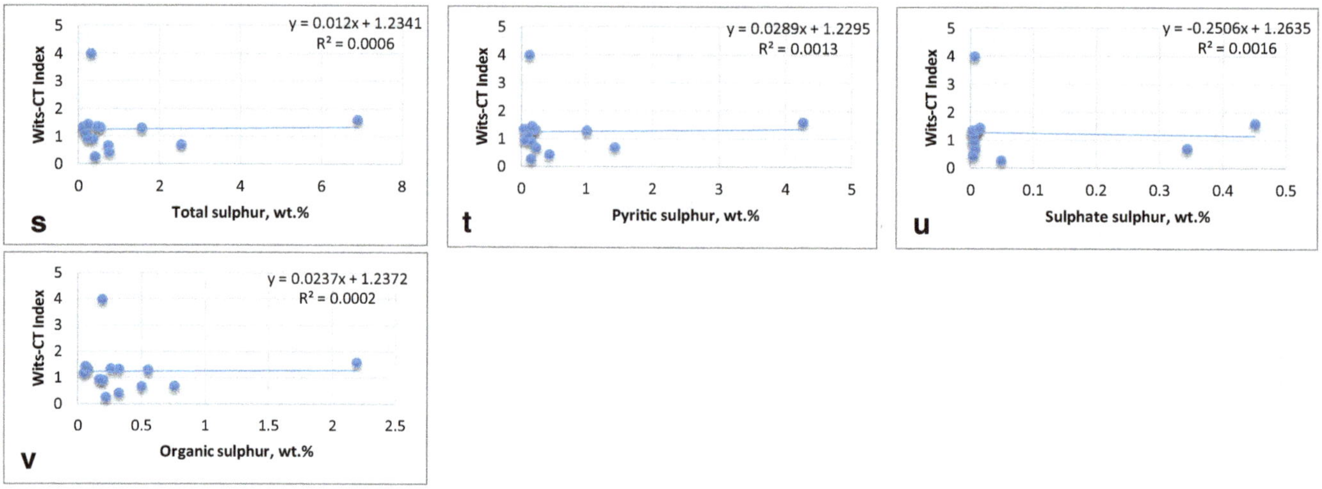

Fig. 6 continued

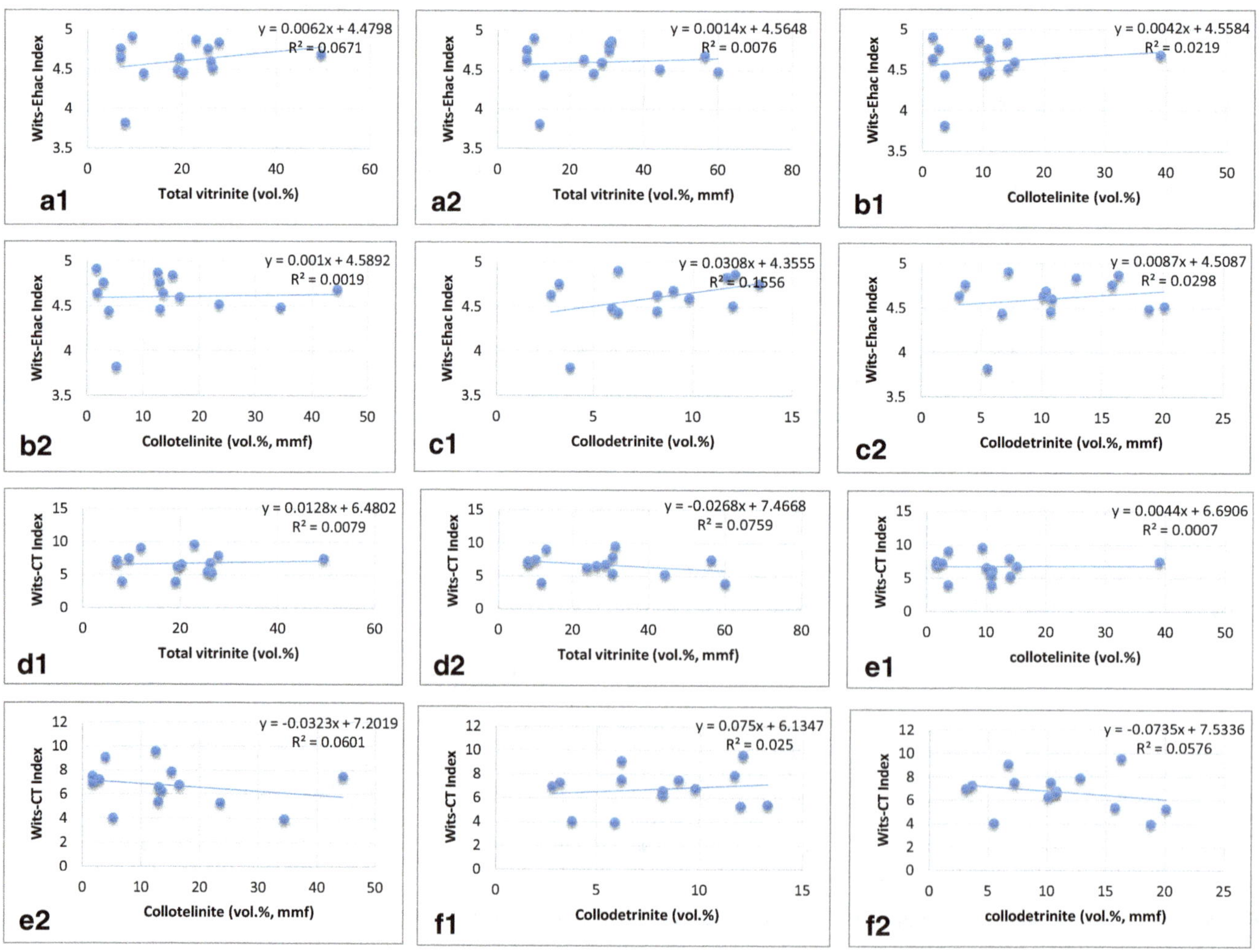

Fig. 7 Influence of vitrinite and its group on spontaneous combustion of coals

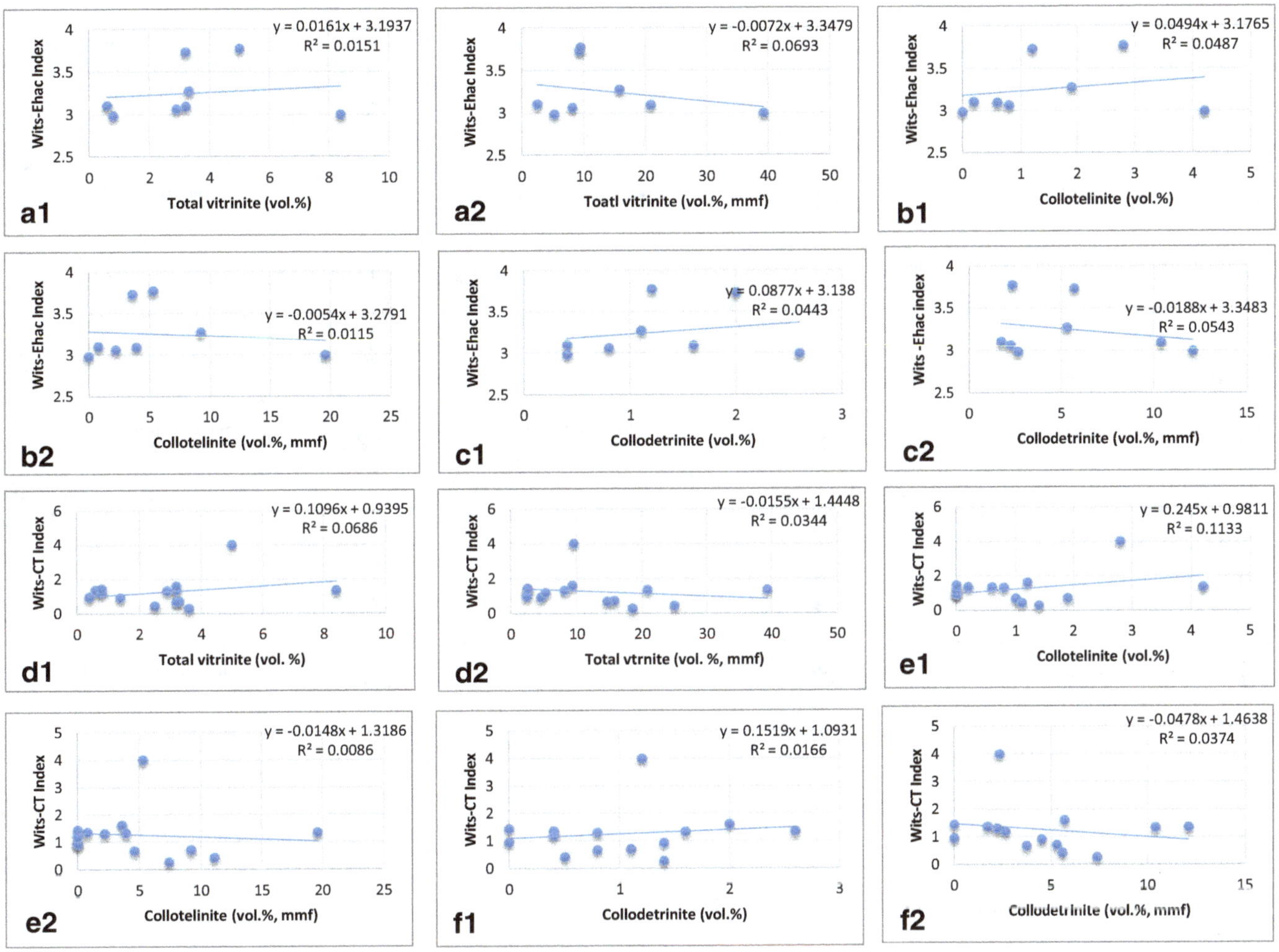

Fig. 8 Influence of vitrinite and its group on spontaneous combustion of coal-shales

relationship between carbon content and spontaneous combustion liability index for coals and coal-shales seems to be strong. The value for coals and coal-shales have a strong effect size. There is a positive relationship between the carbon content and spontaneous combustion liability index for both the coals and coal-shales. This means that as the carbon content increases, the oxidation capacity of coals and coal-shales are more likely to increase in general. The increasing carbon content affects the oxidation potential of both coals and coal-shales by causing a relatively high spontaneous combustion liability.

There is a positive relationship between the hydrogen content and spontaneous combustion liability index for both the coals and coal-shales. This means that as the hydrogen content increases, the oxidation capacity of coals and coal-shales is more likely to increase in general. The increasing hydrogen content affects the oxidation potential of both coals and coal-shales by causing a relatively high spontaneous combustion liability. The R-squared value for the hydrogen content is 0.5703 for coals and 0.7742 for coal-shales (Figs. 5p and 6p). The relationship between the hydrogen content and spontaneous combustion liability index for coals and coal-shales are a strong positive.

The R-squared value for the nitrogen content is 0.7026 for coals and 0.8953 for coal-shales (Figs. 5q and 6q). The relationship between the nitrogen content and spontaneous combustion liability index for coals and coal-shales are a strong positive. This means that as the nitrogen content increases, the oxidation potential of coals and coal-shales is more likely to increase in general.

Figures 5r and 6r show the effect of calculated oxygen on the spontaneous combustion liability of the coals and coal-shales obtained from the linear regression analysis. The evaluation of the calculated oxygen indicated a negative correlation for the coals and a positive correlation for the coal-shales. As the correlation (Fig. 5r) displays R-squared of 0.0103 for coals, it shows that the linear model does not fit well. It is different for the coal-shales.

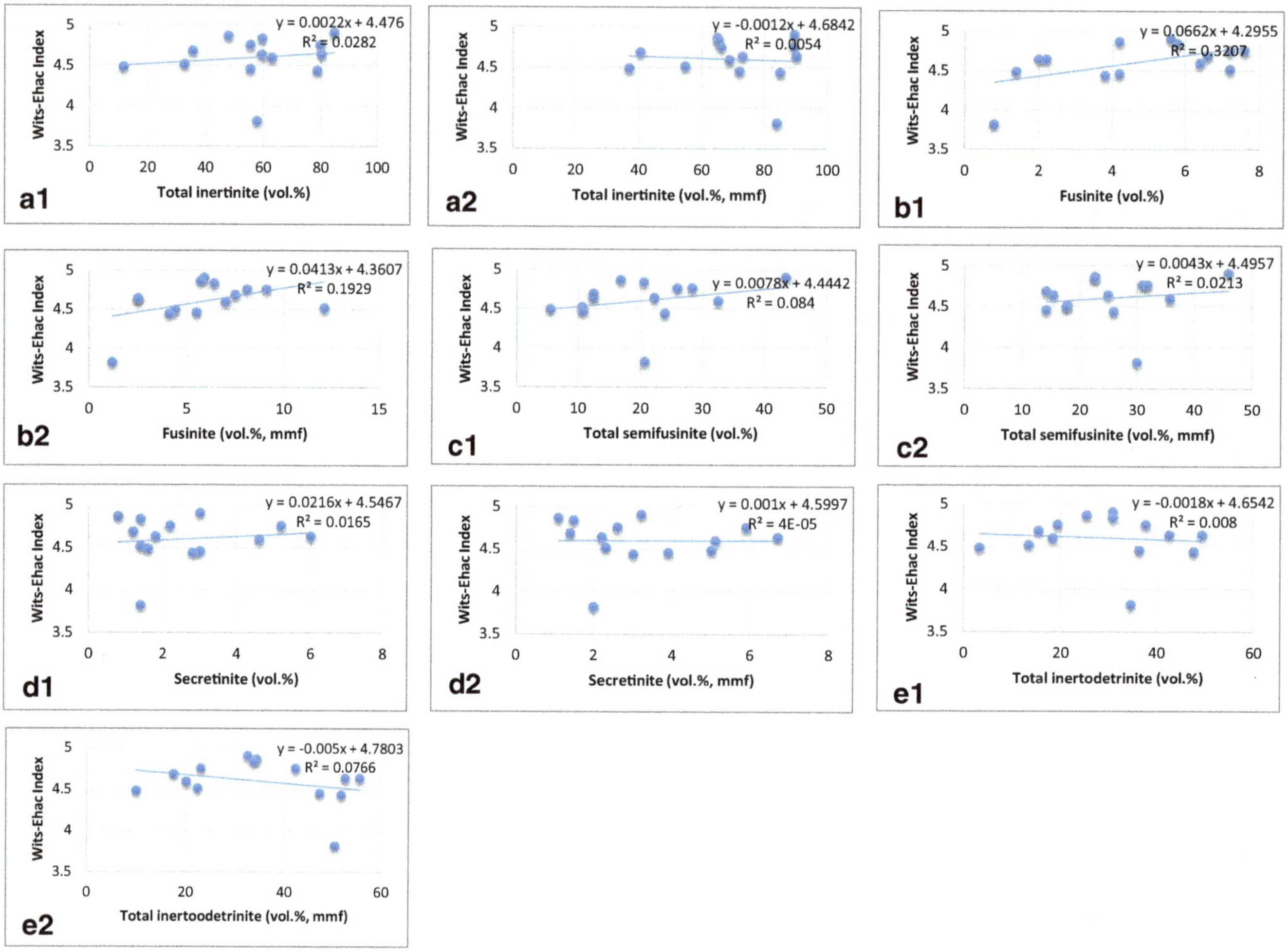

Fig. 9 Influence of inertinite and its group on spontaneous combustion of coals

Here, the R-squared is a bit higher, 0.1052, than the coals. The coals and coal-shales have a very weak effect size.

The evaluation of the total sulphur indicated a negative correlation for the coals and a positive correlation for the coal-shales (Figs. 5h and 6h). This is in-line with the study reported on coal by Singh and Demirbilek, (1987). As the correlation displays an R-squared value of 0.0047 for coals (Fig. 5h), it shows that the linear model does not fit well. It is different for the coal-shales. Here, the R-squared value is a bit higher, 0.3353, than the coals. The value of coals has a very weak effect size, while the value for coal-shales have a moderate effect size.

The R-squared value for pyritic sulphur is 0.0076 for coals and 0.3254 for coal-shales (Figs. 5i and 6i). This is in-line with the study reported on coal by Singh and Demirbilek (1987). The evaluation of the pyritic sulphur indicated a positive correlation coefficient for both the coals and coal-shales. This means that as the pyritic sulphur increases, the oxidation potential of coals and coal-shales is more likely to increase in general. The scatter plots show that the corresponding spontaneous combustion liability values display a wide variation. The value for coals have a very weak effect size, while coal-shales have a moderate effect size. The linear regression analysis was quite similar to those obtained from the total sulphur content data.

The R-squared value for sulphate sulphur is 0.5806 for coals and 0.2878 for coal-shales (Figs. 5u and 6j). The value for coals have a strong effect size, while the value for coal-shales have a moderate effect size. The evaluation of the sulphate sulphur indicated a negative correlation for the coals and a positive correlation for the coal-shales. This means that as sulphate sulphur increases for the coals, the spontaneous combustion liability decreases, while for the coal-shales as the pyritic sulphur increases, the oxidation potential is more likely to increase in general. This is in-line with the study reported on coal by Singh and Demirbilek (1987). The scatter plots show that the corresponding spontaneous combustion liability values display a wide

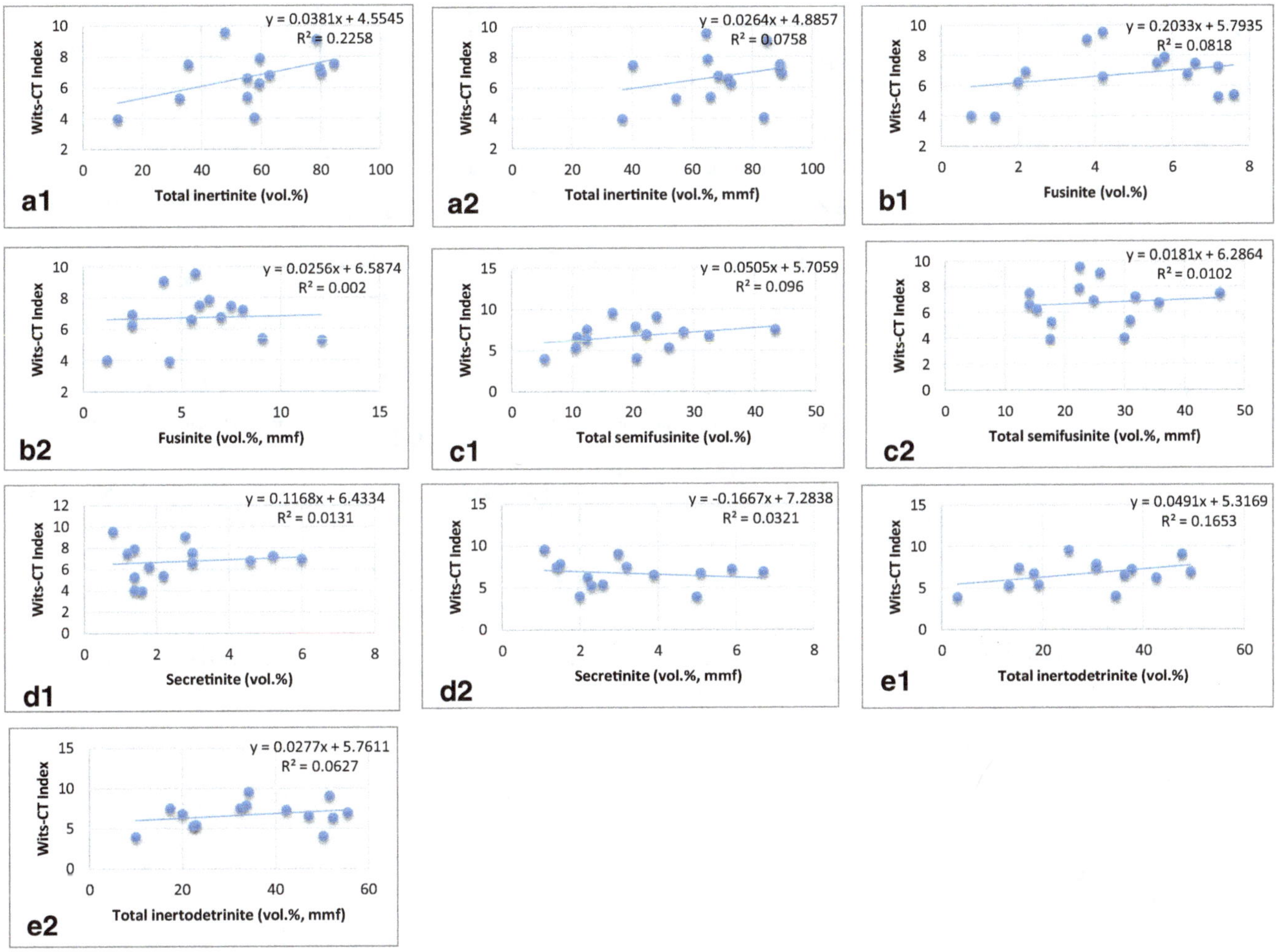

Fig. 10 Influence of inertinite and its group on spontaneous combustion of coals

variation which is quite similar to that obtained from the total sulphur and pyritic sulphur data.

The evaluation of the organic sulphur indicated a negative correlation for the coals and a positive correlation for the coal-shales. This means that as organic sulphur increases for the coals, the spontaneous combustion liability decreases, while for the coal-shales as the organic sulphur increases, the oxidation potential is more likely to increase in general. This is in-line with the study reported on coal by Singh and Demirbilek (1987). The R-squared value for organic sulphur is considered as 0.0849 for coals and 0.3519 for coal-shales (Figs. 5v and 6k). The value for coals have a very weak effect size, while the value for coal-shales have a moderate effect size.

The graphs of the major macerals (vitrinite, inertinite and liptinite) and spontaneous combustion liability index indicates that there may be relationships even though the R-squared values and correlation coefficients are very weak in most cases. A similar study is reported by Eroglu (1992). The analysis showed that the relationships between the various maceral groups and the oxidation potential may be significant when the maceral constituents were unpacked either in the mineral matter free basis (mmf) or inclusive of mineral matter. The results of the linear regression analysis obtained indicated that the various maceral groups have either a positive or negative effect size (Figs. 7, 8, 9, 10, 11, 12, 13, 14, 15 16). It was found that the majority of the vitrinite and its group (collotelinite and collodetrinite) and liptinite and its group (sporinite) in both cases have a very weak effect size. Inertinite macerals (total inertodetrinite, total semifusinite, fusinite and secretinite) shows a better linear relationship to oxidation potential more than the vitrinite and liptinite macerals. The study indicated that the oxidation potential of coals and coal-shales may be influenced by the amount of each maceral composition. This is in-line with the studies reported on coal by (Falcon 1987; Scott and Glasspool 2007). Petrographically observable mineral matter in the coal-shales show a better linear relationship to oxidation potential than the coals (Figs. 14e and 16e).

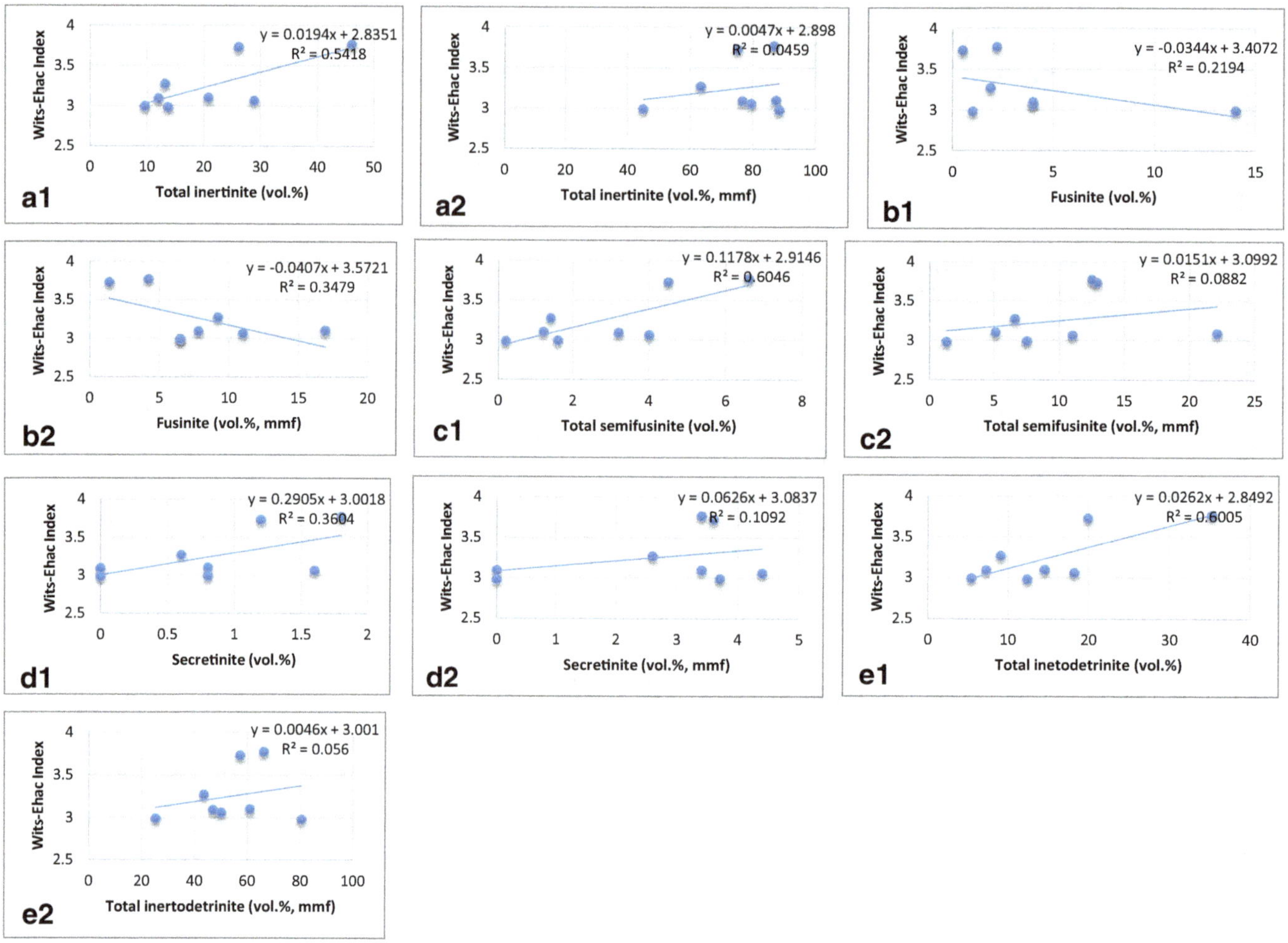

Fig. 11 Influence of inertinite and its group on spontaneous combustion of coal-shales

The study showed that there is a tendency of an increase in intrinsic properties such as moisture, volatile matter, ash, carbon, total sulphur, calculated oxygen, pyritic sulphur, organic sulphur, inertinite macerals and petrographically observable mineral matter in both coals and coal-shales but this seems to be more pronounced for coal-shales than for the coals, while coals seem to be more pronounced in terms of hydrogen, nitrogen and sulphate sulphur content than coal-shales. The inconsistencies (i.e. an independent variable with positive correlation coefficient with the Wits-Ehac Index may have a negative correlation with the Wits-CT Index) in the correlation coefficients between the dependent (Wits-Ehac Index and Wits-CT Index) and independent (some of the intrinsic properties) variables cannot allow accurate determination of the major intrinsic properties affecting spontaneous combustion liability. Similar studies reported different correlation coefficients between some intrinsic factors (moisture, volatile matter, carbon, nitrogen, inertinite, vitrinite, exinite, total sulphur, pyritic sulphur, organic sulphur and calorific value) and the spontaneous combustion liability indices (Kaymakci and Didari 2002; Singh and Demirbilek 1987). The inconsistencies in the trend of correlation coefficients necessitated the developed models (Table 8). It is known that different methods have been used by previous studies to measure the spontaneous combustion liability of coals but no standard method has been considered the best in predicting spontaneous combustion liability (Stott 1980; Wade et al. 1987; Cliff et al. 1998; Beamish et al. 2001). This study investigated the factors affecting spontaneous combustion liability of coals and coal-shales using two different spontaneous combustion test methods (the Wits-Ehac Index and the Wits-CT Index) and evaluates the factor affecting these materials using statistical analysis. The two liability indices have been used to predict the spontaneous combustion liability of these materials and the results are in-line (Onifade and Genc 2018a, b). Limited studies have been conducted to establish relationships between two or more different spontaneous combustion liability indices and selected intrinsic properties (Singh and Demirbilek

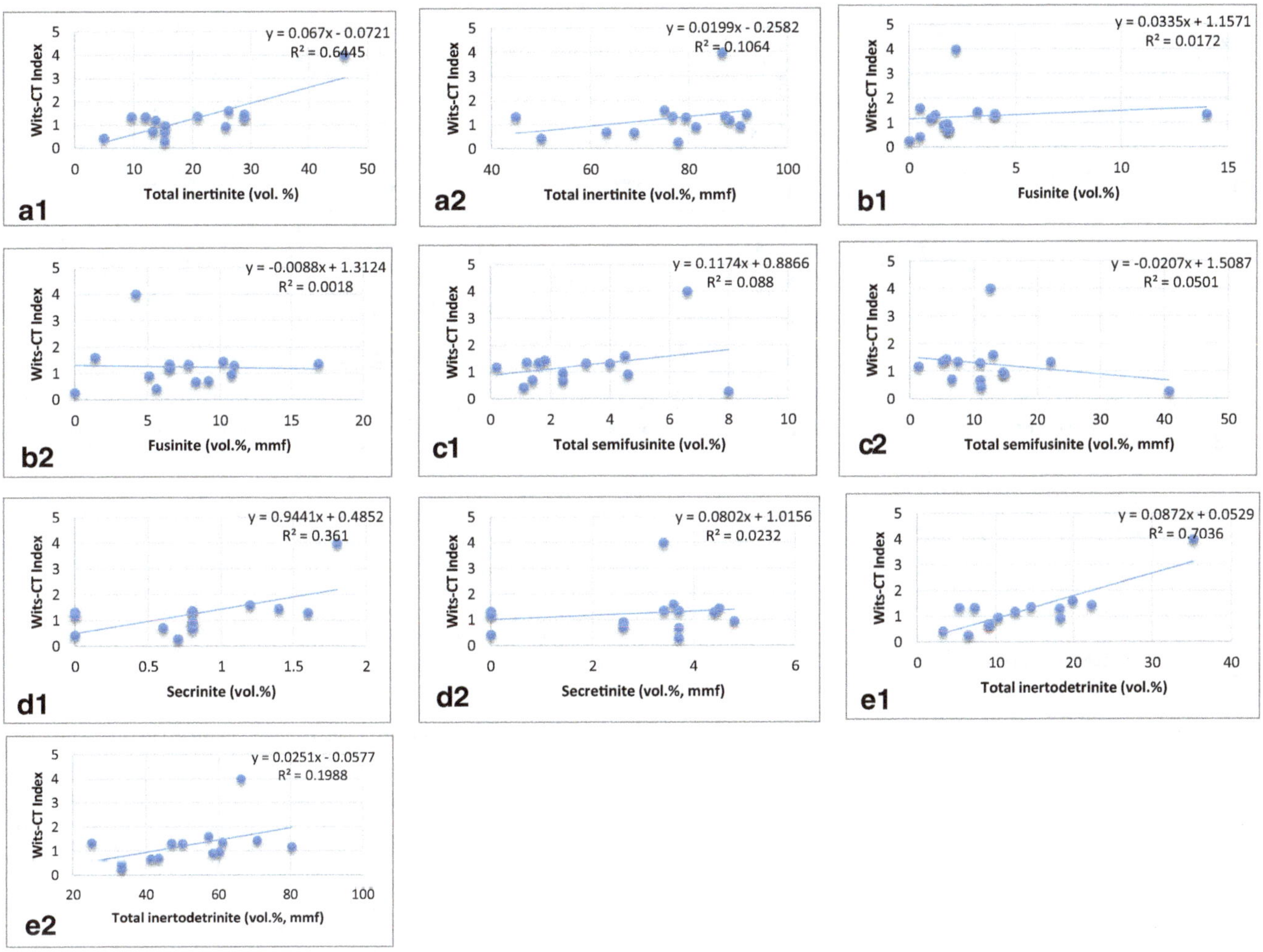

Fig. 12 Influence of inertinite and its group on spontaneous combustion of coal-shales

1987). This study established that some intrinsic properties show different linear trends (i.e. positive or negative correlation coefficients) with spontaneous combustion liability indices due to either the method of calculating the value of each index or the range of values assigned to represent each index for risk classification (see Wits-Ehac test and Wits-CT test procedures).

3.4 Multiple regression analysis

The study indicated that the use of single variable equations obtained from the linear regression analysis to predict the spontaneous combustion liability of coals and coal-shales is an unreliable method (Tables 2, 3, 4, 5, 6, 7). This is because some properties show different trends with the spontaneous combustion liability indices (i.e. similar property may be either positive or negative on the two liability indices which is not supposed to be so in a real sense). The lack of significant relationships between all the dependent and independent variables cannot allow separation of the main intrinsic properties influencing spontaneous combustion liability. It is more likely that the intrinsic properties have an influence on one another to some extent during low-temperature oxidation and this dependence of intrinsic properties determine the spontaneous combustion liability. This implies that the most favourable conditions for spontaneous combustion to occur likely depends on the interrelation of intrinsic properties. This necessitated the use of multiple regression to establish interrelationships between the intrinsic properties and spontaneous combustion liability results. The multiple regression analysis was used to derive the predictive models for spontaneous combustion liability of coals and coal-shales. The predictive models were derived by using 12 independent variables. The multiple regression analysis showed that some of the variables were highly correlated with other independent variables. The best model which provides the highest R-squared values and correlation

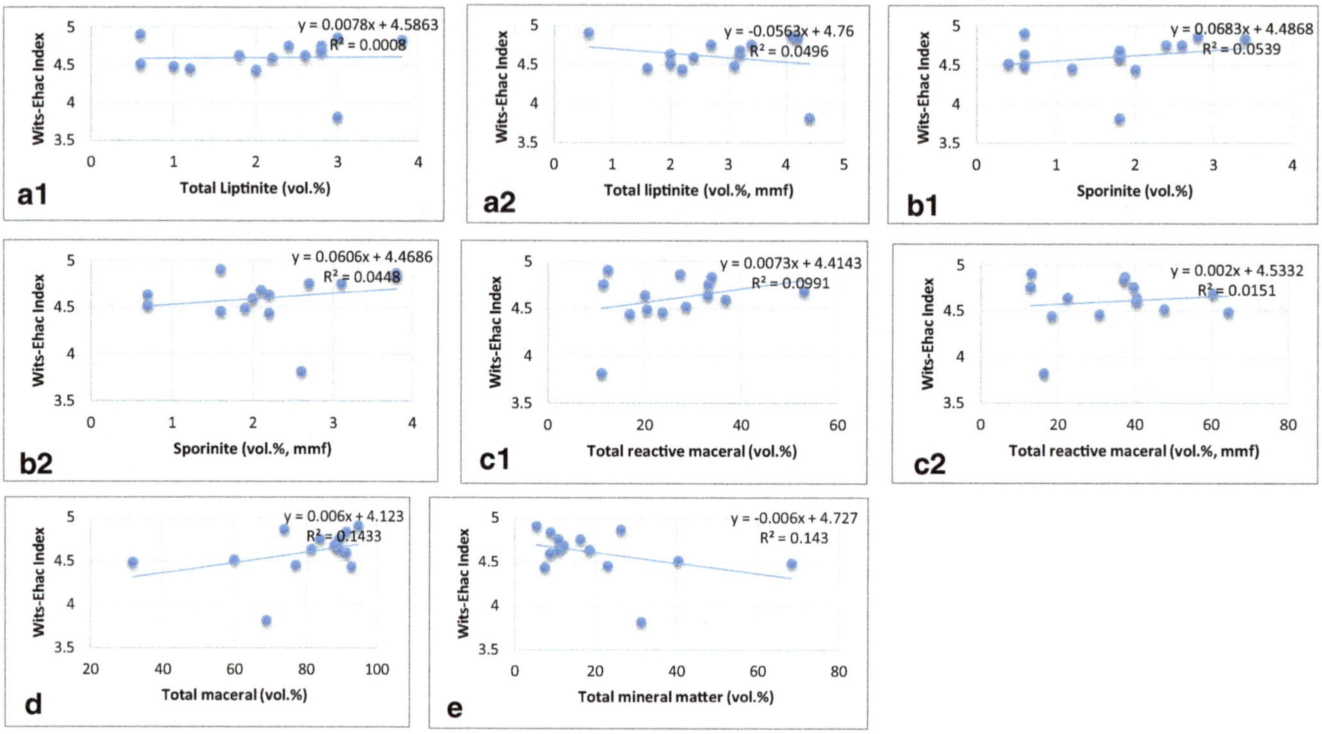

Fig. 13 Influence of liptinite, other macerals, petrographically observable mineral matter on spontaneous combustion of coals

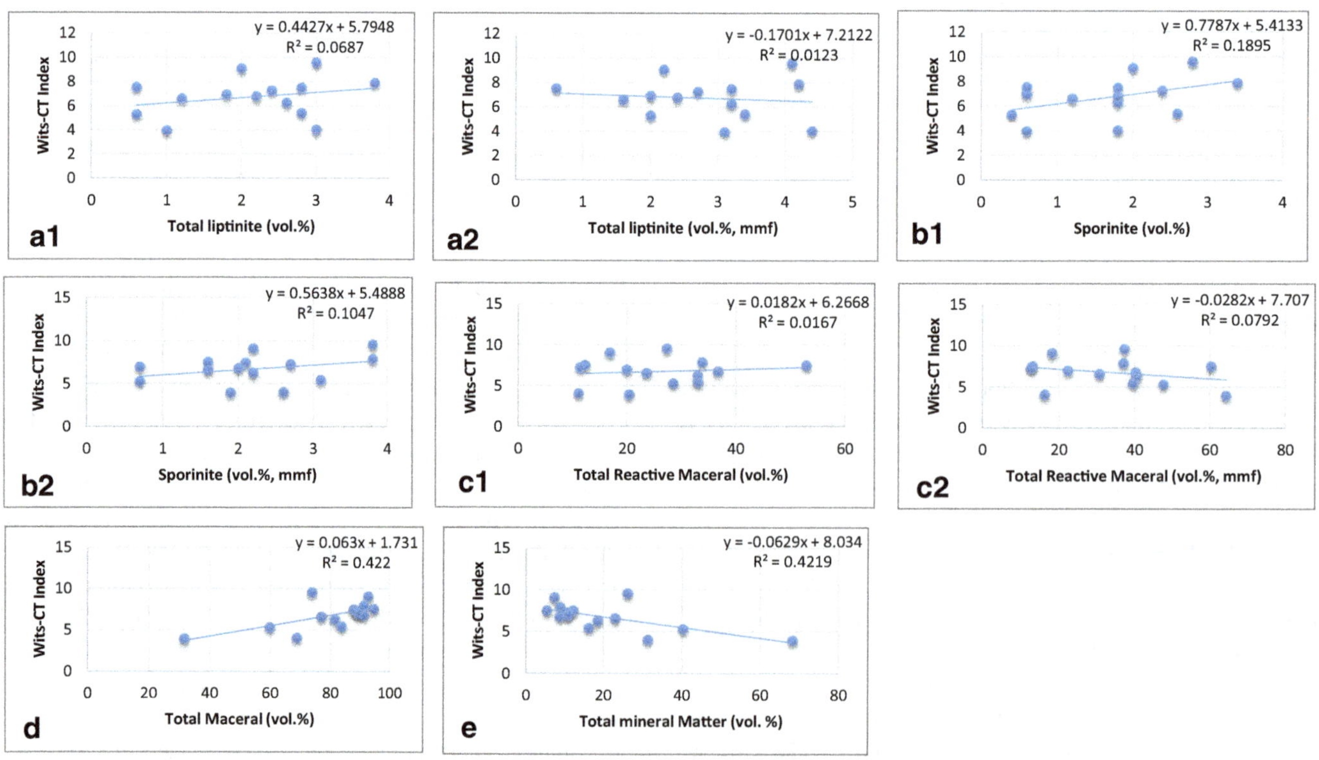

Fig. 14 Influence of liptinite, other macerals, petrographically observable mineral matter on spontaneous combustion of coals

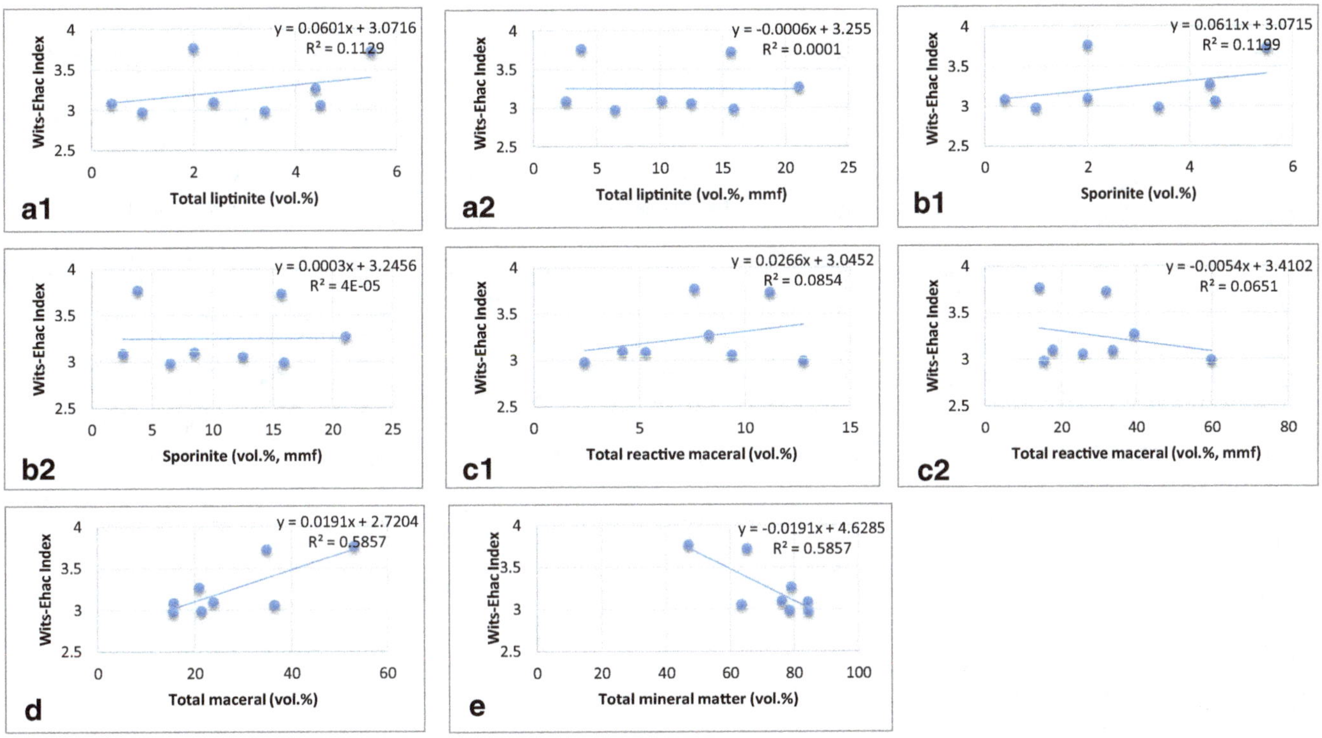

Fig. 15 Influence of liptinite, other macerals and petrographically observable mineral matter on spontaneous combustion of coal-shales

coefficient and lowest standard error of estimate was found through some multiple regression calculations (Table 8).

A statistical standard form from the multiple regression results was selected to analyse and evaluate the influence of selected intrinsic properties on the spontaneous combustion liability of coals and coal-shales. This statistical approach indicated that the changes in spontaneous combustion liability index of coals and coal-shales may be due to an intrinsic parameter, while others are considered. The best prediction for spontaneous combustion liability of 14 coals and 14 coal-shales has been identified in the four models (Table 8). The developed models show that high correlation coefficients and the standard error of estimates values have been reduced to acceptable limits. The multiple regression analysis indicates that the Wits-Ehac Index and intrinsic properties show significant correlation coefficients of 0.939 and 0.998, R-squared values of 0.882 and 0.997, and low standard error of estimates of 0.336 and 0.094 for coals and coal-shales respectively. The Wits-CT Index with intrinsic properties shows significant correlation coefficients of 0.992 and 0.999, R-squared values of 0.983 and 0.999 and higher standard error of estimate of 0.772 and 0.123 for coals and coal-shales respectively. It was found that the models developed for the Wits-Ehac Index gives lower R-squared values, coefficient of correlations and standard error of estimate while, the Wits-CT Index shows higher R-squared values and correlation coefficients and standard error of estimates than the Wits-Ehac Index in both cases. However, considering the index with higher R-squared values and correlation coefficients, it was found that the Wits-CT Index is more suitable to predict the spontaneous combustion liability, while considering the index with low standard error of estimates, the Wits-Ehac Index is also suitable. Hence, multiple regression analysis on the experimental data indicates that the two liability indices can be used to obtain reliable results for spontaneous combustion prediction. Establishing a correlation between liability index and intrinsic properties for coals and coal-shales is essential for a better understanding of the role of coal-shales in spontaneous combustion. The models have shown that spontaneous combustion takes place due to the combined effect of different intrinsic factors.

3.5 Validation of the models

The results of the actual Wits-Ehac and Wits-CT Index for coals and coal-shales are presented in Tables 8 and 9.

The predicted spontaneous combustion liability indices (Wits-Ehac and Wits-CT Index) have been validated with the actual Wits-Ehac Index and the Wits-CT Index and the results are in-line (Tables 9 and 10). Twenty-eight samples evaluated in this paper with an additional four samples (3 coals- KCA, KCB, KCC and KCD; and one coal-shale-KSA) have confirmed the reliability of the developed

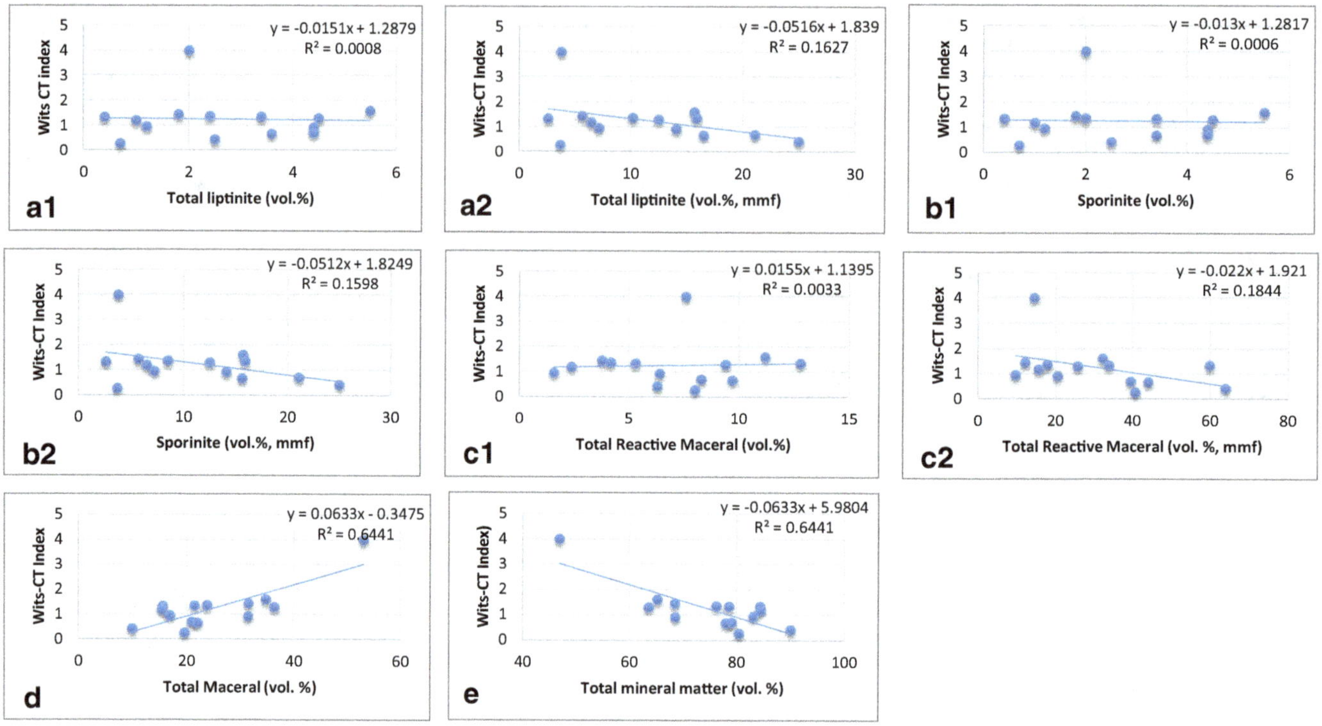

Fig. 16 Influence of liptinite, other macerals, petrographically observable mineral matter and its group on spontaneous combustion of coal-shales

Table 2 Relationships between independent (proximate and ultimate analysis, wt%-ad) and dependent variables for coal samples

Independent variables	Range	Dependent variables correlation coefficients	
		Wits-Ehac index	Wits-CT index
Moisture	1.60–2.50	(0.0637)	(0.5847)
Volatile matter	16.7–26.9	0.5164	0.3273
Ash	13.7–48.4	(0.6884)	(0.8660)
Carbon	36.1–69.7	0.6572	0.8565
Hydrogen	2.55–4.21	0.6616	0.7552
Nitrogen	0.85–1.63	0.6945	0.8382
Oxygen	5.65–10.4	(0.0686)	(0.1016)
Sulphur	0.59–5.30	(0.0115)	(0.0687)
Pyritic sulphur	0.13–4.13	0.0869	0.0553
Sulphate sulphur	0.01–0.42	(0.6422)	(0.7620)
Organic sulphur	0.28–1.09	(0.1787)	(0.2914)

models. The Wits-Ehac of coal-shales SC, SG, SH, SI, SJ and SM that could not be obtained by the actual Wits-Ehac Index due to their low reactivity were successfully obtained with the predicted models (Table 10). The models present a high level of confidence as they produced results similar to the actual liability indices.

4 Conclusion

The statistical analysis showed that the spontaneous combustion liability of coals and coal-shales can be determined by a model consisting of various intrinsic factors. The predicted spontaneous combustion liability indices (Wits-Ehac and Wits-CT Index) have been validated with the actual Wits-Ehac Index and the Wits-CT Index and the results are in-line. The developed models enable one to identify the most important factors affecting spontaneous combustion and establishes relationships between them. The intrinsic properties of the samples have been identified to be the main factors contributing towards spontaneous combustion in coal mines. However, the study shows that no single factor is sufficient to influence the coal and coal-shale spontaneous combustion liability but a cumulative effect of favourable parameters may decide whether a coal or coal-shale will be liable to spontaneous combustion or not. The influence of the intrinsic properties of coal-shales in relation to coals could be a tool to predict spontaneous combustion risk in coal mines. The results provide quantitative descriptions and showed the relationships between the dependence of spontaneous combustion liability indices and intrinsic properties. The linear regression analysis shows that both liability indices indicate better linear relationships with some of the intrinsic factors and thus,

Table 3 Relationships between independent (petrographic analysis, vol%) and dependent variables for coal samples

Independent variables	Range	Dependent variables correlation coefficients	
		Wits-Ehac index	Wits-CT index
Total vitrinite	7.0–49.4	0.2591	0.0890
Total vitrinite (mmf)	7.9–60.0	0.0874	(0.2755)
Collotelinite	1.6–39.0	0.1481	0.0255
Collotelinite (mmf)	1.7–44.4	0.0439	(0.2451)
Collodetrinite	2.8–13.3	0.3944	0.1580
Collodetrinite (mmf)	3.1–18.8	0.1725	(0.2401)
Total inertinite	11.7–84.6	0.1679	0.4752
Total inertinite (mmf)	36.9–90.1	(0.0735)	0.2754
Fusinite	0.8–7.6	0.5663	0.2860
Fusinite (mmf)	1.2–12.1	0.4392	0.0447
Secretinite	0.8–6.0	0.1284	0.1143
Secretinite (mmf)	1.1–6.7	0.0064	(0.1791)
Total semifusinite	5.5–43.4	0.2898	0.3098
Total semifusinite (mmf)	14.1–45.9	0.1460	0.1010
Total inertodetrinite	3.2–49.4	(0.0894)	0.4065
Total inertodetrinite (mmf)	10–55.5	(0.2768)	0.2504
Total liptinite	0.6–3.8	0.0280	0.2621
Total liptinite (mmf)	0.6–4.4	(0.2228)	(0.1108)
Sporinite	0.4–3.4	0.2321	0.4353
Sporinite (mmf)	0.7–3.8	0.2116	0.3235

Table 4 Relationships between independent (total reactive maceral, total maceral and total mineral matter vol.%) and dependent variables for coal samples

Independent variables	Range	Dependent variables correlation coefficients	
		Wits-Ehac index	Wits-CT index
Total reactive maceral	11.2–53.0	0.3148	0.1292
Total reactive maceral (mmf)	12.8–64.4	0.1229	(0.2814)
Total maceral	31.7–94.6	0.3785	0.6496
Total mineral matter	5.4–68.4	(0.3782)	(0.6496)

identifies the major intrinsic factors affecting their liability towards spontaneous combustion. It was found each intrinsic property has an influence on spontaneous combustion liability and their strength varies from one another. Also, a definite positive or negative correlation coefficient exists between the intrinsic factors and the spontaneous combustion test results. The multiple regression of spontaneous combustion liability indices on 12 independent variables has developed acceptable and accurate predictive models as indicated by R-squared values, the coefficient of correlations and standard error of estimates. The models

Table 5 Relationships between independent (proximate and ultimate analysis, wt%-ad) and dependent variables for coal-shale samples

Independent variables	Range	Dependent variables correlation coefficients	
		Wits-Ehac index	Wits-CT index
Moisture	0.8–1.7	0.7715	0.4507
Volatile matter	8.5–16.6	0.6389	0.6901
Ash	51.5–88.7	(0.8352)	(0.9448)
Carbon	2.66–33.7	0.7962	0.9630
Hydrogen	0.75–2.87	0.5795	0.8799
Nitrogen	0.08–0.96	0.6446	0.9462
Oxygen	5.01–11.85	(0.3212)	0.3243
Sulphur	0.12–6.90	0.5791	0.0242
Pyritic sulphur	0.04–4.26	0.5704	0.0367
Sulphate sulphur	0.01–0.45	0.5365	(0.0403)
Organic sulphur	0.05–2.19	0.5933	0.0147

Table 6 Relationships between independent (petrographic analysis-vol%) and dependent variables for coal-shale samples

Independent variables	Range	Dependent variables correlation coefficients	
		Wits-Ehac index	Wits-CT index
Total vitrinite	0.4–8.4	0.1230	0.2620
Total vitrinite (mmf)	2.4–39.3	(0.2632)	(0.1854)
Collotelinite	0–4.2	0.2206	0.3366
Collotelinite (mmf)	0–19.6	(0.1073)	(0.0928)
Collodetrinite	0–2.6	0.2105	0.1289
Collodetrinite (mmf)	0–12.1	(0.2330)	(0.1933)
Total inertinite	4.9–46.1	0.7360	0.8029
Total inertinite (mmf)	44.9–91.7	0.2143	0.3262
Fusinite	0–4.0	(0.4038)	0.1313
Fusinite (mmf)	0–16.9	(0.5898)	(0.0427)
Secretinite	0–1.8	0.6003	0.6009
Secretinite (mmf)	0–4.8	0.3304	0.1523
Total semifusinite	0.2–8.0	0.7776	0.2967
Total semifusinite (mmf)	1.3–40.7	0.2970	(0.2238)
Total inertodetrinite	3.3–35.2	0.7749	0.8388
Total inertodetrinite (mmf)	25.2–80.5	0.2367	0.4460
Total liptinite	0.4–5.5	0.3360	(0.0276)
Total liptinite (mmf)	2.6–25.0	(0.0113)	(0.4033)
Sporinite	0.4–5.5	0.3463	(0.0238)
Sporinite (mmf)	2.6–25.0	0.0059	(0.3997)

Table 7 Relationships between independent (total reactive maceral, total maceral and total mineral matter vol%) and dependent variables for coal-shale samples

Independent variables	Range	Dependent variables correlation coefficients	
		Wits-Ehac index	Wits-CT index
Total reactive maceral	1.6–12.8	0.2923	0.0575
Total reactive maceral (mmf)	9.6–63.9	(0.2552)	(0.4294)
Total maceral	9.9–53.1	0.7653	0.8026
Total mineral matter	46.9–90.1	(0.7653)	(0.8029)

derived from the different samples produced high correlation coefficients and low standard error of estimates. The use of the formulas derived may provide reliable predictions for the spontaneous combustion liability of coals and coal-shales. The influence of the intrinsic properties affecting the spontaneous combustion of coals and coal-shales may be evaluated more effectively by these models than by the linear regression. The results obtained from this study may be used as a reliable tool to support previous works on the role of intrinsic properties on the spontaneous

Table 9 Actual and predicted Wits-Ehac and Wits-CT index for coals

Samples	Wits-Ehac index		Wits-CT index	
	Actual	Predicted	Actual	Predicted
CA	4.64	4.82	6.29	5.88
CB	4.64	4.67	6.96	6.90
CC	4.52	4.52	5.31	5.32
CD	4.60	4.64	6.80	6.71
CE	4.76	4.63	5.42	5.73
CF	4.49	4.49	3.97	3.98
CG	4.91	4.82	7.53	7.74
CH	4.69	4.70	7.51	7.49
CI	3.82	3.83	4.05	4.03
CJ	4.46	4.45	6.61	6.63
CK	4.44	4.62	9.10	8.98
CL	4.87	4.85	9.59	9.64
CM	4.76	4.63	7.27	7.47
CN	4.84	4.79	7.91	8.03
KCA	6.07	6.03	8.87	8.57
KCB	5.41	5.25	8.55	8.78
KCC	5.67	5.67	10.51	10.83
KCD	6.44	6.41	9.01	9.26

Table 8 Predictive formulas derived by multiple regression analysis

Model no.	Equations derived by multiple regression	RSV	R	SEE
All coals (A)	$WE = 557.4939 - 6.6733M_{ad} - 0.0256VM_{ad} - 7.3322A_{ad} - 7.2405C_{ad} - 7.5779H_{ad} - 8.4405N_{ad} - 7.1169S_{ad} - 7.1643O_c + 1.7441\,V + 1.8823L + 1.7419I + 1.7577MM$	0.882	0.939	0.336
All coals (B)	$WC = 5780.205 - 43.1771M_{ad} - 0.3858\,VM_{ad} - 44.1651A_{ad} - 43.7848C_{ad} - 47.8671H_{ad} - 35.8312N_{ad} - 43.9474S_{ad} - 43.79O_c - 13.7768V - 12.3551L - 13.8127I - 13.6081MM$	0.983	0.992	0.772
All coal-shales (C)	$WE = -6159.64 + 61.9497M_{ad} + 1.0644VM_{ad} + 64.5252A_{ad} + 64.5374C_{ad} + 68.2129H_{ad} + 55.0262N_{ad} + 63.5985S_{ad} + 63.0291O_c - 2.9441\,V - 2.9167L - 2.907I - 2.9037MM$	0.997	0.998	0.094
All coal-shales (D)	$WC = 4302.218 - 37.9739M_{ad} - 0.5712VM_{ad} - 38.1227A_{ad} - 37.9071C_{ad} - 40.9502H_{ad} - 29.4017N_{ad} - 37.2958S_{ad} - 37.2423O_c - 4.8458\,V - 5.0309L - 4.9672I - 4.888MM$	0.999	0.999	0.123

M_{ad} is air-dried moisture (wt%), VM_{ad} is air-dried volatile matter (wt%), A_{ad} is air-dried ash (wt%), C_{ad} is air-dried carbon (wt%), H_{ad} is air-dried hydrogen (wt%), N_{ad} is air-dried nitrogen (wt%), S_{ad} is air-dried sulphur (wt%), O_c is calculated oxygen (wt%), V is total vitrinite (vol%), L is total liptinite (vol%), I is total inertinite (vol%), MM is mineral matter (vol%), WE is Wits-Ehac Index, WC is Wits-CT index, RSV is R-squared values, R is correlation coefficient, and SEE is standard error of estimate

Table 10 Actual and predicted Wits-Ehac and Wits-CT index for coal-shales

Samples	Wits-Ehac index		Wits-CT index	
	Actual	Predicted	Actual	Predicted
SA	3.09	3.07	1.33	1.37
SB	3.08	3.08	1.30	1.28
SC	–	2.56	0.91	0.97
SD	3.27	3.31	0.70	0.67
SE	3.73	3.72	1.60	1.61
SF	3.10	3.07	1.36	1.40
SG	–	2.49	0.67	0.63
SH	–	2.35	0.27	0.28
SI	–	2.58	0.95	0.88
SJ	–	2.63	0.42	0.42
SK	2.98	2.98	1.18	1.18
SL	2.99	2.99	1.34	1.35
SM	–	2.54	1.44	1.43
SN	3.77	3.79	3.99	3.97
KSA	3.49	3.50	4.40	4.38

combustion of coals and coal-shales. It has been found that an accurate characterisation of the degree of oxidation could be a useful tool to measure the relative liability of carbonaceous materials toward spontaneous combustion.

Acknowledgements This work was conducted in the context of coal-shale spontaneous combustion in the Witbank coalfields, South Africa and was financially sponsored by Coaltech. The authors wish to express gratitude to Coaltech and the staff of the selected coal mines for their supports. The work presented here is part of a Ph.D. research study in the School of Mining Engineering at the University of the Witwatersrand.

References

ASTM (D-3173-17a) (2017) Standard test methods for moisture in the analysis sample of coal and coke. ASTM International, West Conshohocken, PA. www.astm.org

ASTM (D-3174-11) (2011) Standard test method for ash in the analysis sample of coal and coke from coal and standard classification of coals by rank. ASTM International, West Conshohocken, PA. www.astm.org

ASTM (D-3175-17) (2017) Standard test method for volatile matter in the analysis sample of coal and coke. ASTM International, West Conshohocken, PA. www.astm.org

ASTM (D3176-15) (2015) Standard practice for ultimate analysis of coal and coke and standard test methods for total sulfur in the analysis sample of coal and coke. ASTM International, West Conshohocken, PA, www.astm.org

Banerjee SC (1985) Spontaneous combustion of coal and mine fires. Oxford and IBH Publishing, New Delhi

Beamish BB, Arisoy A (2008) Effect of mineral matter on coal self-heating rate. Fuel 87:125–130

Beamish BB, Blazak DG (2005) Relationship between ash content and R70 self-heating rate of callide coal. Int J Coal Geol 64:126–132

Beamish BB, Barakat MA, St George JD (2001) Spontaneous combustion propensity of NewZealand coals under adiabatic conditions. Int J Coal Geol 45:217–224

Carras J, Young B (1994) Self-heating of coal and related materials: models, application and test methods. Prog Energy Combust Sci 20(1):1–15

Cliff D, Davis R, Bennet A, Galvin G, Clarkson F (1998) Large scale laboratory testing of the spontaneous combustibility of Australia coals. In: Proceedings of Queensland mining industry health & safety conference, Yeppoon, pp 175–179

Deborah JR (2016) Statistics for dummies, 2nd edition, pp 408

Draper NR, Smith H (1998) Applied regression analysis. Wiley-interscience. ISBN 0-471-17082-8

Dullien F (1979) Porous media fluid transport and pore structure. Academic Press, London, p 79

Eroglu HN (1992) Factors affecting spontaneous combustion liability index. Ph.D Thesis, University of the Witwatersrand Johannesburg, South Africa

Falcon LM (1987) The petrographic composition of Southern African Coals in relation to friability, hardness and abrasive indices. Inst Min Metall 87:323–336

Falcon RMS (2004) The constitution of coal and its inherent capacity to self-heat as applied to an integrated spontaneous combustion risk. In: Final proceedings of the international conference in spontaneous combustion. Fossil fuel foundation and SABS, Johannesburg, 8–9

Genc B, Cook A (2015) Spontaneous combustion risk in South African coalfields. J S Afr Inst Min Metall 115:563–568

Genc B, Onifade M, Cook A (2018) Spontaneous combustion risk on South African coalfields: Part 2. In: Proceedings of the 21st International Coal Congress of Turkey ''ICCET'' April 11–13, 2018, Zonguldak, Turkey, pp 13–25

Glant SA, Slinker BK (1990) Primer of applied regression and analysis of variance, ISBN 0-07-023407-8. McGraw-Hill, New York

Gouws MJ, Wade L (1989a) The self-heating liability of coal: predictions based on simple indices. Min Sci Technol 9:75–80

Gouws MJ, Wade L (1989b) The self-heating liability of coal: predictions based on composite indices. Min Sci Technol 9:81–85

Humphreys D, Rowlands D, Cudmore JF (1981) Spontaneous combustion of some Queensland coals. In: Proceedings of the fifth ignitions, explosions and fires in coal mines symposium. Australasian Institute of Mining and Metallurgy, Melbourne, pp 1–19

ISO 12902 (200)1 Determination of total carbon, hydrogen and nitrogen

Kaymakci E, Didari V (2002) Relations between coal properties and spontaneous combustion parameters. Turk J Eng Environ Sci 26(1):59–64

Kim AG (1977) Laboratory studies on spontaneous combustion of coal. IC 8756. US Bureau of Mines

Kucuk A, Kadioglu Y, Gulaboglu MS (2003) A study of spontaneous combustion characteristics of a Turkish lignite: particle size, moisture of coal, humditiy of air. Combust Flame 133(3):255–261

Moghtaderi B, Dlugogorski BZ, Kennedy EM (2000) Effects of wind flow on self-heating characteristics of coal stockpiles. Trans IChemE 78(6):445–453

Nimaje DS, Tripathy DP (2016) Characterization of some Indian coals to assess their liability to spontaneous combustion. Fuel 163:139–147

Onifade M, Genc B (2018a) Establishing relationship between spontaneous combustion liability indices. In: Proceedings of the 21st international coal congress of Turkey ''ICCET'' April 11–13, Zonguldak, Turkey, pp 1–11

Onifade M, Genc B (2018b) Prediction of the spontaneous combustion liability of coal and coal-shale using statistical analysis. In: 6th regional conference of Society of mining professors, pp 63–82

Onifade M, Genc B (2018c) A review of spontaneous combustion studies—South African context. Int J Min Reclamation Environ. https://doi.org/10.1080/17480930.2018.1466402

Onifade M, Genc B (2018d) Spontaneous combustion of coals and coal-shales. Int J Min Sci Technol. https://doi.org/10.1016/j.ijmst.2018.05.013

Onifade M, Genc B, Carpede A (2018) A new apparatus to establish the spontaneous combustion propensity of coals and coal-shales. Int J Min Sci Technol. https://doi.org/10.1016/j.ijmst.2018.05.012

Panigrahi DC, Sahu HB (2004) Classification of coal seams with respect to their spontaneous combustion susceptibility—a neural network approach. Geotech Geol Eng 22:457–476

Panigrahi DC, Saxena VK (2001) An investigation into spontaneous combustion characteristics of coals using differential thermal analysis. In: Proceedings of the 7th international mine ventilation congress, EMAG, Krakow, 17–22 June 2001, pp 495–500

Ren TX, Edwards JS, Clarke D (1999) Adiabatic oxidation study on the propensity of pulverized coal to spontaneous combustion. Fuel 78:1611–1620

Sahu HB, Panigrahi DC, Mishra NM (2005) Assessment of spontaneous combustion susceptibility of coal seams by experimental techniques: a comparative study. In: International symposium on advances in mining technology and management, pp 459–465

SANS, 7404 Parts 1–5 Methods for the petrography analysis of coals. Part 2: Methods of preparing coal samples

Scott AC, Glasspool IJ (2007) Observations and experiments on the origin and formation of inertinite group macerals. Int J Coal Geol 70(1–3):53–66

Singh RN, Demirbilek S (1987) Statistical appraisal of intrinsic factors affecting spontaneous combustion of coal. Min Sci Technol 4(2):155–165

Smith AC, Miron Y, Lazzara P (1988) Inhibition of spontaneous combustion of coal. Report of Investigation RI 9196. Bureau of Mines, US

Stigler SM (1989) Francis Galton's account of the invention of correlation. Stat Sci 4(2):73–79

Stott JB (1980) The spontaneous combustion of coal and the role of moisture transfer, US: Final Report. Bureau of Mines, US

Stracher GB, Taylor TP (2004) Coal fires burning out of control around the world: thermodynamic recipe for environmental catastrophie. Int J Coal Geol 59(1):7–17

Uludag S, Phillips HR, Eroglu HN (2001) Assessing spontaneous combustion risk in South African coal mining by using a GIS tool. In: 17th international mining conference and exhibition, Turkey, pp 243–249

Wade L (1989) The propensity of South African coals to spontaneously combust. Ph.D. Thesis, Department of Mining Engineering, University of the Witwatersrand, Johannesburg

Wade L, Gouws MJ, Phillips HR (1987) An apparatus to establish the spontaneous combustion propensity of South African coals. In: Proceedings of the symposium on safety in coal mines, CSIR, Pretoria. pp 7.1–7.2

8

Mechanical properties and acoustic emission characteristics of thick hard roof sandstone in Shendong coal field

Huigui Li[1,2] · **Huamin Li**[1,2]

Abstract The mechanical properties and acoustic emission characteristics of thick hard roof sandstone were investigated. Samples were taken from the 30.87-m thick sandstone roof in a mine in the Shengdong coal field, China. Firstly, the composition and microscopic characteristics were analyzed by XRD and FE-SEM, respectively. Moreover, the indirect tensile test, uniaxial compression test, three axis compression experiment and AE test are carried out by using RMT-150C mechanics experiment system with DS5-8B AE test system. The experiment results indicate that the main framework particles of sandstone are quartz and feldspar, and mainly quartz. Cements are mainly pyrite, kaolinite, chlorite and zeolite cross needle, clinochlore, and clay minerals. The microstructure of sandstone is very dense, with few pores and high cementation degree. The tensile strength, compressive strength and elastic modulus of sandstone are 4.825, 85.313 MPa, 13.814 GPa, respectively, so the sandstone belongs to hard rock. The AE cumulative counts of sandstone can be divided into three phases: relatively flat growth period, rapid growth period and spurt period. The signal strength of AE waveform can be used as a warning signal. In the tensile fracture zone, the warning value is 0.4 mV, and in the compression shear failure zone, it is 4 mV. The numbers of cumulative counts of AE under different stress conditions have obvious difference. Moreover, the growth of cumulative counts of acoustic emission is more obvious when the stress is more than 60% of the peak stress.

Keywords Mechanical · Acoustic emission (AE) · Sandstone · X-ray diffractometer (XRD) · Field emission scanning electron microscope (FE-SEM)

1 Introduction

The Daliuta coal mine is a large modern mine in the Shenhua coal group, with approved production capacity of 21,700,000 tons. The Shenhua coal group is located in the northern part of Shaanxi Province and composed of the Daliuta and Huojitu coal mines. Daliuta has mine area of 189.9 km^2, coal reserves of 2.32 billion tons, and recoverable reserves of 1.53 billion tons. The roof of the 5–2 coal seam is composed of sandstone with an average thickness of 30.87 m. In order to prevent the roof failure due to weighting over great extent, pre splitting blasting for the thick sandstone in the open cut hole was carried out. However, the initial pressure occurs until the panel advances to the 95 m. Such a large area of the roof once falling will cause great harm. This is a serious threat to the safety and efficiency of production on the working face. Therefore, it is urgent to find out the internal cause of roof failure due to weighting over great extent, so that the corresponding control measures can be put forward. According to the formation of this kind of mine pressure, the rock mechanical properties of the overlying strata are

✉ Huigui Li
lihuigui1221@163.com

[1] School of Energy Science and Engineering, Henan Polytechnic University, Jiaozuo 454000, Henan, People's Republic of China

[2] Opening Project of Key Laboratory of Deep Mine Construction, Henan Polytechnic University, Jiaozuo 454000, Henan, People's Republic of China

the major influential factors. Moreover, acoustic emission monitoring can monitor and predict this kind of dynamic disaster. Therefore, it has engineering and theoretical significance to study the mechanics and acoustic emission characteristics of the rock.

Domestic and foreign scholars have well studied the mechanical properties and AE characteristics of rock and made outstanding contributions to the advancement of rock mechanics and acoustics. The structural plane has an important influence on the mechanical and acoustic emission characteristics of rock. Scholars carried out laboratory experiments and numerical simulations on rocks with structural plane. Meanwhile, the relationship among the structural plane, mechanical properties, and acoustic emission characteristics was also established (Liu et al. 2012; Wang et al. 2013; Han and Zhang 2014; Su et al. 2014; Yang et al. 2014; Zhu et al. 2015). Some scholars have studied the mechanics and damage characteristics of sandy mudstone, metamorphic quartz sandstone and silty sandstone under triaxial compression and triaxial unloading confining pressure (Fang et al. 2015; Deng et al. 2016; Guo et al. 2016). The mechanical properties and acoustic emission characteristics of sandstone under impact loading were studied by (Li 2014; Zhang et al. 2014; Li et al. 2015a, b; Liu et al. 2015). Besides, they explored the effect of strain rate on the mechanical properties and acoustic emission characteristics of sandstone under impact loading. Water also has an important effect on the physical properties, mechanical properties, and acoustic emission characteristics of rock. The effects of water on the physical and mechanical properties and acoustic emission characteristics of silty sandstone, coal and sandstone were analyzed by other researchers (Chen et al. 2013; Deng et al. 2014; Gao et al. 2015a, b; Zhang et al. 2015, 2016), and the results indicate that water can significantly alter the wave velocity, compressive strength, elastic modulus, failure pattern and acoustic emission characteristics of coal and rock. The effect of pore pressure and confining pressure on short-term and creep mechanical behavior of saturated and dry red sandstone were studied by Yang et al. (2015). The effect of strain rate on the mechanical behavior of sandstone with different grain sizes was studied by Wasantha et al. (2015). The microscopic damage and dynamic mechanical characteristics of rock under the freeze–thaw environment were studied by Zhou et al. (2015).

However, studies on the mechanical properties and acoustic emission characteristics of thick hard roof sandstone are scarce. In order to explore the internal cause of dynamic disasters such as the formation of shock, and take corresponding control measures, the mechanical properties and acoustic emission characteristics of the thick hard roof sandstone in the Shendong coal field were studied experimentally and the results are analyzed and discussed in this paper.

2 Experimental

2.1 Sample preparation

The panel of 52307 extracts 5–2 coal seam. The roof of the 5–2 coal seam is composed of sandstone with an average thickness of 30.87 m, as shown in Table 1. In order to study the mechanical properties and acoustic emission characteristics of the sandstone, the method of core drilling was used in the process of sampling. All the collected samples were wrapped by foam, and then put into a core box for protection during the transportation. According to the requirements of regulation (the Ministry of Water Resources of P. R. China 2008), the Brazil split and the uniaxial and triaxial compression tests samples should be processed into $\phi 25$ mm × 25 mm, $\phi 50$ mm × 100 mm respectively.

2.2 Testing equipment

Uniaxial compression test, indirect tensile test and triaxial compression test were carried out on a RMT-150C mechanical testing machine. The system is composed of main control computer, hydraulic controller, three axial pressure sources, digital controller, manual controller, hydraulic actuator and hydraulic source. The acoustic emission test was carried out with a DS5 acoustic emission monitor. The system has 8 channels and can monitor the external parameters such as temperature and pressure. Moreover, the system can identify the location of acoustic emission and record the waveform characteristics in the whole experiment process.

2.3 Testing procedure

(1) AE test

The AE test is applied with the mechanics experiment simultaneously. The experiment system diagram is shown

Table 1 Lithological characteristics of the seam roof

Lithology	Thick (m)	Depth (m)
Fine sandstone	30.87	175.62
Siltstone	0.60	176.22
Medium sandstone	0.70	176.92
Siltstone	2.41	179.33
5–2 coal	7.04	186.37

in Fig. 1. In the AE test, two probes are symmetrically placed in the middle of the sample. In order to maintain the full contact of the sensor and the sample, the sensor is coated with a coupling agent in the test process. During the experiment, the AE system and the mechanics experiment system begin to collect the data simultaneously. The external parameter of the AE test is 40 dB, the threshold value is 50 mv, and the sampling frequency is 3 MHz.

(2) Mechanics experiment

The indirect tensile test adopts force control at the loading rate of 0.2 kN/s, which is repeated for 5 samples. And the uniaxial compression test adopts displacement control at the loading rate of 0.005 mm/s, which is repeated for 3 samples.

Triaxial compression experiment is controlled by displacement. Firstly, loading confining pressure to 5,10,15,20 and 25 MPa by the loading rate of 0.5 MPa/s, under the condition of hydrostatic pressure which turns to be $\sigma_1 = \sigma_2 = \sigma_3$. Then continue loading axial compression by the loading rate of 0.01 mm/s while keeping confining pressure constant till the specimens get destroyed. During the test, the axial loading and acoustic emission testing are kept simultaneously.

3 Composition and microstructure characteristics

3.1 Analysis of mineral composition

Many methods have been used to examine the microfabric of rock (Sachan and Penumadu 2007; Yao et al. 2016). In this study, the method of X-ray diffractometer is used to study the mineral composition of the roof sandstone, and the results are shown in Table 1 and Fig. 2. Table 2

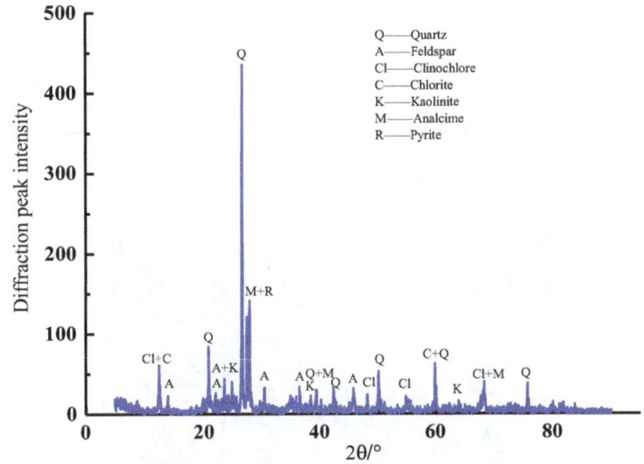

Fig. 2 X-ray diffraction analysis of roof sandstone

indicates that the framework grain of thick hard roof sandstone is quartz and feldspar, constituting about 69.7% of the total mass. Cements mainly have pyrite, kaolinite, chlorite, zeolite cross needle and clinochlore, and the main cement components are clay minerals constituting about 20.8% of the total mass.

3.2 Microstructure characteristics

The microstructure of samples was analyzed by a field emission scanning electron microscope. The images of the roof sandstone are enlarged 100, 200, 500, 1000, 2000 and 5000 times. The SEM images of 500 times and 2000 times are shown in Fig. 3. The figure indicates that the sandstone has a very dense structure, fuzzy boundaries, few pores, highest degree of cementation, abundant clay minerals and mineral Si surrounded very closely by clay minerals.

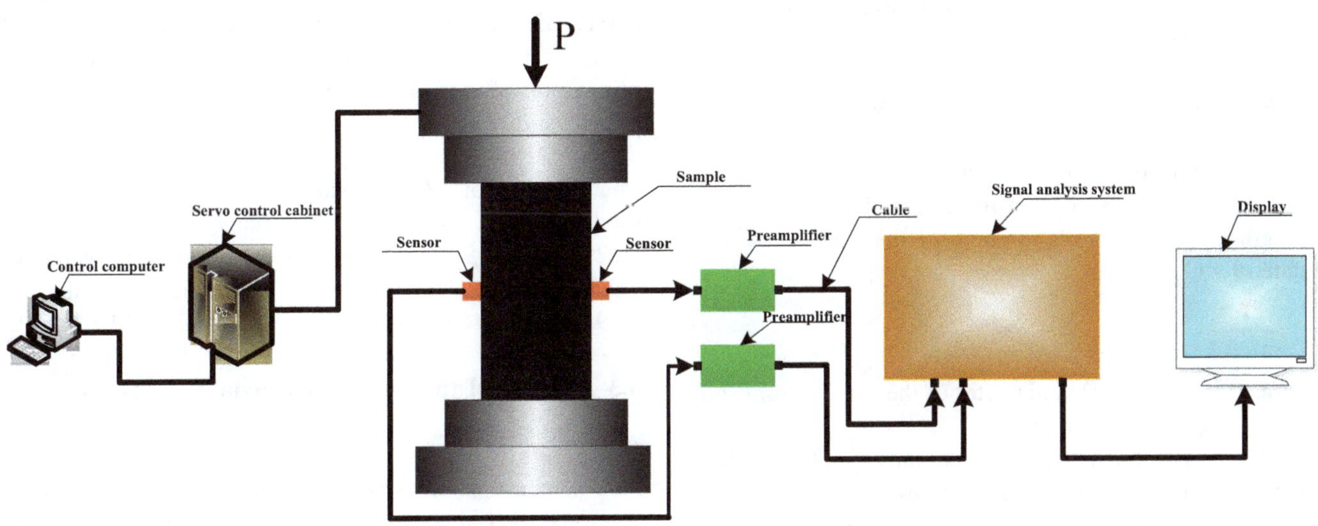

Fig. 1 Figure of test system

Table 2 Analysis result of component

Sample	Mineral components (%)						
	Quartz	Feldspar	Kaolinite	Pyrite	Chlorite	Clinochlore	Analcime
Sandstone	42.6	27.1	13.5	1.7	0.1	7.2	7.9

Magnification ×500

Magnification ×2000

Fig. 3 Scanning electron microscope results of roof sandstone

4 Mechanical properties

In order to study the mechanical properties of thick roof sandstone, the indirect tensile test, uniaxial compression test and triaxial compression test were carried out, respectively (Fig. 4).

4.1 Analysis of the results of indirect tensile test

The damage of roof is not only caused by compression and shear failure, but also by tensile failure. The yield tensile strength and peak tensile strength of the sample are calculated by (1).

$$R = \frac{2P}{\pi DL} \tag{1}$$

where, P is load, N; D is diameter of the sample, mm; L is the height of the sample, mm.

Table 3 lists the mechanical parameters of sandstone under indirect tension. Figure 4 shows the stress–strain curves of sandstone under indirect tension. The tensile strength of roof sandstone ranges from 4.360–5.228 MPa, with the average of 4.825 MPa. The peak strain of roof sandstone is in the range of 1.028×10^{-3}–1.165×10^{-3}.

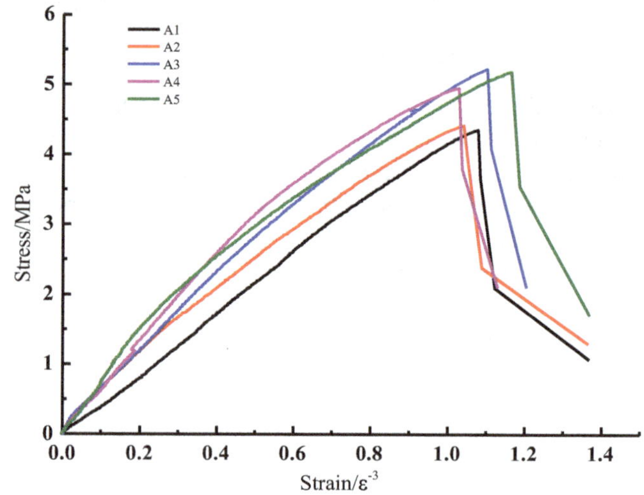

Fig. 4 Stress–strain curves of sandstone under indirect tension

4.2 Analysis of the results of uniaxial compression test

Uniaxial compression tests were carried out to obtain the mechanical parameters (such as the compressive strength, elastic modulus and Poisson's ratio) of the samples of roof sandstone. The test results and derived mechanical parameters are shown in Table 4. Figure 5 shows the stress–strain curves of sandstone under uniaxial compression. Table 4 indicates that the uniaxial compressive strength and elastic of modulus of roof sandstone are relatively high. The range of peak stress is 81.779–89.175 MPa with an average value of 85.313 MPa. According to the definition in regulation (the Ministry of Water Resources of P. R. China 2008), sandstone is a hard rock. The range of peak strain is 7.229×10^{-3}–8.331×10^{-3}, with an average value of 7.929×10^{-3}. The range of elasticity modulus is 12.951–15.155 GPa, with an average value of 13.814 GPa.

4.3 Analysis of the results of triaxial compression test

Five standard samples were chosen to carry out the conventional triaxial compression test to obtain the parameters of cohesion and internal friction angle. According to the

Table 3 Mechanical parameters of sandstone in indirect tension

Serial number	Diameter (mm)	Height (mm)	Loading rate (kN/s)	Tensile strength (MPa)	Strain (10^{-3})
A1	49.80	25.30	0.2	4.360	1.078
A2	49.82	25.92	0.2	4.413	1.041
A3	49.82	26.36	0.2	5.228	1.103
A4	49.92	26.02	0.2	4.939	1.028
A5	49.90	27.90	0.2	5.183	1.165

requirements in regulation, five specimens were tested under the confining pressure of 5, 10, 15, 20 and 25 MPa. Displacement control was adopted in the experiment. The rate of axial loading was 0.002 mm/s and the rate of confining pressure loading was 0.5 MPa/s. The condition of hydrostatic pressure ($\sigma_1 = \sigma_2 = \sigma_3$) was adopted at the initial stage of triaxial compressive loading. Table 5 shows the mechanical parameters of sandstone under different confining pressure. The variables σ_1, σ_3, E and ε_0 represent the peak stress, confining pressure, modulus of elasticity and peak strain, respectively. Figure 6 shows the stress–strain curve of roof sandstone under different confining pressure. Based on the mechanical parameters under different confining pressures, the cohesive force and the internal friction angle were derived to be 23.389 MPa and 32°, respectively.

Figures 5 and 6 indicate that the test process of uniaxial and triaxial compression can be divided into the

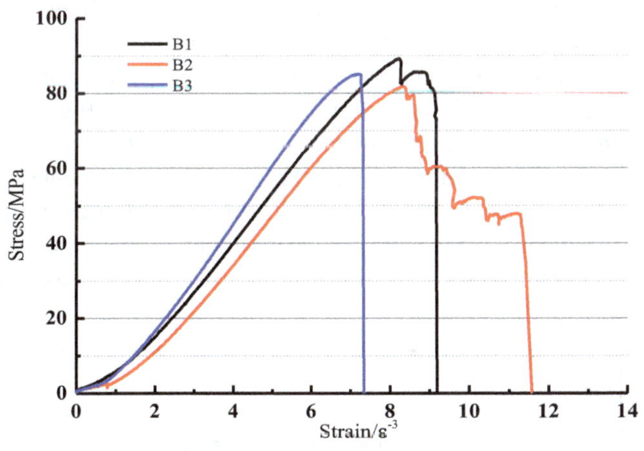

Fig. 5 The stress–strain curves of sandstone under uniaxial compression

compaction phase, linear elastic phase, yield phase and failure phase. In uniaxial compression, the compression phase is relatively long, and the yield phase is short. After the peak stress is reached, the stress drops rapidly, suggesting the brittle characteristics of roof sandstone. However, with the increase of confining pressure, the stage of compaction becomes shorter and the peak strain increases gradually. After the peak stress is reached, the stress drops at a lower rate, and the sandstone is gradually transformed from brittle to ductile.

5 Acoustic emission characteristics

AE is a phenomenon caused by partial fracture of rock material and rapid release of energy in the form of elastic wave (Gao et al. 2015a, b; Li et al. 2015a, b). In the fracturing process of thick hard roof sandstone in the Shendong coal field, energy will be quickly released in the form of elastic wave. Therefore, the characteristics of acoustic emission may provide precursor information during the fracturing process of hard roof and help to the monitor and predict potential dynamic disaster of hard roof.

In this study, the AE signals were monitored in the indirect tensile test, uniaxial compression test and triaxial compression test. The AE counts, cumulative counts of AE and waveform characteristics were obtained as the results of the acoustic emission test.

5.1 Analysis of acoustic emission count and cumulative count characteristics

Figures 7, 8, 9 show the relationship curves among the acoustics emission count, cumulative count, and stress in the roof sandstone during the course of the indirect tensile

Table 4 Mechanical parameters of sandstone under uniaxial compression

Serial number	Diameter (mm)	Length (mm)	Loading rate(kN/s)	Stress (MPa)	Strain (10^{-3})	Elasticity modulus (GPa)	Poisson's ratio
B1	49.70	99.82	0.005	89.175	8.227	13.336	0.249
B2	50.20	100.70	0.005	81.779	8.331	12.951	0.273
B3	50.02	100.40	0.005	84.984	7.229	15.155	0.239

Table 5 Mechanical parameter of sandstone under different confining pressure

Serial number	Diameter (mm)	Length (mm)	σ_3 (MPa)	σ_1 (MPa)	Strain (10^{-3})	E (GPa)
C1	50.02	98.86	5	103.663	10.563	13.813
C2	50.06	100.40	10	114.277	9.388	14.874
C3	50.02	100.38	15	136.319	12.637	12.816
C4	50.02	99.70	20	138.507	13.832	12.166
C5	50.02	100.10	25	173.503	15.446	15.771

test, uniaxial compression test and triaxial compression test. The acoustic emission characteristics during the indirect tensile test, uniaxial compression test and triaxial compression test can be divided into three phases: (1) relative flat growth of the cumulative counts of acoustic emission, (2) rapid growth of the cumulative counts of acoustic emission and (3) spurt of the cumulative counts of acoustic emission. The first phase is particularly evident in the indirect tensile test: acoustic emission appears rarely, and the peak of acoustic emission counts is particularly small. However, acoustic emission appears frequently, and the peak of acoustic emission counts is particularly large during the triaxial compression test. This phenomenon is caused by confining pressure. In the second phase, the frequency of acoustics emission counts appears to gradually increase, and the peak count of acoustics emission gradually increases. In the third phase, the frequency of acoustic emission counts appears very high. Moreover, the peak of acoustic emission counts is larger than those in the first two phases. Especially on the eve of fracture, the acoustic emission count will appear a peak. The peak value of acoustic emission count of sandstone in the uniaxial compression is especially obvious.

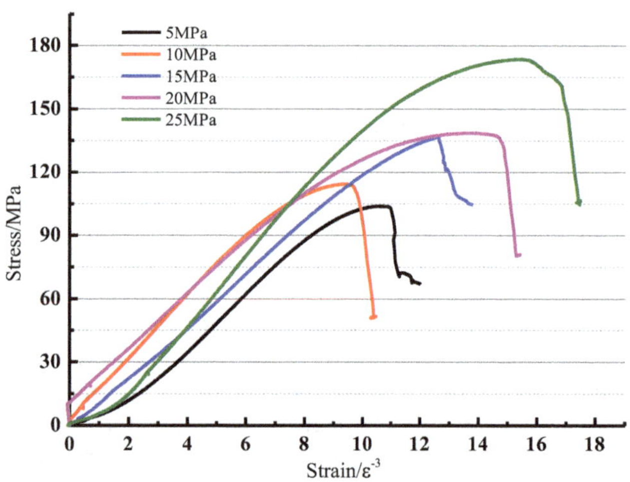

Fig. 6 Stress-strain curves of sandstone under different confining pressure

5.2 Analysis of the characteristics of acoustic emission waveform

In order to study the precursor information of acoustic emission during the fracturing of hard roof, the characteristics of the acoustic emission waveform in the indirect tensile test, uniaxial compression test and triaxial compression test were investigated. It can be seen in Figs. 10, 11 that in the phases with relatively flat growth and rapid growth of cumulative counts, the intensity of the acoustic emission signal is relatively small. However, after entering the spurt phase, the signal intensity gradually increases. In the indirect tensile test, the increase of acoustic emission

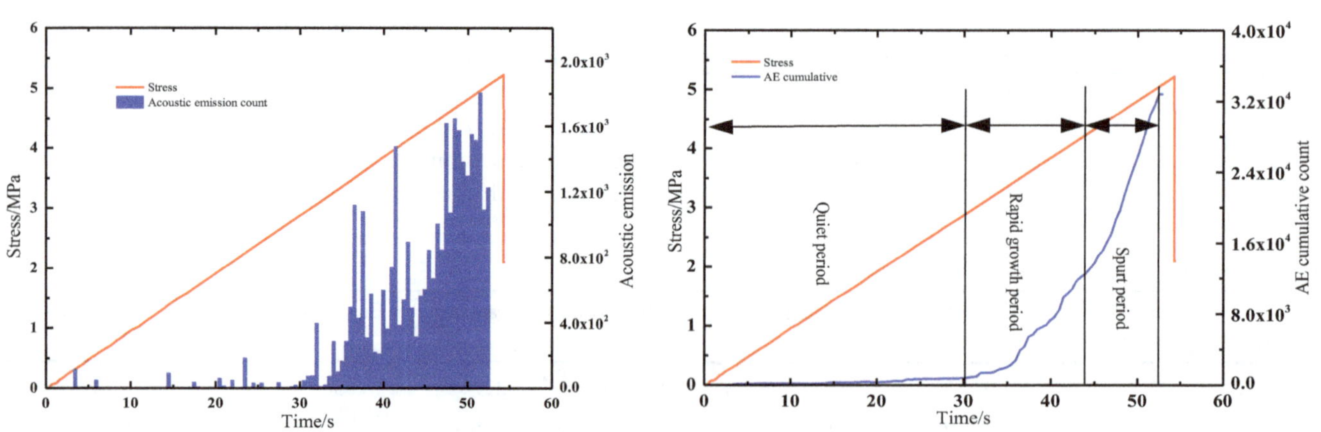

Fig. 7 Relationship among AE counts, cumulative counts, stress with time of sandstone in indirect tensile test

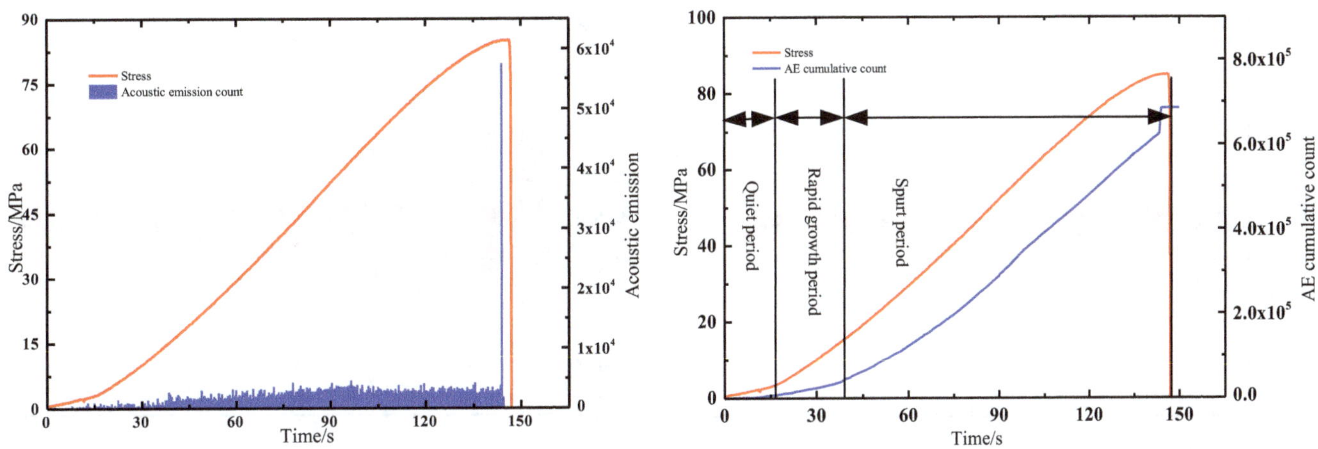

Fig. 8 Relationship among AE counts, cumulative counts, stress with time of sandstone in uniaxial compression test

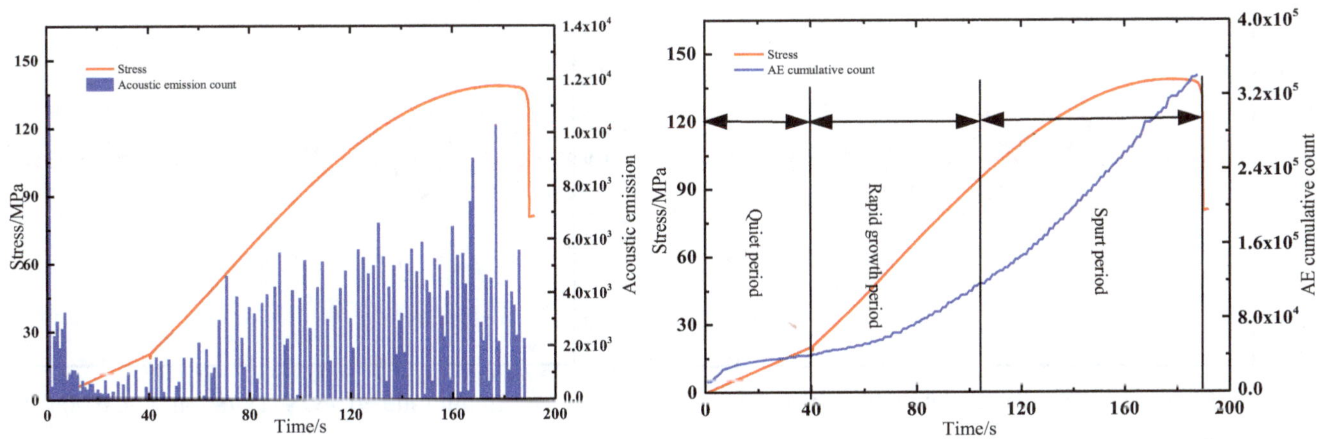

Fig. 9 Relationship among AE counts, cumulative counts, stress with time of sandstone in triaxial compression test

signal intensity is not obvious, but it is particularly obvious during the uniaxial compression test. The peak signal intensity is 10 times of those in the first two phases. In particular, right before the occurrence of fracture, the peak signal intensity is 25 times of those in the first two phases. Therefore, the signal intensity of the warning line can be set to be ±0.4 mV in the tensile fracture zone. When the absolute value of the signal intensity is more than 0.4 mV, early warning signal is sent out. The signal intensity of the warning line can be set to ±4 mV in the compression shear fracture zone. When the absolute value of the signal intensity is more than 4 mV, early warning signal is sent out.

5.3 Analysis of the characteristics of acoustic emission under different stress conditions

The acoustic emission characteristics vary under different stress conditions, and knowledge about the pattern of variation may help to find proper precursor information. The whole process of acoustic emission count is divided into 10 stages to analyze the acoustic emission characteristics of roof sandstone under different stress stages, the stress of 0%–10%, 10%–20%, 20%–30%, 30%–40%, 40%–50%, 50%–60%, 60%–70%, 70%–80%, 80%–90% and 90%–100%, respectively. Table 6 and Fig. 12 show that the acoustic emission count of roof sandstone has great difference under different stress conditions. In the indirect tensile test, before 60% peak tensile strength is reached, the frequency of acoustic emission count appears very low. The range of acoustic emission counts is 0.15%–1.99%. In the uniaxial compression test, before 60% peak tensile strength is reached, the number of acoustic emission counts is less than 10% of the whole process. The range of acoustic emission counts is 3.04%–9.65%. In the triaxial compression test, the frequency of acoustic emission counts appears very high in the first stage. However, from the second stage to the 80% peak stress, the number of acoustic emission counts is less than 10% of the whole process.

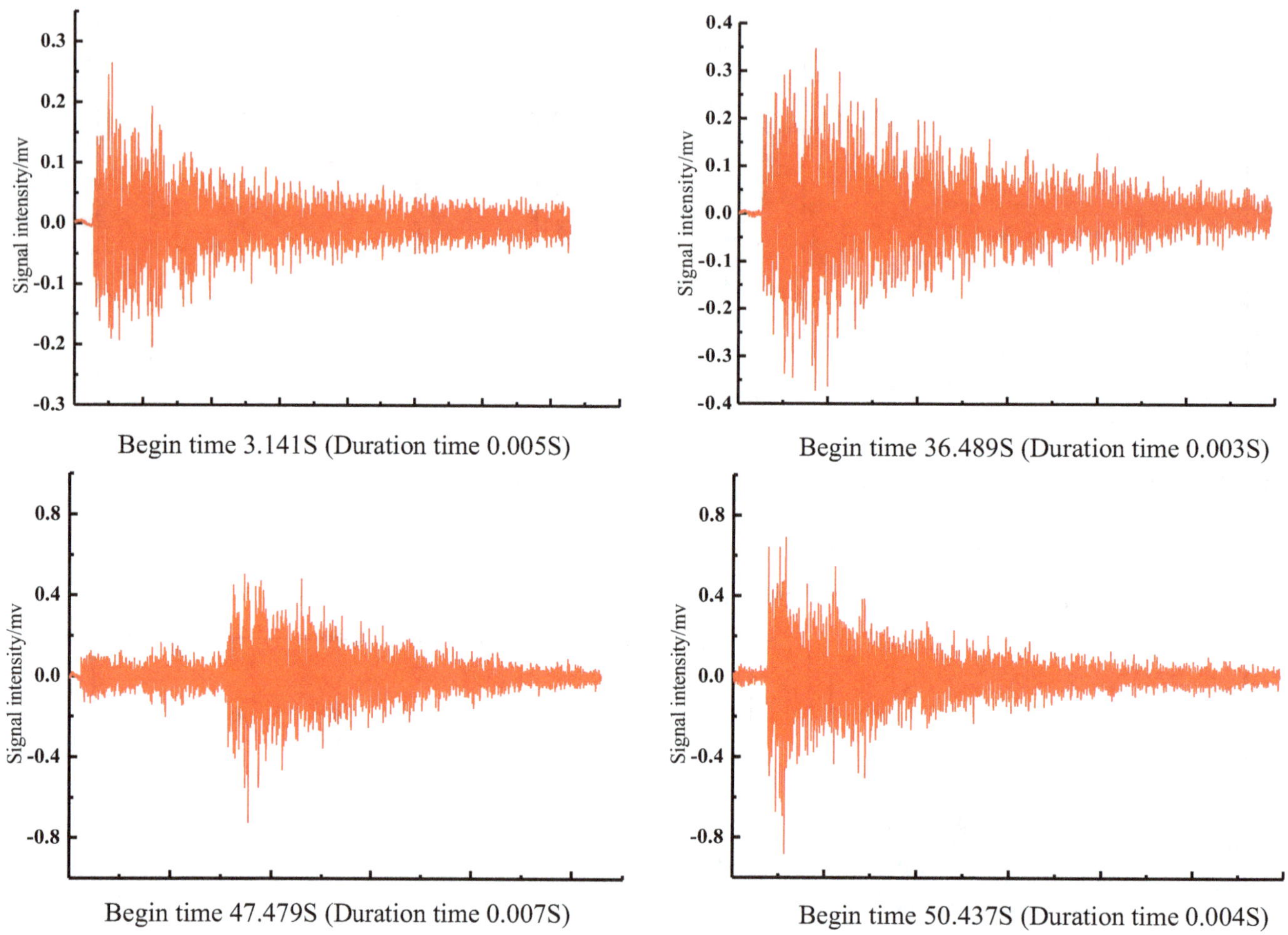

Fig. 10 Waveform characteristics of acoustic emission in different stages of roof sandstone under the indirect tensile test

6 Engineering significance

Effective monitoring, prediction, prevention and control measures of roof dynamic disasters have always been a hot and difficult research topic. At present, displacement and pressure monitoring are commonly used to alert the dynamic disaster of roof. The operation method is simple and easy to master, but the method needs to consume a lot of manpower and material resources. Moreover, continuous and real-time monitoring can not be realized. However, the characteristics of rock mass structures can be reflected in the number of acoustic emission counts, cumulative counts and waveform characteristics in the fracturing process of rock mass. Continuous and real-time monitoring can be carried out by using the AE inspect. Therefore, acoustic emission monitoring provides another way for the prevention of roof dynamic disasters. According to the analyses in Sects. 5.1, 5.2 and 5.3, the AE counts, cumulative counts, AE signal intensity and the acoustic emission cumulative counts under different stress conditions are closely related to the fracturing process of the roof sandstone. Therefore, a prediction system of roof dynamic disaster is established based on the analyses in Sects. 5.1, 5.2 and 5.3, and the technical route is shown in Fig. 13.

The monitoring and prevention technology of roof power disaster mainly includes the following steps:

(1) Reasonable selection of the location of the measuring point;
(2) To determine that the test point is in the tensile fracture zone or compressive shear failure zone;
(3) Based on the AE counts, determine whether the partition is reasonable:
(4) This is mainly based on the number of the acoustic emission counts and the frequency of occurrence to determine the partition is reasonable. If the acoustic emission count in the tensile fracture zone is basically greater than 1000, the division is not reasonable.
(5) Based on the monitoring of acoustic emission parameters, determine whether it is more than the warning value, if more than the warning value, send out an early warning signal.

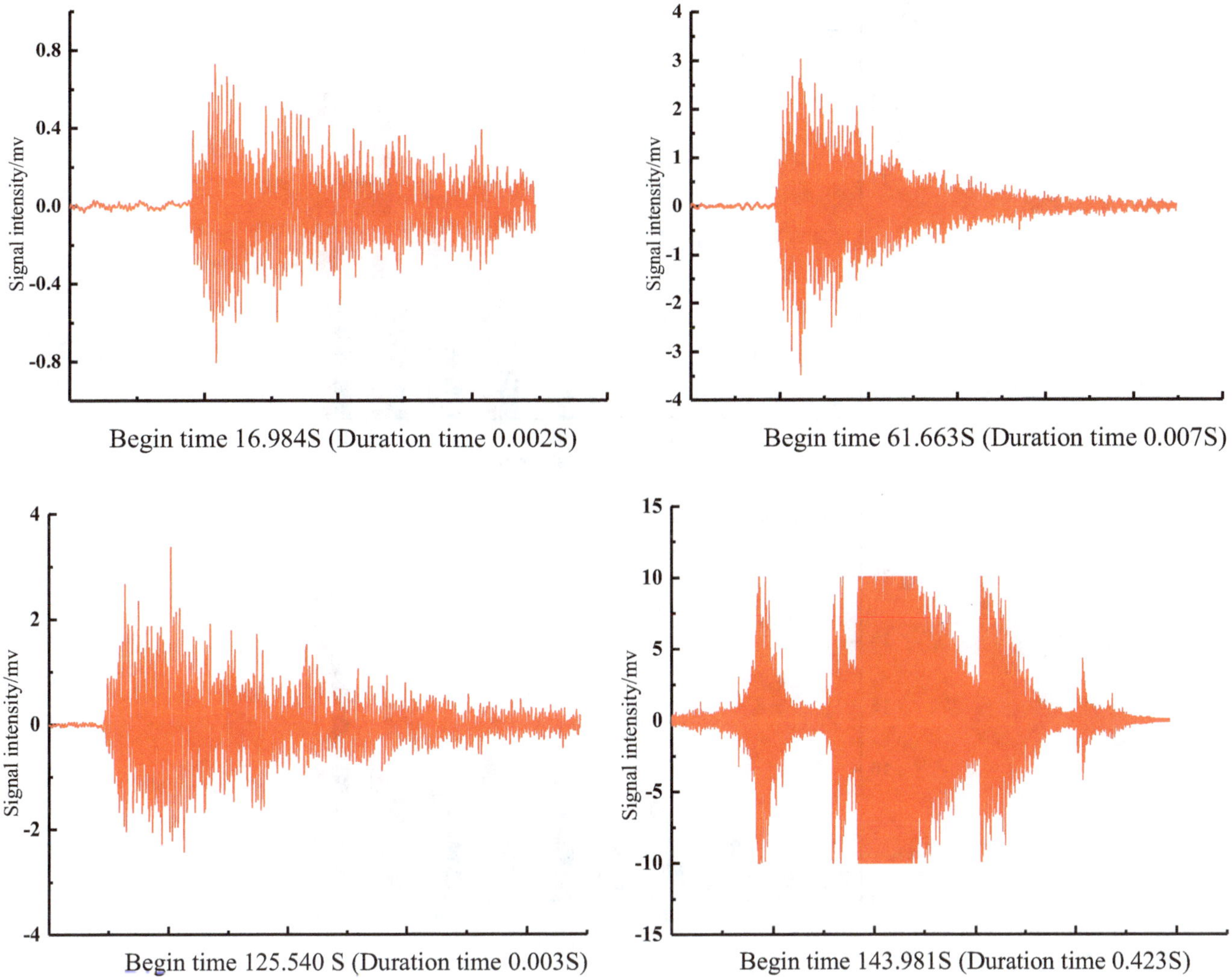

Fig. 11 Waveform characteristics of acoustic emission in different stages of roof sandstone under the uniaxial compression test

Table 6 Acoustic emission characteristics of sandstone under different stress stages

Stress phase (%)	Indirect tensile test		Uniaxial compression test		Triaxial compression test	
	AE count	Percent (%)	AE count	Percent (%)	AE count	Percent (%)
0–10	130	0.40	20875	3.04	36360	11.46
10–20	49	0.15	30923	4.50	8122	2.56
20–30	103	0.31	47347	6.90	5841	1.84
30–40	174	0.53	53075	7.73	9351	2.95
40–50	289	0.88	60937	8.88	15495	4.88
50–60	653	1.99	66284	9.65	21262	6.70
60–70	4359	13.27	77996	11.36	23524	7.41
70–80	6482	19.74	68875	10.03	26008	8.19
80–90	10054	30.62	77877	11.34	44543	14.03
90–100	10546	32.11	182360	26.56	126865	39.97

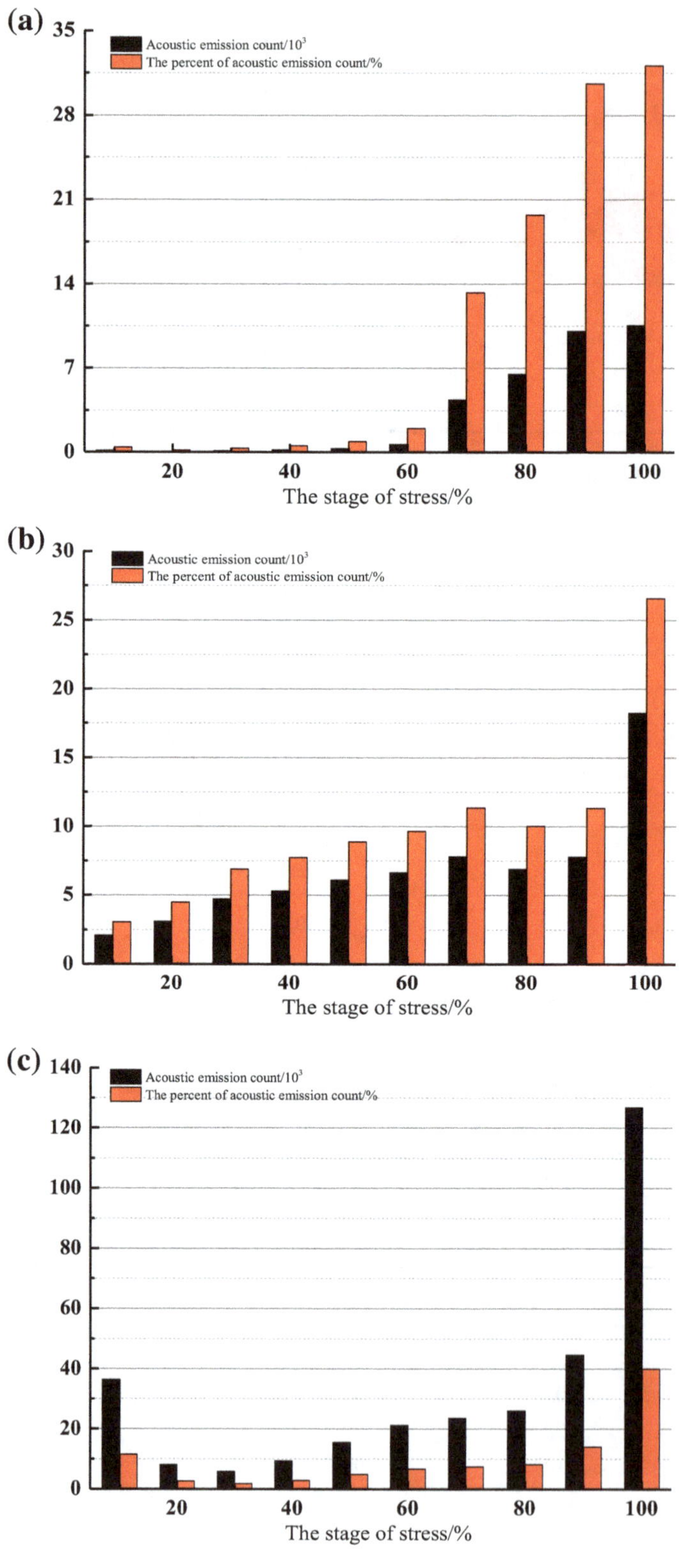

Fig. 12 Acoustic emission characteristics of sandstone under different stress stages. **a** Indirect tensile test, **b** uniaxial compression test, **c** triaxial compression test

(6) Determine whether the pre-warning information is reliable according to the early warning, field situation and other monitoring information. If you think the warning information is true and reliable, immediately start the prevention and control measures. If you think the warning information is not reliable

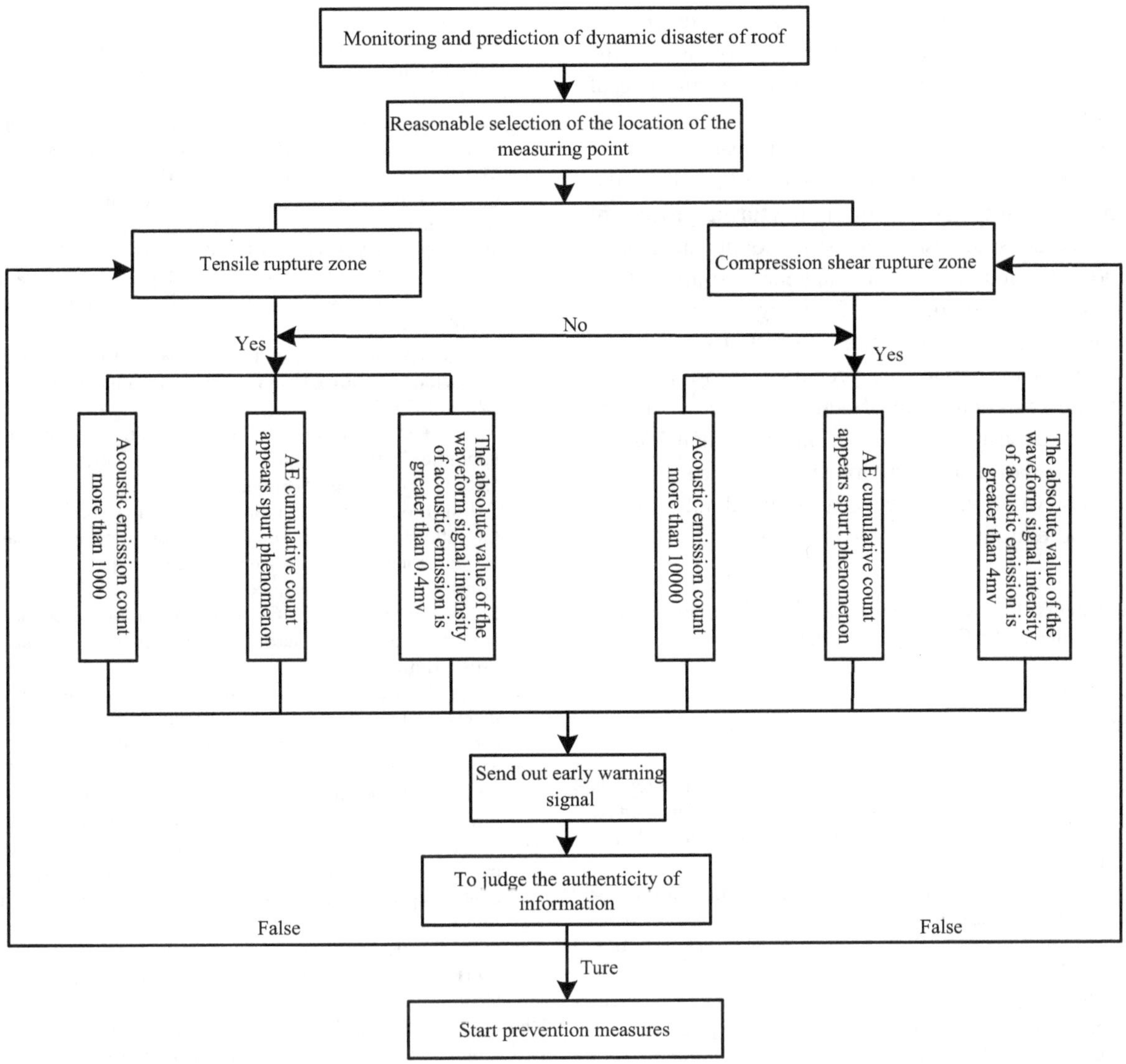

Fig. 13 Technical route of the predicting and forecasting on the thick hard roof dynamic disaster based on acoustic emission characteristics

after the judgment, release the pre-warning information and return to the monitor.

7 Conclusions

In order to explore the internal causes of dynamic disaster of thick hard roof sandstone and the prediction method of dynamic disaster, samples of thick hard roof sandstone were taken as the research object. The composition and microstructure of sandstone were investigated by using SEM and X-ray diffractometer. Finally, the mechanical and acoustic emission characteristics of sandstone were studied by using a RMT-150C mechanical testing machine and acoustic emission monitor. The result of analyses show that

(1) The tensile strength, uniaxial compressive strength, elastic modulus and Poisson's ratio of the roof sandstone are 4.825, 85.313 MPa, 13.814 GPa and 0.254, respectively. The cohesive strength and internal friction angle of the roof sandstone are 23.389 MPa and 32, respectively. The roof sandstone is a hard rock.

(2) The cumulative counts of AE for the roof sandstone in the indirect tensile test, uniaxial compression test and triaxial compression test can be divided into three phases including the relatively flat growth period, the rapid growth period and the spurt period.

(3) The intensities of AE signal in the tensile failure zone and in the compression shear failure zone have obvious difference. In the indirect tensile test, the signal intensity of acoustic emission waveform is

relatively low. However, the signal intensity of acoustic emission waveform is relatively large in the uniaxial compression test. Before the occurrence of fracture, the peak signal intensity of acoustic emission waveform is 0.8 and 10 mV, respectively.

(4) In the tensile fracture zone, the precursor information of acoustic emission in the fracturing process of roof includes acoustic emission count more than 1000, AE cumulative counts appear to spurt and the absolute value of the waveform signal intensity of acoustic emission is greater than 0.4 mv. In the tensile fracture zone, the precursor information of acoustic emission in the fracturing process of roof includes acoustic emission count more than 10,000, AE cumulative counts appear to spurt and the absolute value of the waveform signal intensity of acoustic emission is greater than 4 mV.

Acknowledgments This research is funded by the Chinese Natural Science Committee under Grant No. U1261207.

References

Chen TF, Xu JY, Liu S, Zhi LP (2013) Experimental study on ultrasonic and mechanical properties of sandstone influenced by water-saturation and high temperature. Chin J Undergr Space Eng 9(6):1236–1241

Deng HF, Yuan XF, Li JL, Luo Q, He M (2014) Fracture mechanics characteristics and deterioration mechanism of sandstone under reservoir immersion interaction. Earth Sci J China Univ Geosci 39(1):108–114

Deng HF, Zhou ML, Li JL, Hu YY, Xiao ZY, Wang Y (2016) Experimental research on unloading triaxial rheological mechanical properties of sandy mudstone. Rock Soil Mech 37(2):315–322

Fang JY, Dang FN, Xiao YT, Ding WH, Fang JX (2015) Quantitative study on the CT test process of siltstone under triaxial compression. Chin J Rock Mech Eng 34(10):1976–1984

Gao BB, Li HG, Li HM (2015a) Study on acoustic emission and fractal characteristics of different damage types of rock. Chin J Undergr Space Eng 11(2):358–363

Gao BB, Li HG, Li HM, Li L, Su CD (2015b) Acoustic emission and fractal characteristics of saturated coal samples in the failure process. J Min Saf Eng 32(4):665–670

Guo XF, Yin JM, Yan EC, Zhang LJ (2016) Mechanical property test at different unloading velocities of metamorphic quartz sandstone. J Shanghai Jiao Tong Univ 50(1):78–83

Han TC, Zhang J (2014) Numerical simulation of acoustic emission for defective rock. Chin J Rock Mech Eng 33(suppl):3198–3204

Li M (2014) Research on rupture mechanisms of coal measures sandstone under high temperature and impact load. China Univ Min Technol China, Xuzhou **(Chinese)**

Li DY, Qiu JD, Li XB (2015a) Experimental study on dynamic tensile and compressive properties of bedding sandstone under impact loading. Chin J Rock Mech Eng 34(10):2091–2097

Li HM, Li HG, Gao BB, Jiang DJ, Feng JF (2015b) Study of acoustic emission and mechanical characteristics of coal samples under different loading rates. Shock Vib 2015:1–11

Liu JP, Xu SD, Li YH, Dong LB, Wei J (2012) Studies of AE time-space evolution characteristics during failure process of rock specimens with prefabricated holes. Chin J Rock Mech Eng 31(12):2538–2547

Liu XL, Li XB, Hong L, Yin TB, Rao M (2015) Acoustic emission characteristics of rock under impact loading. J Cent South Univ 22:3571–3577

Sachan A, Penumadu D (2007) Identification of microfabric of kaolinite clay mineral using X-ray diffraction technique. Geotech Geol Eng 25(6):603–616

Su HJ, Jing HW, Zhao HH, Wang YC (2014) Experimental study on the influence of longitudinal fissure on mechanics characteristic of sandstone. J Min Saf Eng 31(4):644–649

The Ministry of Water Resources of the People's Republic of China (2008) GB50487-2008 Code for Water Resources and Hydropower Engineering Geological Investigation. China Planning Press, Beijing

Wang SY, Sloan SW, Tang CA (2013) Three- dimensional numerical investigations of the failure mechanism of a rock disc with a central or eccentric hole. Rock Mech Rock Eng 47(6):1–21

Wasantha PLP, Ranjith PG, Zhao J, Shao SS, Permata G (2015) Strain rate effect on the mechanical behavior of sandstones with different grain sizes. Rock Mech Rock Eng 48:1883–1895

Yang SQ, Huang YH, Jing HW, Liu XL (2014) Discrete element modeling on fracture coalescence behavior of red sandstone containing two unparallel fissures under uniaxial compression. Eng Geol 178:28–48

Yang SQ, Jing HW, Cheng L (2015) Influences of pore pressure on short-term and creep mechanical behavior of red sandstone. Eng Geol 179:10–23

Yao QL, Li XH, Pan F, Weng T, Wang G (2016) Deformation and failure mechanism of roadway sensitive to stress disturbance and its zonal support technology. Shock Vib 2016:1–14

Zhang YB, Mao XB, Tao J, Li M (2014) Impact of strain rate on mechanical property of sandstone under impact load. Blasting 31(3):42–46

Zhang AB, Zhang YB, Liu XX, Zhang P (2015) Test study on water affected to physical mechanics performances of muddy siltstone. Coal Sci Technol 43(8):67–71

Zhang C, Zhang AB, Zhang YB, Liu XX, Chen L (2016) Acoustic emission and damage evolution of dry and water-saturated. J Liaoning Tech Univ (Nat Sci) 35(7):705–711

Zhou KP, Li B, Li JL, Deng HW, Bin F (2015) Microscopic damage and dynamic mechanical properties of rock under freeze-thaw environment. Trans Nonferrous Metals Soc China 25:1254–1261

Zhu TT, Jing HW, Su HJ, Ying Q (2015) Experimental investigation on mechanical behavior of sandstone with coupling effects under uniaxial compression. J China Coal Soc 40(7):1518–1525

9

Comparison of pore characteristics in the coal and shale reservoirs of Taiyuan Formation, Qinshui Basin, China

Yu Liu[1,2] · Yanming Zhu[1,2]

Abstract Coal and shale are both unconventional gas reservoirs. Comparison of pore characteristics in shale and coal would help understand organic pore structure in shale and investigate co-exploration of shale gas and coalbed methane in coal bearing strata. In this study, five shale samples and three coal samples of Taiyuan Formation were collected from Qinshui Basin, China. High pressure mercury injection, scanning electronic microscopy, and fractal theory have been used to compare pore characteristics in shale and coal. The results show that pore volumes in coal are much larger than that in shale, especially pores 3–100 nm. In coal, there are many semi-closed pores in micro pores (<10 nm) and transition pores (10–100 nm). On the contrary, micro pores and transition pores are mainly with open pores in shale. The fractal curves show that pores larger than 65 nm in coal and shale reservoir both have obvious self-similarity and the fractal dimension values in shale and coal are similar. But the fractal characteristics of pores smaller than 65 nm in shale reservoir are quite different from that in coal.

Keywords Comparison · Coal · Shale · Organic pores · Fractal

1 Introduction

Shale and coal seam are both unconventional gas reservoir for shale gas and coal bed methane. Shale gas and coalbed methane have the similar storage mechanism, and they are both mainly composed of free gas and adsorbed gas (Crosdale et al. 1998; Curtis 2002; Pan and Connell 2015; Pan and Wood 2015). Shale gas and coalbed methane are storied in pores, free gas staying in the center of pores and adsorbed gas adsorbed on the surface of pores (Zhang et al. 2012; Pan and Connell 2015). Pores in coal are all organic pores, while pores in shale having different types such as organic pores, interparticle pores and intraparticle pores (Loucks et al. 2012). Among them, the organic pores have a great significance in storage of shale gas (Loucks et al. 2009). Also the methane adsorption of shale is controlled by the organic matter (Zhang et al. 2012). Pore system in shale is more complex than that of coal because shales have both organic pores and inorganic pores (Loucks et al. 2012). For the heterogeneity of shale, it is difficult to study the organic pores and inorganic pores respectively. By comparing pores in shale and coal, it would help understand the organic pores characteristics in shale. Also in the marine- terrestrial sedimentary environment, organic shale is usually adjacent to the coal in the vertical. In view of this, many scholars put forward the idea of co-exploration of shale gas and coalbed methane in coal bearing strata (Zou et al. 2010; Qin et al. 2014). It is necessary to study the differences and similarity of pores in shale and coal by comparing them.

Benefit from exploration and development of coalbed methane and shale gas, studies on the pores in shale and

✉ Yu Liu
liuyu19901213@163.com

1 Key Laboratory of Coal Methane Resource & Reservoir Formation Process of Ministry of Education, China University of Mining & Technology, Xuzhou 221008, Jiangsu, China

2 School of Resource and Earth Science, China University of Mining & Technology, Xuzhou 221116, Jiangsu, China

coal have attracted much attention. In recent years, there have been many achievements on pores of coal and shale reservoir and the pore size distribution, pore types, fractal characteristics have been understood (Yao and Liu 2012; Xue et al. 2012; Clarkson et al. 2013; Zhang et al. 2014; Wang et al. 2014; Dong et al. 2015; Okolo et al. 2015; Pan et al. 2015; Sakhaee-Pour and Bryant 2015; Xiong et al. 2015). Coal and shale both have a wide pore size range, from nm scale to μm scale (Wang et al. 2014; Li et al. 2015). High pressure mercury injection experiment is one of the most powerful methods to characterize the pore structure in both shale and coal (Okolo et al. 2015). Also, scanning electron microscope (SEM) experiment makes the pores in shale and coal visual and brings great help for studies of pores.

Recently, fractal is commonly used to investigate the pore structure and surface irregularities of shale and coal (Mahamud and Novo 2008; ; Zhang et al. 2011; Pan et al. 2016; Li et al. 2015; Sun et al. 2015; Yang et al. 2014). Fractal theory is primary used to study nonlinear and irregular characteristics of shapes (Xie 1996; Pan et al. 2016). Mercury porosimetry and gas adsorption methods are the common methods to investigate the fractal feathers of pore structure of coal and shale (Bu et al. 2015; Yang et al. 2014). The investigation results of fractal methods on coal and shale showed that fractal theory is an effective tool to study pores structure in shale and coal (Bu et al. 2015; Pan et al. 2016), and the methane adsorption capacity of shale also has relations with the fractal dimension (Pan et al. 2016).

Few previous studies were focused on the comparison of pores in shale and coal reservoir, but it would help understand the organic pores in shale and be necessary in the study of co-exploration of shale gas and coalbed methane in coal bearing strata. Besides, pore structure of marine- terrestrial shale has not been understood clearly. In this study, we collected coal and shale samples from the coal bearing strata in Qinshui Basin, China. High pressure mercury and fractal analysis have been used to compare the pore characteristic in coal and shale reservoir.

2 Samples and experimental

2.1 Samples

Five shale samples and three coal samples were chosen from Qinshui Basin in China in the study (Fig. 1). Qinshui Basin is located in the southeast of Shanxi Province and it is the most successful area for coalbed methane exploitation in China (Fig. 1). Qinshui Basin is a synclinore, in which fault is not developed.

The Taiyuan Formation coal and shale in Qinshui Basin are both formed in the marine-continental transitional environment and they have the same type of kerogen (III). But there is a big difference between coal and shale in the composition of the material content. Clay content is the highest in shale, with average of 52.4 %, followed by the quartz content. TOC of the shale is mainly within 1.5 %–5.0 %. In coal, Organic matter is main component, and the content is above 90 %.

All of the shale samples are core samples and all of the coal samples were collected on the mining working face. The sample information is shown in Table 1. The shale and coal samples are all in stage of maturity to over-maturity.

2.2 Experiment

2.2.1 High pressure mercury injection experiment

Because of the nonwetting between mercury and coal, pressure is needed to make mercury into pores. The higher the pressure is, the smaller of the pore size which the mercury could enter will be. So we can measure the amount of mercury of a certain pressure to get the volume of corresponding pore size. The model of high pressure mercury injection experimental instrument is Auto Pore IV 9500 V1.09.

As the upper limit of pressure value is 6000 psi, the lower limit value of the measured pore size is 3.0 nm. The experimental samples are all the bulk samples, about 0.1–1 cm^3 in size. The quality of coal samples is 1.5–2.5 g, that of shale samples being 3–4 g. Samples need to be dried at 70 °C for more than 5 h before the experiment.

2.2.2 Scanning electron microscope (SEM)

Coal and shale samples for scanning electron microscope experiment are bulk samples. The SEM experiment was performed in Chinese University of Petroleum (Beijing). The experimental instrument is FEI Quanta 200. Some of the shale samples are polished by focused ion beam (FIB-SEM), while coal and some shale samples are not. Samples have been prepared by gold spray treatment before the experiment. When observing under scanning electron microscope, energy dispersive spectrometer (EDS) has been combined to judge the pore types.

2.2.3 Pore classification

In this study, we used the classification classified by Fu (2001): micropores <10 nm, transition pore being 10–100 nm, mesopores being 100–1000 nm, macropores are larger than 1000 nm.

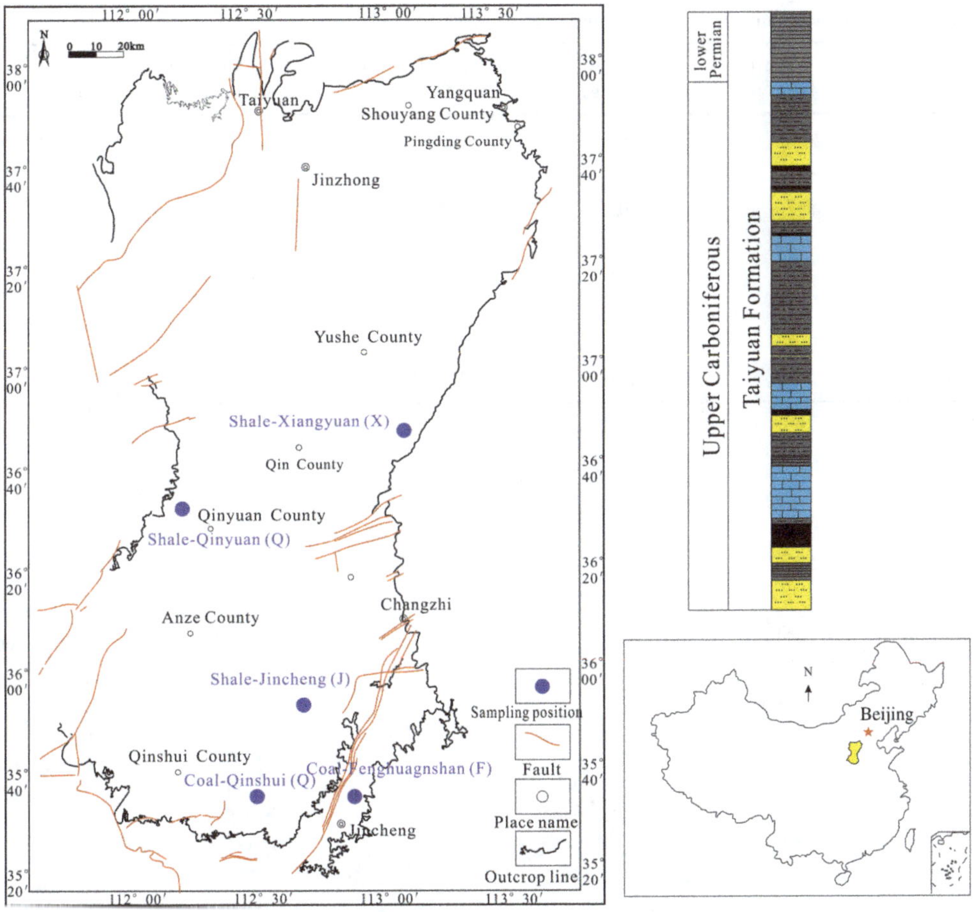

Fig. 1 The sampling position and the typical stratum histogram of Taiyuan Formation in Qinshui

Table 1 Sample information

Sample number	Lithology	TOC (%)	*Ro*	Kerogen type	Clay content (%)	Quartz content (%)
J-1	Shale	3.44	2.92	III	51	29
Q-1	Shale	1.72	–	III	58	30
Q-2	Shale	2.12	1.72	III	43	30
X-1	Shale	3.84	1.91	III	47	28
X-2	Shale	2.58	–	III	–	–
F-1	Coal	–	3.3 (Liu 2007)	III	–	–
F-2	Coal	–	3.3 (Liu 2007)	III	–	–
Q-1	Coal	–	2.4–2.8	III	–	–

Note The reflectance of vitrinite in coal sample Q-1 is from region data

3 Results and discussion

The high pressure mercury injection experimental results show that the porosity of coal samples in Qinshui Basin is 5.76 % on average, significantly larger than that of shale samples (1.65 %) (Table 2). Coal reservoir and shale reservoir of Taiyuan Formation in Qinshui Basin are both mainly with pores smaller than 100 nm (Fig. 2). Pores whose throat is 1000–10,000 nm are not developed in shale and coal reservoir. Mesopores (100–1000 nm) is well developed in coal reservoirs, but these pores are not developed in shale. Figure 2 also shows that in the stage of 3–100 nm, the number of pores in coal is significantly larger than that in shale. That is because organic pores are

Table 2 Data of high pressure mercury injection of shale and coal

Sample number	Lithology	Porosity (%)	Volume (mL/g)	Specific area (m^2)	Median pore diameter (volume) (nm)	Average pore diameter (nm)
F-1	Coal	4.66	0.04	19.67	7.5	7.2
F-2	Coal	6.84	0.06	27.29	9.2	8.1
Q-1	Coal	5.79	0.05	23.09	10	8.4
J-1	Shale	1.88	0.01	2.38	191.1	15
Q-1	Shale	1.74	0.01	3.2	13.6	10.2
Q-2	Shale	1.17	0.01	1.55	26.2	13.7
X-1	Shale	2.41	0.01	6.03	7.9	7.7
X-2	Shale	1.4	0.01	2.34	16.4	10.9

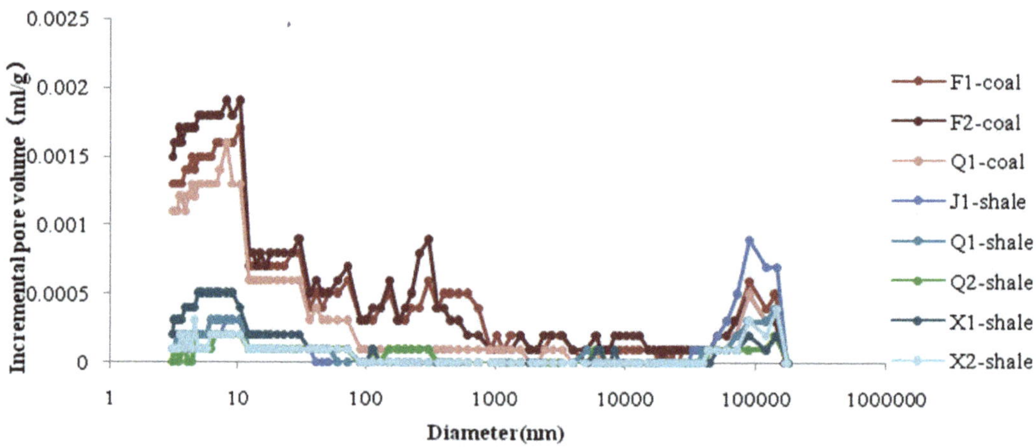

Fig. 2 Pore structure distribution of coal and shale in Qinshui Basin

mainly 3–100 nm, and organic matter content in coal is much larger than that in shale. The increase of organic matter in coal brings much more organic pores, and these pores make the porosity of coal samples much larger than that of shale.

It should be noted that when testing the nanometer pores, the test pressure would become larger than 30 MPa and the high pressure mercury injection experiment would overestimate the pore volume for the compressibility in coal and shale. Okolo et al. (2015) has studied effects of coal compressibility on the pore structure distribution from mercury injection experiment. It was showed that the pore volume would be overestimated by about 20 %–25 %. Also, the pores could be damaged as the test pressure becomes larger than 30 MPa. The effects of damage in pore and compressibility are similar to coal and shale samples. Considering these factors, it is also clear that the pore volume of coal samples is much larger than that of shale samples, especially for pores smaller than 100 nm.

Coal and shale have great differences in genetical types of pores. As the coal is the organ-enriched reservoir, absolute advantage of organic matter content makes that the pores in coal are mainly with organic pores. In the shale, clay, quartz and other minerals content accounted for the major. In the shale reservoir, pores larger than 100 nm are mainly mineral intergranular pores, solution pores and clay pores, and some organic pores (Fig. 3a, b, d).

The organic matter type and maturity in coal and shale of Taiyuan Formation in Qinshui Basin is similar. Large numbers of organic gas pores (mainly transition pores and micropores) exit in both the coal reservoir and shale reservoir. Those gas pores in coal and shale are very similar in pore size and shape (Fig. 3d, e, f, g). For the same quality of coal and shale, the content of organic matter in the coal reservoir is 20–30 times that in shale reservoir. Given the similarity of the gas pores in coal reservoir and shale reservoir, the porosity of the 3–100 nm organic pores in coal reservoir should be 20–30 times of organic pores in shale reservoir. But in fact, the total volume and surface area of 3–100 nm pores in coal reservoir is about 10 times of that in shale, and this reflects that there are certain amounts of inorganic pores in shale. By calculating, the

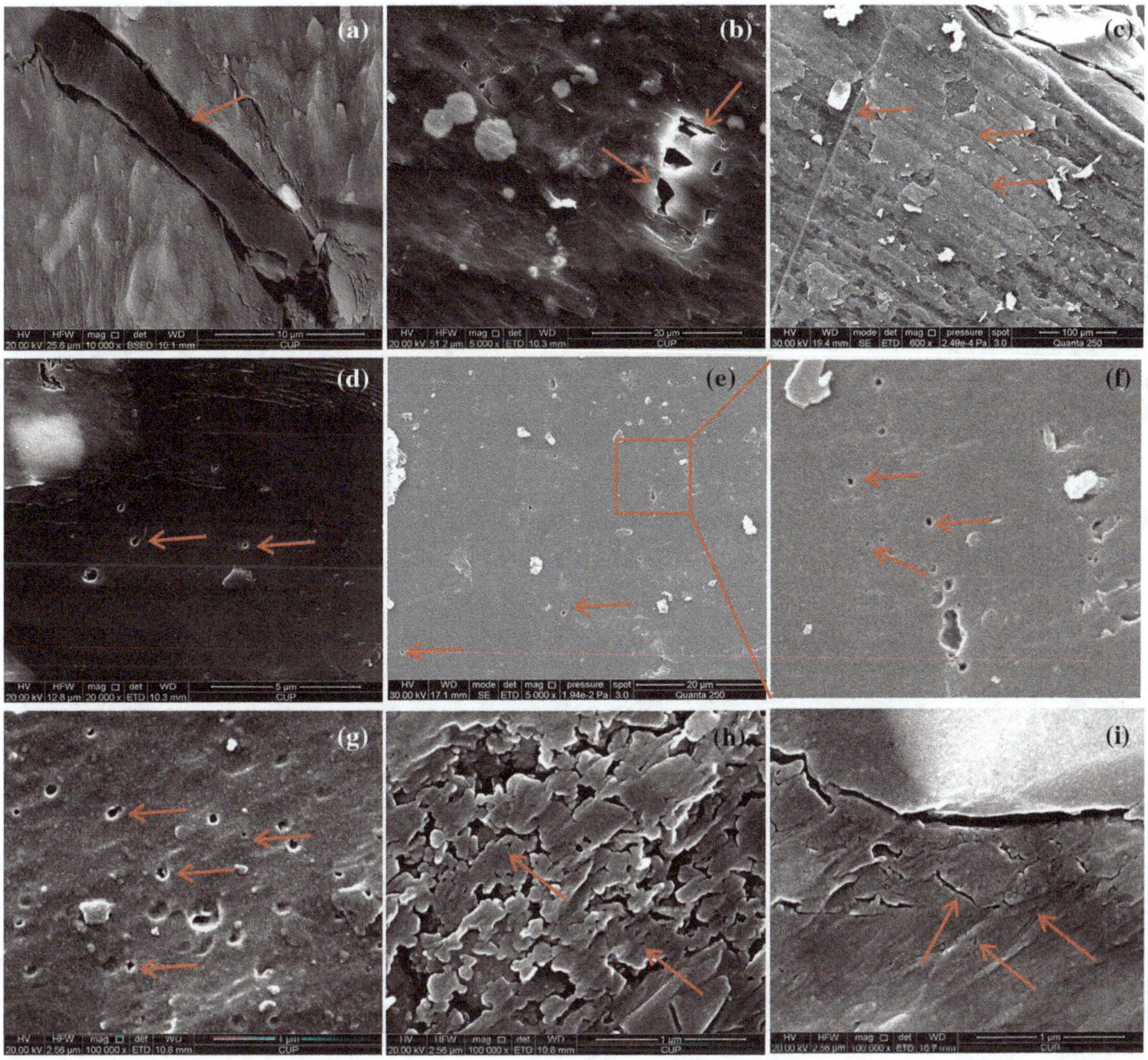

Fig. 3 Pore of shale reservoir in the Taiyuan Formation from Qinshui Basin. *a* shale, FIB-SEM, intergranular pores. *b* shale, FIB-SEM, mineral dissolution pores *c* coal, SEM, cleats are well developed in the coal reservoir. *d* shale, FIB-SEM, organic pores. *e* coal, SEM, organic gas pores, familiar to organic pores in shale showed on d. *f* coal, SEM, organic gas pores, magnified part of c. *g* shale, FIB-SEM, organic pores, familiar to organic gas pores in coal reservoir showed on f. *h* shale, FIB-SEM, clay pores. *i* shale, FIB-SEM, clay pores

organic pores accounted for about 30 %–50 % of total 3–100 nm pores. Also we can observe many clay pores which are smaller than 100 nm by using SEM (Fig. 3h, i).

3.1 Mercury injection curves and mercury withdrawal curves

Mercury intrusion curves and mercury evacuation curves are often used in previous studies to reflect the pore shape of the pores in coal and shale reservoir (Chen et al. 2013; Qin 1994). Figures 4 and 5 show the mercury injection curves and mercury withdrawal curves of the coal and shale samples.

The pore morphology would control the mercury injection- mercury withdrawal curves of the pores (Fig. 6). The differences in pore morphology of coal and shale samples would make the mercury injection-mercury withdrawal curves different. From Figs. 4 and 5, we can find that the mercury injection curves and mercury withdrawal curves of coal are quite different from that of shale, especially in the micro and transition pores stage. In micro and transition pores stage, mercury injection curve almost overlap mercury withdrawal curve in the coal reservoir, and this shows that there are a lot of semi-closed pres in the micro and transition pores (Qin 1994) (Fig. 6). But in the shale reservoir, the mercury injection curves are far from the mercury withdrawal curves (Fig. 5). At the beginning, the mercury injection curves show convex, and then have a sharp decline in 10 nm or so. It reflects that the pores at these stages in shale

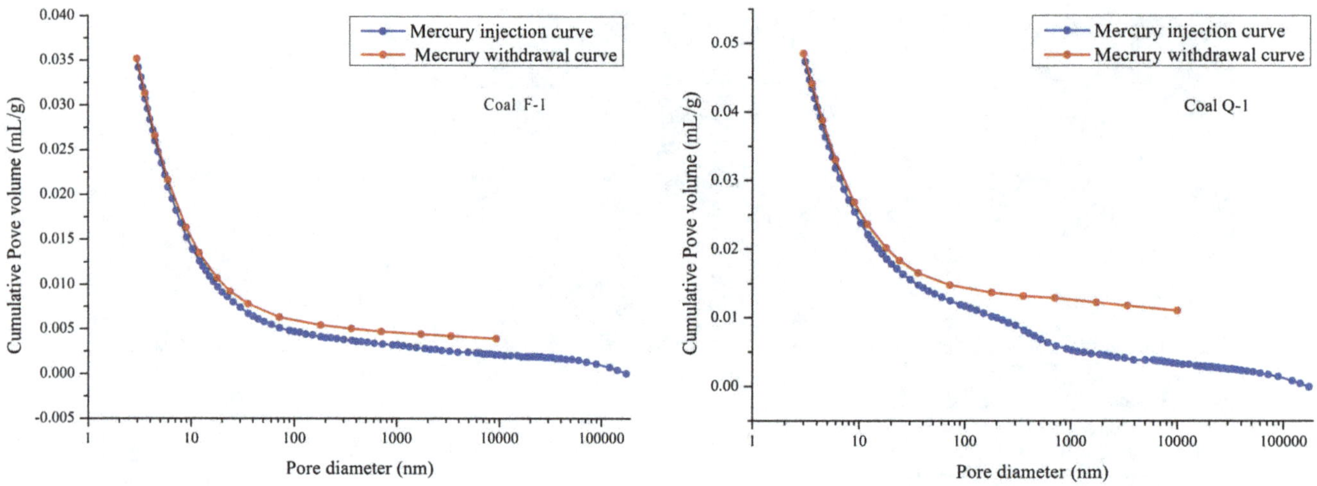

Fig. 4 Mercury injection curve and mercury withdrawal curve of coal samples

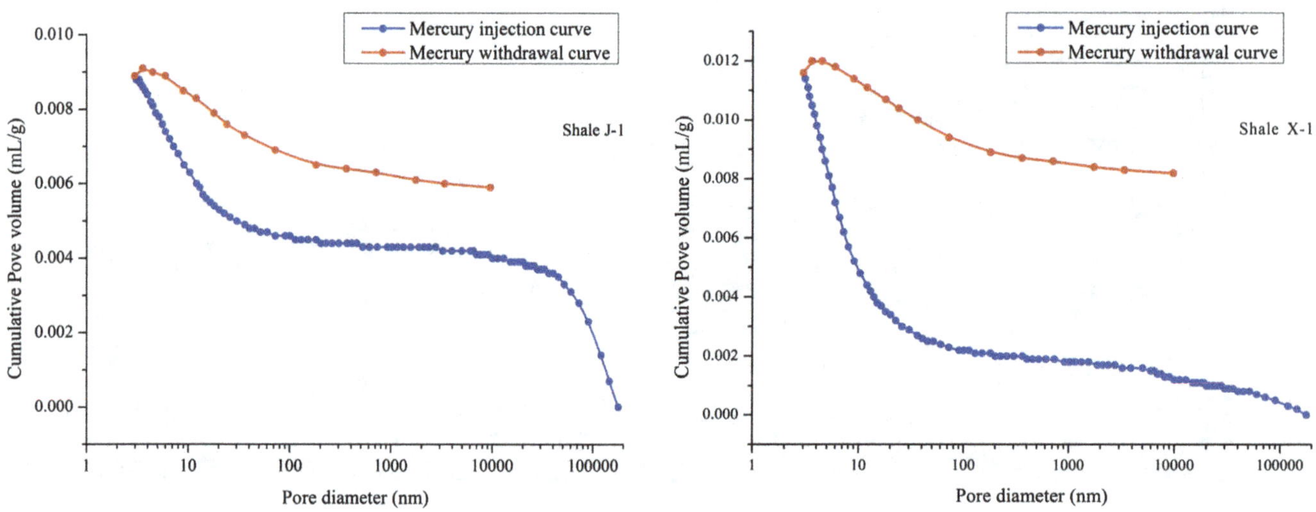

Fig. 5 Mercury injection curve and mercury withdrawal curve of shale samples

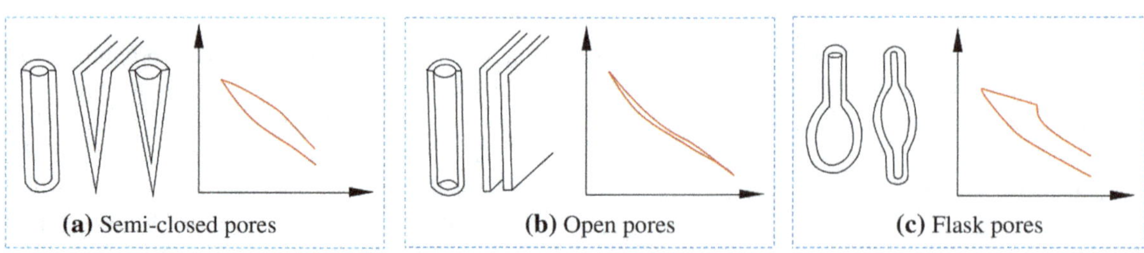

Fig. 6 Pore morphology and mercury injection-mercury withdrawal curves (Qin 1994)

reservoir are mainly open pores, and also have many "flask" pores around 10 nm (Qin 1994). At the stage of larger than 100 nm, the mercury withdrawal curves of coal reservoir and shale reservoir all have certain distance from the mercury injection curves.

3.2 Fractal characteristics of pores in coal and shale reservoir

Coal and shale are both heterogeneous porous reservoir. For the heterogeneity of the shale and coal, fractal theory is introduced to study the pore characteristics of coal and shale (Mahamud and Novo 2008; Yang et al. 2014). In this study, high pressure mercury injection data of shale and coal samples was used to investigate the fractal features. The model used in the study is Menger sponge model, and the specific process of deduction and calculation are as follows (Xie 1996):

Assuming the length of initial element cube unit is R, divide the initial cube into m^3 small same cubes, then select a rule to remove part of the small cubes, and the number of cubes left would be $N1$; repeat the above operation on the smaller cubes. Then the side length of cubes decreases, and the cube number increases. After the operation for K times, the side length of small cube $r = R/mk$, and the number of $N_k = N_1 k$, at the same time:

$$N_k = (R/r_k)^D = C_1/r_k^D = C_1\ r_k^{-D} \tag{1}$$

There into:

$$D = \lg N_1/\lg m,\ C_1 \text{ is a Constant} \tag{2}$$

$$V_k = N_k\ C_2\ r_k^3 = C_1\ C_2\ r_k^{3-D} \tag{3}$$

In the mercury injection experiment, pressure P (R) and aperture r satisfy the Washburn equation (Eq. 4):

$$P(r) = -2\sigma \cos\theta/r \tag{4}$$

In the formula, θ is mercury and solid medium contact angle (140 about) σ is the surface tension of mercury (0.48 N/m).In the mercury injection experiment, mercury injection volume under certain pressure (dVP (R)) is equal to the total pore volume (V (k)) at certain size:

$$dV(k) = dV_{P(r)} \tag{5}$$

$$dV_{P(r)}/dP(r) \propto [r/P(r)] \cdot r^{2-D} \tag{6}$$

$$D = 4 + lg(dV_p/dP(r))/lg(P(r)) \tag{7}$$

Through counting the mercury injection data for dVp $(R)/d$ $(P$ $(R))$ and dP (R), $lg(dVp$ $(R)/P$ $(R))$ − lg(dP (R)) diagram can be made, and the slope (K) of the curve can be calculated, and then the pore fractal dimension D would be obtained (Eq. (7)).

From Figs. 7 and 8 it could be found that the pore fractal curves of shale are similar to those of coal, and they all can be divided into two stages. But the boundary points of the two stages in shale and coal are different. The boundary points of coal reservoir fractal are in the interval of 54–85 nm, with an average of 65 nm, and that of shale reservoir pore fractal curve are smaller, 18–40 nm, with an average of 28 nm.

The pores larger than 65 nm in coal and shale reservoir both have obvious self-similarity: the degree of association between the pore fractal curves and the fitting line is larger than 0.9. The fractal dimension values of pores in coal are 2.93–3.14, and fractal dimension values of pores in shale are 2.99–3.12. The fractal dimension values of shale are similar to that of coal, and this reflects that surface roughness of pores larger than 65 nm in both coal and shale is very similar.

Based on the fractal principle, the calculated fractal dimension for this model can't be larger than 3, but some calculated results shows the fractal dimensions are larger than 3. The reasons of these phenomena have been discussed in previous studies (An et al. 2011; Wang and He 1996). On the one hand, as the pores measured by mercury injection experiment are all connected pores, closed pores in the reservoir cannot be measured. But the fractal model used in the calculation (Menger sponge model) contains not only connected pores, but also closed pores. Thus the fractal dimension calculated by using the mercury injection data has slight deviations compared to the practical dimension. On the other hand, in the mercury injection experiments, with decrease of test aperture, the mercury pressure increases. In this case, smaller pores would be overestimated for the compressibility of coal and shale samples. As fractal dimension value is very close to 3, slight deviations would make the fractal dimension calculated larger than 3.

The fractal characteristics of pores smaller than 65 nm in the coal reservoir are very different from that in shale reservoir. In the coal reservoirs, pores between 28 and 65 nm do not have self-similarity, but the self-similarity of

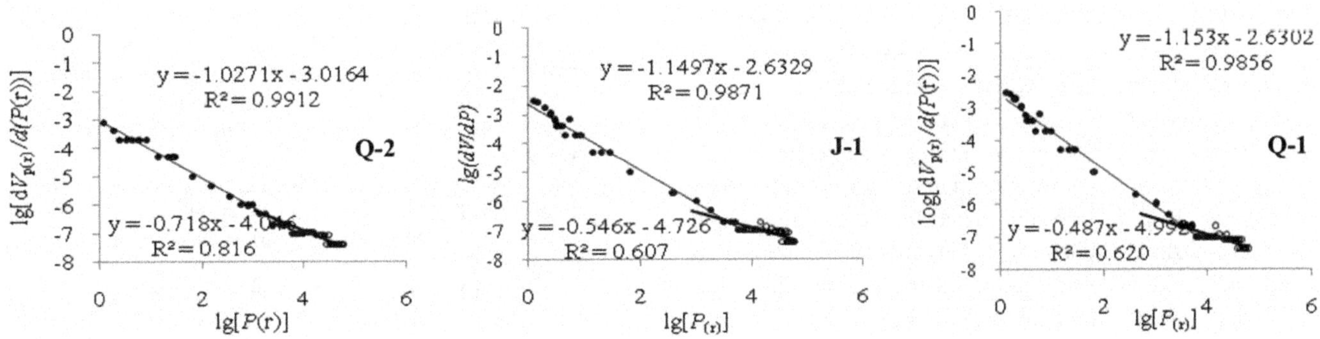

Fig. 7 Pore fractal curve of shale reservoir

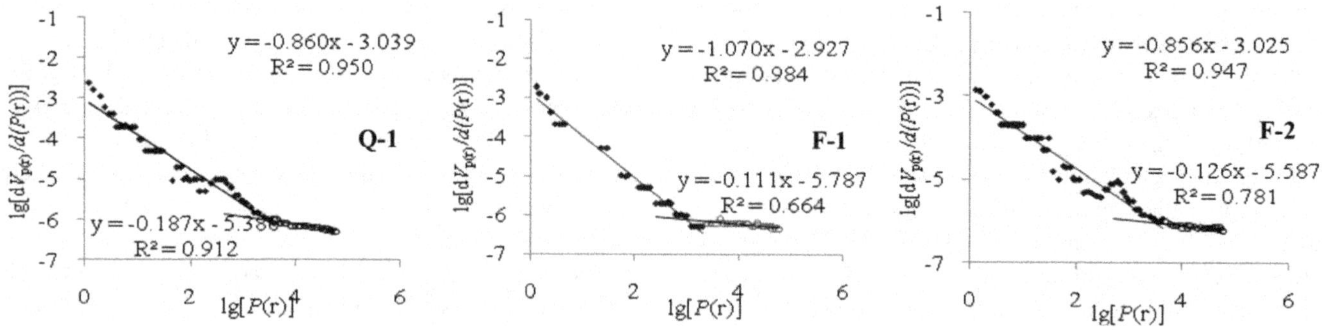

Fig. 8 Pore fractal curve of coal reservoir

shale pores in this stage is still obvious. This reflects that pores 28–65 nm in coal and shale reservoir have differences in shape. When the pore size is smaller than 28 nm, the fractal curves of shale are similar to that of coal samples. The reasons are that although the pores of the 28–65 nm in shale reservoir have a lot of organic pores, but inorganic pores still occupies a large part of the total pores. With the decrease in pore size, the proportion of organic pores increases continuously. When the pores size is smaller than 28 nm, pores in shale are mainly with organic pores, and the fractal curves of shale are similar to that of coal samples.

4 Conclusions

The porosity of coal samples from Qinshui Basin is 5.76 % on average, and it is significantly larger than that of shale samples (1.63 %). Pore structure distribution of coal and shale samples show that pores smaller than 100 nm in coal are far more than that in shale.

Mercury injection curves and mercury withdrawal curves of shale samples are quite different from that of coal samples. In micro and transition pores stage, mercury injection curve almost overlap mercury withdrawal curve in the coal reservoir. There are a lot of semi closed pores in micro pores and transition pores in coal, but in shale micro pores and transition pores are mainly with open pores.

Pores larger than 65 nm in coal and shale reservoir have similar fractal dimension values. But the fractal characteristics of pores smaller than 65 nm in coal are different from that in shale. That is because inorganic pores still occupy part of the pores smaller than 65 nm in shale. With the decrease in pore size, the proportion of organic pores increases in shale. When the pores size is smaller than 28 nm, pores in shale are mainly with organic pores, and the fractal curves of shale are similar to that of coal samples.

Acknowledgments The authors thank the National Science Foundation of China (41472135), the Research and Innovation Project for College Graduates of Jiangsu Province (KYLX15_1396), the Scientific Research Foundation of the Key Laboratory of Coalbed Methane Resources and Reservoir Formation Process, Ministry of Education (China University of Mining and Technology) (No. 2015-04) for the support of the research.

References

An S, Sang S, Li Y, Wang G (2011) Study on pore fractal characteristics of high-rank coal reservoirs in Southern Qinshui Basin. Coal Geol China 23(2):17–21

Bu H, Ju Y, Tan J, Wang G, Li X (2015) Fractal characteristics of pores in non-marine shales from the Huainan coalfield, eastern China. J Nat Gas Sci Eng 24:166–177

Chen S, Xia X, Qin Y, Wang Y (2013) Classification of pore structures in shale gas reservoir at the Longmaxi Formation in the south of Sichuan Basin. J China Coal Soc 38(5):760–765

Clarkson CR et al (2013) Pore structure characterization of North American shale gas reservoirs using USANS/SANS, gas adsorption, and mercury intrusion. Fuel 103:606–616

Crosdale PJ, Beamish BB, Valix M (1998) Coalbed methane sorption related to coal composition. Int J Coal Geol 35:147–158

Curtis JB (2002) Fractured shale-gas systems. AAPG Bull 86(11):1921

Dong Q, Zhang H, Zhu Z (2015) Evolution of structure properties during Zhundong coal pyrolysis. Procedia Eng 102:4–13

Fu X (2001) Physical simulation of multiphase medium coal rock properties and numerical simulation. University of Mining and Technology, Xuzhou

Li W, Liu H, Song X (2015) Multifractal analysis of Hg pore size distributions of tectonically deformed coals. Int J Coal Geol 144–145:138–152

Liu F (2007) The characteristics of coal reserviors and evaluation of coalbed methane enrichment and high-productivity in QinShui Basin of ShanXi Province. Chengdu University of Technology, Chengdu

Loucks RG, Reed RM, Ruppel SC, Jarvie DM (2009) Morphology, genesis, and distribution of nanometer-scale pores in siliceous mudstones of the Mississippian Barnett Shale. J Sediment Res 79(12):848–861

Loucks RG, Reed RM, Ruppel SC, Hammes U (2012) Spectrum of pore types and networks in mudrocks and a descriptive classification for matrix-related mudrock pores. AAPG Bull 96(6):1071–1098

Mahamud MM, Novo MF (2008) The use of fractal analysis in the textural characterization of coals. Fuel 87(2):222–231

Okolo GN, Everson RC, Neomagus HWJP, Roberts MJ, Sakurovs R (2015) Comparing the porosity and surface areas of coal as measured by gas adsorption, mercury intrusion and SAXS techniques. Fuel 141:293–304

Pan Z, Connell LD (2015) Reservoir simulation of free and adsorbed gas production from shale. J Nat Gas Sci Eng 22:359–370

Pan Z, Wood DA (2015) Coalbed methane (CBM) exploration, reservoir characterisation, production, and modelling: a collection of published research (2009–2015). J Nat Gas Sci Eng 26:1472–1484

Pan J, Zhu H, Hou Q, Wang H, Wang S (2015) Macromolecular and pore structures of Chinese tectonically deformed coal studied by atomic force microscopy. Fuel 139:94–101

Pan J et al (2016) Micro-pores and fractures of coals analysed by field emission scanning electron microscopy and fractal theory. Fuel 164:277–285

Qin Y (1994) Micropetrology and structural evolution of high-bank coals in P.R.China. China University of Mining and Technology press, Beijing

Qin Y, Liang J, Shen J, Liu Y, Wang C (2014) Gas logging shows and gas reservoir types in tight sandstones and shales from Southern Qinshui Basin. J China Coal Soc 39(8):1559–1565

Sakhaee-Pour A, Bryant SL (2015) Pore structure of shale. Fuel 143:467–475

Sun W, Feng Y, Jiang C, Chu W (2015) Fractal characterization and methane adsorption features of coal particles taken from shallow and deep coalmine layers. Fuel 155:7–13

Wang E, He X (1996) Fractal desription of coal seam porosity and crack system and its application. J Fuxin Min Inst (Nat Sci) 15(4):407–410

Wang Y, Zhu Y, Chen S, Li W (2014) Characteristics of the nanoscale pore structure in Northwestern Hunan Shale gas reservoirs using field emission scanning electron microscopy, high-pressure mercury intrusion, and gas adsorption. Energy Fuels 28(2):945–955

Xie H (1996) Introduction to fractal petrology. Science Press, Beijing

Xiong J, Liu X, Liang L (2015) Experimental study on the pore structure characteristics of the Upper Ordovician Wufeng Formation shale in the southwest portion of the Sichuan Basin, China. J Nat Gas Sci Eng 22:530–539

Xue G, Liu H, Li W (2012) Deformed coal types and pore characteristics in Hancheng coalmines in Eastern Weibei coalfields. Int J Min Sci Technol 22(5):681–686

Yang F, Ning Z, Liu H (2014) Fractal characteristics of shales from a shale gas reservoir in the Sichuan Basin, China. Fuel 115:378–384

Yao Y, Liu D (2012) Comparison of low-field NMR and mercury intrusion porosimetry in characterizing pore size distributions of coals. Fuel 95:152–158

Zhang S, Yang S, Cheng J, Zhang B, Lu C (2011) Study on relationships between coal fractal characteristics and coal and gas outburst. Procedia Eng 26:327–334

Zhang T, Ellis GS, Ruppel SC, Milliken K, Yang R (2012) Effect of organic-matter type and thermal maturity on methane adsorption in shale-gas systems. Org Geochem 47:120–131

Zhang S, Tang S, Tang D, Huang W, Pan Z (2014) Determining fractal dimensions of coal pores by FHH model: problems and effects. J Nat Gas Sci Eng 21:929–939

Zou C, Dong D, Wang S, Li J (2010) Geological characteristics and resource potential of shale gas in China. Pet Explor Dev 37(6):641–653

10

Modeling of time dependent subsidence for coal and ore deposits

Ryszard Hejmanowski[1]

Abstract Coal and ore underground mining generates subsidence and deformation of the land surface. Those deformations may cause damage to buildings and infrastructures. The environmental impact of subsidence will not be accepted in the future by the society in many countries. Especially there, where the mining regions are densely urbanized, the acceptance of the ground deformations decreases every year. The only solution is to limit the subsidence or its impact on the infrastructure. The first is not rentable for the mining industry, the second depends on the precise subsidence prediction and good preventing management involved in the mining areas. The precision of the subsidence prediction depends strictly on the mathematical model of the deformation phenomenon and on the uncertainty of the input data. The subsidence prediction in the geological conditions of the raw materials used to be made on the basis of numerical modeling or the stochastic models. A modified solution of the stochastic model by Knothe will be presented in the paper. The author focuses on the precise description of the deposit shape and on the time dependent displacements of the rock mass. A two parameters' time function has been introduced in the algorithm.

Keywords Subsidence prediction · Influence function · Modeling in time

1 Introduction

Specialists dealing with the protection of mining areas have recently discussed the issue of reliability and higher accuracy of deformation occurrence predictions (Kowalski 2005; Niedojadło 2008; Hejmanowski and Malinowska 2009). Such considerations stem from the increasing financial and social costs of damage done by this type of mining activity. The reliability issue is mainly connected with the quality of geologic-mining data (Naworyta and Sroka 2005; Stoch 2005), whereas higher accuracy of predictions may be related with the appropriate model of deposit and accuracy of the estimated parameters of the calculation model. Bearing in mind the popularity of Knothe theory and its common applicability in the Polish mining, the discussion may be focused on model parameters a and r (or interchangeably "tgβ"), and also additionally horizontal displacement parameter B (Budryk 1953) and time coefficient c (Knothe 1953). Over the years of study of deformations and predictions of their occurrence, the author noticed that deposit models used for calculations are frequently simplified (Knothe 2005), which results in considerable inaccuracies of the obtained results. The discrepancies of predictions and the actual deformations on the surface clearly indicate that such simplifications should never be made. The influence of a simplified description of deposit on the accuracy of model calculations is presented in this paper. An exemplary model employing author's prediction system MODEZ (Hejmanowski and Kwinta 2009) is also presented. This program can be used for accurate modeling of deposition in a reservoir and the mining process itself.

✉ Ryszard Hejmanowski
hejman@agh.edu.pl

[1] AGH University of Science and Technology, Cracow, Poland

2 Classic modeling of displacements in a function of time

The most popular model of rock mass displacement in Poland and in many other countries all over the world is Knothe model worked out in 1953 (Knothe 1953), called the influence function model or stochastic model.

In this case, the vertical displacements of any point in the rock mass or on the surface can be modeled with this dependence:

$$S(x,y,z) = \frac{ag}{r_z^2}\iint \varphi(x,y)\mathrm{d}x\mathrm{d}y \tag{1}$$

where $\varphi(x,y)$ is the influence function in Knothe model, which will be simplified to the form: $\varphi(x,y) = \mathrm{e}^{-\pi\frac{x^2+y^2}{r_z^2}}$, for a point located at the origin of the coordinates system, a is the extraction coefficient which defines the way in which mining voids should be filled; g is the thickness of the deposit under production; and r_z is the radius of major influences in Knothe model, at a level z over the mined deposit. For $z = H$, this will be a dependence of the angle of major impact zone β and depth of extraction H (Fig. 1):

$$r(H) = H\mathrm{ctg}\beta \tag{2}$$

and then

$$r_z = r_H\left(\frac{z}{H}\right)^n \tag{3}$$

where H is the depth of extraction (m); β is the angle of major impact zone (grad), and n is the rock mass parameter, characterizing propagation of deformations in the space above the deposit. The n value does not matter for the terrain surface, though according to many authors $n < 1, 0$ (Krzysztoń 1965; Drzęźla 1979; Preuße 1990).

This model (1) was originally worked out for predicting static states of vertical displacements, however, later Knothe supplemented it by the function of time. This function was written in the form of a linear differential equation describing the subsidence rate as a function difference of final and momentary subsidence:

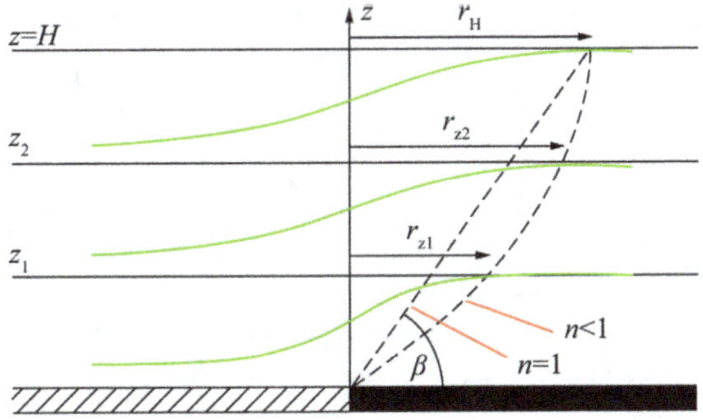

Fig. 1 Mining impact dispersed in the rock mass

$$\frac{\delta S(t)}{\delta t} = c[S_K(t) - S(t)] \tag{4}$$

where $S_k(t)$ is the subsidence of a point when the mining has been stopped at a moment t; $S(t)$ is the actual subsidence of a point at a moment t, and c is the time parameter-proportionality coefficient in Eq. (4), characterizing delaying properties of the rock mass as far as mining impacts effects appearing in time [1/year] are concerned.

A general solution of (4) is (5):

$$S(t) = \exp(-ct)\int_0^t \exp(-c\tau)S_K(\tau)c\mathrm{d}\tau \tag{5}$$

This model well describes subsidence of the analyzed point of the terrain surface after the mining operations on a given mining field (or fragment of the lot) are finished. It can be hardly applied for the advancing longwall fronts, as in this case calculations should be made for fragments (strips) of the lot. The author of this model also noticed this regularity and suggested a solution for the advancing extraction front (Knothe 1984). A similar application of Knothe model for the calculation of dynamic subsidence of terrain was presented by a group of Chinese scientists (Hu et al. 2011).

If part of the deposit has been mined and depleted instantly (i.e. $S_K(t)$ = constants and $S(t)_{t=0} = 0$), the solution takes the form of Eq. (4):

$$S(t) = S_K(t)(1 - \mathrm{e}^{-ct}) \tag{6}$$

As observed earlier (Kwinta et al. 1996), this function has a drawback, i.e. the subsidence is described from the moment of $t = 0$, when the subsidence rate is the maximum. This shortcoming can be liquidated by simulating the advancing extraction front in the form of strips (for longwall coal extraction), or deposit elements in the form of a cuboid.

3 Numerical modeling of displacements in time-reservoir elements model

This method was used in the calculation model based on the division of reservoir into reservoir elements (Hejmanowski 1993). In this case, the subsidence of an arbitrary point caused by mining activity can be written as:

$$\Delta S_{zi,j}(d_{i,j}, t_i) = V_i(t)\varphi_z(d_{i,j}, H_i) \tag{7}$$

and mining-induced subsidence of the entire mining lot or part of the reservoir is a sum of elementary impacts:

$$S_{zj}(t) = \sum_{i=1}^{N}\Delta S_{zi,j}(d_{i,j}, t_i) \tag{8}$$

where N is the number (quantity) of exploitation elements; i is the number of a subsequent individual exploitation element; j is the number of the (individual) calculation point; $V_i(t)$ is the volume of the post-exploitation elementary subsidence bowl at the time t, due to mining of the layer volume in the ith element of exploitation; H_i is the local mining depth in (m); t is the time difference between exploitation date of every reservoir element and date of the prediction (years); $d_{i,j}$ is the (horizontal) distance between the point calculation, and the centre of the exploitation element (m); and $\varphi_z(d_{i,j}, H_i)$ is the influence function, given with the formulae:

$$\varphi_z(d_{i,j}, H_i) = \frac{1}{r_z^2}\exp\left(-\pi\frac{d_{i,j}^2}{r_z^2}\right) \tag{9}$$

The model of reservoir division is presented in figure (Fig. 2). The horizontal dimension (square) is adjusted to the distance of the reservoir from the calculation level, accounting for the accuracy issue. The size of such elements is considerable, i.e. length L usually ranges from 30 to 60 m for extraction depth to 1000 m and predictions made for the surface. The height of the element corresponds with the thickness of the mined reservoir g_i.

3.1 Modeling of geology and mining technology

The division of reservoirs into elements has a very important utility aspect. All attributes connecting with geologic conditions and technological parameters of mining, which are locally variable, can be attributed to elements. In this situation the reservoir element may bring about information about the reservoir, which is especially significant for the prediction of deformation occurrence, and also spatial analyses and for evaluation of mining hazards.

3.1.1 Deposit thickness variable

The average thickness of extracted deposit layer is a typical generalization assumed in the traditional prediction models. It is not very important for coal seams, though in the metal ore deposits the thickness may vary significantly (Fig. 3).

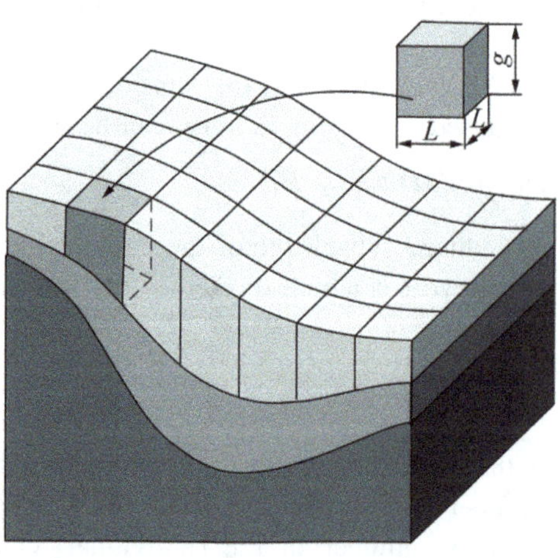

Fig. 2 Exemplary division of reservoir into elements

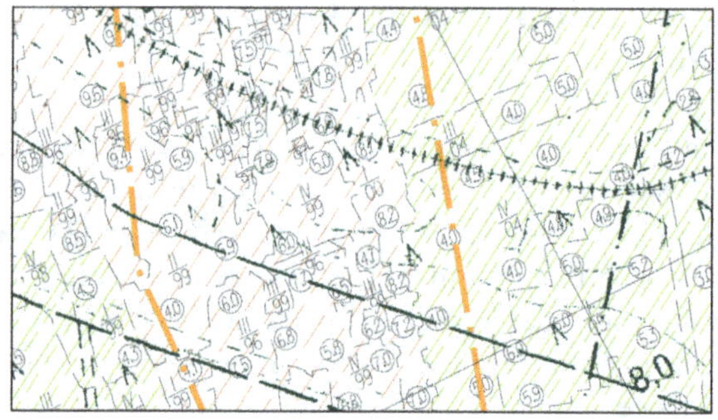

Fig. 3 Exemplary thickness of a copper ore layer [thickness given in (m)]

The reservoir element model allows including this attribute in displacement calculations for thus diversified thicknesses of mined layers (Fig. 4). Consequently, the picture of the summary subsidence basin will vary, depending on the varying thickness of deposit.

3.1.2 Mixed mining system

The accuracy of displacement predictions is known to depend on the correct parametrization of the mathematical model. The parameters should correspond to the local geologic-mining conditions. The extraction coefficient a and angle of major impact β were the most important parameters in Knothe model, on the basis of which the presented modified model was worked out. In view of the planned extraction system, the extraction coefficient can also be simulated for the locally varying conditions, e.g. hydraulic backfilling (Fig. 5). In the presented example the room and pillar system with roof deflection is predicted for extraction coefficient of *ca.* 0.6. The system including hydraulic backfill better minimizes displacements and the extraction coefficient usually ranges between 0.2 and 0.3.

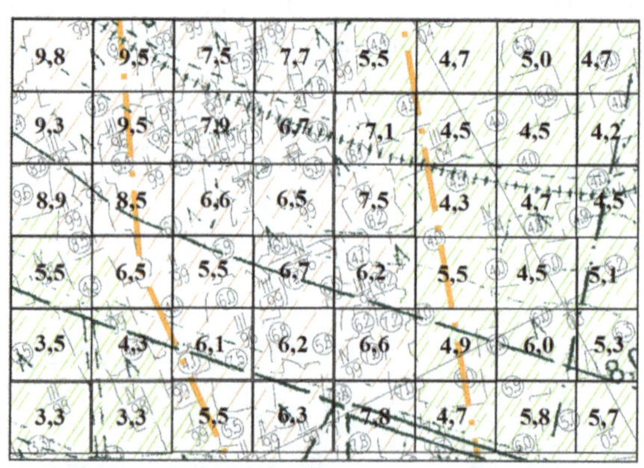

Fig. 4 Reservoir elements with 'reservoir thickness g_i' attribute in (m)

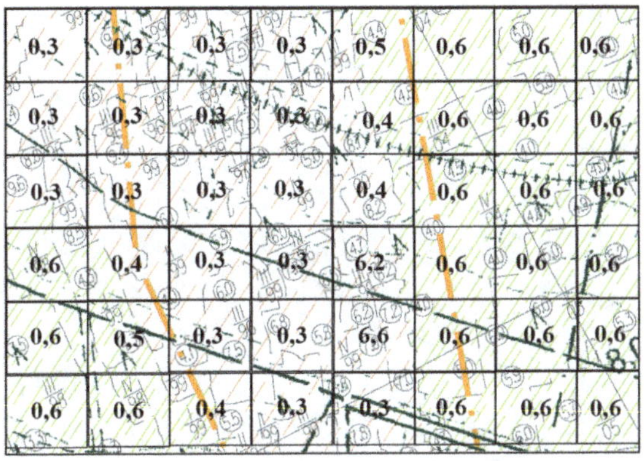

Fig. 5 Reservoir elements with 'extraction coefficient a_i' attribute

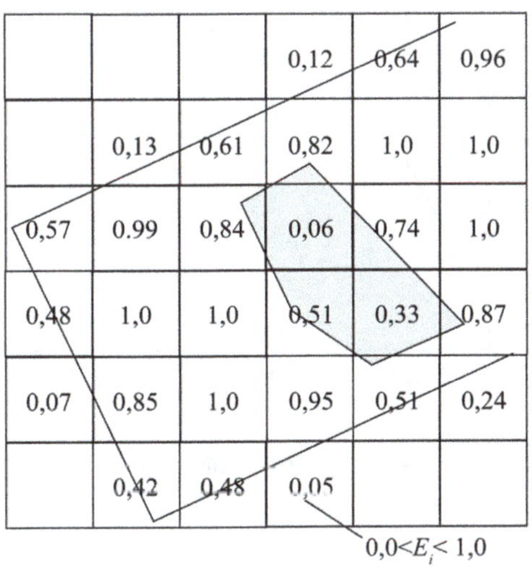

Fig. 6 Modeling low-mineral pillar in mining area (excluded from predictions)

3.1.3 Varying abundance of deposit and its modeling

The reservoir elements model allows for very accurate modeling of local changes in the ore resources thanks to which they can be selectively extracted. The influence of a given fragment of reservoir on the rock mass displacements can be modelled by introducing information about the surface of the planned mining area within the reservoir element as compared to the total reservoir element area. Figure 6 illustrates a reservoir pillar which will not be extracted. Numbers corresponding to the percentage of mined area of a given reservoir element are introduced in particular cells of the model. Thanks to this the volume of an arbitrary reservoir element, i.e. its influence on rock mass displacement, can be calculated from dependence (10):

$$V_i^E = E_i \cdot L_i^2 \cdot g_i \qquad (10)$$

where V_i^E is the volume of reservoir element to be mined; E_i is the ratio of surface of mining area to total surface in

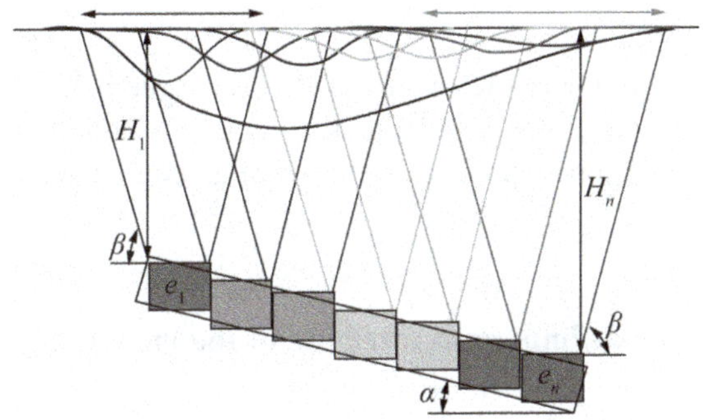

Fig. 7 Asymmetry of summaric basin for inclined deposit layers

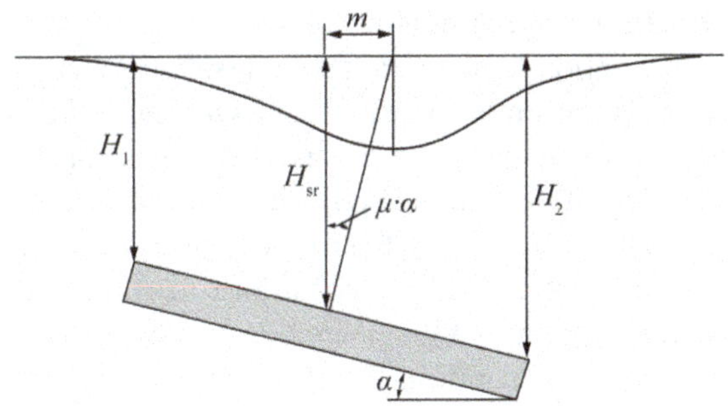

Fig. 8 Modeling of subsidence basin deviation caused by deposit inclination

a given reservoir element; L_i is the length of the edge of reservoir element; and g_i is the local thickness of deposit layer to be mined.

3.1.4 Tectonic disturbances and reservoir tilt

The biggest problem in modeling displacements is usually stratigraphy and tectonics of the reservoir formation. Unlike the presented model, classic models are frequently insufficiently flexible for modeling tectonic disturbances or variable reservoir tilt, e.g. within the coal longwall itself. Another advantage of the improved model is the local height datum as well as local direction and angle of dip of the deposit. Figure 7 reveals the way in which the difference of depth of deposition is accounted for. This affects the mining impact area of particular elementary basins on the surface. The summaric subsidence basin, being a sum of elementary basins, will show a visible asymmetry in the cross section parallel to the dip of deposit. This effect is observed in practice and no asymmetric influence functions need to be applied.

Additional information about the angle of inclination of deposit layers and the direction of the dip of the layer can be used for deviating the basin (m_i) i.e. shifting elementary basin towards the dip by a value calculated from dependence (11) (Fig. 8).

$$m_i = H_i \cdot \mathrm{tg}(\mu\alpha) \tag{11}$$

where α is the angle of dip of deposit layer or roof formation (grad), $0 < \alpha < 50$ grad; μ is the deviation coefficient determined empirically for local geologic conditions, $\mu < 1$.

3.2 Time as important attribute of the model

Most importantly, the division of a reservoir into elements allows one to very accurately describe the mining development in time, and so to simulate the advancement of the extraction longwall or development of the room and pillar field, even for varying mining rates.

The extraction time of a given reservoir element as compared to the time, for which the subsidence has been calculated, is a unique pair of variables (dates) in the calculation algorithms. An arbitrary function of time can be used for this pair of variables to obtain a result which would be closest to the real conditions of mining deformation propagation in a given area. In this way the course of time in which manifests itself the mining impact at a given calculation point will automatically account for the space and time conditions referred to in the model.

It seems obvious that the surface deformation rate reaches the maximum after some time, but begins from the preliminary phase when the distance of the extraction from the given point on the surface is relatively big. Thus Knothe time model (4–6) will not be a suitable tool for describing this type of displacements. A far better solution will be the two-parameter function of time proposed by Sroka and Schober (Schober and Sroka 1983). This function, whose properties were discussed in detail (Kwinta et al. 1996) relates the cause of deformation, i.e. convergence of working with the subsidence basin. The volume of elementary basin needed in the forecast of displacements (7) is expressed with the formula (12).

$$V_i(t) = V_i^E \cdot \theta(t) \tag{12}$$

where $\theta(t)$ is the function of time.

$$\theta(t) = 1 + \frac{\xi}{\vartheta - \xi} e^{-\vartheta t} - \frac{\vartheta}{\vartheta - \xi} e^{-\xi t} \tag{13}$$

where ξ is the convergence coefficient characterizing delaying properties of roof rocks in the workings area, (year^{-1}); ϑ is the coefficient characterizing delaying properties of the cap rock in displacement propagation, (year^{-1}).

These two time parameters characterize time in which appear all mining impacts on the surface it can be assumed that they are the same as time coefficient introduced by Knothe. Then:

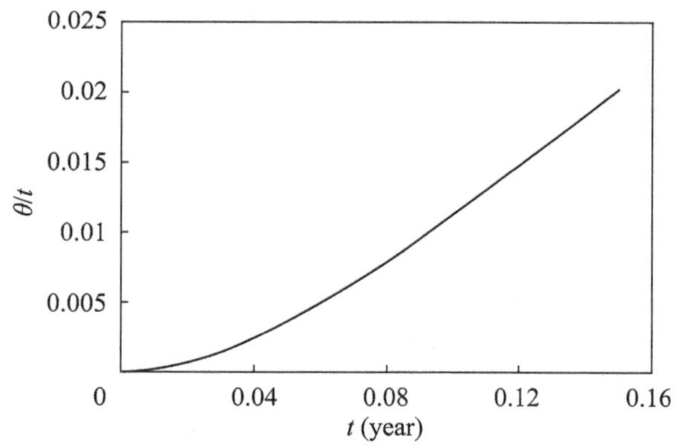

Fig. 9 Schober-Sroka two-parameter function of time (0–0.15 year)

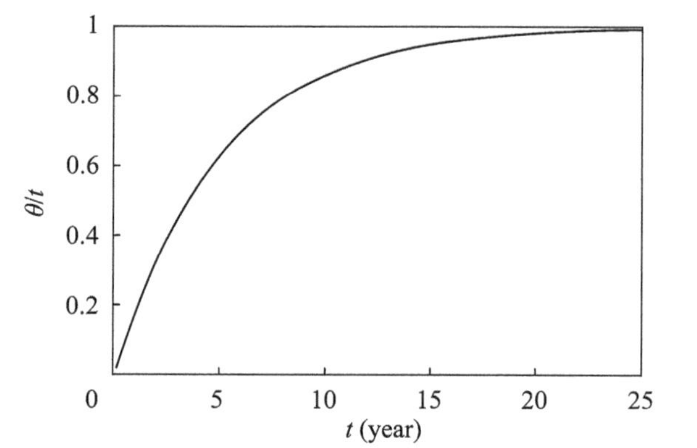

Fig. 10 Schober-Sroka two-parameter function of time (0.15–25.0 years)

$$\frac{1}{c} = \frac{1}{\vartheta} + \frac{1}{\xi} \tag{14}$$

For traditional hard coal and ore extraction, the values of these coefficients take the following values:

(1) For hard coal (goaf system), $15 < \xi < 50$;
(2) For copper ores (room and pillar), $0.2 < \xi < 1.5$, $1 < \vartheta < 15$.

Graphic characteristics of the presented function of time are presented for exemplary values of time coefficient in figures (Figs. 9, 10).

The function of time has zero rate and acceleration at the initial stage of movement. Then the displacement rate increases to the maximum in the linear phase of movement, delays, and the rate lowers at the vanishing movement stage. These properties of the function of time show to its adequacy to the actual conditions of displacement propagation in the rock mass as compared to Knothe function. Nonetheless the two-parameter function requires a much broader scope of studies to determine both these parameters. For the sake of determining the convergence coefficient, convergence measurements need to be performed in similar geologic and mining conditions, which is not always possible.

4 Concluding remarks

The problem of modeling displacements in the coal or ore rock mass shows considerable advantages of a model based on Knothe influence function model. Modifications in the article in Knothe model refer to the space and time. The division of reservoir has the following advantages:

(1) this model gives much better possibility of including local variability of geologic properties of the rock mass and deposit as compared to Knothe model,
(2) properties characteristic of the designed extraction system (backfill, working convergence rate, leaving off pillars and unmined deposit remains) can be accounted for,
(3) time-modeling of extraction in planned reservoir lots, accounting for delays in displacements propagation in the rock mass.

Utilization of this model and author's computer software MODEZ (Hejmanowski and Kwinta 2009; Hejmanowski and Zespol (2009)) makes prediction of the rock mass deformations and displacements much easier in the coal and copper ore mining conditions. The introduction of this software to thc mining practice will increase the accuracy of predictions.

References

Budryk W (1953) Wyznaczanie wielkości poziomych odkształceń terenu. Archiwum Górnictwa i Hutnictwa, t.1, z.1, Warszawa

Drzęźla B (1979) Zmienność zasięgu wpływów w górotworze, Przegląd Górniczy, nr 10, Katowice

Hejmanowski R (1993) Zur Vorausberechnung förderbedingter Bodensenkungen über Erdöl- und Erdgaslagerstätten. Praca doktorska, TU Clausthal

Hejmanowski R, Zespół I (2009) Dokładność prognoz deformacji powierzchni terenu górniczego "Lubin I" i "Małomice I", w aspekcie zmian dynamiki ruchów powierzchni w odniesieniu do prognoz krótko- i długookresowych. Praca nauk-badawcza na zlecenie KGHM Polska Miedź SA. (niepublik.)

Hejmanowski R, Kwinta A (2009) System prognozowania deformacji Modez. Materiały konferencji X Dni Miernictwa Górniczego i Ochrony Terenów Górniczych", Kraków

Hejmanowski R, Malinowska A (2009) Evaluation of reliability of subsidence prediction based on spatial statistical analysis. Int J Rock Mech Min Sci 46:432–438

Hu QF, Cui XM, Wang G, Wang MR, Ji YX, Xue WW (2011) Key technology of predicting dynamic surface subsidence based on Knothe time function. J Softw 6(7):1273–1280

Knothe S (1953) Równanie profilu ostatecznie wykształconej niecki osiadania. Archiwum Górnictwa i Hutnictwa, t.1, z.1, Warszawa

Knothe S (1984) Prognozowanie wpływów eksploatacji górniczej. Wyd. Śląsk

Knothe S (2005) Asymetryczna funkcja rozkładu wpływów eksploatacji górniczej w ośrodku zmieniającym swoje własności. Archiwum Górnictwa 50(4):401–415

Kowalski A (2005) O błędach prognozy wskaźników deformacji powierzchni spowodowanych błędami parametrów teorii. Materiały konferencji, VIII Dni Miernictwa Górniczego i Ochrony Terenów Górniczych, Ustroń

Krzysztoń D (1965) Parametr zasięgu niecek osiadania w ośrodku sypkim, Archiwum Górnictwa, tom X, zesz.1, str. 29–51, Kraków

Kwinta A, Hejmanowski R, Sroka A (1996) A time function analysis used for the prediction of rock mass subsidence.//96 International symposium on mining science and technology, Xuzhou, pp 16–18

Naworyta W, Sroka A (2005) Ocena niepewności prognozy wskaźników deformacji terenu metodą symulacji geostatystycznej. Materiały konferencji, VIII Dni Miernictwa Górniczego i Ochrony Terenów Górniczych, Ustroń

Niedojadło Z (2008) Problematyka eksploatacji złoża miedzi z filarów ochronnych szybów w warunkach LGOM. Rozprawy i monografie nr 177. Wydawnictwo AGH, Kraków

Popiołek E, Ostrowski J (1981) Próba ustalenia głównych przyczyn rozbieżności prognozowanych i obserwowanych poeksploatacyjnych wskaźników deformacji. Ochrona Terenów Górniczych nr 58, Katowice

Preuße A (1990) Markscheiderische analyse und prognose der vertikalen beanspruchungvon Schachtsäulen im Einwirkungsbereich untertägigen Steinkohlenabbaus. Praca doktorska, TU Clausthal, Clausthal-Zellerfeld

Schober F, Sroka A (1983) Die Berechnung von Bodenbewegungen uber Kavernen unter Berucksichtigung des zeitlichen Konvergenz und Gebirsverhaltens. Kahli und Steinsaltz Bd.8, Heft 10 **(in German)**

Stoch T (2005) Wpływ warunków geologiczno-górniczych eksploatacji złoża na losowość procesu przemieszczeń i deformacji powierzchni terenu. Akademia Górniczo-Hutnicza w Krakowie, Kraków (praca doktorska)

11

Plant growth and their root development after inoculation of arbuscular mycorrhizal fungi in coal mine subsided areas

Yinli Bi[1] · Yanxu Zhang[1] · Hui Zou[1]

Abstract Coal mining often cause serious land degradation, soil erosion, and desertification affecting growth of the local vegetation, especially the roots. Arbuscular mycorrhizal fungi (AMF) inoculation is considered a potential biotechnological tool for mined soil remediation because mycorrhizal fungi could improve plant growth environment, especially under adverse conditions due to their good symbiosis. A field experiment was conducted to study the ecological effects of AMF (*Funneliformis mosseae*, *Rhizophagus intraradices*) on the growth of *Amygdalus pedunculata* Pall. and their root development in the regenerated mining subsidence sandy land. The reclamation experiment included four treatments: inoculation of *Funneliformis mosseae* (F.m), inoculation of *Rhizophagus intraradices* (R.i), combined inoculation of F.m and R.i and non-inoculated treatment. Root mycorrhizal colonization, plant height, crown width, soil moisture, root morphology and certain soil properties were assessed. The results showed that AMF improved the shoot and root growth of *Amygdalus pedunculata* Pall., and significantly increased root colonization after 1 year of inoculation. Available phosphorus content, activities of phosphatase as well as electrical conductivity in soil rhizosphere of all the three inoculation treatments were higher than that of the non-inoculated treatment. AMF increased the quantity of bacteria and fungi in soil rhizosphere compared with the non-inoculated treatment. Our study indicates that revegetation with AMF inoculum could influence plant growth and root development as well as soil properties, suggesting that AMF inoculation can be effective method for further ecological restoration in coal mine subsided areas.

Keywords Arbuscular mycorrhizal fungi · Plant growth · Root development · Soil remediation · *Amygdalus pedunculata* Pall.

1 Introduction

China is the world's largest coal producing and consuming country (Li et al. 2011). Although coal brings huge economic benefits to China, coal mining causes numerous environmental problems, such as drastic reduction in the number of biological communities in the reconstructed soil induced by any disturbance in coal mining, plant root damage caused by mining subsidence, reduced topsoil fertility, water–fertilizer disharmony in soil, and pollution by filling materials (Zang et al. 2010; Tripathi et al. 2009). Simultaneously, the vegetation coverage is gradually decreasing and thus severely restricts the sustainable development of eco-environment in Western China. Previous studies demonstrated that microbial activity is a key factor affecting the functioning of all terrestrial systems (Quadros et al. 2016).

Soil microbiota communities play crucial role in maintaining the soil ecological balance as well as the sustainability of both natural ecosystem and agro-ecosystem. Beneficial rhizospheric microbe-plant interactions have a great influence on plant health and soil quality (Wu et al.

✉ Yinli Bi
ylbi88@126.com

[1] State Key Laboratory of Coal Resources and Safe Mining, China University of Mining and Technology (Beijing), Beijing 100083, China

2008). Among these beneficial rhizospheric microbes, arbuscular mycorrhizal fungi (AMF) and plant-growth promoting rhizobacteria (PGPR) can be considered. Arbuscular mycorrhizal fungi are obligate symbionts that colonise the roots of most cultivated plant species. Mycorrhizal symbiosis can be found in nearly all types of ecological situations and most plant species are able to form this symbiosis naturally (Gong et al. 2012). These associations occur naturally when plantlets are transplanted into the field, favouring plant development by increasing nutrient uptake, growth rates and hormonal activities (del Carmen Jaizme-Vega et al. 2006; Chandanie et al. 2009). Mycorrhizae may also increase plant tolerance to stress conditions such as salinity and drought (Latef and He 2011), heavy metals (Tian et al. 2013), root soil borne pathogens (Miller and Jastrow 1992) and the improvement of soil structure (Matthiasc et al. 2010).

Amygdalus pedunculata Pall. is a native tree specie with quick adaptability and high economic and ecological value, which is one of the economic plants used as ecological restoration pioneering species in Shendong coal mine areas (Ma et al. 2006). Studies concerning *Amygdalus pedunculata* Pall. mainly focus on the physiological and ecological characteristics, plant breeding methods, resources development and utilization, etc., mechanisms and effects of AMF on it has rarely been studied (Ma et al. 2006; Xu and Chu 2013; Chen et al. 2013; Cao et al. 2016). In this study, the impact of two Glomus species on the plant growth and their root development in coal mining subsided land were studied.

2 Materials and methods

2.1 Study site description

The study was conducted at Daliuta mining area located at the transition zone between Loess Plateau in Northern Shaanxi and the Mu Us Desert. Its altitude ranges from 1000 to 1250 m and has a semiarid continental monsoon climate. The average annual temperature is about 8.4 °C with the highest temperature of 38 °C and the lowest of − 28 °C. Mean annual precipitation is 441.2 mm, which is concentrated between June and September. Annual average evaporation is 2211.2 mm. The area has thick loosened aquifer, and the soil is sandy (Bi and Zhang 2014).

2.2 Experimental materials

Two AM fungi were used in the study. *Funneliformis mosseae* (F.m)was cultured under maize with 90% of root colonisation, and *Rhizophagus intraradices*(R.i)was cultured under maize with 84% of root colonization. They are two kinds of AMF with wide application and strong ability to adapt to the coal mine environment. The tested plant is *Amygdalus pedunculata* Pall. under field conditions. They were planted in April 2012. For the inoculation treatments, 50 g of sagent were inoculated near by the root after digging out the root carefully in the following July at the beginning of the nursery phase. After microbial inoculants were evenly applied on the plants, soil was filled back and compacted.

2.3 Experimental design

The experiment was conducted by a completely random design method to 4 treatments: (1) non-inoculated as the control treatment(CK); (2) single inoculation of F.m, (3) single inoculation of R.i, and (4) combined inoculation of F.m and R.i. The soil samples were collected in September 2013, 1 year after inoculation. Each treatment was replicated ten times.

2.4 Experimental methods

After choosing standard plants randomly in each small quadrat and removing the surface litter around plant roots, the rhizosphere soil in depths of 0–20 cm was collected into sterile plastic bags, frozen and quickly transported back to the laboratory for analysis. The soil samples were divided into two parts, one was sieved through a mesh of < 2 mm in diameter, and preserved at 4 °C for analysis of microbial biomass and enzymatic activity; the other was naturally air-dried, and sieved through a mesh of < 1 mm in diameter for analysis of physical–chemical properties. Microbial populations were conducted by dilution plate method (Shen et al. 1999). The medium for actinomycesis named improved Gao compound medium, including soluble starch 2g, KNO_3 0.1g, K_2HPO_4 0.05g, $MgSO_4{\cdot}7H_2O$ 0.05g, NaCl 0.05g, $FeSO_4{\cdot}7H_2O$ 0.001g, agar 1.8g, deionized water 100 mL, pH 7.4–7.6; the medium for fungi is named Martin—Bangladesh red medium, including Rose Bengal agar 3.5g, deionized water 100mL; the medium for bacteria is named beef extract peptone agar medium including beef extract (0.3%), peptone (0.5%) and agar (1.5%) in water. Root morphology was measured by CI-600. Fine roots were cleared in 10% KOH, stained with 0.05% w/v trypan blue for 24 h, and then were cut into 1.5-cm-long segments (Phillips and Hayman 1970). Mycorrhizal infection rate was measured by the glass slide method. Mycorrhizal infection rate(%) = formation of twigs of the number of root segments/number of the measured root segments × 100.

Soil pH and electrical conductivity (Ec) (soil:deionized water 1:5) were determined with a pH meter and electrical conductivity meter, respectively. Soil acid phosphatase activity was measured by the improved method of Tabatabai and Brimner (Zhao and Jiang 1986). Available phosphorus by the $NaHCO_3$ extraction colorimetric method (Bao 1998).

2.5 Statistical analysis

All data were subjected to one-way analysis of variance using SPSS program. Means were compared by Duncan's test at the 5% level.

3 Results

3.1 Plant growth and root morphology

The impact of different Glomus species on the plant growth of *Amygdalus pedunculata* Pall. under field conditions is shown in Fig. 1. Compared with the non-inoculated control treatment, the plant height of single inoculation of R.i, single inoculation of F.m, co-inoculation of F.m and R.i increased by 40%, 21%, 19% respectively, and the plant crown width increased by 26%, 10%, 11%. Analysis of variance showed that single inoculation of R.i and single inoculation of F.m significantly increased plant height compared to the control treatment. Single inoculation of R.i proved better than the single inoculation of F.m and the co-inoculation of F.m and R.i, while there was no significant difference between single inoculation of F.m and the co-inoculation of F.m and R.i. There were no difference in the crown width between inoculated and non-inoculated control treatment. The data indicated that AM fungi can enhance plant growth of *Amygdalus pedunculata* Pall., in favor of revegetation in the mining subsidence area for ecological recovery.

The impact of different *Glomus* species on the root growth of *Amygdalus pedunculata* Pall. is shown in Table 1. Compared with the non-inoculated control treatment, single inoculation of R.i and single inoculation of F.m increased the root projected area and the number of root tips significantly, and single inoculation of R.i increased the root surface area significantly, while there was no significant difference in the average root diameter and volume.

3.2 Mycorrhizal colonization rate

Mycorrhizal colonization rate of different treatments for *Amygdalus pedunculata* Pall. is shown as Fig. 2. The average percentage of root colonization of three treatments was 78%, higher than that of the non-inoculated control treatment (59%), indicating that all the three *Glomus* species can increase root colonization. Among the three inoculation treatments, the percentage of root colonization by the single inoculation of F.m was the highest (91%), and the combined inoculation of F.m and R.i was the second (75%), and the single inoculation of R.i was the lowest (67%).

3.3 Soil water content

Soil water content in different treatments was shown in Fig. 3. One year after inoculation, soil water content was not significantly affected. The soil water content of single inoculation of F.m, single inoculation of R.i, and the co-inoculation of F.m and R.i was 6.64%, 5.97%, 5.38% respectively, higher than that of the non-inoculated control treatment (4.71%), indicating that three inoculation treatments all had certain effect on increasing soil moisture, probably due to the increasement of soil organic matter and soil water capacity. There was no significant difference between these treatments.

3.4 Soil pH and soil electrical conductivity

Soil pH in different treatments for *Amygdalus pedunculata* Pall. after inoculation 1 year is shown in Fig. 4. There were some differences for different Glomus species, that is the soil pH of single inoculation of R.i and the co-inoculation of F.m and R.i were both higher than that of the non-inoculated control treatment, while the soil pH of single inoculation of F.m was lower than that of the non-

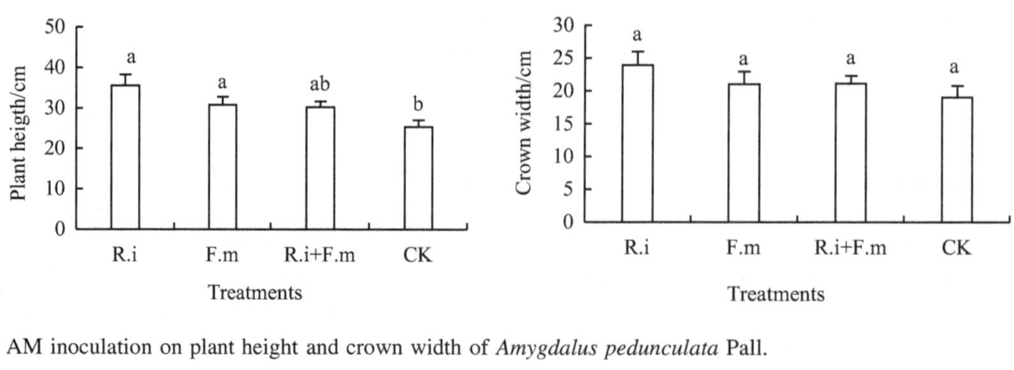

Fig. 1 Effect of AM inoculation on plant height and crown width of *Amygdalus pedunculata* Pall.

Table 1 Effect of AM inoculation on root morphology of *Amygdalus pedunculata* Pall.

Treatment	Total root length (cm)	Total projected area (cm^2)	Root surface area (cm^2)	Average root diameter (mm)	Root volume (cm^3)	Number of root tips
R.i	182.94a	4.25a	13.34a	0.27a	0.11a	107a
F.m	210.45a	7.85a	19.53ab	0.29a	0.21a	178a
R.i + F.m	188.56ab	4.03a	12.65ab	0.27a	0.10a	78ab
CK	121.80b	3.15b	9.89b	0.26a	0.07a	92b

Note Dates are expressed as mean of ten replicates, and different letters in each variable denote means that are significantly different ($P < 0.05$)

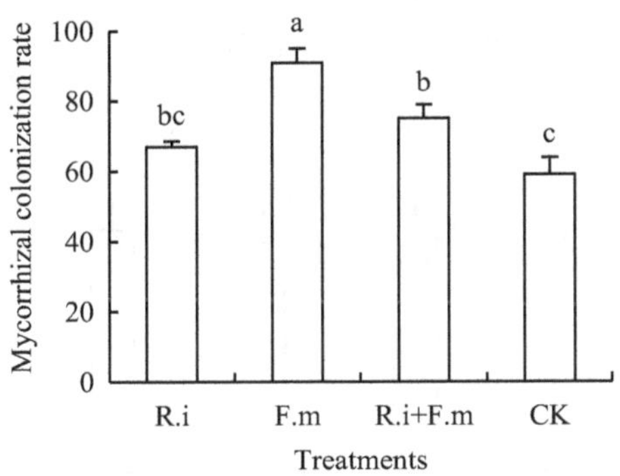

Fig. 2 Effect of AM inoculation on mycorrhizal colonization rate of *Amygdalus pedunculata* Pall.

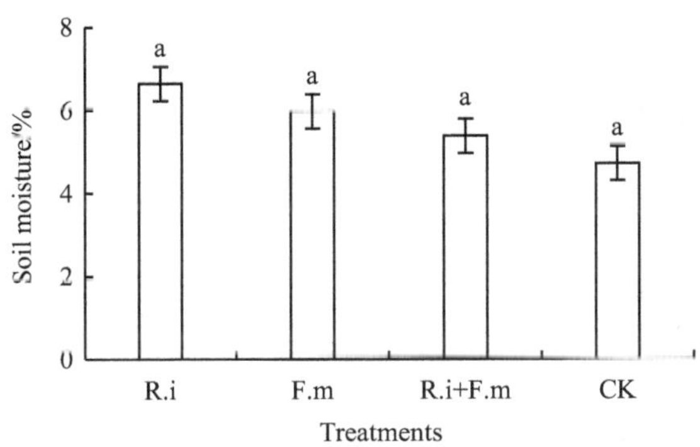

Fig. 3 Effect of AM inoculation on soil moisture of *Amygdalus pedunculata* Pall.

inoculated control treatment; there was no significant difference between single inoculation of R.i and the co-inoculation of F.m and R.i. The electrical conductivity for different treatments is shown in Fig. 4. We can see that all the three inoculation treatments were higher than the non-inoculated control treatment, indicating that inoculating AM fungi can increase the soil electrical conductivity.

3.5 Soil available phosphorus and acid phosphatase

The effect of different treatments on phosphorus and acid phosphatase activity are shown in Fig. 5. Phosphorus content of all the three inoculation treatments were higher than that of the non-inoculated treatment, indicating that inoculating AM fungi can increase soil available phosphorus, which is vital to the growth and health of plants, and favors the vegetation reconstruction in mining subsidence area. Soil acid phosphatase plays an important role in the transformation and bioavailability of organic phosphorus into inorganic phosphorus. The activity of soil acid phosphatase for different treatment is shown in Fig. 5. The activities of phosphatase of all the three inoculation treatments showed higher than the non-inoculated treatment.AM fungi accelerated the activities of phosphatase in soil rhizosphere, which can promote the transformation of soil insoluble phosphorus into plant available phosphorus, increasing soil available phosphorus.

3.6 Soil microbe populations

The effect of different treatments on soil microbe populations was shown in Fig. 6. The quantity of bacteria and fungi of all the three inoculation treatments tended to be higher than that of the non-inoculated control treatment, indicating that AM fungi increased the quantity of bacteria and fungi, especially bacteria. The population of different

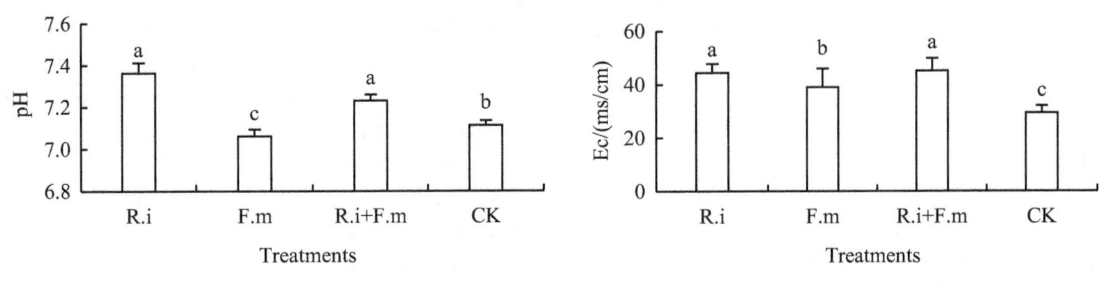

Fig. 4 Effect of AM inoculation on soil pH and Ec of *Amygdalus pedunculata* Pall.

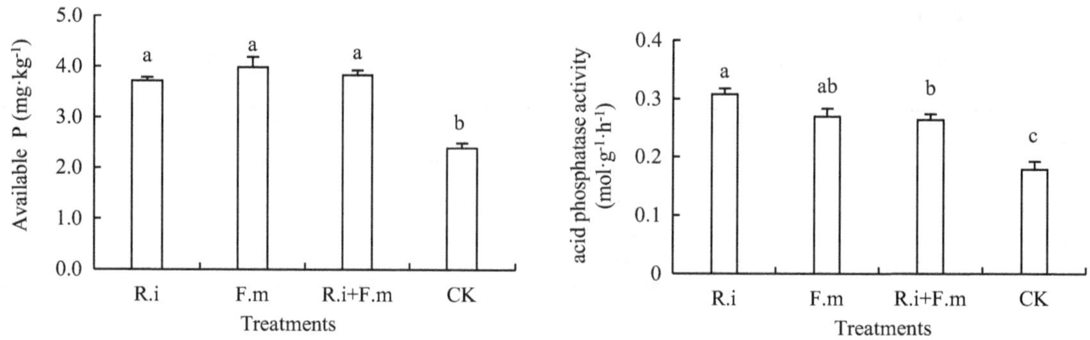

Fig. 5 Effect of AM inoculation on soil acid phosphatase and available phosphorus of *Amygdalus pedunculata* Pall.

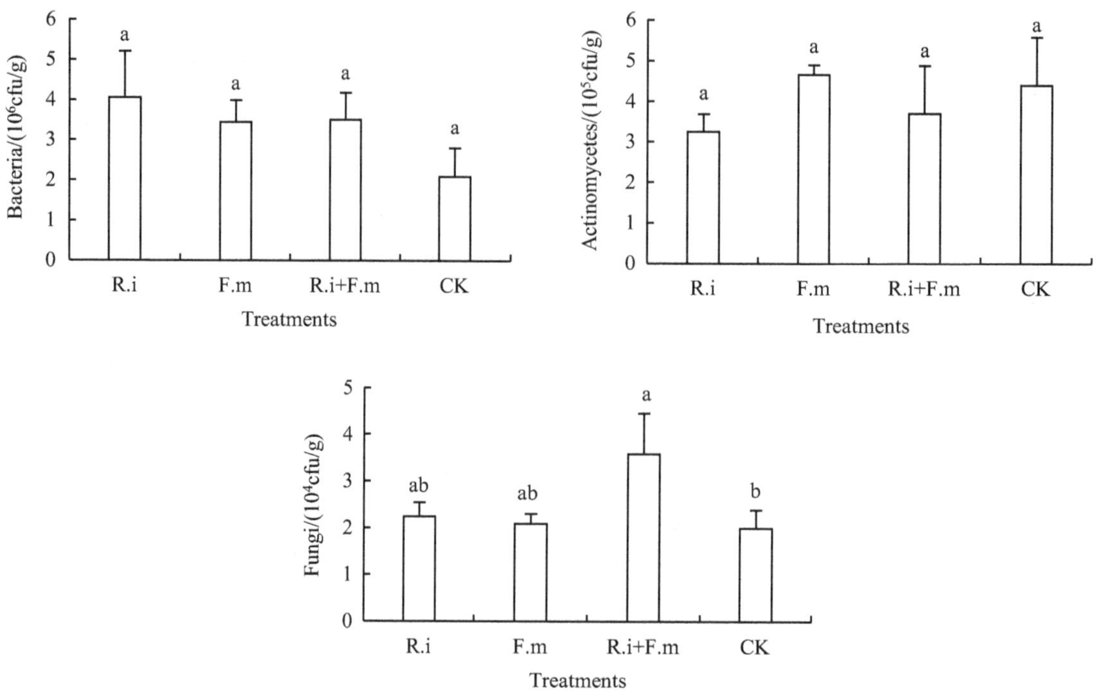

Fig. 6 Effect of AM inoculation on soil microbe populations of *Amygdalus pedunculata* Pall.

soil microbes tended to decrease in the following order: bacteria (about 80% of the total number of microbes) > actinomycetes (about 10%–20% of the total number of microbes) > fungi (about 0.2%–0.7% of the total number of microbes). The study area was located at southeast Mu Us desert with poor soil quality, the increase in soil bacteria and fungi can improve the soil microbial and nutritional environment, conducive to the plant growth and local ecology recovery.

4 Discussion

Mycorrhizas are symbiotic associations established between certain soil fungi and most vascular plants. They play a good role in optimizing plant fitness and soil quality (Barea et al. 2011). In this study, different kinds of fungi species could infect plant roots of *Amygdalus pedunculata* Pall., which indicate that good symbiosis between AM fungi and *Amygdalus pedunculata* Pall. has been formed. Mycorrhizal colonization rate of different treatments were higher than that of the non-inoculated control treatment, which showed artificial inoculation to be effective in increasing root mycorrhizal colonization. In addition, the infection rate of different treats were different, showing that host plant and fungi select each other and its colonization rate may be probably influenced by host plants and AM fungal isolates.

Mycorrhizal symbiosis could improve the resilience of plant communities against environment stresses, including nutrient deficiency, drought and soil disturbance and promote the growth rate of the plant. After inoculation for

1 year, plant height of inoculation treats presented significantly higher than that of control treatment. At the same time, certain indicators of the roots of inoculated plant were higher than that of no-inoculated plant. Rapid growth of plant is beneficial to the local ecological restoration, especially for ecological fragile mining areas. These results are consistent with former research (Wang et al. 2014).

Water is a key factor of plant growth, especially for the arid mining area where water loss is serious. Mycorrhiza can enhance nutrient acquisition and water absorption of host plants by extending hyphae beyond the nutrient depletion zone of roots (Wu et al. 2015). Through the development of hyphae network from mycorrhizal roots into the surrounding soil, hyphae directly enmesh soil particles. Besides AMF could release glomalin-related soil protein (GRSP) into soils, which could improve soil structure and improve water retention of rhizosphere (Bedini et al. 2009; Matthiasc et al. 2010). In this study, rhizosphere soil moisture content increase than control, probably due to the increase of soil organic matter and soil water holding capacity. After mycorrhizal inoculation, plant rhizosphere environment has also changed soil pH value and conductivity changed after mycorrhizal inoculation for 1 year, which may be due to the secretion caused a certain effect on rhizosphere soil biochemical reaction after mycorrhizal inoculation. Meanwhile, the content of available P for different inoculation treatments showed higher than that of non-inoculated treatment, which may be the result of higher soil acid phosphatase activities, which could promote the transformation of soil insoluble phosphorus into plant available phosphorus.

Microorganisms in the soil are critical in decomposing organic residues and recycling soil nutrients, and is one of the indicators of soil quality evaluation. Soil microorganism quantity and enzyme activity changes because of mining activities (Quadros et al. 2016). Arbuscular mycorrhizal inoculation can significantly improve the survival environment of rhizosphere microbes, is conducive to the improvement of the quality of the soil and the growth of plants. Research conclusions of our study were consistent with Camilla (Camilla et al. 2003).

5 Conclusion

Coal mining subsidence causes numerous environmental problems which have negative effects on vegetation growth. In this study, we found that AMF inoculation could promote the shoot growth and the root development of *Amygdalus pedunculata* Pall. The AMF inoculated plant showed higher root mycorrhizal colonization, soil acid phosphatase and available phosphorus as well as soil electrical conductivity compared to control. Single inoculation of R.i had good performance in plant height, soil moisture, and soil microbe population quantity; while single inoculation of F.m had good performance in soil available phosphorus concentration and soil fungi quantity. Associated inoculation of F.m and R.i can increase the soil electric conductivity and enhance soil acid phosphatase activity. AMF inoculation can be effective method for further ecological restoration in fertile environment in coal mine subsided areas.

Acknowledgements The study was financially supported by the National Natural Science Foundation of China (51574253) and the National Key Research and Development Program of China (2016YFC0501106).

References

Bao SD (1998) Soil agricultural chemistry analysis. China Agriculture Press, Beijing, pp 56–83

Barea JM, Palenzuela J, Cornejo P, Sánchezcastro I, Navarro-fernández C, Lopézgarcía A (2011) Ecological and functional roles of mycorrhizas in semi-arid ecosystems of southeast spain. J Arid Environ 75(12):1292–1301

Bedini S, Pellegrino E, Avio L, Pellegrini S, Bazzoffi P, Argese E et al (2009) Changes in soil aggregation and glomalin-related soil protein content as affected by the arbuscular mycorrhizal fungal species *Glomus mosseae* and *Glomus intraradices*. Soil Biol Biochem 41(7):1491–1496

Bi YL, Zhang YX (2014) Role of the different planting age of seabuckthorn forests to soil amelioration in coal mining subsidence land. Int J Coal Sci Technol 1(2):192–197

Camilla W, Søren Christensen I, Jakobsen AK Müller, Sørensen SJ (2003) The mycorrhizal fungus (*Glomus intraradices*) affects microbial activity in the rhizosphere of pea plants (*Pisum sativum*). Soil Biol Biochem 35(10):1349–1357

Cao ZW, Wang C, Zhang JB, Bi GY, Bi Y, Wen BY, Li WY, Li YX (2016) Growth of seedlings and morphological characteristics of leaves for different provenance of introduced *Amygdalus pedunculata*. Prot For Sci Technol 158(11):1–3

Chandanie WA, Kubota M, Hyakumachi M (2009) Interactions between the arbuscular mycorrhizal fungus *Glomus mosseae* and plant growth-promoting fungi and their significance for enhancing plant growth and suppressing damping-off of cucumber (*Cucumis sativus* l.). Appl Soil Ecol 41(3):336–341

Chen Q, Li C, Fang ZG, Chen B, Shen YH (2013) Preparation of biodiesel from desert *Amygdalus pedunculata* pall oil. J Northwest Univ (Nat Sci Edn) 43(2):229–232

del Carmen Jaizme-Vega M, Rodríguez-Romero AS, Núñez LAB (2006) Effect of the combined inoculation of arbuscular mycorrhizal fungi and plant growth-promoting rhizobacteria on papaya (*Carica papaya* l.) infected with the root-knot nematode *Meloidogyne incognita*. Fruits 61(3):151–162

Gong MG, Tang M, Zhang QM, Feng X (2012) Effects of climatic and edaphic factors on arbuscular mycorrhizal fungi in the rhizosphere of *Hippophae rhamnoides*, in the loess plateau, china. Acta Ecol Sin 32(2):62–67

Latef AAHA, He CX (2011) Effect of arbuscular mycorrhizal fungi on growth, mineral nutrition, antioxidant enzymes activity and fruit yield of tomato grown under salinity stress. Sci Hortic 127(3):228–233

Li W, Chen LQ, Zhou TJ, Zhang K, Long LI (2011) Research

progress of soil quality in china mining subsidence area. Coal Sci Technol 05:125–128

Ma XW, Guo CH, Luo M (2006) Influence of endocarp, salt and water stress on the seed germination of *Amygdalus pedunculata*. J Northwest For Univ 21(4):69–72

Matthiasc R, Noorf M, Evaf L, Pedrom A (2010) Mycelium of arbuscular mycorrhizal fungi increases soil water repellency and is sufficient to maintain water-stable soil aggregates. Soil Biol Biochem 42(7):1189–1191

Miller RM, Jastrow JD (1992) The role of mycorrhizal fungi in soil conservation. ASA Special Publication, Madison, pp 29–44

Phillips JM, Hayman DS (1970) Improved procedures for clearing roots and staining parasitic and vesicular-arbuscular mycorrhizal fungi for rapid assessment of infection. Trans Br Mycol Soc 55(1):158–161

Quadros PDD, Zhalnina K, Davis-Richardson AG, Drew JC, Menezes FB, Camargo FADO et al (2016) Coal mining practices reduce the microbial biomass, richness and diversity of soil. Appl Soil Ecol 98:195–203

Shen P, Fan XR, Li GB (1999) Microbiology experiment. Higher Education Press, Beijing, pp 123–128

Tian Y, Zhang HH, Meng XY, Wang J, Hu JW, Sun GY (2013) Effects of arbuscular mycorrhizal fungi (Glomus mosseae) on growth and photosynthesis characteristics of *Lolium merenne* L. under Cd contaminated soil. Acta Agrestia Sin 21(1):135–141

Tripathi N, Singh RS, Singh JS (2009) Impact of post-mining subsidence on nitrogen transformation in southern tropical dry deciduous forest, India. Environ Res 109(3):258–266

Wang JG, Gao XM, Bai SL, Zheng R, Liu M (2014) Effects of arbuscular mycorrhiza on the drought tolerance of *Prunus mongolica*. J Arid Land Resour Environ 12:138–142

Wu QS, Xia RX, Zou YN (2008) Improved soil structure and citrus growth after inoculation with three arbuscular mycorrhizal fungi under drought stress. Eur J Soil Biol 44(1):122–128

Wu QS, Srivastava AK, Cao MQ, Wang J (2015) Mycorrhizal function on soil aggregate stability in root zone and root-free hyphae zone of trifoliate orange. Arch Agron Soil Sci 61(6):813–825

Xu XQ, Chu JM (2013) Analysis on development potential and countermeasures of *Amygdalus pedunculata* industry. For Resour Manag 2(1):22

Zang YT, Wang J, Ding GD, Gao Y, He X, Yan L, He Z, Na Q, Gong P, Ren Y (2010) Variation of physico-chemical properties of aeolian sandy soil at coal mining subsidence and its evaluation. Acta Pedol Sin 47(2):262–269

Zhao LP, Jiang Y (1986) Determination of the method of soil phosphatase activity. Chin J Soil Sci 17(3):138–142

12

Presenting an engineering classification system for coal spontaneous combustion potential

Amir Saffari[1] · **Farhang Sereshki**[1] · **Mohammad Ataei**[1] · **Keramat Ghanbari**[2]

Abstract The phenomenon of coal spontaneous combustion is one of the common hazards in coal mines and also one of the important reasons for the loss of coal in piles and mines. Based on previous researches, different types of coals have different spontaneous combustion characteristics. For coal loss prevention, a measure is necessary for prediction of coal spontaneous combustion. In this study, a new engineering classification system called ''Coal Spontaneous Combustion Potential Index (CSCPI)'' is presented based on the Fuzzy Delphi Analytic Hierarchy Process (FDAHP) approach. CSCPI classifies coals based on their spontaneous combustion capability. After recognition of the roles of the effective parameters influencing the initiation of a spontaneous combustion, a series of intrinsic, geological, and mining characteristics of coal seams are investigated. Then, the main stages of the implementation of the FDAHP method are studied and the weight of each parameter involved is calculated. A classification list of each parameter is formed, the CSCPI system is described, and the engineering classifying system is subsequently presented. In the CSCPI system, each coal seam can be rated by a number from 0 to 100; a higher number implies a greater ease for the coal spontaneous combustion capability. Based on the CSCPI system, the propensity of spontaneous combustion of coal can be classified into three potential levels: low, medium, and high. Finally, using the events of coal spontaneous combustion occurring in one of the Iranian coal mines, Eastern Alborz Coal Mines, an initial validation of the mentioned systematic approach is conducted. Comparison of the results obtained in this study illustrate a relatively good agreement.

Keywords Coal · Classification · Coal Spontaneous Combustion Potential Index (CSCPI) · Fuzzy Delphi Analytic Hierarchy Process (FDAHP) · Eastern Alborz Coal Mines

1 Introduction

Coal mining is a very intricate system and process. The rough working conditions and the hazardous environment are the most important factors that affect a coal mining process.

The hazards of underground mining are critical factors, which should be considered in the design and planning step of coal mines. Some significant hazards in underground coal mining can be summarized as subsidence, outburst, and spontaneous combustion. Therefore, it is necessary to accurately identify the risks involved and to find the ways to forecast, prevent, and control them.

Coal is a combustible material, which is applicable to a variety of oxidation scenarios with conditions ranging from the atmospheric temperature to the ignition temperature. One of the most frequent and serious causes of coal fires is spontaneous combustion or self-heating.

Spontaneous combustion is an oxidation reaction that occurs without an external heat source. This process changes the internal heat profile of the material, leading to

✉ Amir Saffari
amirsaffari5710@yahoo.com

[1] Faculty of Mining, Petroleum and Geophysics Engineering, Shahrood University of Technology, Daneshgah Blvd., P.O. Box 3619995161, Shahrood, Iran

[2] HSE Management of Eastern Alborz Coal Mines Company, Shahrood, Iran

a rise in the temperature. This can eventually lead to an open flame and burning of the material (Akgun and Arisoy 1994; Carras and Young 1994; Ren et al. 1999; Nugroho et al. 2000a; Wang et al. 2003; Smith and Glasser 2005; Beamish and Arisoy 2008b).

Oxidation of coal starts with exothermic chemical reactions, and it can be described as a process including three sequential steps. These steps are as what follow (Mohalik et al. 2009; Yuan and Smith 2012):

(1) Physical adsorption of oxygen on the coal surface;
(2) Chemical absorption (chemisorption), which leads to the formation of coal-oxygen complexes and oxygenated carbon-species;
(3) Chemical reaction.

The chemical reaction (step 3) breaks down the less stable coal-oxygen complexes, and results in the formation of gaseous products, typically carbon monoxide (CO), carbon dioxide (CO_2), and water vapor (H_2O).

Wang et al. (2003) also agreed that coal oxidation is a complicated process involving four phenomena (see Fig. 1), namely:

(1) Oxygen transport to the surfaces of coal particles;
(2) Oxygen transport within coal pores;
(3) Chemical interaction between coal and O_2;
(4) Release of heat and emission of gaseous products.

The inspection of events is a useful part of the risk assessment process in that it provides some information of the noxious impacts of spontaneous combustion, which include:

(1) Mortality of personnel;
(2) Psychic perturbations in survivors of calamity;
(3) Pollution of the air, water, and soil in the vicinity of the burning coal field; this is a hazard to the ecosystem of the region;

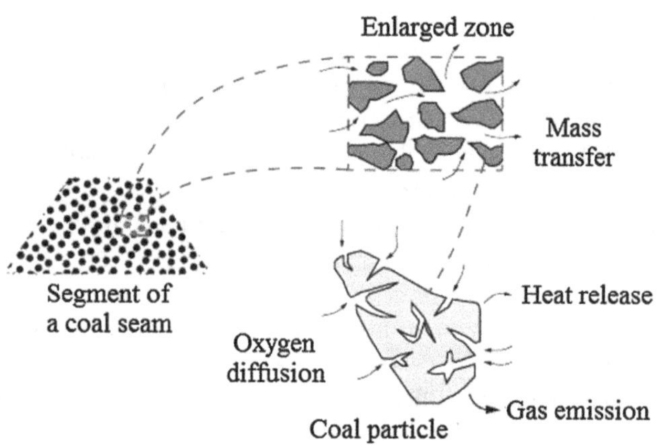

Fig. 1 Fundamental phenomena occurring in coal oxidation process (Wang et al. 2003)

(4) Emissions from coal spontaneous combustion not only pollute the local atmosphere but also add substantial amounts of the greenhouse gases (CO_2 and CH_4) along with SO*x*, NO*x*, H_2S, and CO;
(5) Subsidence of land surface;
(6) Climate change and its contribution to global warming;
(7) Mine closures;
(8) Loss of equipment;
(9) Loss of production;
(10) Useless loss of a non-renewable energy resource;
(11) Loss of popularity and market position;
(12) Costs of therapeutic and recovery measures.

As mentioned earlier, one of the risks that is a substantial amount occurring in the coal mines is ''spontaneous combustion''. Other terminologies of ''Coal Spontaneous Combustion'' are ''Coal Self-burning'', ''Coal Self-heating'' and ''Coal Self-ignition''. Coal spontaneous combustion can occur in underground mines, open pit mines, abandoned mines, coal storage locations, waste dumps, and during transports, especially on ocean-going vessels.

Evaluation of the coal spontaneous combustion hazards in coal mines should start in the first stage of design and carry on during their whole lifecycle, even after mine closure because the coal spontaneous combustion risks still remain as safety, economic, and environmental issues (Michaylov 2002).

In order to evaluate the spontaneous combustion capability of each coal seam, the effects of the important parameters involved should be recognized. The factors that appear to have an influence on the development of a spontaneous combustion event include, but are not limited to, pyrite content, ash content, humidity, coal rank, etc. These effective parameters that have specific effects on the spontaneous combustion phenomenon have different impacts from one to another. For this purpose, it is very important to know the qualitative and quantitative impacts of each parameter. The best method for surveying the coal spontaneous combustion capability is using a classification system based on the above-mentioned factors.

So far, many studies, mentioned in Table 1, have been carried out about intrinsic, geological, and mining characteristics of coal which, affecting coal spontaneous combustion, but until now, a comprehensive classification approach considering all the parameters affecting coal spontaneous combustion capability has not been developed. Also through analyzing the historical records of coal mine fires, it can be seen that the hazards of coal spontaneous combustion are influenced not only by the intrinsic characteristics of coal but also by a multiplicity of factors like geological and mining characteristics. To modify the traditional idea, the prediction of coal spontaneous

Table 1 The most famous and important studies with their used parameters

References	R	P	Mo	A	PS	GC	MSE	EM	AR	DS
Kuchta et al. (1980)	●									
Banerjee (1985)			●		●					
Ghosh (1986)		●								
Mitchell (1990)							●			
Arisoy and Akgun (1994)			●							
Bhat and Agarwal (1996)			●							
Ren et al. (1999)			●		●					
Sujanti and Zhang 1999	●				●					
Sujanti et al. (1999)	●				●	●				
Nugroho et al. (2000a, b)					●					
Jones (2001)			●							
Kadioğlu and Varamarz (2003)			●							
Kucuk et al. (2003)			●		●					
Beamish (2005)	●									
Beamish and Blazak (2005)				●						
Beamish and Jabouri (2005)						●				
Beamish et al. (2005)	●			●						
Ren et al. (2005)								●		
Cao et al. (2007)										●
Nelson and Chen (2007)	●	●								
Ramlu (2007)						●				
Singh et al. (2007)	●	●	●							
Beamish and Arisoy (2008a)	●			●						
Beamish and Sainsbury (2008)				●						
Beamish and Schultz (2008)			●							
Bo-tao et al. (2009)	●								●	
Beamish and Beamish (2010)			●							
Beamish and Beamish (2011)			●							
Beamish and Beamish (2012)			●							
Beamish et al. (2012)		●								
Beamish et al. (2013)	●									
Sasaki et al. (2014)	●		●							
Arisoy and Beamish (2015)		●	●							
Deng et al. (2015)		●								

R rank, *P* pyrite, *Mo* moisture, *A* ash, *PS* particle size, *GC* gas content, *MSE* multi-seam extraction, *EM* extraction method, *AR* advanced rate, *DS* deep seam

combustion should not be based on the unilateral identification of coal spontaneous combustion tendency, but on a comprehensive hazard evaluation system, which contains different factors and complicated interactions.

In this paper, according to a review of all the latest studies on the coal spontaneous combustion subject, the important, effective, and applicable parameters were identified and selected. It requires effort to make a comprehensive classification system that is an intelligent method for evaluation of coal spontaneous combustion potential in underground coal mines, especially mines extracted by the long wall method mining.

The most important step involved in providing a comprehensive classification system for evaluation of a phenomenon with a certain number of parameters is the determination of the weight of each parameter. The illustrious methods used for weighting parameters include the multi-criteria decision-making (MCDM) methods, rock engineering system (RES) method, and experimental-comparison method.

MCDM deals with the problem of helping the decision-maker to choose the best alternative according to several criteria (Chen and Klein 1997; Valls and Torra 2000).

Due to the fact that, depending on the problem, decision makers use different types of criteria (e.g. numerical, qualitative or stochastic values), different specializations of these methods have been developed in order to deal with these types of values (Valls and Torra 2000).

One of the newest processes used for weighting parameters in the MCDM methods is Fuzzy Delphi Analytic Hierarchy Process (FDAHP). The main objective of this paper is to present a new system for evaluating the factors affecting the coal spontaneous combustion capability using FDAHP. In this process, the opinions and experiences of a large number of experts are used, and thus it has a high efficiency. In this paper, according to the ability and reputation of FDAHP to determine the significance of each parameter in influencing the hazards and weighting parameters in the classification system, it is being used. This research work considers the coal spontaneous combustion classification as a group decision problem and applies the fuzzy logic theory as the tool for weighting calculations. Finally, a new classification system named ''Coal Spontaneous Combustion Potential Index (CSCPI)'' is presented.

2 Fuzzy Delphi Analytic Hierarchy Process (FDAHP)

Analytic hierarchy process (AHP) is a multiple-criteria decision-making (MCDM) technique that is suitable to deal with complex systems related to making a choice among several alternatives, and provides a comparison of the considered option, firstly proposed by Saaty (1980). It is a tool used to combine the qualitative and quantitative factors in the selection of a process, and is used for setting priorities in a complex, unanticipated, multi-criteria, and problematic situation. As Saaty has stated ''AHP is a measurement theory that prioritizes the consistency of the judgmental data provided by a group of decision-makers''. Furthermore, AHP provides a flexible and easy way to understand the analysis of complicated problems (Saaty 1980, 1990). The main advantage of AHP is its ability to handle complex and ill-structured problems that cannot be usually handled by rigorous mathematical models. AHP incorporates the evaluations of all decision-makers into a final decision without having to elicit their utility functions on subjective and objective criteria by pairwise comparisons of the alternatives (Saaty 1980).

The traditional AHP method is problematic in that it uses an exact value for the decision-maker's opinion in a

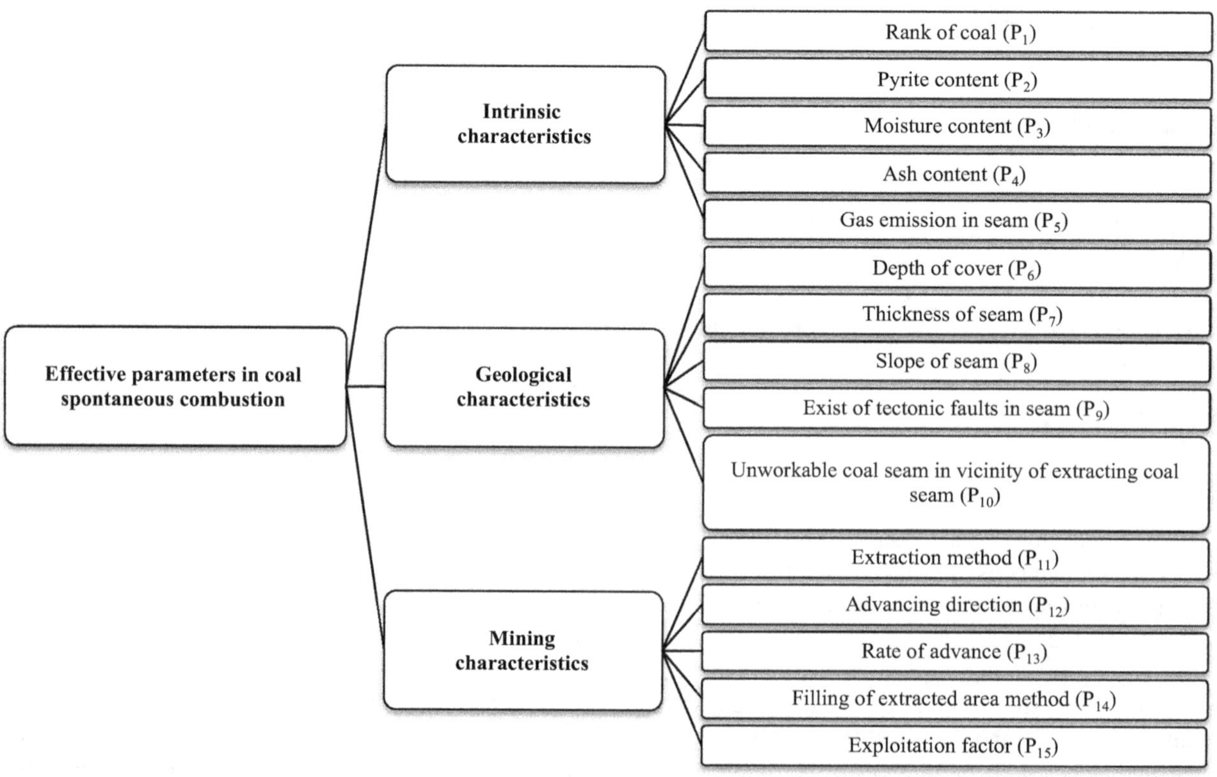

Fig. 2 Important characteristics influencing coal spontaneous combustion

Table 2 Fundamental scale of absolute numbers (Saaty 1980)

Intensity of importance	Definition
1	Equal importance
3	Moderate importance
5	Strong importance
7	Very strong importance
9	Extreme importance
2, 4, 6, 8	Preferences between above intervals

comparison of alternatives (Wang and Chen 2007). The AHP method is often criticized for the use of unbalanced scale of judgments and its inability to adequately handle the inherent uncertainty and imprecision in the pairwise comparison process (Deng 1999). To overcome all these shortcomings, FDAHP was developed for solving the hierarchical problems. Decision-makers usually find that it is more confident to give interval judgments than fixed value judgments.

The Delphi method is a technique used for structuring an effective group communication process by providing feedback of contributions of information and assessment of group judgments to enable individuals to re-evaluate their judgments. Since its development in the 1960s at Rand Corporation, the Delphi method has been widely used in various fields (Hoseinie et al. 2009; Mikaeil et al. 2013; Soltanmohammad et al. 2013).

On the other hand, the Delphi method uses the crisp number and a mean to become the evaluation criterion; these shortcomings might distort the experts' opinion. In order to deal with the fuzziness of human participants'

Table 3 A sample of questionnaire form main characteristics

No.	Main characteristics	Importance of each parameter				
		VW (1)	W (3)	M (5)	S (7)	VS (9)
1	Intrinsic characteristics					
2	Geological characteristics					
3	Mining characteristics					

VW very weak importance, *W* weak importance, *M* moderate importance, *S* strong importance, *VS* very strong importance

Table 4 A sample of questionnaire form intrinsic characteristics

No.	Intrinsic characteristics	Importance of each parameter				
		VW (1)	W (3)	M (5)	S (7)	VS (9)
1	Rank of coal					
2	Pyrite content (%)					
3	Moisture content (%)					
4	Ash content (%)'					
5	Gas emission in seam (ton/m^3)					

VW very weak importance, *W* weak importance, *M* moderate importance, *S* strong importance, *VS* very strong importance

Table 5 A sample of questionnaire form geological characteristics

No.	Geological characteristics	Importance of each parameter				
		VW (1)	W (3)	M (5)	S (7)	VS (9)
1	Depth of cover					
2	Thickness of seam					
3	Slope of seam					
4	Exist of tectonic faults in seam					
5	Unworkable coal seam in vicinity of extracting coal seam					

VW very weak importance, *W* weak importance, *M* moderate importance, *S* strong importance, *VS* very strong importance

Table 6 A sample of questionnaire form mining characteristics

No.	Mining characteristics	Importance of each parameter				
		VW (1)	W (3)	M (5)	S (7)	VS (9)
1	Extraction method					
2	Advancing direction					
3	Rate of advance					
4	Filling of extracted area method					
5	Exploitation factor					

VW Very Weak importance, *W* Weak importance, *M* Moderate importance, *S* Strong importance, *VS* Very Strong importance

Table 7 Experts' responses for main characteristics

Experts	1	2	3	4	5	6	7	8	9	10	11	12	13	14	15
Experts' responses for main characteristics															
Intrinsic characteristics	9	7	9	9	9	7	9	9	5	9	9	9	7	9	9
Geological characteristics	5	7	7	9	5	7	5	5	7	9	7	5	5	5	3
Mining characteristics	7	5	5	7	9	7	7	7	7	7	9	7	7	7	7

Table 8 Experts' responses for intrinsic characteristics

Experts	1	2	3	4	5	6	7	8	9	10	11	12	13	14	15
Experts' responses for intrinsic characteristics															
Rank of coal	9	7	9	9	9	9	9	7	7	9	7	7	7	7	9
Pyrite content	9	9	7	7	7	9	9	9	5	9	7	7	5	9	7
Moisture content	7	7	7	9	9	5	5	7	7	5	9	7	5	9	7
Ash content	7	7	7	9	7	5	5	5	7	5	5	7	5	9	5
Gas emission in seam	9	7	7	9	9	9	9	7	9	7	7	7	5	9	5

Table 9 Experts' responses for geological characteristics

Experts	1	2	3	4	5	6	7	8	9	10	11	12	13	14	15
Experts' responses for geological characteristics															
Depth of cover	9	7	9	9	7	7	9	9	9	9	7	7	9	9	7
Thickness of seam	3	7	5	5	7	3	7	7	7	5	5	7	3	7	7
Slope of seam	3	5	5	5	5	1	5	1	5	1	5	3	3	5	1
Exist of tectonic faults in seam	7	9	5	9	5	5	9	7	9	7	7	5	5	7	9
Unworkable coal seam in vicinity of extracting coal seam	3	3	5	9	9	7	7	3	7	3	7	7	5	9	5

Table 10 Experts' responses for mining characteristics

Experts	1	2	3	4	5	6	7	8	9	10	11	12	13	14	15
Experts' responses for mining characteristics															
Extraction method	9	7	7	9	9	7	9	7	7	9	7	9	7	9	9
Advancing direction	7	7	5	9	7	9	7	9	7	5	7	7	5	5	7
Rate of advance	7	7	9	9	7	9	9	7	7	7	7	7	7	7	7
Filling of extracted area method	7	7	5	9	9	7	7	7	7	7	9	7	7	7	7
Exploitation factor	7	7	7	9	7	7	7	7	7	7	7	5	7	7	7

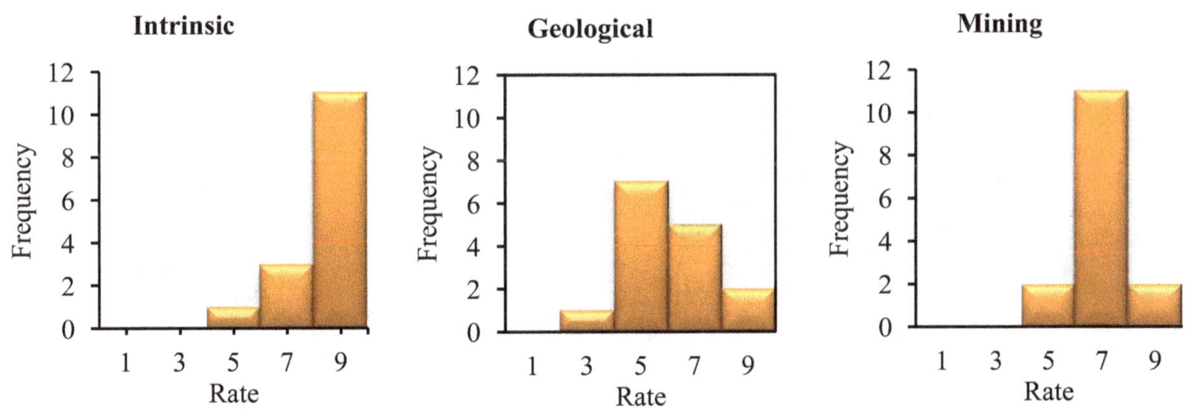

Fig. 3 Histograms of opinions of experts about main characteristics

judgments in the traditional Delphi method, Ishikawa et al. (1993) positioned "fuzzy set theory" proposed by Zadeh (1965, 1999) into the Delphi method to improve the time-consuming problems such as the convergence of the experts' options presented by Hwang and Lin (1987). The fuzzy Delphi method (FDM) is a methodology in which the subjective data of the experts is transformed into the quasi-objective data using the statistical analysis and fuzzy operations (Liu and Chen 2007a, b). The main advantages of FDM are that it can reduce the number of surveys to save time and cost, and it also includes the individual attributes of all experts (Kaufmann and Gupta 1988; Liu and Chen 2007a, b; Hoseinie et al. 2009). Thus it can effectively determine the weighting of each parameter.

It is known that group decision-making is a very important and powerful tool to accelerate the consensus of various opinions from the experts who are experienced in practices (Liu and Chen 2007a, b; Hoseinie et al. 2009). In this paper, FDM has been used to combine the opinions of experts. The main steps of the FDAHP method has been described in Sect. 3.1 with solving the problem in this research work.

The ultimate weights of each characteristic is obtained by multiplying the weight underhand characteristics in the weight overhand characteristics listed in Table 19 and shown in Fig. 8.

3 Methodology

3.1 Development of Fuzzy Delphi Analytic Hierarchy Process (FDAHP) method application for classification of coal spontaneous combustion potential

In order to develop a new classification for assessing the coal spontaneous combustion capability, five main steps must be taken into account as what follow.

3.1.1 Selection of most important parameters affecting coal spontaneous combustion potential (review of previous studies)

In the first step, identification of the parameters responsible for the occurrence of risks in the case of coal spontaneous combustion is necessary. According to the available literature and studies on the coal spontaneous combustion subject, a total of 15 important and influential parameters

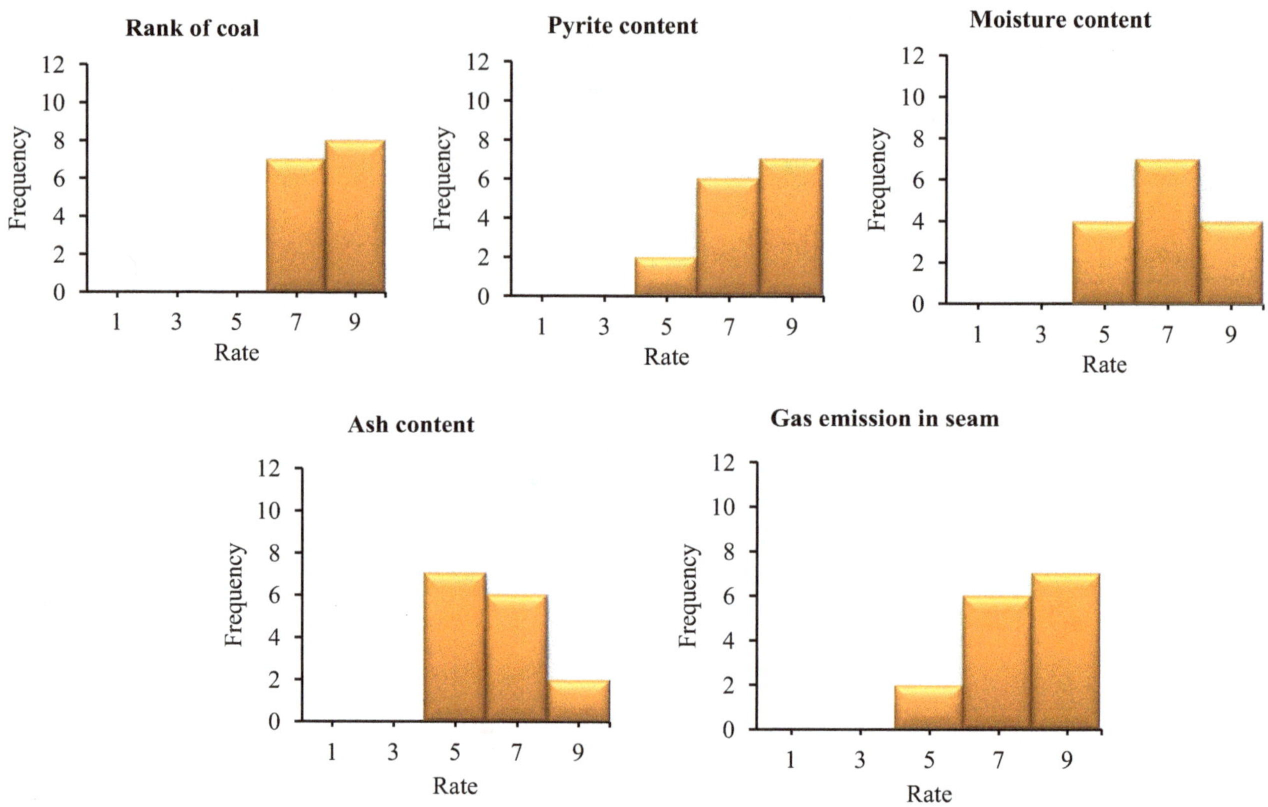

Fig. 4 Histograms of opinions of experts about intrinsic characteristics

were identified for the coal spontaneous combustion potential and classification, listed in Fig. 2.

Generally speaking, the important characteristics that influence the coal spontaneous combustion capability can be classified into three major categories: intrinsic, geological, and mining characteristics. Finally, as a result of this preview, all the mentioned parameters were selected for use in the FDAHP evaluation system and development of the CSCPI classification.

On the development of a new classification system, 3 basic principles were taken into consideration:

(1) The minimum number of parameters was used for classification.
(2) Utilization of the parallel and overlapping parameters was avoided.
(3) Using the parameters that did not have measurement abilities was avoided.

3.1.2 Sending questionnaire to experts and experts' responses

In the second step, after identification of the parameters affecting the coal spontaneous combustion, in order to use the FDAHP method and determine the weight of each parameter, technical questionnaires were prepared and sent to the experts. They were asked to mark in the questionnaires the importance of each parameter by Likert 5 point scale. In order to use the questionnaire data in the FDAHP method, for each important level, an intensity number from 1 to 9 was assigned (Table 2) based on the Saaty's method (Saaty 1980); questionnaire forms are shown in Tables 3, 4, 5, 6. The experts' responses (completed questionnaires) are shown in Tables 7, 8, 9 10. These opinions were used for the input FDAHP method.

The experts' opinion histograms (rates) about each parameter are illustrated in Figs. 3, 4, 5, 6. As it can be seen, the intrinsic characteristics have the highest frequency of rate 9. It shows that they are the most important characteristics of coal spontaneous combustion potential from the viewpoint of the experts' opinions.

3.1.3 Calculation of relative fuzzy weights of parameters

In the third step, relative fuzzy weights of parameters are calculated. The main steps are described as following.

3.1.3.1 Establishment of pairwise comparison matrices In order to establish the main pairwise comparison matrix using the FDAHP method, it is necessary to have the comparison matrix of the parameters based on each

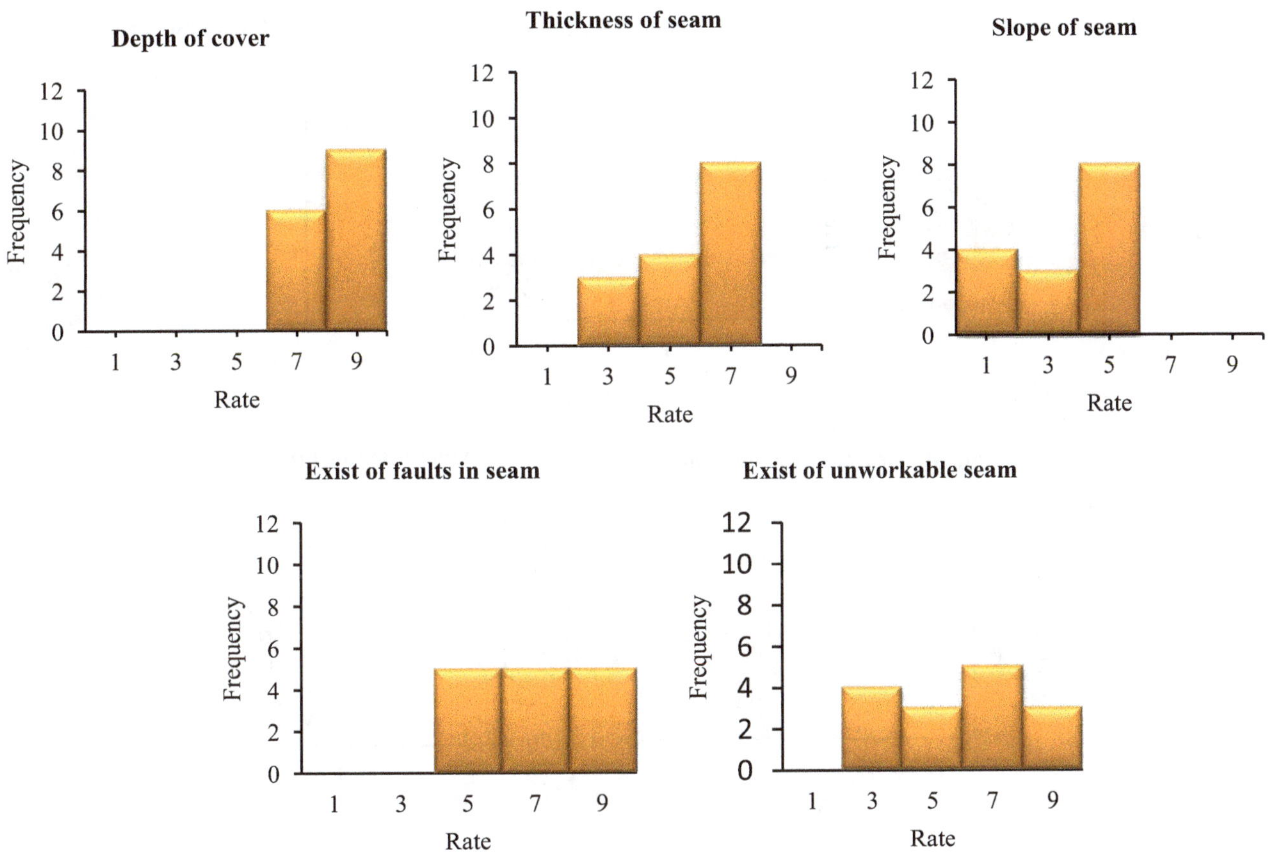

Fig. 5 Histograms of opinions of experts about geological characteristics

expert's opinion. For this purpose, according to the questionnaire, 4 comparison matrices were established for each expert (matrices 1, 2, 3, and 4 for main, intrinsic, geological, and mining characteristics, respectively; totally 60 comparison matrices).

Let $C_1, C_2, \ldots, C_n$ denote the set of elements, while a_{ij} represents a quantified judgment on a pair of elements C_i and C_j. The relative importance of two elements is obtained from the division rate of C_i on C_j based on the questionnaire. This yields an $n \times n$ matrix $\boldsymbol{A}$, as in Eq. (1) (Hoseinie et al. 2009):

$$\boldsymbol{A} = [a_{ij}] = \begin{array}{c} \\ C_1 \\ C_2 \\ \vdots \\ C_n \end{array} \begin{array}{c} \begin{array}{cccc} C_1 & C_2, & \ldots, & C_n \end{array} \\ \begin{bmatrix} 1 & a_{12} & \cdots & a_{1n} \\ 1/a_{12} & 1 & \cdots & a_{2n} \\ \vdots & \vdots & \ddots & \vdots \\ 1/a_{1n} & 1/a_{2n} & \cdots & 1 \end{bmatrix} \end{array} \quad (1)$$

In this research work, there were four group characteristics (main, intrinsic, geological, and mining characteristics with 3, 5, 5, and 5 elements, respectively) and 15 experts. Therefore, there were fifteen 3×3, fifteen 5×5, fifteen 5×5, and fifteen 5×5 pairwise comparison matrices for the main, intrinsic, geological, and mining characteristics, respectively. Due to very high calculations, their details are avoided. The matrices were established for the following calculations.

3.1.3.2 Establishment of major pairwise comparison matrix For establishing the major pairwise comparison matrix in the FDAHP method and calculation of the relative fuzzy weights of the decision elements, three steps should be done, as follow (Liu and Chen 2007a, b; Hoseinie et al. 2009):

(a) Computation of triangular fuzzy numbers (TFNs); $\tilde{a}_{ij}$. In this work, the TFNs (as shown in Fig. 7 and Eq. (2)), which represent the pessimistic, moderate, and optimistic estimates were used to represent the opinions of the experts about each parameter.

$$\tilde{a}_{ij} = (\alpha_{ij}, \delta_{ij}, \gamma_{ij}) \quad (2)$$

$$\alpha_{ij} = \mathrm{Min}(\beta_{ijk}), \quad k = 1, 2, \ldots, \lambda \quad (3)$$

$$\delta_{ij} = \left(\prod_{k=1}^{\lambda} \beta_{ijk} \right)^{1/k}, \quad k = 1, 2, \ldots, \lambda \quad (4)$$

$$\gamma_{ij} = \mathrm{Max}(\beta_{ijk}), \quad k = 1, 2, \ldots, \lambda \quad (5)$$

where, $\alpha_{ij} \leq \delta_{ij} \leq \gamma_{ij}, \alpha_{ij}, \delta_{ij}, \gamma_{ij} \subseteq [1/9, 1] \cup [1, 9]$ (obtained from Eqs. (3)–(5); α_{ij} indicates the lower

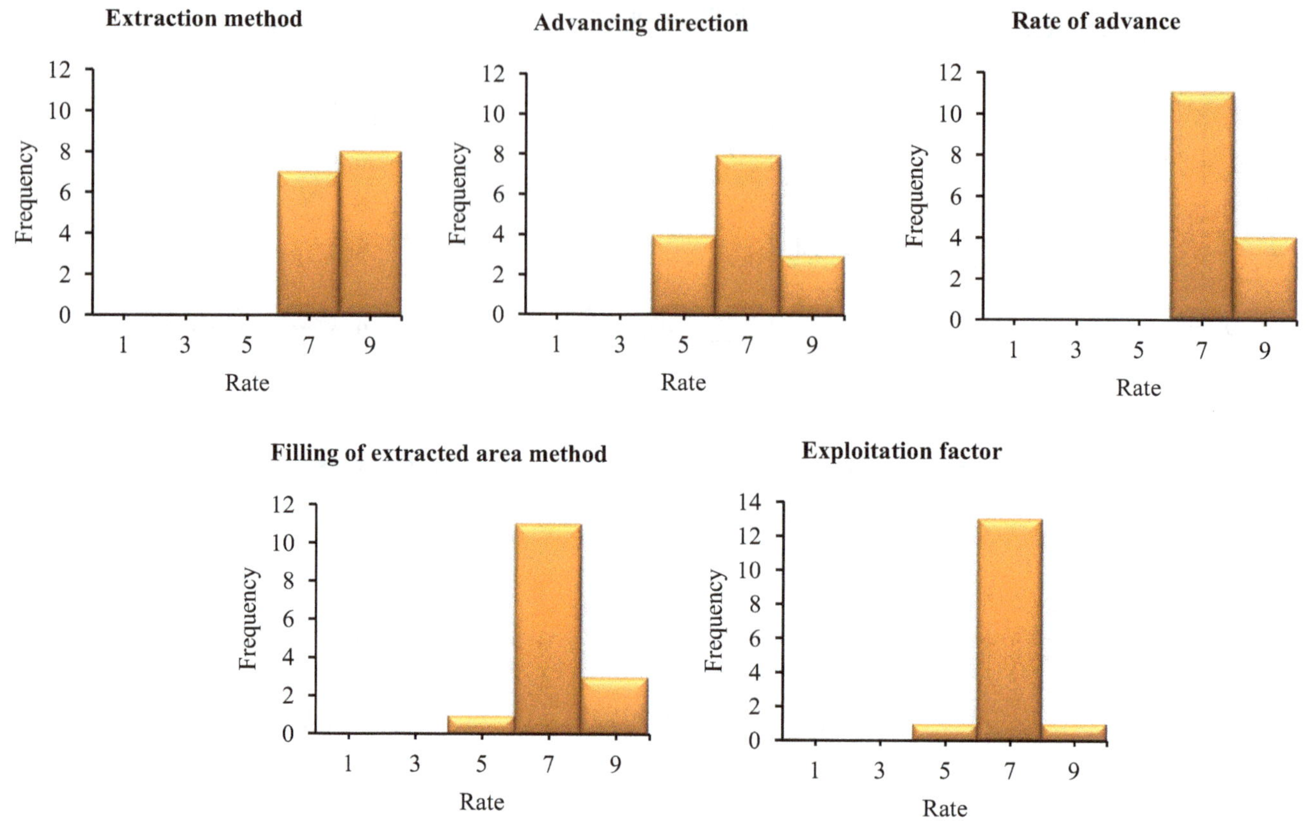

Fig. 6 Histograms of opinions of experts about mining characteristics

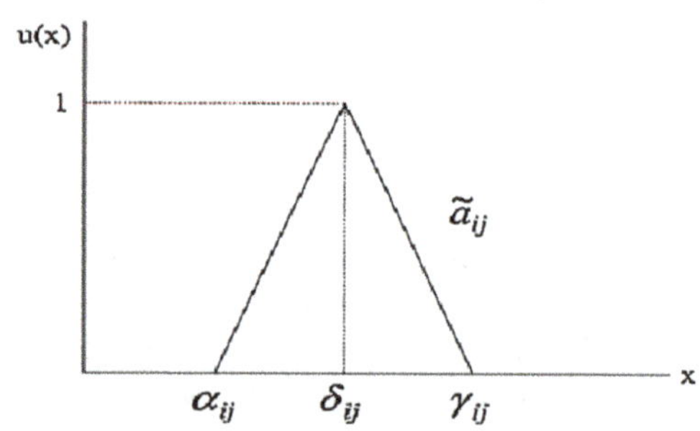

Fig. 7 Membership function of fuzzy Delphi method

bound, γ_{ij} indicates the upper bound, β_{ijk} indicates the relative intensity of importance of expert k between parameters i and j, and λ is the number of experts in the decision-making.

b) Following the above outlines, a fuzzy positive reciprocal matrix $\tilde{\boldsymbol{A}}$ can be calculated as Eq. (6):

$$\tilde{A} = [\tilde{a}_{ij}], \ \tilde{a}_{ij} \times \tilde{a}_{ji} \approx 1, \ \forall i,j = 1,2,\ldots n$$

or

$$\tilde{A} = \begin{bmatrix} (1,1,1) & (\alpha_{12},\delta_{12},\gamma_{12}) & (\alpha_{13},\delta_{13},\gamma_{13}) \\ (1/\gamma_{12},1/\delta_{12},1/\alpha_{12}) & (1,1,1) & (\alpha_{23},\delta_{23},\gamma_{23}) \\ (1/\gamma_{13},1/\delta_{13},1/\alpha_{13}) & (1/\gamma_{23},1/\delta_{23},1/\alpha_{23}) & (1,1,1) \end{bmatrix} \tag{6}$$

(c) Calculation of relative fuzzy weights of the evaluation factors according to Eq. (7):

$$\tilde{\boldsymbol{Z}}_i = \left[\tilde{a}_{ij} \otimes \cdots \otimes \tilde{a}_{in}\right]^{1/n},$$
$$\tilde{\boldsymbol{W}}_i = \tilde{Z}_i \otimes (\tilde{Z}_i \oplus \cdots \oplus \tilde{Z}_n)^{-1}$$

where $\tilde{a}_1 \otimes \tilde{a}_2 = (\alpha_1 \times \alpha_2, \delta_1 \times \delta_2, \gamma_1 \times \gamma_2)$; the symbol $\otimes$ here denotes the multiplication of fuzzy numbers, and the symbol $\oplus$ denotes the addition of fuzzy numbers. $\tilde{\boldsymbol{W}}_i$ is a row vector in consist of a fuzzy weight of the ith factor($\tilde{\boldsymbol{W}}_i = (\omega_1, \omega_2, \cdots,$

Table 11 Major pairwise comparison matrix between 3 main characteristics

Item	Intrinsic characteristics (IC)	Geological characteristics (GC)	Mining characteristics (MC)
IC	(1.000, 1.000, 1.000)	(0.714, 1.408, 3.000)	(0.714, 1.189, 1.800)
GC	(0.333, 0.710, 1.400)	(1.000, 1.000, 1.000)	(0.429, 0.845, 1.400)
MC	(0.556, 0.841, 1.400)	(0.714, 1.184, 2.333)	(1.000, 1.000, 1.000)

$\omega_n), i = 1, 2, \cdots, n)$. The defuzzification (changing the fuzzy number to a usual number) is based on the geometric average method (Eq. (8)):

$$W_i = \left(\prod_{i=1}^{3} \omega_j\right)^{1/3} \quad (8)$$

In this research work, 4 major pairwise comparison matrices, as shown in Tables 11, 13, 15, and 17, and the fuzzy and usual final weight of each group, as shown in Tables 12, 14, 16, and 18, were established by the procedure described.

Table 19 and Fig. 8 show that the parameters rank of coal, gas emission in seam, pyrite content, depth of cover, moisture content, extraction method, rate of advance, and filling of extracted area method have the highest weights in the system, and so a small change in these parameters will affect, to a considerable extent, the behavior of the system.

3.1.4 *Classification and rating of parameters*

In the fourth step, the rating of the parameter values was carried out based upon their effects on the coal spontaneous combustion. Totally, six classes of rating, from 0 to 5, were considered, where 0 shows the best case (most favorable condition) and the maximum value that indicates the worst position (most unfavorable condition).

In the case of coal spontaneous combustion, the rating of each parameter is presented in Table 20. The ranges of parameters in Table 20 were proposed based on the results obtained by other researchers (mentioned in Table 1).

Table 12 Fuzzy and usual final weight of 3 main characteristics

Item	$\tilde{Z}_i$ (α, δ, γ)	$\tilde{W}_i$ (Fuzzy weight)	Usual weight
Characteristics			
Intrinsic	(0.79906, 1.18732, 1.75441)	(0.1780, 0.3919, 0.8530)	0.3906
Geological	(0.52276, 0.84352, 1.25146)	(0.1164, 0.2785, 0.6085)	0.2704
Mining	(0.73485, 0.99848, 1.48378)	(0.1637, 0.3296, 0.7214)	0.3390
$(\tilde{Z}_i \oplus \cdots \oplus \tilde{Z}_n)^{-1}$	(0.22273, 0.33011, 0.48622)		$\Sigma - 1.000$

Table 13 Major pairwise comparison matrix between 5 intrinsic characteristics

Item	Rank of coal (RC)	Pyrite content (PC)	Moisture content (MC)	Ash content (AC)	Gas emission (GE)
RC	(1.000, 1.000, 1.000)	(0.778, 1.064, 1.400)	(0.778, 1.170, 1.800)	(0.778, 1.294, 1.800)	(0.778, 1.064, 1.800)
PC	(0.714, 0.940, 1.286)	(1.000, 1.000, 1.000)	(0.714, 1.100, 1.800)	(0.714, 1.216, 1.800)	(0.556, 1.000, 1.400)
MC	(0.556, 0.855, 1.286)	(0.556, 0.909, 1.400)	(1.000, 1.000, 1.000)	(1.000, 1.106, 1.800)	(0.556, 0.909, 1.400)
AC	(0.556, 0.773, 1.286)	(0.556, 0.822, 1.400)	(0.556, 0.904, 1.000)	(1.000, 1.000, 1.000)	(0.556, 0.822, 1.000)
GE	(0.556, 0.940, 1.286)	(0.714, 1.000, 1.800)	(0.714, 1.100, 1.800)	(1.000, 1.216, 1.800)	(1.000, 1.000, 1.000)

Table 14 Fuzzy and usual final weight of 5 intrinsic characteristics

Item	$\tilde{Z}_i$ (α, δ, γ)	$\tilde{W}_i$ (Fuzzy weight)	Usual weight
Characteristics			
Rank of coal	(0.81787, 1.11350, 1.52191)	(0.1182, 0.2218, 0.4171)	0.2220
Pyrite content	(0.72656, 1.04696, 1.42286)	(0.1050, 0.2086, 0.3899)	0.2044
moisture Content	(0.70281, 0.95195, 1.35311)	(0.1016, 0.1896, 0.3708)	0.1926
Ash content	(0.62486, 0.86067, 1.12475)	(0.0903, 0.1714, 0.3082)	0.1684
Gas emission	(0.77713, 1.04696, 1.49621)	(0.1123, 0.2086, 0.4100)	0.2126
$(\tilde{Z}_i \oplus \cdots \oplus \tilde{Z}_n)^{-1}$	(0.14453, 0.19920, 0.27403)		$\Sigma = 1.000$

Table 15 Major pairwise comparison matrix between 5 geological characteristics

Item	Depth (D)	Thickness (T)	Slope (S)	Faults (F)	Unworkable seam (US)
D	(1.000, 1.000, 1.000)	(1.000, 1.507, 3.000)	(1.400, 2.769, 9.000)	(0.778, 1.196, 1.800)	(0.778, 1.483, 3.000)
T	(0.333, 0.664, 1.000)	(1.000, 1.000, 1.000)	(1.000, 1.838, 7.000)	(0.429, 0.794, 1.400)	(0.429, 0.984, 2.333)
S	(0.111, 0.361, 0.714)	(0.143, 0.544, 1.000)	(1.000, 1.000, 1.000)	(0.111, 0.432, 1.000)	(0.143, 0.535, 1.667)
F	(0.556, 0.836, 1.286)	(0.714, 1.260, 2.333)	(1.000, 2.315, 9.000)	(1.000, 1.000, 1.000)	(0.556, 1.239, 3.000)
UC	(0.333, 0.675, 1.286)	(0.429, 1.016, 2.333)	(0.600, 1.868, 7.000)	(0.333, 0.807, 1.800)	(1.000, 1.000, 1.000)

Table 16 Fuzzy and usual final weight of 5 geological characteristics

Item	$\tilde{Z}_i$ (α, δ, γ)	$\tilde{W}_i$ (Fuzzy weight)	Usual weight
Characteristics			
Depth	(0.96731, 1.49228, 2.70864)	(0.0959, 0.2829, 0.9150)	0.2928
Thickness	(0.57199, 0.99038, 1.87000)	(0.0567, 0.1877, 0.6317)	0.1895
Slope	(0.19066, 0.53886, 1.03549)	(0.0189, 0.1021, 0.3498)	0.0881
Faults	(0.73904, 1.24749, 2.40822)	(0.0732, 0.2365, 0.8136)	0.2425
Unworkable seam	(0.49112, 1.00656, 2.06775)	(0.0487, 0.1908, 0.6985)	0.1872
$(\tilde{Z}_i \oplus \cdots \oplus \tilde{Z}_n)^{-1}$	(0.09911, 0.18955, 0.33782)		$\Sigma = 1.0000$

Table 17 Major pairwise comparison matrix between 5 mining characteristics

Item	Extraction method (EM)	Advancing direction (AD)	Rate of advance (RA)	Filling method (FM)	Exploitation factor (EF)
EM	(1.000, 1.000, 1.000)	(0.778, 1.189, 1.800)	(0.778, 1.069, 1.286)	(0.778, 1.112, 1.400)	(1.000, 1.150, 1.800)
AD	(0.556, 0.841, 1.286)	(1.000, 1.000, 1.000)	(0.556, 0.899, 1.286)	(0.714, 0.935, 1.286)	(0.714, 0.967, 1.400)
RA	(0.778, 0.935, 1.286)	(0.778, 1.112, 1.800)	(1.000, 1.000, 1.000)	(0.778, 1.040, 1.800)	(1.000, 1.075, 1.400)
FM	(0.714, 0.899, 1.286)	(0.778, 1.070, 1.400)	(0.556, 0.962, 1.286)	(1.000, 1.000, 1.000)	(0.714, 1.034, 1.400)
EF	(0.556, 0.870, 1.000)	(0.714, 1.034, 1.400)	(0.714, 0.930, 1.000)	(0.714, 0.967, 1.400)	(1.000, 1.000, 1.000)

Table 18 Fuzzy and usual final weight of 5 mining characteristics

Characteristics	$\tilde{Z}_i$ (α, δ, γ)	$\tilde{W}_i$ (Fuzzy weight)	Usual weight
Extraction method	(0.86003, 1.10218, 1.42286)	(0.1323, 0.2200, 0.3670)	0.2203
Advancing direction	(0.69094, 0.92662, 1.24369)	(0.1063, 0.1850, 0.3208)	0.1849
Rate of advance	(0.86003, 1.03073, 1.42286)	(0.1323, 0.2058, 0.3670)	0.2155
Filling method	(0.73904, 0.99112, 1.26505)	(0.1137, 0.1979, 0.3263)	0.1944
Exploitation factor	(0.72656, 0.95846, 1.14407)	(0.1118, 0.1913, 0.2951)	0.1849
$(\tilde{Z}_i \oplus \cdots \oplus \tilde{Z}_n)^{-1}$	(0.15388, 0.19964, 0.25796)		$\Sigma = 1.0000$

3.1.5 Definition of CSCPI and vulnerability index ranges

In the fifth step, rating of Coal Spontaneous Combustion Potential index (CSCPI) for each coal seam can be calculated according to Eq. (9) (modified after Hudson 1992).

$$CSCPI_j = \sum_{i=1}^{15} a_i \frac{P_{ij}}{P_{\mathrm{Max}i}} \tag{9}$$

where, i refers to parameters (1 to 15); j refers to number of seams; a_i is the weighting of each parameter (%) (obtained from Table 19); P_{ij} is the rating assigned to different

Table 19 Ultimate weights of characteristics in coal spontaneous combustion potential resulting from FDAHP method

Parameter	Ultimate weight	Ultimate weight (%)
Rank of coal	0.0867	8.67
Pyrite content	0.0798	7.98
Moisture content	0.0752	7.52
Ash content	0.0658	6.58
Gas emission in seam	0.0830	8.30
Depth of cover	0.0792	7.92
Thickness of seam	0.0512	5.12
Slope of seam	0.0238	2.38
Exist of tectonic faults in seam	0.0656	6.56
Unworkable coal seam in vicinity of extracting coal seam	0.0506	5.06
Extraction method	0.0747	7.47
Advancing direction	0.0627	6.27
Rate of advance	0.0730	7.30
Filling of extracted area method	0.0659	6.59
Exploitation factor	0.0627	6.27

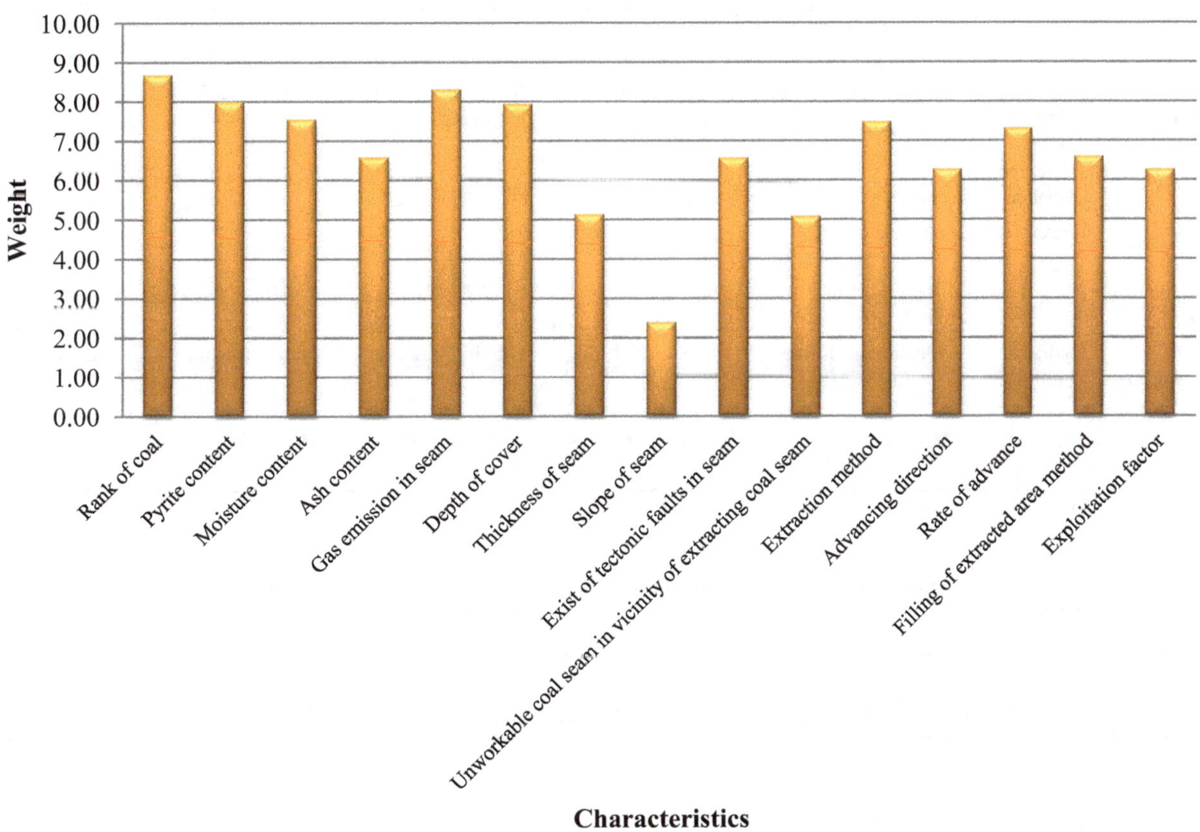

Fig. 8 Histogram ultimate weights of characteristics in coal spontaneous combustion potential resulting from FDAHP method

Table 20 Proposed ranges and ratings for effective parameters in coal spontaneous combustion

Parameter	Suggested ranges and ratings					
	0	1	2	3	4	5
Rank of coal	Anthracite groups	Bituminous groups	Lignite	Sub-bituminous groups	–	–
Pyrite content (%)	$P_2<1$	$1\leq P_2<2$	$2\leq P_2<5$	$5\leq P_2<8$	$8\leq P_2<10$	$P_2\geq 10$
Moisture content (%)	$P_3<3$	$3\leq P_3<5$	$5\leq P_3<7$	$7\leq P_3<12$	$12\leq P_3<24$	$P_3\geq 24$
Ash content (%)	$P_4\geq 60$	$40\leq P_4<60$	$20\leq P_4<40$	$10\leq P_4<20$	$P_4<10$	–
Gas emission in seam (m^3/ton)	$P_5\geq 15$	$10\leq P_5<15$	$5\leq P_5<10$	$P_5<5$	–	–
Depth of cover (m)	$50\leq P_6<200$	$200\leq P_6<400$	$400\leq P_6<600$	$600\leq P_6<800$	$P_6<50$ $P_6\geq 800$	–
Thickness of seam (m)	$P_7<1.2$	$1.2\leq P_7<2.4$	$2.4\leq P_7<3.2$	$3.2\leq P_7<5$	Thick seam with multi-cut	
Slope of seam (degree)	$P_8<15$	$15\leq P_8<30$	$30\leq P_8<45$	$45\leq P_8<60$	$60\leq P_8\leq 90$	–
Exist of tectonic faults in seam	Not available	Less than 2 small fault/500 m	More than 2 small fault/500 m	Big fault with great failures	–	–
Unworkable coal seam in vicinity of extracting coal seam	Not available	Spacing of 2 seams is more than 30 times extraction seam thickness	Spacing of 2 seams is between 20-30 times extraction seam thickness	Spacing of 2 seams is between 10-20 times extraction seam thickness	Spacing of 2 seams is between 5-10 times extraction seam thickness	Spacing of 2 seams is less than 5 times extraction seam thickness
Extraction method	Mechanized long wall with shearer loader	Short wall mining	Semi-mechanized long wall with Plow	Room and pillar mining	Traditional long wall	
Advancing direction	Retreat	Advance	–	–	–	–
Rate of advance (m/d)	$P_{13}\geq 5$	$4\leq P_{13}<5$	$3\leq P_{13}<4$	$2\leq P_{13}<3$	$1\leq P_{13}<2$	$P_{13}<1$
Filling of extracted area method	Hydraulic filling	Pneumatic filling	Mechanical filling	Gravity stowing	Hand stowing	Caving method
Exploitation factor	$P_{15}\geq 80$	$65\leq P_{15}<80$	$50\leq P_{15}<65$	$30\leq P_{15}<50$	$P_{15}<30$	–

Table 21 Classification of coal spontaneous combustion potential index (modified after Mazzoccola and Hudson 1996)

Risk significance	Low	Medium	High
Category	I	II	III
Coal Spontaneous Combustion Potential index (CSCPI)	$0\leq CSCPI>33$	$33\leq CSCPI>66$	$66\leq CSCPI\leq 100$

classes of parameter i value, different for different seams j (obtained from Table 20); P_{Maxi} is the maximum value rating of parameter i (obtained from Table 20); $P_{\text{Max}i}$ is for normalization by dividing with the maximum rating; $CSCPI_j$ is the coal spontaneous combustion potential index of each seam; the maximum value of the index is 100 that refers to the most unfavorable conditions, and the minimum index is 0 that refers to the most favorable conditions.

Vulnerability index is in a range of 100 points that can be divided into five or three areas. In this range, high levels of classification, indicating the vulnerability of

Fig. 9 Location map showing position of Tazareh coal mines in Eastern Alborz, Iran

inappropriate conditions, and by reducing the amounts of index, improved conditions (Mazzoccola and Hudson 1996). In this study, Based on the CSCPI system, the propensity of spontaneous combustion of coal can be classified into three potential levels: low, medium and high risk significance, as shown in Table 21.

3.2 Case study: Eastern Alborz Coal Mines

The eastern Alborz coal mines are located in the Alborz Mountains. They include two major mining areas including Tazareh (including Pashkalat, Kelariz, and Razmja mines) and Olang. In order to classify the coal spontaneous combustion potential in the Eastern Alborz coal mines by means of the FDAHP approach, the Pashkalat, Kelariz, and Razmja coal fields of Tazareh were selected (Fig. 9).

4 Results

Each seam of coal fields should be rated to calculate the intrinsic, geological, and mining characteristic-related CSCPI. The rating of each seam is presented in Table 22. It should be noted that obtaining the intrinsic, geological, and mining characteristics of each seam under study involves field sampling and laboratory works. For short, the details are avoided, and just the category is provided, assigned according to Table 20.

After calculating CSCPI for each seam, based on value ranges which presented in Table 21, each coal seam was classified. The proposed classification is given in Table 23. It is noteworthy that at present this classification and indexing in order to determine the scope of the index on each class, of the evidence and the events occurred in each class in the past has been used. The comparison demonstrated relatively good concordance. That indicate use of a systematic approach in analyzing of the potential of coal spontaneous combustion seams in large scale and in the issues of multiple factors can be very useful.

5 Discussion and conclusions

One of the biggest challenges to safety, economic, and environment in coal mines is coal spontaneous combustion, which depends on many parameters. Therefore, prediction of coal spontaneous combustion considering all the effective parameters is very difficult.

In this research work, after primary studies, the most important parameters affecting coal spontaneous combustion (15 parameters) in underground coal mines were selected, the whole Fuzzy Delphi Analytic Hierarchy Process (FDAHP) method was formulated the weight of characteristics affecting the coal spontaneous combustion was evaluated, and the effect of each parameter on the spontaneous combustion potential was estimated. In this paper, a new classification was developed based on the fuzzy Delphi analytic hierarchy process (FDAHP) technique and using the intrinsic, geological, and mining characteristics, named CSCPI. The CSCPI method can be widely used in the evaluation of coal spontaneous combustion studies and industrial applications.

Information has an important practical use, and, for instance, has implications on site characterization since it allows the designer to identify parameters that should be characterized in more detail in any particular case. For example, the results obtained showed that the parameter "rank of coal" had the highest expected interaction with the system. Similarly, "gas emission in seam", "pyrite content", "depth of cover", "moisture content", "extraction method", "rate of advance", and "filling of extracted area method" have the highest weight in the system, and so a small change in these parameters will affect, to a considerable extent, the behavior of the system.

In this study, the rating of each coal seam based on CSCPI was calculated. The CSCPI results indicate that the K5 seam of Pashkalat Coal Mines was in the high risk state (Category III). Comparison between the results of CSCPI and spontaneous combustion events shows that there is a relatively good concordance. Thus the resulting

Table 22 Calculation of CSCPI for seams of coal fields

Item	Rating of parameters															
Weight of each parameter $a_i(\%)$	8.67	7.98	7.52	6.58	8.30	7.92	5.12	2.38	6.56	5.06	7.47	6.27	7.30	6.59	6.27	$\sum_{i=1}^{15} a_i = 100$
P_{Maxi}	3	5	5	4	3	4	4	4	3	5	4	1	5	5	4	–
Parameters Seam	P_1	P_2	P_3	P_4	P_5	P_6	P_7	P_8	P_9	P_{10}	P_{11}	P_{12}	P_{13}	P_{14}	P_{15}	CSCPI
Pashkalat																
P18	1	0	0	2	3	0	0	1	0	0	4	1	5	5	1	44.3
P15	1	2	0	2	3	0	0	2	0	0	4	1	5	5	1	48.1
P10	1	1	0	2	3	1	0	1	1	0	4	1	5	5	1	50.0
P3	0	0	0	2	3	1	0	2	0	3	4	1	5	5	1	47.0
K25	0	0	0	2	3	3	0	1	1	0	4	1	5	5	1	49.5
K23	0	0	0	2	3	3	0	2	1	0	4	1	5	5	1	50.1
K21	0	0	0	2	3	3	0	1	1	0	4	1	5	5	1	49.5
K19U	0	2	0	1	3	3	0	2	1	0	4	1	5	5	1	51.7
K19L	0	2	1	2	3	3	0	1	1	0	4	1	5	5	1	54.2
K14	0	0	0	2	3	3	0	2	1	0	4	1	5	5	1	50.1
K13	0	2	0	2	2	3	0	2	1	0	4	1	5	5	1	50.5
K12	0	2	0	1	3	3	0	1	1	0	4	1	5	5	1	51.1
K10	0	2	0	4	3	3	2	3	3	0	4	1	5	5	1	64.1
K9	0	0	0	2	2	3	0	2	0	0	4	1	5	5	1	45.2
K8	0	2	0	2	3	3	0	2	1	0	4	1	5	5	1	53.3
K6	0	2	0	2	3	4	0	2	1	0	4	1	5	5	1	55.3
K5	0	3	0	4	3	4	4	3	3	0	4	1	5	5	1	70.3
Razmja																
K21	0	1	0	2	2	0	0	2	0	0	4	1	5	5	1	40.8
K20	0	2	1	2	2	0	0	2	2	0	4	1	5	5	1	48.3
K19	0	0	1	2	2	0	1	2	0	3	4	1	5	5	1	45.0
K17	0	2	0	2	1	0	2	2	0	4	4	1	5	5	1	46.3
K10	0	4	1	2	1	1	3	2	0	0	4	1	5	5	1	50.2
K8	0	2	2	2	1	1	2	2	2	0	4	1	5	5	1	51.6
K5	0	2	3	2	2	1	3	2	2	0	4	1	5	5	1	57.1
Kelariz																
P18	1	1	0	2	3	0	0	4	0	0	4	1	5	3	1	45.0
P15	1	2	0	2	3	1	0	4	0	0	4	1	5	3	1	48.6
P10	0	1	0	2	3	1	1	4	0	0	4	1	5	3	1	45.4
K28	0	0	0	2	3	3	0	4	0	0	4	1	5	3	1	46.5
K25	0	0	0	2	3	3	1	4	0	0	4	1	5	3	1	47.8
K23	0	1	0	2	3	3	1	4	0	0	4	1	5	3	1	49.4
K19	0	2	0	2	3	3	1	4	0	0	4	1	5	3	1	50.9
K10	0	3	0	2	3	3	1	4	0	0	4	1	5	3	1	52.5
K8	0	2	0	2	2	3	0	4	0	0	4	1	5	3	1	46.9
K6	0	2	0	2	3	4	0	4	0	0	4	1	5	3	1	51.6
K5	0	2	0	3	3	4	1	4	0	0	4	1	5	3	1	54.6
K4	0	2	0	2	3	4	0	4	0	0	4	1	5	3	1	51.6

Table 23 Final ranking obtained for seams of coal fields

Seam	CSCPI	Category	Risk description	Status of coal spontaneous combustion
Pashkalat				
P18	44.3	II	Medium	No spontaneous combustion.
P15	48.1			
P10	50.0			
P3	47.0			
K25	49.5			
K23	50.1			
K21	49.5			
K19U	51.7			
K19L	54.2			
K14	50.1			
K13	50.5			
K12	51.1			
K10	64.1			
K9	45.2			
K8	53.3			
K6	55.3			
K5	70.3	III	High	In 2006, spontaneous combustion occurred.
Razmja				
K21	40.8	II	Medium	No spontaneous combustion.
K20	48.3			
K19	45.0			
K17	46.3			
K10	50.2			
K8	51.6			
K5	57.1			
Kelariz				
P18	45.0	II	Medium	No spontaneous combustion.
P15	48.6			
P10	45.4			
K28	46.5			
K25	47.8			
K23	49.4			
K19	50.9			
K10	52.5			
K8	46.9			
K6	51.6			
K5	54.6			
K4	51.6			

classification by this method is reliable and could present a comprehensive view of the situation about the case study.

The results of this research work indicate that utilization of the systematic approach in analyzing the coal spontaneous combustion seams potential in a large scale can be very useful.

Acknowledgment The authors wish to thank the Eastern Alborz coal mines company management and the anonymous experts for their appreciable cooperation and support in this research work.

Funding Funding was provided by Shahrood University of Technology.

References

Akgun F, Arisoy A (1994) Effect of particle size on the spontaneous heating of a coal stockpile. Combust Flame 99:137–146

Arisoy A, Akgun F (1994) Modelling of spontaneous combustion of coal with moisture content included. Fuel 73:281–286

Arisoy A, Beamish B (2015) Mutual effects of pyrite and moisture on coal self-heating rates and reaction rate data for pyrite oxidation. Fuel 139:107–114

Banerjee SC (1985) Spontaneous Combustion of Coal and Mine Fire. Oxford & IBH Publishing, Central Mining Research Station, Dhanbad

Beamish BB (2005) Comparison of the R70 self-heating rate of New Zealand and Australian coals to Suggate rank parameter. Int J Coal Geol 64:139–144

Beamish BB, Arisoy A (2008a) Effect of intrinsic coal properties on self-heating rates. In: 12th U.S./North American mine ventilation symposium (The Society of Mining, Metallurgy and Exploration Inc.), Littleton, USA, pp 149–153

Beamish BB, Arisoy A (2008b) Effect of mineral matter on coal self-heating rate. Fuel 87:125–130

Beamish BB, Beamish R (2010) Benchmarking moist coal adiabatic oven testing. Coal operators' conference, University of Wollongong, Australia, pp 264–268

Beamish BB, Beamish R (2011) Experience with using a moist coal adiabatic oven testing method for spontaneous combustion assessment. Coal operators' conference, University of Wollongong, Australia, pp 380–384

Beamish B, Beamish R (2012) Benchmarking coal self-heating using a moist adiabatic oven test. In: Proceedings of the fourteenth US/ North American mine ventilation symposium, Salt Lake City, pp 423–427

Beamish BB, Blazak DG (2005) Relationship between ash content and R70 self-heating rate of Callide Coal. Int J Coal Geol 64:126–132

Beamish BB, Jabouri I (2005) Factors affecting hot spot development in bulk coal and associated gas evolution. Coal operators' conference, University of Wollongong, Australia, pp 187–193

Beamish BB, Sainsbury W (2008) Development of a site specific self-heating rate prediction equation for a high volatile bituminous coal. Coal operators' conference, University of Wollongong, Australia, pp 161–165

Beamish BB, Schultz TJ (2008) Moisture content impact on the self-heating rate of a highly reactive subbituminous coal. Coal operators' conference, University of Wollongong, Australia, pp 155–160

Beamish BB, Blazak DG, Hogarth LC, Jabouri I (2005) R70 Relationships and their interpretation at a mine site. Coal operators' conference, University of Wollongong, Australia, pp 183–185

Beamish BB, Lin Z, Beamish R (2012) Investigating the influence of reactive pyrite on coal self-heating. Coal operators' conference, University of Wollongong, Australia, pp 294–299

Beamish BB, McLellan P, Endara H, Turunc U, Raab M, Beamish R (2013) Delaying spontaneous combustion of reactive coals through inhibition. Coal operators' conference, University of Wollongong, Australia, pp 221–226

Bhat S, Agarwal PK (1996) The effect of moisture condensation on the spontaneous combustibility of coal. Fuel 75:1523–1532

Botao Q, Qingguo S, Deming W, Leilin Z, Qin X (2009) Analysis and key control technologies to prevent spontaneous coal combustion occurring at a fully mechanized caving face with large obliquity in deep mines. J Min Sci Technol 19:446–451

Cao D, Fan X, Guan H, Wu C, Shi X, Jia Y (2007) Geological models of spontaneous combustion in the Wuda coalfield, Inner Mongolia, China. Geol Soc Am 18:23–30. doi:10.1130/2007.4118(03)

Carras JN, Young BC (1994) Self-heating of coal and related materials: models, application and test methods. Prog Energy Combust Sci 20:1–15

Chen C, Klein CM (1997) An efficient approach to solving fuzzy MADM problems. Fuzzy Sets Syst 88:51–67

Deng H (1999) Multi criteria analysis with fuzzy pair-wise comparison. Int J Approx Reason 21:215–231

Deng J, Ma X, Zhang Y, Li Y, Zhu W (2015) Effects of pyrite on the spontaneous combustion of coal. Int J Coal Sci Technol 2(4):306–311

Ghosh R (1986) Spontaneous combustion of certain Indian coals-some physico-chemical considerations. Fuel 65(8):1042–1046

Hoseinie SH, Ataei M, Osanloo M (2009) A new classification system for evaluating rock penetrability. Int J Rock Mech Min Sci 46:1329–1340

Hudson J (1992) Rock engineering systems: theory and practice. Ellis Horwood, Chichester

Hwang CL, Lin ML (1987) Group decision making under multiple criteria. Springer, Berlin

Ishikawa A, Amagasa M, Shiga T, Tomizawa G, Tatsuta R, Mieno H (1993) The max-min Delphi method and fuzzy Delphi method via fuzzy integration. Fuzzy Sets Syst 55:241–253

Jones JC (2001) On the extrapolation of results from oven heating tests for propensity to self-heating. Combust Flame 124(1–2): 334–336

Kadioğlu Y, Varamaz M (2003) The effect of moisture content and air drying on spontaneous combustion characteristics of two Turkish lignites. Fuel 82(13):1685–1693

Kaufmann A, Gupta MM (1988) Fuzzy mathematical models in engineering and management science. North-Holland, Amsterdam

Kuchta JM, Rowe VR, Burgess DS (1980) Spontaneous Combustion Susceptibility of US Coals. U.S. Dept. of the Interior, Bureau of Mines Report of Investigation 8474

Kucuk A, Kadioglu Y, Gulaboglu MS (2003) A study of spontaneous combustion characteristics of a Turkish lignite: particle size, moisture of coal, humidity of air. Combust Flame 133(3):255–261

Liu YC, Chen CS (2007a) A new approach for application of rock mass classification on rock slope stability assessment. Eng Geol 89:129–143

Liu YC, Chen CS (2007b) A new approach to classification base on association rule mining. J Decis Support Syst 42:674–689

Mazzoccola DF, Hudson JA (1996) A comprehensive method of rock mass characterization for indicating natural slope instability. Q J Eng Geol 29:37–56

Michaylov M (2002) Expert system for assessment of risk from spontaneous combustion. Min Miner Process 44–45:27–35

Mikaeil R, Ozcelik Y, Yousefi R, Ataei M, Hosseini SM (2013) Ranking the sawability of ornamental stone using Fuzzy Delphi and multi-criteria decision-making techniques. Int J Rock Mech Min Sci 58:118–126

Mitchell DW (1990) MINE FIRES Prevention, Detection, Fighting. Maclean Hunter Publishing Company, Chicago

Mohalik N, Singh R, Singh V, Tripathi D (2009) Critical appraisal to assess the extent of fire in old abandoned Coal Mine Areas-

Indian Context. Coal Operators' Conference, University of Wollongong, Australia, pp 271-280

Nelson MI, Chen XD (2007) Survey of experimental work on the self-heating and spontaneous combustion of coal. Geol Soc Am Rev Eng Geol 18:31–83. doi:10.1130/2007.4118(04)

Nugroho YS, McIntosh AC, Gibbs BM (2000a) Low-temperature oxidation of single and blended coals. Fuel 79:1951–1961. doi:10.1016/S0016-2361(00)00053-3

Nugroho YS, McIntosh AC, Gibbs BM (2000b) On the prediction of thermal runaway of coal piles of differing dimension by using a correlation between heat release and activation energy. Twenty-Eighth Symposium (International) on Combustion: Pittsburgh. The Combustion Institute, Pennsylvania, pp 2321–2327

Ramlu MA (2007) Mine disasters and mine rescue, 2nd edn. chapter 1, mine fires. Universities Press, Hyderabad, p 2–15

Ren TX, Edwards JS, Clarke D (1999) Adiabatic oxidation study on the propensity of pulverized coals to spontaneous combustion. Fuel 78:1611–1620

Ren TX, Balusu R, Humphries P (2005) Development of innovative goaf inertisation practices to improve coal mine safety. Coal operators' conference, University of Wollongong, Australia, pp 315–322

Saaty TL (1980) The analytic hierarchy process. McGraw- Hill, New York

Saaty TL (1990) The analytic hierarchy process series. University of Pittsburg, Pittsburg

Sasaki K, Wang Y, Sugai Y, Zhang X (2014) Numerical modelling of low rank coal for spontaneous combustion. Coal operators' conference, University of Wollongong, Australia, pp 344–349

Singh AK, Singh RVK, Singh MP, Chandra H, Shukla NK (2007) Mine fire gas indices and their application to Indian underground coal mine fires. Int J Coal Geol 69(3):192–204

Smith MA, Glasser D (2005) Spontaneous combustion of carbonaceous stockpiles. Part II: factors affecting the rate of the low-temperature oxidation reaction. Fuel 84:1161–1170

Soltanmohammad N, Modiri M, Fathi Hafashjani K (2013) Ranking important factors influencing organizational strategy in selecting distribution channels with the approach of FMCDM. Uncertain Supply Chain Manag 1:253–263

Sujanti W, Zhang DK (1999) A laboratory study of spontaneous combustion of coal: the influence of inorganic matter and reactor size. Fuel 78(5):549–556

Sujanti W, Zhang DK, Chen XD (1999) Low-temperature oxidation of coal studied using wire-mesh reactors with both steady-state and transient methods. Combust Flame 117:646–651

Valls A, Torra V (2000) Using classication as an aggregation tool in MCDM. Fuzzy Sets Syst 115:159–168

Wang TC, Chen YH (2007) Applying consistent fuzzy preference relations to partnership selection. Int J Manag Sci 35:384–388

Wang H, Dlugogorski BZ, Kennedy EM (2003) Coal oxidation at low temperatures: oxygen consumption, oxidation products, reaction mechanism and kinetic modeling. J Prog Energy Combust Sci 29:487–513

Yuan L, Smith AC (2012) The effect of ventilation on spontaneous heating of coal. J Loss Prev Process Ind 25:131–137

Zadeh LA (1965) Fuzzy sets. Inf Control 8:338–353

Zadeh LA (1999) Some reflections on the anniversary of Fuzzy Sets and Systems. Fuzzy Sets Syst 100:1–3

13

Subsidence rules of underground coal mines for different soil layer thickness: Lu'an Coal Base as an example, China

Haifeng Hu[1] · **Xugang Lian**[1]

Abstract Damage caused by underground coal mining is a serious problem in mining areas in China; therefore, studying and obtaining the rules of ground movement and deformation under different geological conditions is of great importance. The numerical software ANSYS was used in this study to simulate mining processes under two special geological conditions: (1) thick unconsolidated soil layer and thin bedrock; (2) thin soil layer and thick bedrock. The rules for ground movement and deformation for different soil layer to bedrock ratios were obtained. On the basis of these rules, a prediction parameter modified model of the influence function was proposed, which is suitable for different values of unconsolidated soil layer thickness. The prediction results were verified using two sets of typical field data.

Keywords Numerical simulation · Ground movement and deformation · Prediction of ground subsidence · Subsidence rules · Ratio of soil layer to bedrock

1 Introduction

Underground coal mining causes deformation and movement of strata and ground surface. Some protected objects, such as houses and railways, around the coal panel would be damaged if a mining plan was carried out without considering the consequences. Mining subsidence has been studied for more than 60 years and some general rules have been obtained. Strata with different properties will follow different deformation rules. In this paper, we focused on strata consisting of two types of components: bedrock lying above a coal seam and the unconsolidated soil layer. The ratio of soil layer to bedrock is very important. When the ratio is large, the soil layer is thick and the bedrock is thin; if the ratio is small, the soil layer is thin and the bedrock is thick. The numerical simulation software ANSYS was used to analyze the ground deformation and movements of underground mining for different values of the soil layer to bedrock ratio. A prediction model using an influence function was used in this paper to model the subsidence for different soil layer thickness.

Numerical simulation is an effective way to analyze geotechnical engineering mechanics. A number of previous studies of mining subsidence have used numerical simulation software. For example, on the basis of the specific geological and mining conditions of certain coal panels in China, a series of numerical models for different unconsolidated layer thicknesses were established using the simulation software FLAC3D. The relationship between unconsolidated layer thickness and surface movement was studied (Dai et al. 2011). Based on mining under thick unconsolidated layers and the specific geological conditions, a numerical model of strip mining under thick unconsolidated layers was established, and the rules of ground movement for this condition were established (Xu and Li 2012). A numerical model using the finite difference method to simulate the safety problems of a dam near the Wutong coal mine area induced by underground mining was constructed. Four scenarios were simulated in the Wutong coal mine area, the best choice being the model in

✉ Haifeng Hu
tyhhf65@163.com

1 College of Mining Engineering, Taiyuan University of Technology, Taiyuan 030024, China

which the nearest distance from the boundary of the surface movement area to the edge of the foundation is 35 m (Xu and Kulatilake 2013). By studying the regularity of strata movements in deep mining, a coupled model involved unconsolidated layer and bedrock of deep mining strata composite medium was obtained. Deep mining was simulated using FLAC3D, and the coupling relationship between unconsolidated layer and bedrock was analyzed in strata composite layer (Yang and Chen 2011). Coal mine subsidence was simulated using the CSIRO Exploration and Mining software COSFLOW. In addition to the major benefits compared to conventional continuum numerical models of sedimentary rock, this software is able to simulate rock breakage and slip efficiently as well as separation along bedding planes (Guo et al. 2004). ANSYS was used to simulate the dynamic changes and tendency of overburden movement and bed development during strip mining and stopping under railways. The results showed that movement and rheological changes in key strata are some of the major causes of surface subsidence and induced deformation of surface railways (Wang et al. 2008). Using actual project data, the rules governing displacement and deformation damage of overlying strata and the surface were obtained using FLAC3D, and a reasonable strip mining method and interrelated parameters were presented for the specific geological and mining conditions of coal. In addition, some research has been carried out using an influence function prediction model in conditions with a thick soil layer (Ji and Yu 2011). The mechanism of surface subsidence caused by pressure consolidation of loose layers has been explored. If the loose layer is significantly thick, it is inevitable that pressure consolidation of a water-bearing loose layer or deformation of the soil body will cause additional stress in the mining soil body, and dissipation of the excess pore water pressure will result in consolidation. Using the surface subsidence features of thick loose layers in mining areas, a correction model has been applied using the probability integral method to predict subsidence with satisfactory predictive results (Zhang et al. 2012). Based on the conditions of the Huainan coal mining area in China, by combining physical simulation, theoretical analysis and experimental data, the subsidence mechanism induced by underground mining of thick unconsolidated soil layer was studied, and a prediction model for this condition was established (Zhou et al. 2014). Quantitative prediction of mining subsidence under thin bedrock and a thick unconsolidated layer was obtained by means of artificial neural networks (ANN). The improved neural network was used for modeling and predicting mining subsidence. The ANN output can reflect changes in the trends of ground movement and deformation. The forecast results are in good agreement with actual observations (Yang and Xia 2013).

Previous studies have dealt with the general rules of ground movement for different strata conditions based on simulation software and analysis of field data, which obviously addresses the influence of bedrock and the unconsolidated layer. However, previous research did not directly transfer the rules of mining under unconsolidated layer and bedrock to the classic influence function of mining subsidence prediction, and the research results cannot be used to predict the value of surface movement and deformation. In this study, different ratios of soil layer and bedrock are added to the influence-function prediction method: this study is the first time that this has been done. In this paper, the ground movement rules for different geological conditions are studied in Sects. 1 and 2 and the prediction model were described in Sect. 3.

2 Numerical simulation of two typical geological conditions of underground coal mines in Lu'an Coal Base

2.1 Thick soil layer and thin bedrock

In the numerical simulation based on the Sima coal mine in Southern Shanxi, China, the key parameters of the selected geological section were: mining depth 200 m, mining height 6.7 m, bedrock thickness 20 m, soil layer thickness 180 m, and a horizontal coal seam.

The model set up was as follows: model size 600 m × 216.7 m, extracted zone size 164 m × 6.7 m, and mining height 6.7 m. At the bottom boundary of the model, a displacement of zero was applied in the vertical direction, and at the left and right boundaries a displacement value of zero was applied in the horizontal direction. A gravity value of 9.8 m/s^2 was applied over the whole model. The physical and mechanical parameters of the direct roof and the strata above the coal seam were obtained from rock and mechanical tests of field drill cores.

Figures 1 and 2 show the horizontal and vertical displacement contours in bedrock and the unconsolidated soil layer after extraction. Figures 3 and 4 show the horizontal and vertical displacement curves of the ground after extraction. From Figs. 2 and 4, the results of the model are that the ground maximum subsidence is 20.327 m under gravity loading. After the coal has been extracted, the ground maximum subsidence is 26.564 m, meaning that the practical maximum ground subsidence is 6.237 m in the middle of the goaf. As illustrated in Fig. 3, the maximum horizontal movement is 1.344 m, which occurs at both sides of the goaf approximately 120 m from the goaf center, and approximately 38 m outside the boundary of the goaf. In general, the maximum ground horizontal movement occurs immediately above the inflexion point of

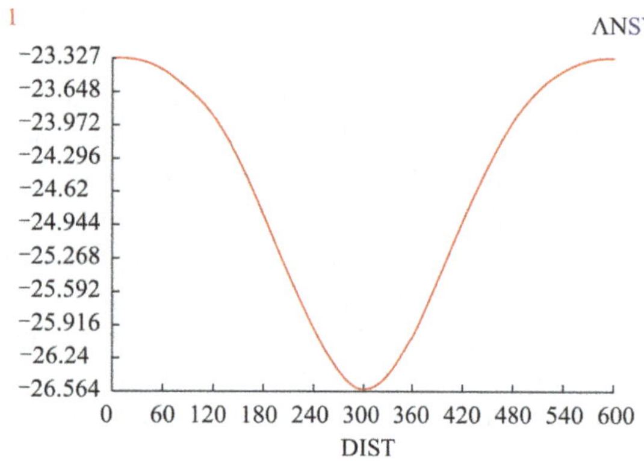

Fig. 1 Curve of ground *vertical* displacement after extraction in the Sima coal mine

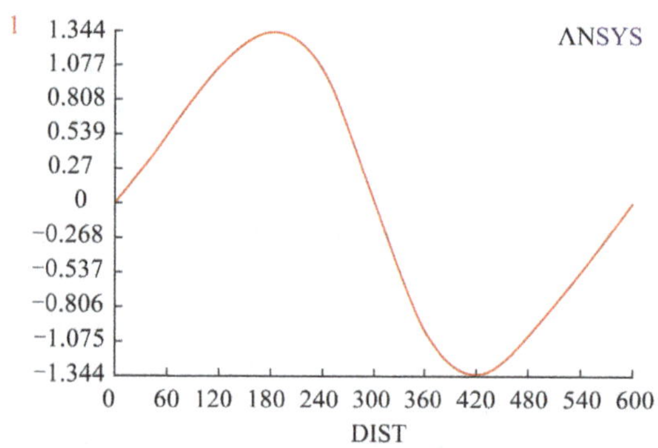

Fig. 2 Curve of ground *horizontal* displacement after extraction in the Sima coal mine

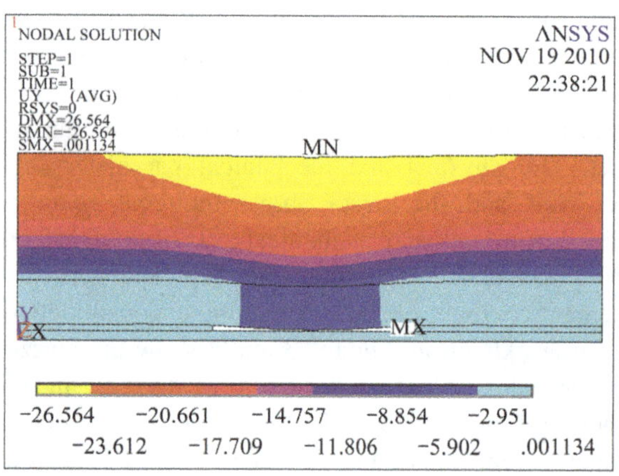

Fig. 3 Contours of *vertical* displacement after extraction in the Sima coal mine

the ground subsidence basin. Because the inflexion point moves outside of the goaf, the ground movement within the basin is gentle and the range of movement is increased. Thus, compared with mining in normal geotechnical conditions, in conditions with a thick soil layer and thin

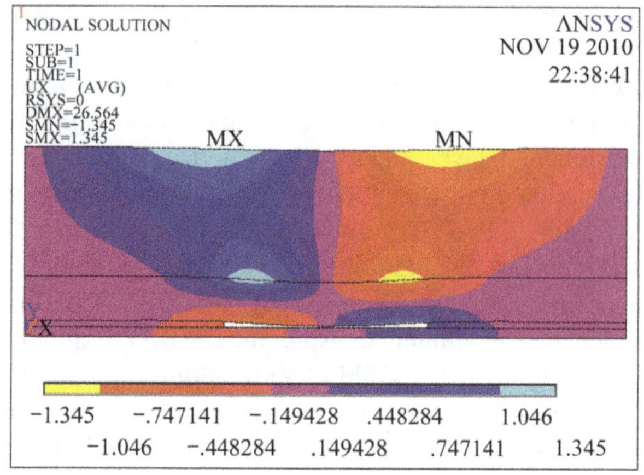

Fig. 4 Contours of *horizontal* displacement after extraction in the Sima coal mine

bedrock, both the range of ground movement and the maximum subsidence are increased.

2.2 Thin soil and thick bedrock

This model was based on the Wuyang coal mine in southern Shanxi Province, China. A section of the geology of this mine was used in the simulation. The key parameters of the selected geological section were as follows: mining depth 242 m, mining height 6 m, bedrock thickness 212 m, soil layer thickness 30 m, and the coal seam was horizontal.

The model size was 800 m × 268 m, the extracted zone size was 180 m × 6 m, and the mining coal seam height was 6 m. The boundary limitation and gravity setup of this model were the same as those of the model described in Sect. 2.1.

As illustrated in Figs. 5 and 6, after coal extraction the maximum ground subsidence above the goaf is 1.397 m. Subtracting the subsidence before extraction caused by gravity loading yields a practical maximum ground

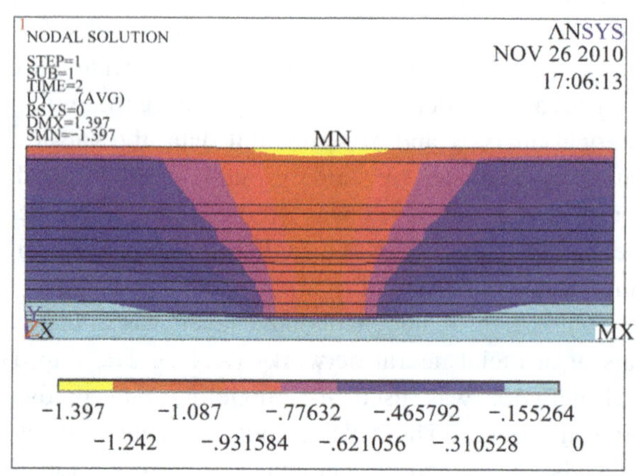

Fig. 5 Contours of vertical displacement after extraction in the Wuyang coal mine

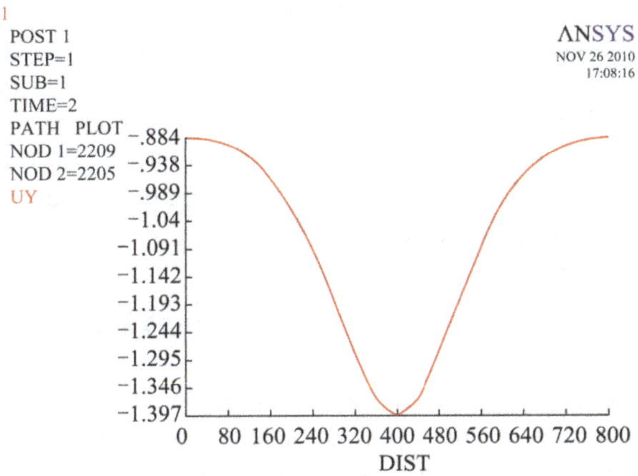

Fig. 6 Curve of ground *vertical* displacement after extraction in the Wuyang coal mine

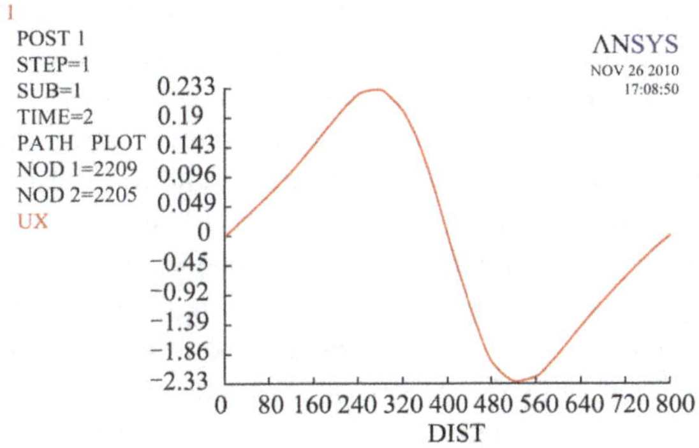

Fig. 8 Curve of ground *horizontal* displacement after extraction in the Wuyang coal mine

subsidence of 0.508 m. As illustrated in Fig. 5, the vertical displacement contour is steeply inclined which is different from the results for the thick soil layer and thin bedrock shown in Fig. 1. In the Wuyang coal mine, the surface soil layer is very thin or thinned out and the slope of the surface basin is steeply inclined.

Figures 7 and 8 showed the maximum horizontal movement is 0.233 m and is located approximately 30 m outside the boundary of the goaf. The range of significant influence is much smaller than that of the model described in Sect. 2.1.

3 Ground mining subsidence rules for different soil layer and bedrock thickness

3.1 Simulation of different thicknesses of unconsolidated soil layer

When the bedrock thickness is constant, the ground movement and deformation rules change as the thickness of the soil layer alters. To study the rules of ground subsidence in conditions of different soil layer thickness, models were established with constant bedrock thickness and variable soil layer thickness. The length of the panel was 200 m, the mining coal seam height was 7 m, the bedrock thickness was 80 m, and the thickness of the unconsolidated soil layer ranged from 20 to 200 m in 20 m increments.

The subsidence and horizontal displacement curves of ground points for different soil layer thicknesses are drawn in Figs. 9 and 10.

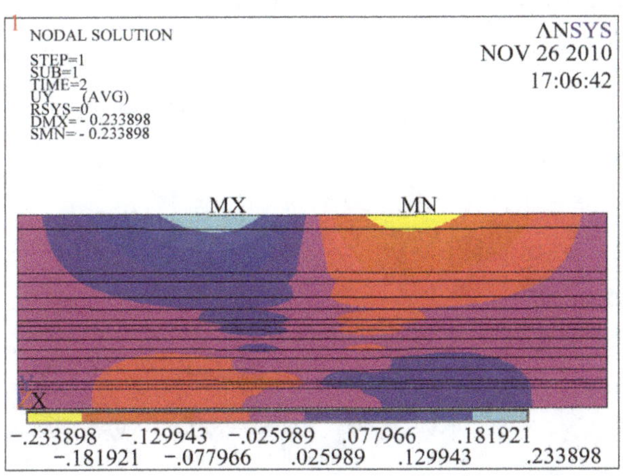

Fig. 7 Contours of *horizontal* displacement after extraction in the Wuyang coal mine

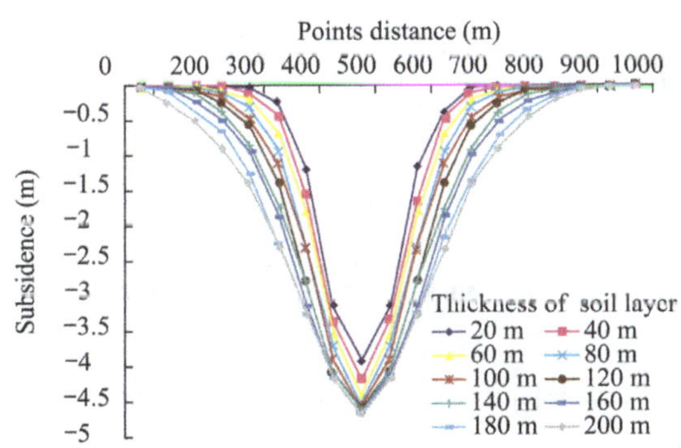

Fig. 9 Ground subsidence curves for different soil layer thickness

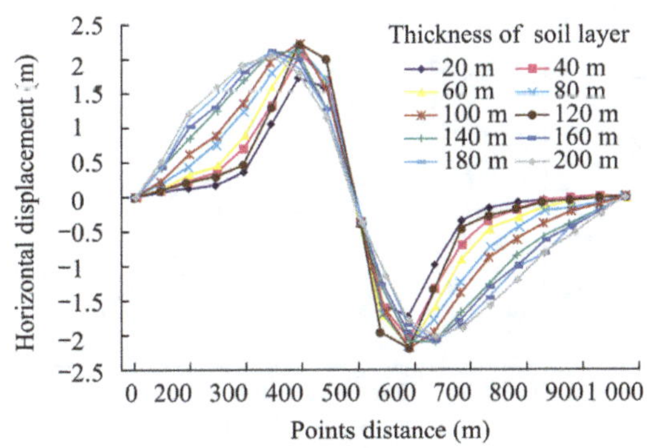

Fig. 10 Ground *horizontal* displacement curves for different soil layer thicknesses

When the bedrock of the model is of constant thickness, as the soil layer thickness or the soil layer to bedrock ratio increases, the subsidence curve changes from steeply inclined to flat, the range of movement deformation expands, and the amount of subsidence and the subsidence factor both increase (Fig. 9). Figure 10 shows that the position of maximum horizontal displacement is not changed, but the values outside the boundaries of goaf are increased. The maximum subsidence point is in the centre of the goaf, and the points of maximum horizontal displacement are at the two side boundaries of the goaf.

The curves of maximum subsidence and horizontal displacement with different soil layer thicknesses (Figs. 11 and 12) reveal some rules of ground movement and deformation.

(1) With increasing soil layer thickness, the ground maximum subsidence describes a logarithmic curve (Fig. 11): for soil layers 20–100 m thick the maximum subsidence increases very quickly, but for thicknesses greater than 100 m the maximum subsidence increases very slowly.

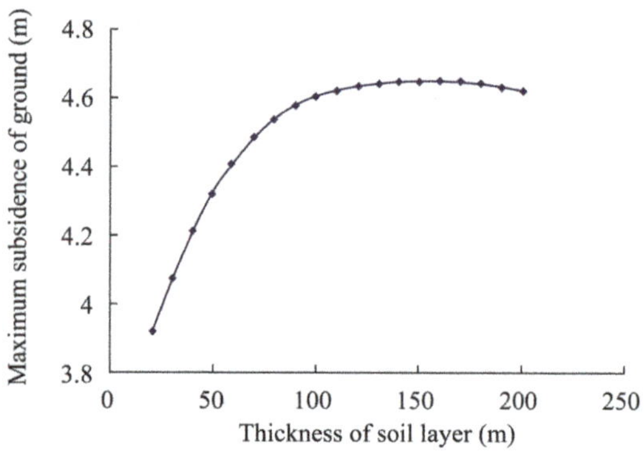

Fig. 11 Relationship between maximum subsidence and soil layer thickness

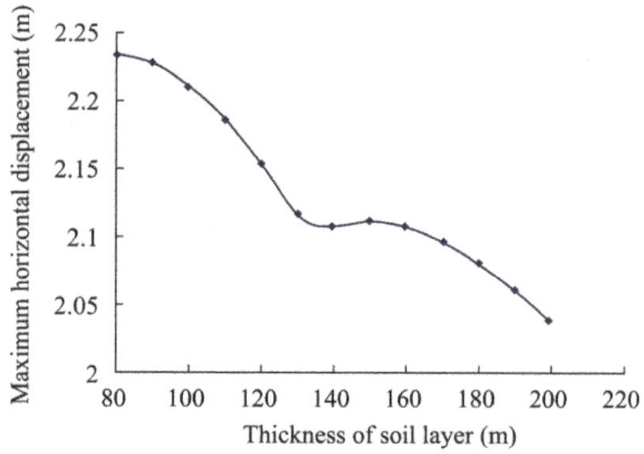

Fig. 12 Relationship between maximum displacement and soil layer thickness

(2) With increasing soil layer thickness, the ground maximum horizontal displacement forms a periodic linearly declining curve (Fig. 12): for soil layer thicknesses between 20 and 130 m, the displacement values decline linearly; between 130 and 160 m, the curve flattens out; and after 160 m, the values again show a linear decrease.

3.2 Simulation of different bedrock thicknesses

When the thickness of the unconsolidated soil layer is constant, the ground deformation and movement rules alter with changes in the bedrock above the coal seam. To study the subsidence rules for different thicknesses of bedrock, a numerical simulation model with constant soil layer thickness and changing bedrock thickness was set up. The model assumptions were that the mining coal seam height was 7 m, the panel length was 200 m, the thickness of the unconsolidated soil layer was 20 m, and the bedrock thickness was 80–160 m in increments of 10 m.

The curves of ground subsidence and horizontal displacement plotted against bedrock thickness are shown in Figs. 13 and 14. Figures 15 and 16 contain the curves of

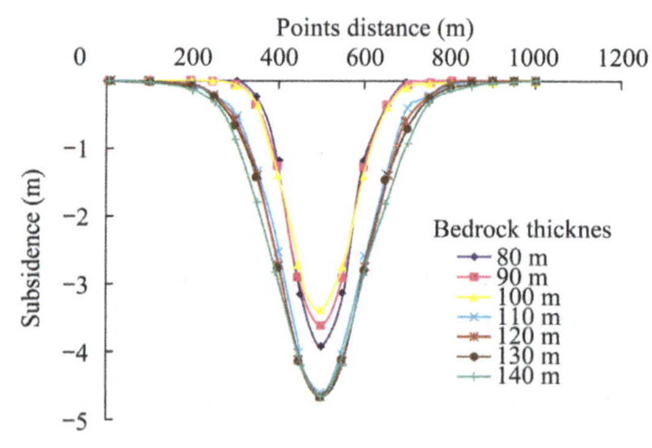

Fig. 13 Ground subsidence curves for different bedrock thicknesses

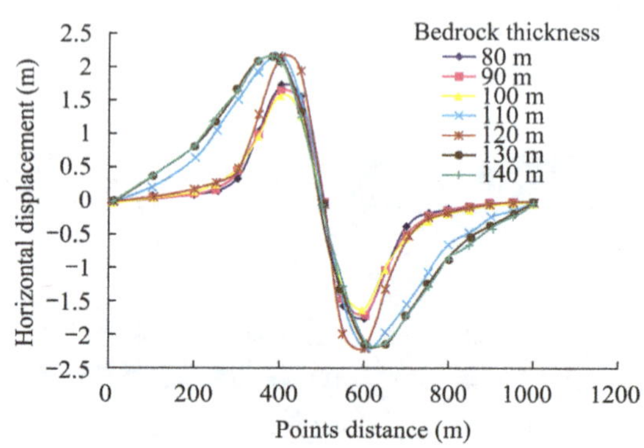

Fig. 14 Ground *horizontal* displacement curves for different bedrock thicknesses

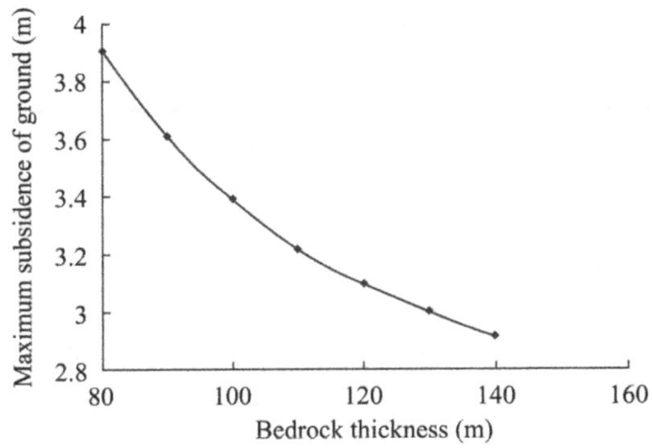

Fig. 15 Relationship between maximum ground subsidence and bedrock thickness

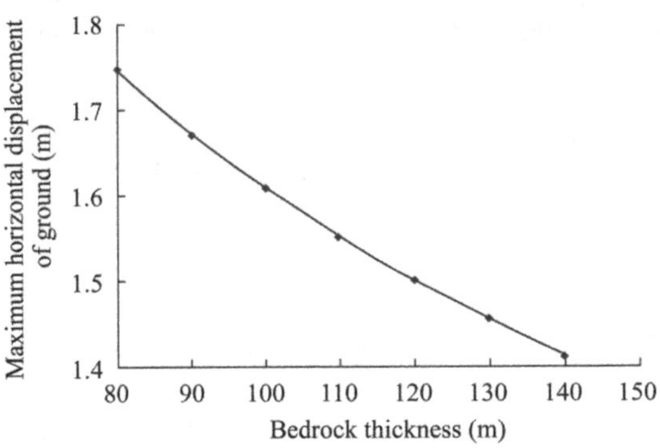

Fig. 16 Relationship between maximum ground displacement and bedrock thickness

maximum subsidence and maximum horizontal displacement plotted against bedrock thickness. The simulation results demonstrate that ground movement and deformation are negatively correlated with bedrock thickness, i.e., when the bedrock is thin the ground movement and deformation are large; when the bedrock is thick the ground movement and deformation are small. The ground maximum subsidence and maximum horizontal displacement decrease as bedrock thickness increases, and the relations between them are negative exponential curves, as illustrated in Figs. 15 and 16.

4 Influence function prediction method on the soil layer to bedrock ratio

The influence function is a general prediction method for strata and ground movement induced by underground coal mining and is suitable for general geological conditions such as when the soil layer is very thin or absent. However, for a very thick soil layer this prediction method has some limitations, for example, it cannot be used to obtain large subsidence factors and a large range of ground movement. The mechanics of mining subsidence are different for the soil layer and for the bedrock.

4.1 Prediction parameters of the influence function modified for a thick unconsolidated soil layer

(1) Main influence radius r

$$r = \frac{H_s}{\tan\varphi} + \frac{H_j}{\tan\beta}, \tag{1}$$

where H_s is the thickness of the unconsolidated soil layer; φ is the displacement angle of the soil layer; H is the bedrock thickness; and $\tan\beta$ is the tangent of the main influence angle.

(2) Prediction parameters

Based on the results of Sects. 2 and 3, it is apparent that the ratio of soil layer to bedrock, i.e., H_s/H_j, is a key factor affecting the ground movement rules. However, because the traditional influence function is hard to predict for the coupled strata of soil and bedrock, we attempted to construct a series of functions incorporating prediction parameters and H_s/H_j. This type of function has the form:

$$K = f\left[\frac{H_s}{H_j}, H_0\right], \tag{2}$$

where K is one of the prediction parameters; H_s/H_j is the soil layer to bedrock ratio; and H_0 is the total mining depth, $H_0 = H_s + H_j$. Each prediction parameter function can be obtained as follows:

(1) Subsidence factor $q = a_1 \frac{H_s}{H_j} + b_1$,
(2) Horizontal displacement factor $b = a_2 \frac{H_s}{H_j} + b_2$,
(3) Tangent of main influence angle $\tan\beta = a_3 \frac{H_s}{H_j} + b_3$,
(4) Propagation angle of extraction $\theta = a_4 \frac{H_s}{H_j} + b_4 - c_4\alpha$,
(5) Deviation of inflection point $s = a_5 \frac{H_s}{H_j} + b_5 H_0$, where a_1 to a_5, b_1 to b_5, and c_4 are constants for a given mining area.

4.2 Case study

On the basis of ground observational data for the Lu'an coal group in southern Shanxi, China (Table 1), the prediction parameter function can be established using regression analysis:

Table 1 Subsidence parameters of observation stations of the Lu'an Group in southern Shanxi, China

Colliery	Panel No.	H_0 (m)	H_s (m)	q	H_s/H_j	b	Tan β	k_i	k_θ
Wuyang	7305 up	212	25	0.72	0.13	0.26	2.50	0.15	1.0
	7305 down	212	25	1.00	0.13	0.33	3.20	0.11	1.0
	7309 down	164	20	0.72	0.14	0.30	2.70	0.13	0.9
	7503	318	30	0.86	0.10	0.33	2.71	0.13	1.0
	7506	329	29	0.82	0.10	0.28	2.70	0.12	0.8
	7511	270	40	0.86	0.17	0.31	2.70	0.14	0.6
Wangzhuang	4326	300	50	0.74	0.20	0.31	2.70	0.11	0.6
Zhengcun	2201	350	62	0.82	0.22	0.25	3.60	0.14	0.6
Sima	1101	242	186	0.94	3.47	0.26	2.70	0.10	0.6
Tunliu	2201	540	40	0.80	0.02	0.30	2.40	0.10	0.7

Notes H_0 is the mining depth of the coal seam; H_s is the thickness of the soil layer above bedrock; H_j is the thickness of bedrock above the coal seam; H_s/H_j is the ratio of soil layer to bedrock; q is the subsidence factor; b is the horizontal displacement factor; tan β is the tangent of the main influence angle; k_i is the factor of deviation of the inflection point; and k_θ is obtained from the function $\theta = 90° - k_\theta\,\alpha$, which is the propagation angle of extraction θ in general geological conditions.

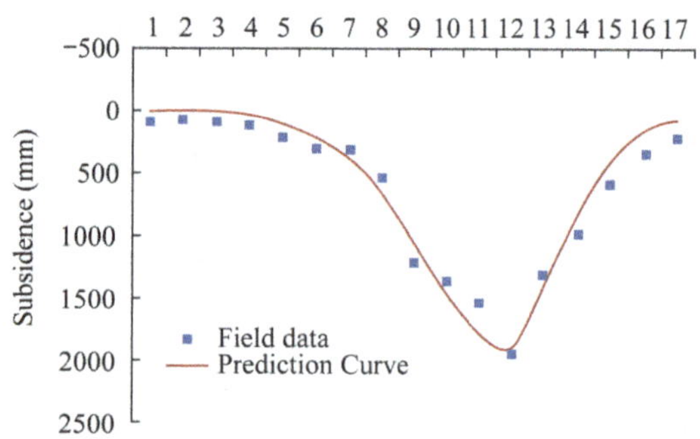

Fig. 17 Difference between predicted and measured subsidence values along the inclined direction of the Wuyang 7506 *panel*

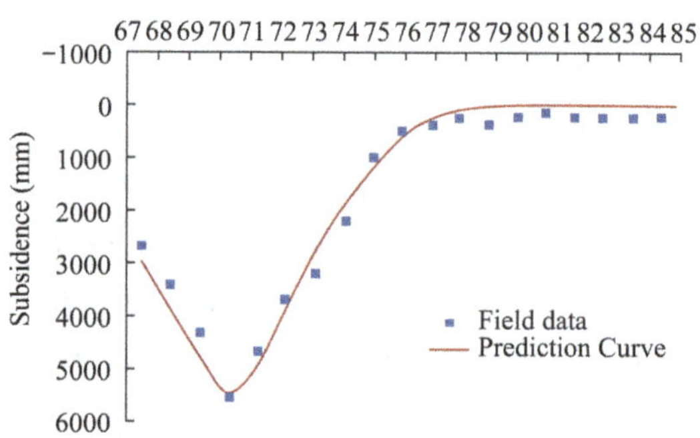

Fig. 18 Difference between predicted and measured subsidence values along the inclined direction of the Sima 1101 panel

$$q = 0.046\frac{H_s}{H_j} + 0.74 \quad b = 0.03\frac{H_s}{H_j} + 0.3333$$

$$\tan\beta = -0.1\frac{H_s}{H_j} + 2.2833$$

$$\theta = 77.9 - 1.407\alpha + 0.45\frac{H_S}{H_j} \quad s = 0.087H_0 + 1.07\frac{H_s}{H_j}$$

Figures 17 and 18 demonstrate that the predicted results are in good agreement with field data for different thicknesses of the unconsolidated soil layer. The prediction parameters showed in Table 2.

The standard deviation m of the predictions can be calculated by:

$$m = \pm\sqrt{\frac{[VV]}{n-1}},$$

where m is the standard deviation of prediction, mm; V is the difference between field data and the predicted value, mm; and n is the number of prediction points.

In general, for accuracy of mining subsidence the relative error m_d was used:

$$m_d = \pm\frac{m}{W_{\max}} \times 100\,\%,$$

where $W_{\max}$ is the maximum subsidence in the field data.

The prediction accuracy of the measurements illustrated in Figs. 17 and 18 are listed in Table 3.

Table 2 Prediction parameters calculated using H_s/H_j

Panel No.	H_O (m)	H_s (m)	H_j (m)	H_s/H_j	q	b	tan β	θ	s
Wuyang 7506	330	30	300	0.10	0.76	0.33	2.27	85.2	28.83
Sima 1101	220	171	49	3.47	1.01	0.44	1.94	86.0	22.27

Table 3 Accuracy of prediction results

Panel No.	Standard deviation (mm)	The maximum subsidence (mm)	Relative error (%)
Wuyang 7506	131	2858	4.76
Sima 1101	276	5565	4.96

From Table 3, the relative error of the revised prediction model is less than 5 %, which is an acceptable result.

5 Conclusions

(1) Based on the geological conditions of the Lu'an coal base in China, two types of mining subsidence simulation were performed: thin bedrock with thick soil layer and thick bedrock with thin soil layer. The results of these two conditions were analyzed.

(2) Two variables, bedrock thickness and soil layer thickness, were studied separately. The results showed that ground deformation will rapidly increase with increasing soil layer thickness, but after a certain value (100 m in this paper), the ground deformation will change little. When the thickness of the soil layer is constant, surface deformation will decline with increasing bedrock thickness.

(3) On the basis of field data from the Lu'an coal base, the prediction parameters of the influence function can be regressed. Two sets of field data were used to verify the correction of the regression parameter function, and a good match was obtained.

Acknowledgments This project was supported by the National Natural Science Foundation of China (41301469) and the Natural Science Foundation of Shanxi Province, China (2014011001-3).

References

Dai H, Li W, Liu Y (2011) Numerical simulation of surface movement laws under different unconsolidated layers thickness. Trans Nonferrous Metals Soc China 21(S3):599–603

Guo H, Chen S, Craig MS (2004) Coal mine subsidence simulation using COSFLOW. The AusIMM Bull 2:50–52

Ji H, Yu X (2011) Research on numerical simulation of mining subsidence. Appl Mech Mater 148:384–387

Wang S, Fu Z, Liao J (2008) Study on dynamic numerical simulation of overburden movement mining under railways. Journal of Coal Science and Engineering (China) 1:29–33

Xu N, Kulatilake PH (2013) Surface subsidence prediction for the Wutong mine using a 3-D finite difference method. Comput Geotech 48:134–145

Xu W, Li Y (2012) Numerical simulation of safe strip mining under thick unconsolidated layers. Appl Mech Mater 182–183:941–944

Yang F, Chen S (2011) Numerical simulation of compound media coupling mechanism of deep mining overburden strata. Trans Nonferrous Metals Soc China 21(S3):631–636

Yang W, Xia X (2013) Prediction of mining subsidence under thin bedrocks and thick unconsolidated layers based on field measurement and artificial neural networks. Comput Geosci 52:199–203

Zhang H, Zhang G, Shen Y (2012) Mechanism and prediction research of the surface mining subsidence of thick loose layers in mining area. Appl Mech Mater 204–208:3488–3493

Zhou D, Wu K, Cheng G (2014) Mechanism of mining subsidence in coal mining area with thick alluvium soil in China. Arab J Geosci 8:1855–1867

14

Comparison of international mine reclamation bonding systems with recommendations for China

Linlin Cheng[1] · **Jeffrey G. Skousen**[2]

Abstract Mine reclamation bonds are used in countries with mineral mining to ensure that reclamation of the mined area is completed. The United States, Canada, and Australia are countries with established mine reclamation bond programs, with the United States coal system having been in place since 1977. China implemented a bonding system in 1998 and by 2013 all 31 provinces had established a system. An effective bonding system must be conditioned on fair and enforceable nationwide reclamation standard, stimulate companies to conduct reclamation by forming economic incentives rather than penalties that become a liability, and allow for full public participation. Based on these principles, this paper compares seven important factors for a successful reclamation bonding system: laws and regulations, administrative authority, bond types, bond size, calculation method, bond release, and public participation. The results show variation in policies and procedures for bonding among countries. Using principles and policies primarily from the United States, China should establish a national reclamation bonding system with legislation that forms a national authority to oversee and enforce reclamation standards and bond requirements. In addition, China can expand bond financial types and strategies, set the size of reclamation bonds at the level of a third-party reclamation cost, and set unified standards for calculation. Phased bond release should be established with specific reclamation criteria for each phase of release. Finally, bonding regulations should clearly identify opportunities for full public participation in the process.

Keywords Mining legislation · Administrative authority · Bond types · Size of bond · Calculation method · Bond release · Public participation

1 Introduction

Mining produces materials that are essential to human needs, but the mining process also disturbs the environment, including the ecosystem and organisms, land surface, and hydrologic balance (Zipper et al. 2011). In the United States alone, from 1930 to 2000, coal mining altered 2.5 million ha of natural landscape, most of that land originally being forested landscapes (MacDonald et al. 2015). In Canada, coal is mined primarily in the three western Provinces of British Columbia, Alberta and Saskatchewan with estimates of 200000 ha of land being disturbed (Fedkenheuer and Macyk 2000). According to the Queensland Resources Council, there will be as much as 120000 ha of land disturbed from mining by 2020 in central Queensland, Australia (Slezak and Robertson 2016). Coal mining in China produced over 2 billion metric tons of raw coal in 2012, with much of that coming from underground mines. Estimates for land surface impacts from subsidence are as high as 1 million ha (Hu 2009).

Mine reclamation is required by the governments of most countries. Before the mining process begins,

✉ Linlin Cheng
chll@cumtb.edu.cn

Jeffrey G. Skousen
jskousen@wvu.edu

1 Department of Surveying and Land Use, School of Geoscience and Surveying Engineering, China University of Mining and Technology (Beijing), D11 Xueyuan Road, Haidian District, Beijing 100083, China

2 Division of Plant and Soil Sciences, West Virginia University, Morgantown, WV 26506-6108, USA

companies develop mining plans that outline the process of land disturbance to obtain the mineral resource. The mining plan also outlines a reclamation procedure to restore the land to a productive post-mining land use (Skousen and Zipper 2014). Common reclamation goals include backfilling and regrading to a specified landform, replacing topsoil, and establishing a vegetation cover to meet the land use goals of the landowner. The mining plan and reclamation plan are the basis for a mining permit issued by the regulatory authority with the objective of mitigating mining impacts to land and water.

As part of the mining permit application, the mining operator is also required to post a reclamation bond. The reclamation bond provides a financial guarantee that the land being disturbed for the operation of the mine will be reclaimed to an acceptable condition. If the operator successfully completes reclamation according to the requirements of the reclamation plan, the reclamation bond is released to the operator. If the operator fails to reclaim the site and all means are exhausted to compel the operator to reclaim the site, then the reclamation bond is forfeited and used to reclaim the land under the supervision of the regulatory authority.

1.1 Reclamation bond types

Reclamation bonds can be provided or purchased in several ways. The most commonly used types are as below.

1. Companies can deposit cash or liquid assets directly into an account that is accessible to the regulatory authority. Assets such as certificates of deposit, bank guarantees, first-lien interests in real estate, and other financial holdings may be deposited.
2. Companies can purchase an insurance policy from a financial surety provider that will pay out the insured amount of reclamation bond if the mining permit is forfeited. The insurance policy is for a specified length of time for which the company pays a premium for the coverage.
3. A surety bond is a three-party contract between a mining company, a surety company, and a regulatory agency. The mining company is obligated to conduct reclamation and the surety company is a secondary source of funds to the mining company. The mining company is still the primary source of funds for reclamation.
4. A trust fund is a financial account managed by a third-party or trustee. The mining company places an agreed-upon amount into a trust fund and a designated trustee manages the funds provided by the mining company. The funds become payable to the regulatory authority if the company fails to perform reclamation.
5. Companies can be self-bonded or provide corporate guarantees. This means that the company relies on its own financial strength to provide assurance for reclamation by meeting certain financial tests established by the regulatory authority without depositing assets into an account directly or the use of a separate surety company.
6. A bond pool is an alternative bonding program where members in the bond pool pay membership and resource extraction fees. The money in the bond pool increases in size with fee collection and investment options. The amount or rate at which a specific member pays into the bond pool is conditional and based on their risk history. Companies may be assigned a rating of A, B, or C, based on their compliance history with permit obligations.

1.2 Factors considered when determining bond amount

Most reclamation bond calculations include estimates of the potential disturbance area and other liabilities such as infrastructure or water treatment. The amount of money for the reclamation bond is usually based on an analysis of anticipated liabilities in the event of forfeiture. Most regulatory authorities have a process for calculating liabilities and thus have a way to determine bond amounts for a given property based on site factors and the nature of the disturbance and the jurisdiction's environmental and land use requirements. The liability analysis is based on several considerations such as the period of operation, size and depth of the disturbance, underground and surface water handling and treatment, equipment and facilities removal and disposal, hazardous waste disposal, landscape restoration and earthwork, topsoil replacement, and revegetation. For example, the United States Forest Service summarized the factors that it uses to determine bond amounts on United States federal lands with the following (Marks et al. 2004): (1) interim operations and maintenance, (2) hazardous materials, (3) water quality, (4) demolition, removal and disposal of uncontaminated structures, equipment and materials, (5) earthwork, (6) revegetation, (7) mitigation, (8) long term operation, maintenance and monitoring.

1.3 Use of bonding

Bonding offers several advantages. It provides an assurance that a disturbance will be reclaimed either by the mining company or the regulator or a combination of both. It also provides the funding or a mechanism for funding that is needed to complete reclamation. One disadvantage of bonding is the added cost to the mining company for setting aside sufficient funds before the disturbance starts. The capital investment of placing a large sum of cash in a bond fund or financial account can put the company in fiscal difficulty from the start. Another disadvantage is that the funds allocated for the bond initially may become insufficient to complete reclamation later when new standards or different reclamation requirements emerge. Therefore, some type of periodic review or recalculation is needed.

Although reclamation bonding may have some disadvantages (Shogren et al. 1993), international experience has proven that the bonding procedure is an effective and reasonable practice to ensure reclamation if a company defaults on permit obligations (McLeod 2000; Cheng 2009). As such, the use of bonding is common in the mining industry worldwide (Gerard 2000; Cheng 2009).

One of the first United States laws requiring a reclamation bond in coal mining was Pennsylvania's Anthracite Strip Mining Law of 1947. Other United States eastern states followed this approach with gradual increases in reclamation requirements during the ensuing 30 years. In 1977, the national Surface Mining Control and Reclamation Act (SMCRA) was passed by the United States Congress, which included strong bond provisions to ensure reclamation occurred if the mining company defaulted on its obligations.

To promote mine reclamation, China implemented a bonding system in 1998 in Zhejiang province. By 2013, all 31 provinces (including autonomous regions or municipalities that are directly under the central government) have established bond systems. But it is relatively new and has not developed some of the provisions found in bonding systems of other countries. Currently, only 25% of land disturbed in China has been reclaimed (Liu 2011), while countries with a longer history of bonding systems such as the United States and Australia have 80% of disturbed lands reclaimed (Luo and Wang 2010). Therefore, the older bonding programs of other countries may provide successful options and practices for reclamation bonds in China.

1.4 Standards for the evaluation of reclamation bonding system

An effective reclamation bonding system should have the following features (Cheng 2009):

1. Must be based on fair and enforceable nationwide reclamation standards for mining companies.

A successful reclamation bonding system should be a 'fair' system. That is, all the mining companies throughout the country should face the same requirements and costs per unit of disturbed area for reclamation. Reclamation bonds help to ensure reclamation of the disturbed area and to enhance public welfare and attitude.

2. Stimulate companies to conduct reclamation by forming economic incentives rather than penalties that become a liability.

Reclamation is no longer viewed as an extra cost or option for mining companies; rather it is a rule that mining companies reclaim the land and undo the damage resulting from the mining operation. This principle reflects the proper ethical philosophy for environmental protection. The main purpose of reclamation bonding is to make the mining companies bear the external cost caused by mining activities and perform reclamation work by themselves, not to collect a huge reclamation fund for the regulatory authority. Therefore, an effective reclamation bonding system should achieve this purpose.

3. Allow for full public participation.

The well-being of society is based on a clean and healthy environment. The public is affected both positively and negatively by mining operations. They receive the benefits of economic development and jobs the company provides, but they also bear the burden of a poor environment in the long term if there is little or no reclamation. Therefore, the public should have the right to participate in mine permitting and the reclamation bonding system. In addition, public participation can create a social consciousness that can be both for and against mining operations, place strong pressure on the credibility and reputation of mining companies, and supervise or demand reclamation performance. Thus, the purpose of reclamation bonding systems can be more easily achieved with public participation.

Important factors of a bonding system are: (1) laws and regulations which are passed to set standards and enforcement, (2) authority or agency that enforces laws and administers the reclamation bond, (3) type of bond to be collected, (4) size of the bond, (5) method of calculation, (6) provisions for release of the bond, and (7) public participation. The objective of this paper is to use the above factors to compare bonding systems in four countries (United States, Australia, Canada, and China) and provide recommendations for a sustainable and effective bonding system in China.

2 Current bonding systems in the United States, Australia, Canada, and China

2.1 The United States

All coal mining operations in the United States are subject to state and federal laws governing reclamation performance and bonding requirements. While laws exist for non-coal resource extraction, regulation of the coal industry is significantly more mature and detailed than that found for the non-coal mining industry (Gorton 2013).Therefore, the bonding system for the United States coal industry will be used as the example.

2.1.1 Laws and regulations

Surface coal mines prior to 1950 in the United States were generally left with little to no reclamation. As government regulations progressed, mining companies were required to backfill the area and plant grasses or trees (Gorman et al. 2001; Skousen et al. 2006). During the ensuing 30 years, more requirements were instituted to promote reclamation and some states had more stringent laws and regulations, and some had a reclamation bond. When the Surface Mining Control and Reclamation Act was passed in 1977 (SMCRA), the United States mining companies were required to post a reclamation bond as part of the mining permit to ensure that the regulatory authority would have sufficient funds to reclaim the site if the company failed to complete the reclamation plan approved in the permit. The Surface Mining Control and Reclamation Act established the legal requirement of reclamation bonds and created a federal agency to implement the provisions of the act (Code of Federal Regulations Title 30, Chapter 25, Section 1259). This federal agency, the Office of Surface Mining Reclamation and Enforcement (OSMRE), has established detailed provisions for reclamation bonds (Office of Surface Mining Reclamation and Enforcement 2000a, b). 'Primacy' means that a state assumes the primary enforcement responsibility. Most coal-producing states have 'primacy' to enforce the Surface Mining Control and Reclamation Act and have enacted their own state statutes, which meet or exceed the national regulatory requirements to control environmental impacts of coal mining activities (Gorton 2013).

2.1.2 Administrative authority

The Office of Surface Mining Reclamation and Enforcement is the federal administrative authority governing the provisions for reclamation bonds. The Office's objective is to ensure that citizens and the environment are protected during mining and that the land is restored to beneficial uses when mining is finished. The headquarters of OSMRE are located in Washington DC with three regional offices in the Appalachian, Mid-Continent, and Western regions. Besides the management of reclamation bonds, it also provides technical training both in the legal and administrative aspects of bonding and in the calculation of reclamation bond amounts.

For states with primacy, reclamation standards and enforcement are administered by their own state reclamation agency. The OSMRE has the right of regular and special inspections in an oversight role. If OSMRE finds that a state regulatory authority has not taken appropriate action or shown good cause, it may issue a federal notice of violation or take other enforcement action directly against the mining company or permit holder (Power 2010).

2.1.3 Bond types

Several types of reclamation bonds are available to companies in the United States: (1) cash; (2) certificates of deposit; (3) first-lien interests in real estate; (4) letters of credit; (5) federal, state, or municipal bonds; (6) investment-grade securities; (7) corporate surety bonds; and (8) self-bonds. Bond types vary from state to state but they must be one of the types recognized by OSMRE (Office of Surface Mining Reclamation and Enforcement 2016). Some states such as Colorado, North Dakota and Wyoming accept any combination of these types of bonds, while Montana does not allow self-bonding. Some states authorize or require mining companies to participate in a "bond pool." Bond pools gained popularity in the 1990s and by 2014, six states had utilized some form of bond pool (Conrad 2014).

2.1.4 Size of the bond

According to the Surface Mining Control and Reclamation Act, the performance bond provides a guarantee that funds will be available to the regulatory authority to complete the approved reclamation plan in the event that the mining company fails to do so (Office of Surface Mining Reclamation and Enforcement 2000a, b). It should be based on, but not limited to, the estimated cost of the reclamation plan submitted by the permit applicant and should reflect the probable difficulty of reclamation, giving consideration to such factors as topography, geology, hydrology, and revegetation. The OSMRE wrote a guidance document called the Handbook for Calculation of Reclamation Bond Amounts (Office of Surface Mining Reclamation and Enforcement 2000a, b). The handbook's bond calculation assumed that the bond amount would be sufficient to engage a third-party contractor to complete the reclamation plan and to reflect the worst-case scenario (or the maximum reclamation cost liability). It also required that the liability calculation be based on at least two estimates from the OSMRE and a third-party reclamation entity, taking into account that unanticipated events may increase the reclamation liability. The minimum amount of the bond for one permitted parcel is $10000.

2.1.5 Bond calculation

The Handbook for Calculation of Reclamation Bond Amounts made a detailed description of the five major steps in bond calculation (Office of Surface Mining Reclamation and Enforcement 2000a, b).

1. Determine the point of maximum reclamation cost liability.
2. Estimate direct reclamation costs such as earthmoving, revegetation, and the removal and demolition of structures not to be retained as part of the post-mining land use.
3. Adjust direct costs for inflation.
4. Estimate indirect reclamation costs, including contractor and equipment mobilization and demobilization charges, contingency allowances, redesign expenses (including surveying, aerial photography, and monitoring in support of this effort), profit and overhead, and contract management fees.
5. Calculate the total bond amount.

The Handbook is used when calculating bonds where there is no state primacy under a Federal lands program or whenever the OSMRE issues a permit for surface mining operations. It also can be used by the OSMRE during oversight of approved state programs as a technical guide when assessing the adequacy of bonds, although the states are not compelled to adhere exactly to the methods in the Handbook.

Many primacy states have their own calculation methods. In Colorado, bonds are calculated on a task by task basis with an automated calculation program. Ohio has developed a Performance Security Spreadsheet that engineers use to estimate the cost of reclamation (Interstate Mining Compact Commission 2013).

2.1.6 Bond release

According to federal regulations that implement the Surface Mining Control and Reclamation Act, the release of the bond can be divided into three phases. Each phase has specific reclamation completion criteria. Phase 1 release may occur after the operator completes the backfilling, regrading, and surface water drainage control of the bonded area in accordance with the approved reclamation plan. After completing this phase, 60% of the bond or collateral for the applicable permit area can be released. Phase 2 release may occur after vegetation has been established on the backfilled and regraded mined area in accordance with the approved reclamation plan. The amount of bond released for Phase 2 is an additional 25% of the total reclamation bond. Phase 3 release may occur after the mining company has successfully completed all mining and reclamation requirements in the permit. After Phase 3, the remaining 15% of the bond is released.

Based on the three-phase release specified by the Act, primacy states may make their own guidance. For example, in Wyoming, bond release includes a total area release, as well as Phase 1, Phase 2, and Phase 3 incremental bond release. Each phase has the performance standards the operator must meet (Krzyszowska-Waitkus and Blake 2011). Some states minimize the use of phased-bond release especially where long term water treatment is a possibility.

2.1.7 Public participation

The Surface Mining Control and Reclamation Act made detailed provisions for public participation in the release of the bond. Citizens can attend public hearings, be aware of mining activities through notices in local news outlets, write or deliver personal comments to regulators, and request visits to the mining area.

2.2 Canada

In Canada, mining companies have been required by law to reclaim all lands disturbed by mining since 1969 (Technical and Research Committee on Reclamation 2010). All the provincial and territorial governments in Canada have legislation in place to provide financial assurance for reclamation in the event the mine operator is unable to do so (Fraser Institute 2015). After the mining company conducts the proper site reclamation, the financial security deposit is returned to the company.

2.2.1 Laws and Regulations

Unlike the United States, laws and regulations for mine reclamation and bonding in Canada reside primarily at a provincial level. For example, the British Columbia Ministry of Energy, Mines and Petroleum Resources sets out criteria and the Mines Act requires a company to estimate the cost of reclamation and establish the bond amount (CCSG Associates 2001). In Alberta, reclamation certificates are required for mining operations as directed by the Environmental Protection and Enhancement Act (Alberta Energy Regulator 2015). In Ontario, the Mining Act of 1990 (Bill 26), Chapter M 14, Part VII covers the requirements and specifies standards, procedures and requirements needed for rehabilitation and closure (World Bank Group 2008).

2.2.2 Administrative authority

No national authority exists in Canada. Each province or territory has its own administrative authority over the bond. For example, bonding in the Yukon Territory is regulated by the Department of Indian Affairs and Northern Development and Yukon Territory Water Board (Searle 2000).

Bonding in Ontario is administered by the Ministry of Northern Development and Mines (World Bank Group 2008), while British Columbia reclamation bonding is administered by the Ministry of Energy and Mines (British Columbia 2015).

2.2.3 Bond types

Several bond types are acceptable in Canada including cash, letters of credit, surety bonds, government bonds, trust funds, company guarantees, investment certificates, self-bonds, and combinations of these. Differences exist among provinces. For example, in British Columbia, corporate guarantees and self-bonds are not acceptable types of financial assurance for sites which may have long term water quality and liability concerns (CCSG Associates 2001; World Bank Group 2008). In Alberta, the Mine Financial Security Program is used to collect financial security from the oil sands and coal industry to protect the public from paying for end-of-life project closure costs. It allows for security deposits to be made that will account for potential risks during the life cycle of the mine.

2.2.4 Size of the bond

The size of the bond differs among provinces. A survey in 2001 found that most Canadian governments did not calculate security bond amounts equal to the full cost of clean-up. No examples were found of a security bond that was sufficient to cover the full costs of remediating a site (CCSG Associates 2001).

2.2.5 Bond calculation

Bond calculation in Canada is generally based on current construction rates and third-party estimates. Contingency surcharges are imposed and range from 10% to 30% add-on depending on the degree of uncertainty. British Columbia and Northwest Territories provide a costing spreadsheet to use and others use a reclamation model or a regulatory guide (British Columbia 2015; Cowan et al. 2010).

2.2.6 Bond release

In Ontario, the bond may be available for on-going rehabilitation. If some of the work is carried out in accordance with legislative requirements, some of the bond may be released considering the current value of the remaining rehabilitation work. After successful mine closure, the bond is released but some funds may be retained for short term monitoring costs or long term care (World Bank Group 2008). In the Northern Territories, securities are deposited with the Department of Indian and Northern Affairs (DIAND) and held until reclamation and closure are completed to the satisfaction of DIAND (Searle 2000).

2.2.7 Public participation

In Ontario, the public has an opportunity to review closure plans, but the security amounts are not open to public review (World Bank Group 2008). Under British Columbia mining legislation, there are no specific legal requirements mandating public participation in setting mine securities or bonding (Fair Mining Collaborative 2015).

2.3 Australia

All Australian states have a requirement to elicit a security deposit by the leaseholder against possible failure to rehabilitate or meet the environmental requirements after mining (McLeod 2000).

2.3.1 Laws and regulations

Laws about bonding vary among Australian states. For example, in Western Australia, provisions for reclamation bonding are encompassed in the Mining Act of 1978. In Queensland, the Mineral Resources Act of 1989 requires that a "security" is deposited prior to a mining title being issued. The Environmental Protection Act of 1994 requires the rehabilitation program to include the proposed amount of the financial surety for larger projects while the Codes of Environmental Compliance require a financial surety for small projects. A single surety can be used to cover the requirement of both the Mineral Resources Act and the Environmental Protection Act (World Bank Group 2008).

2.3.2 Administrative authority

Each state has its own administrative authority for bonding. For example, the Department of Mines and Petroleum of Western Australia administers mining securities on behalf of the Minister, in accordance with the provisions of the Mining Act of 1978. Bonding requirements are guided by departmental policy and administrative procedures (Government of Western Australia 2010). In Queensland, the Department of Mines and Energy is responsible for the receipt and management of both the security under the Mineral Resources Act and the financial surety under the Environmental Protection Act (World Bank Group 2008).

2.3.3 Bond type

Bond types differ among Australian states. Acceptable forms of financial surety in Queensland include cash, bank guarantees, letters of credit, and insurance bonds. In

Western Australia, before 2014, the bond type was usually in the form of a performance bond fully guaranteed by a bank or other acceptable financial institution, which must show the name of the licensee. It should not be in the form of cash. A performance bond is a contract between the Minister for Mines and Petroleum and a third-party of financial standing acceptable to the Minister. The third-party will unconditionally pay an agreed sum to the Minister upon his request following the failure of the mining company to meet environmental commitments (Government of Western Australia 2010). Since July 2014, except in cases of high risk, the majority of mining companies are not required to provide a bond for future approvals. Participation in the bonding fund is compulsory except for mining companies with a rehabilitation liability estimate below a threshold of $50000 (Government of Western Australia 2014).

2.3.4 *Size of the bond*

In Western Australia, the Department of Mines and Petroleum recommended that the performance bond be set at a level to encourage reclamation. It is based on a set of predetermined criteria that relate to the estimated cost of remediation and may not necessarily reflect the actual reclamation cost. The bond covers all the land disturbed and will be reviewed annually and increased or decreased depending on the total area disturbed and the area successfully reclaimed (Government of Western Australia 2014). In Queensland, bonds for exploration and small standard mining projects are based on the total area of disturbance and the risk associated with rehabilitation. The bond amount for a non-standard project is made on a project-specific basis (World Bank Group 2008).

2.3.5 *Bond calculation*

There are large differences among states as to how the bond amount is calculated. In Queensland, the amount of financial assurance is calculated by determining the cost to reclaim the site minus any applicable discounts. To obtain a discount, a set of mandatory criteria must first be met (World Bank Group 2008). In Western Australia, the calculation method for performance bonds uses a minimum bond rate based on the mine's surface area disturbance, with minimum bond rates for small areas (Government of Western Australia 2014).

2.3.6 *Bond release*

In Western Australia, the full bond amount can be returned to the company when the mine closure standards have been met and all environmental regulations have been achieved. There will also be bond reduction if progressive reclamation has been carried out (Government of Western Australia 2014). In Queensland, the financial surety is not available for on-going reclamation. However, when a new mining operation plan is submitted and the reclamation liability is recalculated, work that has been completed will no longer be included in the total. The financial surety remains in place until the Environmental Protection Agency is satisfied that no further claim is likely to be made against it (World Bank Group 2008).

2.3.7 *Public participation*

In Queensland, the government will no longer accept self-bonds as a financial surety instrument because public opinion has no faith in them (World Bank Group 2008). In Victoria, there is no third-party involvement in establishing the financial surety and no process of verification (World Bank Group 2008).

2.4 China

2.4.1 *Laws and regulations*

The first real attention for land reclamation in China was in the 1980s. In 1986, the Land Management Law was passed by the Sixth National People's Congress, which was the first time China addressed reclamation in the law. It specified that the company causing a land disturbance should be responsible for reclamation, recovery, and land utilization after mining. It remains as the only national law on reclamation in China.

The Land Reclamation Regulation (LRR), effective in 1989, was then promulgated by the State Council, which stipulated definitions, planning and management of land reclamation. In 2006, the Ministry of Land and Resources issued the Notice of Strengthening Land Reclamation Management Work of Production and Construction Projects (NSLR), which incorporated a reclamation plan in the mining license and approval process. In 2011, the State Council issued the Land Reclamation Ordinance (LRO), which revised the 1989 Land Reclamation Regulation. Compared with the 1989 Land Reclamation Regulation, the 2011 Land Reclamation Ordinance further defined the responsibility of mining companies, establishing that reclamation plans are part of the permitting process and that reclamation monitoring and supervision should occur. Over the past 30 years, China's reclamation rate of disturbed land has been increased from 1% in 1987 to 25% at present (Liu 2011).

Zhejiang Province was the first province in China to implement a mine reclamation bonding system in 1998 (Cheng and Hu 2008). By 2013, all 31 provinces (including autonomous regions and municipalities that are directly under the central government) had established a reclamation bonding system and over 70000 mines have paid a reclamation bond (Ministry of Land and Resources 2013), which accounts for 80% of the total number of mines.

Three policies were issued that helped to establish mine reclamation bond systems in the 31 provinces of China (Cheng and Hu 2008). The policies were: (1) The Notice on the Comprehensive Rectification and Standardization of Mineral Resources Exploitation Order (NCRS) issued in 2005; (2) Guidance on the Gradual Establishment of the Responsibility Mechanism of Mine Environment Management and Ecological Restoration (GERM) issued in 2006; (3) Provisions on the Protection of Mine Geological Environment (PPMGE) issued in 2009.

Each policy enhanced procedures and helped to further develop a mine reclamation bonding system. For example, the GERM stipulated that the local Ministry of Finance, the department of Land Resources, and the department of Environmental Protection in each province should jointly determine the amount of the reclamation bond based on mineral type, size of mine disturbance, mineral output, sales income of mineral products, the design life of the mine, environmental treatments, ecological recovery rate, and reclamation cost.

Besides the above three national policies, each province has made their own regulations and implementation measures of mine reclamation bond. There are no specific laws and regulations at the national level.

2.4.2 *Administrative authority*

There is no national administrative authority that oversees reclamation standards or bonding. Most provinces stipulate that administrative departments of land and resources at or above the county level calculate, collect, and release bonds.

2.4.3 *Bond types*

There is no national standard for the types of financial instruments that can be used for reclamation bonds in China. Except Shandong Province, reclamation bonds in the other 30 provinces can only be in the form of cash. In 2015, Shandong Province specified that the bond submitted can be in cash or other non-cash guarantee forms such as performance bonds or letters of credit provided by a bank. The non-cash bond in Shandong Province must be submitted to the Department of Land and Resources. The cash bond in most provinces is required to be deposited directly into a special financial or bank account jointly administered by the financial department, the Department of Land and Resources, and the mining company.

2.4.4 *Size of the bond*

Most provinces collect the reclamation bond according to the principle of "not lower than the treatment fee." For example, administrators of mine reclamation bonds in Tianjin municipality, Shandong province, and Anhui province all stipulate that 'the bond collected should not be lower than the treatment fee.'

2.4.5 *Bond calculation*

There is no unified bond calculation method in China. Some provinces calculate the bond based solely or in combination with the mineral type, the mining area, a percentage of sales revenue, and the mineral output. Most provinces use a combination of these criteria to form a basis (Cheng 2009). For example, Guizhou Province determines the amount of bond according to mineral type, the designed mining scale, and the valid period of the mining permit. Zhejiang Province calculates the bond according to the designed mining scale and a scaling coefficient based on the depth of mining disturbance.

2.4.6 *Bond release*

Most provinces stipulate that the bond shall be released at one time after all the reclamation work is finished if the period of the permit is less than 3 years. Bonds for those mines whose permit period is more than 3 years shall be released after each reclamation phase is fulfilled. But most of the provinces do not make detailed reclamation standards for each phase in their regulations or the standards are not specific.

2.4.7 *Public participation*

There is no public participation allowed in the regulations of China's 31 provinces concerning the collection and release of bonds, nor in the enforcement of regulations and reclamation standards.

3 Comparison and discussion

Table 1 summarizes current mine reclamation bonding policies in the United States, Canada, Australia and China.

3.1 Laws and administrative authority

Laws and administrative authority provide that the reclamation bonding system has legal status and is enforced. From the above table, it can be seen that among the four countries, only the United States has a law and administrative authority for reclamation bonding at the national level. There are no laws or administrative authority for reclamation bonds at the national level in Canada and Australia, and each province/state has its own reclamation laws and administrative authority. In China, national laws and regulations and administrative authority for reclamation bonding also do not exist. Rather, each province has its own regulation of bonds and the bond is usually administered by departments of land and resources at or above the county level in each province.

Compared to provincial or state laws and regulations, national laws help to establish a unified and standardized reclamation bonding system, which must be based on fair and enforceable nationwide reclamation standards for mining companies. National laws also have greater weight and binding force for implementing reclamation standards and bond systems. Before the Surface Mining Control and Reclamation Act in the United States, some states had their own laws governing surface mining, but these state laws varied widely, enabling mining operations to relocate to states where regulations were less strict. The passage of the Surface Mining Control and Reclamation Act equalized reclamation standards across states and provided a baseline of procedures for reclamation practices, enforcement, and bonding. A national regulatory authority of bonding (the OSMRE) was also created. Reclamation bonds are required for all active mines throughout the country. If the mine permit applicant cannot submit the bond according to the provisions in the Surface Mining Control and Reclamation Act, supervised by the OSMRE, the company will not be granted a mine permit anywhere in the country.

3.2 Bond type

The United States, Canada, and Australia accept a variety of financial instruments and assets for reclamation bonds including cash, corporate assets, letters of credit, real estate, surety bonds, insurance bonds, trusts, and bond pools. In China, only Shandong Province uses non-cash bonds but has only approved this circumstance since 2015. All other provinces in China accept only cash bonds.

Financial assurance is usually divided into hard and soft assurance (Miller 1998). The hard forms involve cash and company assets that are placed in an account. The soft forms of assurance are based on liens and lines of credit. Even these forms of assurance represent a significant burden in that they reduce the credit rating and borrowing ability of the company (Miller 2005). With the economic burden caused by reclamation bonds, many companies have turned to bond pools where allowed. Reclamation bonds composed of a combination of hard and soft funds may be more acceptable in the future (Searle 2000).

Although a cash bond may be a preferred type for administrative authorities because of low risk as stated above, a desirable bonding system should form economic incentives rather than economic burdens for mining companies (Cheng 2009). There is evidence that the high cost of bonding sharply reduces investment in and by the mining industry. Cash bonds are expected when the credit or strength of the mining company is unknown or questionable, or the permit time is short. But cash bonds impose great liquidity constraint and economic burdens on many companies, particularly small companies, and other types of bonding types should be allowed for more established and reputable companies.

Self-bonding is coming under particular scrutiny. In the United States, self-bonding for coal companies tops $3.3 billion nationwide, which includes $2.3 billion in liabilities (Gruver 2016). As the industry declines, more companies, who were previously strong and solvent, have filed for bankruptcy (Mufson and Warrick 2016). In some cases, money available to cover the cost of reclamation was not posted as a bond, and the company had no financial reserve to complete reclamation and environmental cleanup (Gruver 2016). Therefore, there has been a move away from self-bonding. The states of New Mexico and Nevada in the United States limit self-bonding to no more than 75% of a company's reclamation liability (Conrad 2014).

3.3 Size of bond and calculation method

The core of a reclamation bonding system is the size and the method of calculation for the bond amount. If the amount calculated is too small, there will be a huge capital gap for reclamation and the land will be left unreclaimed or result in a great financial burden to the government and community to reclaim the operation once the mining company forfeits their reclamation obligation. If the size calculated is too large, it will negatively impact the mining company's fiscal capacity and potentially affect mine development (Cheng 2009).

Among the four countries, Canada does not calculate bond amounts equal to the full cost of reclamation and bond amounts in Australia do not necessarily reflect the actual cost. China's level is not lower than the treatment fee. None of these countries have a unified calculation method at the national level. Only the United States calculates bond amounts at a third-party reclamation cost and has unified the calculation method with rules established by the OSMRE.

A preferred mechanism of a reclamation bonding system is that the bond amount be sufficient for a third-party to do the reclamation work and supervised by the administrative authority. The cost should also include all administrative, maintenance, and monitoring costs (World Bank Group

Table 1 Comparison of mine reclamation bonding system in the United States, Canada, Australia and China

Item	United States	Canada	Australia	China
Laws and regulations	Surface Mining Control and Reclamation Act (SMCRA) with accompanying state laws at least as stringent as national law	No specific national laws and regulations. Each province or territory has its own law	No specific national laws and regulations. Each state has its own law	No specific national laws and regulations. Each province has its own
Administrative authority	State mine reclamation agency, with oversight by the federal OSMRE. Non-primacy states are directly administered by OSMRE	No national administrative authority, each province has its own administrative authority	No national administrative authority, each state has its own administrative authority	No national administrative authority, bond is administered by departments of land and resources at or above county level in each province
Bond type	Major bond types include cash, corporate surety bonds, self-bonds, certificates of deposit, first-lien interests in real estate, letters of credit, federal, state, or municipal bonds, investment-grade securities etc Some states use bond pool and trust funds	Bond types differ among provinces and territories, but include cash, letters of credit, corporate surety bonds, government bonds, trust funds, company guarantees, investment-grade securities etc	Bond types differ among states, but include cash, letters of credit, bank guarantees, insurance bonds, etc. Some states use Mining Rehabilitation Fund	Only Shandong province allows non-cash bonds. All other provinces only use cash
Size of the bond	The bond amount would be sufficient to engage a third-party contractor to complete the reclamation plan, and reflects the worst-case scenario	Most provinces and territories do not calculate security bond amounts equal to the full cost of reclamation	The bond does not reflect the actual cost	Not lower than the treatment feeor reclamation cost
Calculation method	Directed by Handbook for Calculation of Reclamation Bond Amounts, written by OSMRE	Differs among provinces and territories	Differs among states	No unified calculation method, differs among provinces
Bond release	Phased-release with clear reclamation criteria	Primarily one-time release at closure	Primarily one-time release at closure	One-time release if permit time is below 3 years. Phased release beyond 3 years, but no criteria for each phase or the criteria is vague
Public participation	Stipulated by SMCRA	Partial public participation	Partial public participation	No public participation

2008). In Canada, the lack of calculating the reclamation bond with a third-party and administration cost analysis has decreased incentives for companies to act responsibly and complete reclamation and closure, and therefore increases the risk to public financial resources and health (CCSG Associates 2001). In China, few mining companies reclaim the disturbed land, or have any of the bond refunded, because the bond amount paid is lower than the reclamation costs (Xiao et al. 2010; Yan et al. 2012). This defeats the original intention of a bonding system which should be to encourage mining companies to fulfill reclamation obligations and have the bond money returned. Bond calculations based on a third-party reclamation cost provides an incentive for a company to complete reclamation work by itself because it can do so less expensively than a third-party (CCSG Associates 2001). A unified bond calculation method can help to ensure the adequacy of bond amounts and will contribute to fairness among mining companies for reclamation.

3.4 Bond release

Release of the bond is the important final key piece of a bonding system. The United States releases mine reclamation bonds according to a three-phase schedule stipulated by the federal regulations that implement the Surface Mining Control and Reclamation Act with each stage having its own clear criteria. In Canada and Australia, bond reduction is allowed for on-going reclamation, but most of the bond is retained until mine closure.

In China, the bond is usually released at the end of reclamation if the permit time is less than 3 years. If the permit time is more than 3 years, bonds can be phase-released but most provinces do not have reclamation criteria for incremental bond release at each stage. For example, in Jilin province, a company can apply for release of bond money when they have partially or fully completed reclamation work, which is determined by the regulatory authority. The authority has 20 days to make an evaluation after which they may release part or all of the bond based on their evaluation of the company's reclamation. There are no specific criteria about the evaluation defined in the law and therefore the release of bonds is difficult to implement in practice.

Phased-bond release can reduce the economic burden on a mining company and encourage and stimulate a more proactive approach to progressive reclamation. It also embodies the principle of concurrent reclamation and promotes the idea that mining companies should reclaim disturbed land actively and timely (Cheng 2010a, b). Reclamation standards should be clearly defined so that the reclamation work is properly done and evaluations can be done in a timely and efficient manner.

3.5 Public participation

Among these four countries, only the United States stipulates public participation in the law. There is no public participation or only partial public participation in the reclamation bonding systems of the other three countries. Public participation allows local citizens and communities to engage in the mining and reclamation process. It allows for greater collection and access of information about the disturbed site, and can promote cooperation and coordination of the mining activity and reclamation procedures. Allowing for public input helps to avoid wrong decisions by the regulatory authority, can reduce administrative and supervision costs, and can promote fairness, justice and accountability (Environmental Defender's Office NSW 2011).

4 Recommendations for China

4.1 Establish a national law to unify and standardize reclamation bonding system

Currently, reclamation bonds in 31 provinces are diverse and regulated by provincial laws. A national law should be established to unify and standardize the content of reclamation bonding system to create a fair and competitive environment for mining enterprises. Since all 31 provinces in China have implemented land reclamation bonds with regulations, refinements can be made for a national law from the procedures already established in provincial laws.

4.2 Create a special national administrative authority with power to implement the reclamation bonding system

An entity like the OSMRE in the United States could be created within the Ministry of Land and Resources, with branches in each province. This special administrative authority would be responsible for the regulation, enforcement, evaluation, collection, calculation, and release of reclamation bonds. Under the direction of this authority, a nationwide unified reclamation bond system could be established to create fair and enforceable standards with realistic reclamation costs for mining companies. The bonds can be based on a per unit reclamation cost.

4.3 Expand bond types

Besides cash, other non-cash bonds and financial instruments should be promoted to lessen the economic burden on mining companies. Expanding the types of bonds available can be applied gradually (Cheng 2010a, b). Initially, some types of less "hard" bonds such as letters of credit, trust funds, deposits of security, or a combination of financial assurances along with cash can be available for companies to use as financial collateral. Later, other "soft" types of financial securities such as corporate assets and investments can be adopted. Self-bonding is not recommended because of the experiences in other countries of company bankruptcies and that no reserves of funds are available to complete the reclamation work (Riley and Isidore 2016; Corkery 2016). Research about bond types and their suitable conditions should also be carried out by academia.

4.4 Set the size of reclamation bond at the level of the third-party reclamation cost and unify the calculation method

The size of reclamation bonds should be set at the level of a third-party reclamation cost to ensure sufficient funding for the administrative authority to complete reclamation and to encourage reclamation and closure by the mining company. The calculation method should also be unified and based on standardized costs for mining companies. The administrative authority should provide guidance and publish the bond calculation methods so that the methods are known and widely available. Related research should also be carried out by academia to discover the most practical and effective ways, and provide theoretical support for bond calculations.

4.5 Unify the phases of bond release and establish detailed reclamation criteria and the corresponding amount released for each phase

There should be a nationally unified release criteria that allows phased-release of bonds. Supporting rules and regulations should also be formulated. Detailed reclamation criteria and the amount released for each phase and procedures of release should be clearly stipulated.

4.6 Stipulate full public participation in legislation

Public participation in reclamation bonding systems should be promoted. All the information on the standards, collection, use and release of the bond should be open to the public. Public evaluation and opinion, methods and procedures of engagement, and channels for communication should be stipulated in legislation to ensure the public can fully participate in the reclamation bonding system.

5 Conclusions

The comparisons in this paper show that the United States has a more mature and unified reclamation bonding system than the other three countries. Although controversies and problems still exist where companies do not secure enough bonds, do not update their liabilities during mining process, the system in the United States has been highly successful in helping companies be responsible for reclamation liabilities. More than 80% of the land disturbed since 1977 has been reclaimed. As a country that implemented mine reclamation bonding system only recently, China can adopt laws and regulations that clearly work well from the experience of other countries and improve many aspects of the reclamation bonding system including the passage of a national law and regulations, creating a national administrative authority, expanding bond types, unifying bond calculation methods to determine the size of bonds, establishing bond release criteria, and allowing for full public participation in the reclamation system. Recommendations for the above aspects outlined in this paper are beneficial for the establishment of a healthy and sustainable bonding system in China and a healthy and competitive mining industry.

Acknowledgements This study was supported by National Natural Science Foundation of China (No. 41401637) and China Scholarship Council (No. 201406435027). This article is scientific article #3305 from the Agricultural and Forestry Experiment Station, West Virginia University, Morgantown, WV.

Compliance with ethical standards

Conflicts of interest The authors declare no conflict of interest.

References

Alberta Energy Regulator (2015) Abandonment and reclamation https://www.aer.ca/abandonment-and-reclamation/reclamation. Accessed 10 Nov 2016

British Columbia (2015) Permitting and reclamation http://www.empr.gov.bc.ca/Mining/Permitting-Reclamation/Pages/default.aspx. Accessed 15 July 2016

CCSG Associates (2001) Financial option for the remediation of mine sites: a preliminary study. Report to Mining Watch Canada, Ottawa, Ontario, 2001. http://www.miningwatch.ca/sites/www.miningwatch.ca/files/financial_options_paper_0.pdf. Accessed 16 Oct 2016

Cheng L (2009) Study on the mode of mine reclamation bond of China. Ph.D. thesis, China University of Mining and Technology (Beijing), Beijing, China **(in Chinese with English abstract)**

Cheng L (2010a) Study on the release of mine land reclamation performance bond in China. In: Proceedings, 2010 IEEE international conference on advanced management science, vol 2, pp 658–661

Cheng L (2010b) Designs of the form and operation mode of mine reclamation bond. Coal Econ Res 30(5):15–18

Cheng L, Hu Z (2008) Study on the land reclamation performance bonds systems in mineral areas in China. China Mining Magazine 17:18–19 **(in Chinese with English abstract)**

Conrad GE (2014) Mine reclamation bonding—from dilemma to crisis to reinvention: what's a state regulator to do? Interstate Mining Compact Commission. February 11, 2014. http://www.imcc.isa.us/EMLF%20Bonding%20Presentation%20Final.pdf. Accessed 16 Oct 2016

Corkery M (2016) Regulators fear $1 billion coal cleanup bill. New York Times. June 6, 2016. http://www.nytimes.com/2016/06/07/business/dealbook/regulators-fear-1-billion-coal-cleanup-bill.html?_r=0. Accessed 16 Oct 2016

Cowan W, Mackasey W, Robertson J (2010) The policy frame work in Canada for mine closure and management of long term liabilities: a guidance document. http://www.abandoned-mines.org/pdfs/PolicyFrameworkCanforMinClosureandMgmtLiabilities.pdf. Accessed 21 Aug 2016

Environmental Defender's Office NSW (2011) Mining law in New South Wales. http://d3n8a8pro7vhmx.cloudfront.net/edonsw/pages/285/attachments/original/1380667551/110628mining_law_discussion_paper.pdf?1380667551. Accessed 16 Oct 2016

Fair Mining Collaborative (2015) Securing the cost of mine clean-up: public participation. http://www.fairmining.ca/code/securing-the-cost-of-mine-clean-up-2/public-participation/. Accessed 21 Aug 2016

Fedkenheuer A, Macyk T (2000) Reclamation of surface mined coal lands in western Canada. In: Barnhisel R, Darmody R, Daniels W (eds) Reclamation of Drastically Disturbed Lands. Agronomy 41. Soil Science Society of America, Madison

Fraser Institute (2015) What happens to mine sites after a mine is closed? Mining facts. http://www.miningfacts.org/Environment/What-happens-to-mine-sites-after-a-mine-is-closed/. Accessed 15 July 2016

Gerard D (2000) The law and economics of reclamation bonds. Resour Policy 26:189–197

Gorman J, Skousen J, Sencindiver J, Ziemkiewicz P (2001) Forest productivity and minesoil development under a white pine plantation versus natural vegetation after 30 years. In: Proceedings, 18th annual meeting, American Society for Surface Mining and Reclamation, 2–7 June, 2001, Albuquerque, NM, pp 103–111

Gorton WT (2013) State and federal reclamation bonding programs: lessons learned and current challenges. In: 2013 National Bonding Workshop, St. Louis, MI. Interstate Mining Compact Commission. http://www.imcc.isa.us/Bill%20Gorton%20Paper.pdf. Accessed 16 Oct 2016

Government of Western Australia (2010) The Mining Rehabilitation Fund—the first two years. Department of Mines and Petroleum. http://www.dmp.wa.gov.au/Documents/Environment/MRF_The_First_Two_Years.pdf. Accessed 16 Oct 2016

Government of Western Australia (2014) Mining Rehabilitation Fund. Fact sheet May 2014 compulsory year. Department of Mines and Petroleum. http://www.dmp.wa.gov.au/Documents/Environment/ENV-MEB-381.pdf. Accessed 21 Aug 2016

Gruver M (2016) Mine environmental risk grows with bankruptcies in big coal. Associated Press. http://bigstory.ap.org/article/2b8ae16e1f8e42e7b1af97584ec1cf99/mine-environmental-risk-grows-bankruptcies-big-coal. Accessed 16 Aug 2016

Hu ZQ (2009) Review and prospect of land reclamation and ecological restoration in China. Sci Tech Rev 27(17):25–29 **(in Chinese with English abstract)**

Interstate Mining Compact Commission (2013) Bonding survey report: coal. August 2013. http://www.imcc.isa.us/BondingSurvey2013Final.pdf. Accessed 16 Oct 2016

Krzyszowska-Waitkus A, Blake C (2011) Tracking bond release at a large Wyoming coal mining operation. In: Proceedings, American Society of Mining and Reclamation. Lexington, Kentucky, pp 338–350

Liu Z (2011) To give land life and vigor - Interpretation of "Land reclamation ordinance" by the director of policy and regulations of the Ministry of Land and Resources. Land Resour Commun 5:12–13

Luo M, Wang J (2010) The Enlightenment to China from the construction of land reclamation system and the research of science and technology in Australia. China Land 4:51–53

MacDonald E, Landhausser S, Skousen J, Franklin J, Frouz J, Hall S, Jacobs D, Quideau S (2015) Forest restoration following surface mining disturbance: challenges and solutions. New For 46:703–732

Marks M, Buchta T, Newman C, Lentz R, Gabardi J, Werner P, DeTar B, Curry T, Doran M (2004) Training guide for reclamation bond estimation and administration. USDA-Forest Service, April 2004. http://www.fs.fed.us/geology/bond_guide_042004.pdf. Accessed 10 Nov 2016

McLeod H (2000) A review of the bond system in Fiji. SOPAC Technical Report 305, December 2000. http://ict.sopac.org/VirLib/TR0305.pdf. Accessed 16 Oct 2016

Miller G (1998) Use of financial surety for environmental purposes. International Council on Metals and the Environment. http://www.icme.com/icme/finsurety.htm. Accessed 10 July 2016

Miller G (2005) Financial assurance for mine closure and reclamation. https://www.icmm.com/document/282. Accessed 15 Oct 2016

Ministry of Land and Resources (2013) 2013 land resources bulletin. http://www.mlr.gov.cn. Accessed 15 July 2016

Mufson S, Warrick J (2016) Can coal companies afford to clean up coal country? Washington Post. April 2, 2016. https://www.washingtonpost.com/business/economy/can-coal-companies-afford-to-clean-up-coal-country/2016/04/01/c175570c-ec73-11e5-a6f3-21ccdbc5f74e_story.html. Accessed 16 Aug 2016

Office of Surface Mining Reclamation and Enforcement (2000a) Handbook for calculation of reclamation bond amounts. USDI-OSMRE, October 2000. http://www.osmre.gov/LRG/docs/directive882.pdf. Accessed 16 Oct 2016

Office of Surface Mining Reclamation and Enforcement (2000b) Chronology of major SMCRA-related events. USDI-OSMRE, October 2000. http://www.osmre.gov/LRG/chronlisting.shtm. Accessed 16 Oct 2016

Office of Surface Mining Reclamation and Enforcement (2016) Reclamation performance bonds. http://www.osmre.gov/resources/bonds/bondsoverview.shtm. Accessed 16 Oct 2016

Power CB (2010) SMCRA primacy, mining permit transfers, ownership and control, and excess reclamation responsibilities: a primer on confusing topics. Energy Miner Law Inst 31:268–297

Riley C, Isidore C (2016) The largest U.S. coal company just filed for bankruptcy. CNN. April 13, 2016. http://money.cnn.com/2016/04/13/news/companies/peabody-coal-bankruptcy/. Accessed 16 Oct 2016

Searle D (2000) Reclamation and bonding in the North. Department of Indian and Northern Affairs (DIAND), Canada. http://www.miningnorth.com/_rsc/site-content/library/reclam%20%20bond%20-%20reprint.pdf

Shogren J, Herrigies J, Govindasamy R (1993) Limits to environmental bonds. Ecol Econ 8:109–133

Skousen J, Zipper C (2014) Post-mining policies and practices in the Eastern USA coal region. Int J Coal Sci Technol 1:135–151

Skousen J, Ziemkiewicz P, Venable C (2006) Tree recruitment and growth on 20-yr-old, unreclaimed surface mined lands in West Virginia. Int J Min Reclam Environ 20:142–154

Slezak M, Robertson J (2016) Full of holes: why Australia's mining boom will leave permanent scars. The Guardian. https://www.theguardian.com/environment/2016/jul/20/full-holes-australia-mining-boom-permanent-scars. Accessed 16 Oct 2016

Technical and Research Committee on Reclamation (2010) Mining and reclamation in British Columbia, Canada http://technology.infomine.com/reviews/reclamationbc/welcome.asp?view=full (accessed 16 Oct 2016)

World Bank Group (2008) Guidance notes for the implementation of financial surety for mine closure. The World Bank Group of the Oil, Gas and Mining Policy Division. http://siteresources.worldbank.org/INTOGMC/Resources/financial_surety_mine.pdf. Accessed 21 Aug 2016

Xiao W, Hu Z, Xu X (2010) Cost definite method of land reclamation in coal mining area. J China Coal Soc 35(S1):175–179

Yan S, Shen W, Zou C (2012) Theory and practice of deposit system for restoration and control of mining environment. Environ Sci Technol 35(6I):423–427

Zipper C, Burger J, Skousen J, Angel P, Barton C, Davis V, Franklin J (2011) Restoring forests and associated ecosystem services on Appalachian coal surface mines. Environ Manag 47:751–765

15

Laboratory testing of a shuttle car canopy air curtain for respirable coal mine dust control

W. R. Reed[1] · Y. Zheng[1] · M. Yekich[1] · G. Ross[2] · A. Salem[3]

Abstract Canopy air curtain (CAC) technology has been developed by the National Institute for Occupational Safety and Health (NIOSH) for use on continuous miners and subsequently roof bolting machines in underground coal mines to protect operators of these machines from overexposure to respirable coal mine dust. The next logical progression is to develop a CAC for shuttle cars to protect operators from the same overexposures. NIOSH awarded a contract to Marshall University and J.H. Fletcher to develop the shuttle car CAC. NIOSH conducted laboratory testing to determine the dust control efficiency of the shuttle car CAC. Testing was conducted on two different cab configurations: a center drive similar to that on a Joy 10SC32AA cab model and an end drive similar to that on a Joy 10SC32AB cab model. Three different ventilation velocities were tested—0.61, 2.0, 4.3 m/s (120, 400, and 850 fpm). The lowest, 0.61 m/s (120 fpm), represented the ventilation velocity encountered during loading by the continuous miner, while the 4.3 m/s (850 fpm) velocity represented ventilation velocity airflow over the shuttle car while tramming against ventilation airflow. Test results showed an average of the dust control efficiencies ranging from 74 to 83% for 0.61 m/s (120 fpm), 39%–43% for 2.0 m/s (400 fpm), and 6%–16% for 4.3 m/s (850 fpm). Incorporating an airflow spoiler to the shuttle car CAC design and placing the CAC so that it is located 22.86 cm (9 in.) forward of the operator improved the dust control efficiency to 51%–55% for 4.3 m/s (850 fpm) with minimal impact on dust control efficiencies for lower ventilation velocities. These laboratory tests demonstrate that the newly developed shuttle car CAC has the potential to successfully protect shuttle car operators from coal mine respirable dust overexposures.

Keywords Shuttle car · Canopy air curtain · Dust · Airflow · Coal mining

1 Introduction

The development of the canopy air curtain (CAC) dates back to the 1970s starting with the initial development of the CAC by the Donaldson Company, Inc. under contract from the U.S. Bureau of Mines (Krisko 1975). This CAC was originally developed for continuous miner operators when continuous mining machines had cabs. The need for a CAC on the continuous miner was eliminated when the cab was removed from the machine design. However, CAC development progressed to include CAC designs for a roof bolting machine to protect roof bolters from respirable coal mine dust (Goodman and Organiscak 2002; Listak and Beck 2012; Reed et al. 2017). This roof bolting machine CAC research continues to the present day.

The findings and conclusions in this report are those of the author(s) and do not necessarily represent of the official position of the National Institute for Occupational Safety and Health. Mention of any company name, product, or software does not constitute endorsement by NIOSH.

✉ W. R. Reed
rreed@cdc.gov

[1] Pittsburgh Mining Research Division, Centers for Disease Control and Prevention, National Institute for Occupational Safety and Health, 626 Cochrans Mill Rd., Pittsburgh, PA 15236, USA

[2] J.H. Fletcher, Inc., Huntington, WV, USA

[3] Marshall University, Huntington, WV, USA

National Institute for Occupational Safety and Health (NIOSH) conducted a study which indicated that coal mine respirable dust overexposures are a concern for shuttle car operators when blowing face ventilation is used to ventilate the continuous miner face while cutting and loading coal. Table 1 summarizes the averages of the coal mine respirable dust exposure of shuttle car operators measured during continuous miner operation—cutting and loading coal—at different mining operations (Potts et al. 2011). In Table 1, straight cuts are defined as the continuous miner cutting straight into the entry. Right and left cuts are defined as the continuous miner cutting or turning a crosscut in the respective direction off the entry.

These exposures occur while the shuttle car operator is operating downwind of the continuous miner, waiting to be loaded with coal. It can be seen, from Table 1, that many of the exposures exceed 1.5 mg/m^3. While these exposures only occur during continuous miner cutting and loading cycles when the shuttle car is downwind of the miner, it can be seen that they may be high enough to result in overexposures.

Research on the CAC is being expanded to include a CAC for shuttle car operators to provide respiratory protection from respirable coal mine dust. Ambient mine air is filtered and blown over the operator through a plenum built into the shuttle car canopy. A new version of the CAC, specifically designed for the shuttle car, has been developed under a NIOSH contract by Marshall University and J.H. Fletcher [contract #200-2015-63485], and is based upon NIOSH design recommendations. NIOSH completed the required laboratory testing of the shuttle car CAC to determine its ability to reduce the shuttle car operators' respirable coal dust exposure. This paper details results of testing the shuttle car CAC in 0.61 m/s (120 fpm), 2.0 m/s (400 fpm), and 4.3 m/s (850 fpm) ventilation airflows. Since results with 4.3 m/s (850 fpm) ventilation airflows were not satisfactory additional tests with modifications to the location and design of the CAC were conducted. Modifications included moving the CAC 22.86 cm (9 in.) forward of operator location and adding a 5.08 cm (2 in.) spoiler. These modifications provided satisfactory results, thus showing that a shuttle car CAC can be a viable dust control device for the protection of shuttle car operators to coal mine respirable dust.

1.1 Testing

The testing of the shuttle car CAC was conducted on a simulated shuttle car cab in an airflow corridor at NIOSH Pittsburgh Mining Research Division (PMRD) to determine effectiveness for dust reduction. The corridor dimensions were 2.29 m (90 in.) high by 1.98 m (78 in.) wide by an 18.9 m (62 ft.) long corridor. Two different shuttle car cab designs were evaluated—a center drive cab similar to that on a Joy 10SC32AA model shuttle car and an end-drive cab similar to that on a Joy 10SC32AB shuttle car. These two shuttle car models were found to be the most commonly used in underground coal mines. These designs are shown in Figs. 1 and 2. The dimensions are approximated from actual measurements of typical shuttle cars at an operating underground coal mine site. During setup, the cabs were placed in the center of NIOSH's longwall gallery return airway at 25.4–30.5 cm (10–12 in.) above the floor as specified by the ground clearance for each shuttle car model. No obstructions to cab openings, such as wheel fenders, caging of cab openings, etc. were simulated.

The CAC (Fig. 3) was attached onto the bottom side of the shuttle car roof (canopy) and centered over the seating area of the operator. For purposes of lab testing, all necessary components for generating airflow, including the blower, drive, and intake filter were set up outside the test area. Figure 4 shows the center-drive cab setup for testing.

Table 1 Average coal mine respirable dust concentrations with 85% confidence intervals, measured at the location of shuttle car operators when the continuous miner cuts and loads cars (Potts et al. 2011)

Mine	Cut depth (m)	Straight cut (mg/m^3)		85% CI	Right cut (mg/m^3)		85% CI	Left cut (mg/m^3)		85% CI
A	6.1	4.13	±	0.70	5.31	±	1.01	NA		NA
	12.2	6.39	±	1.50	3.55	±	0.58	NA		NA
D	6.1	0.75	±	0.36	NA		NA	NA		NA
	9.2	2.73	±	0.68	NA		NA	NA		NA
E	6.1	1.27	±	0.18	1.24	±	0.25	1.50	±	0.35
	12.2	1.15	±	0.23	1.13	±	0.29	1.10	±	0.45
F	6.1	1.77	±	0.24	NA		NA	NA		NA
	9.2	2.13	±	0.25	NA		NA	NA		NA

CI confidence interval, *NA* not available

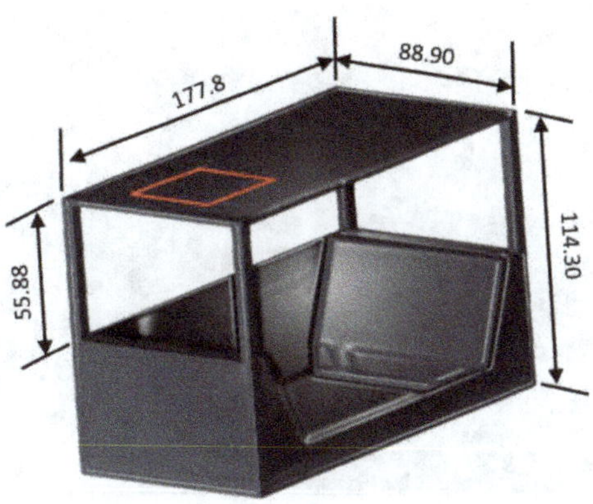

Fig. 1 The center-drive shuttle car cab for the Joy 10SC32AA (dimensions in centimeters). The red outline shows the location of the canopy air curtain underneath the cab roof during testing (drawing by NIOSH)

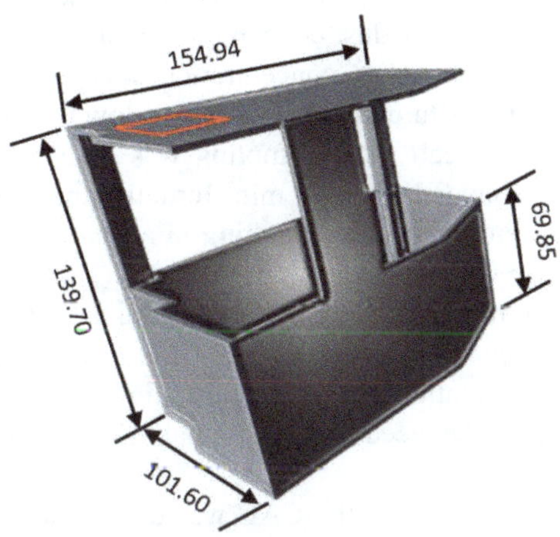

Fig. 2 The end-drive shuttle car cab for the Joy 10SC32AB (dimensions in centimeters). The red outline shows the location of the canopy air curtain underneath the cab roof during testing (drawing by NIOSH)

1.2 Sampling method

Both gravimetric and instantaneous samplers were used for testing the CAC for respirable dust control. Each gravimetric sampler consisted of an Escort ELF® pump, a 10-mm Dorr-Oliver cyclone, and a 37-mm, 5-μm PVC filter. The Escort ELF pumps were calibrated to maintain 2.0 L/min airflow. The Thermo Scientific pDR-1000®was used as the instantaneous sampler. The typical sampling package was comprised of two gravimetric samplers and one instantaneous sampler. One sampling package was placed approximately 0.91 m (3 ft.) upstream of the shuttle car cab and another was placed 0.91 m (3 ft.) downstream of the shuttle car cab. These sampling packages monitored the respirable upstream and downstream dust concentrations to ensure consistent dust concentrations throughout the test.

To test the CAC for dust control effectiveness, the sampling heads of four gravimetric samplers were placed at different locations at approximately 25.4 cm (10 in.) underneath the CAC (Fig. 5). A pDR-1000 was also placed in the center of the four gravimetric samplers underneath the CAC (Fig. 6). Later on during testing of the modifications to the CAC, continuous personal dust monitors (CPDM) Thermo Fisher Scientific Model 3600 were used in place of the gravimetric samplers due ease of obtaining immediate results.

The amount of air supply to the canopy was measured using a hot wire anemometer. Blower velocity measurements were conducted by inserting the anemometer at a port in a 1.52-m (5-ft) length of straight 10.2-cm (4-in.) PVC pipe connected to the blower outlet. The port was located approximately 1 m (40 in.) from the blower outlet to minimize turbulent airflow effects from the blower outlet.

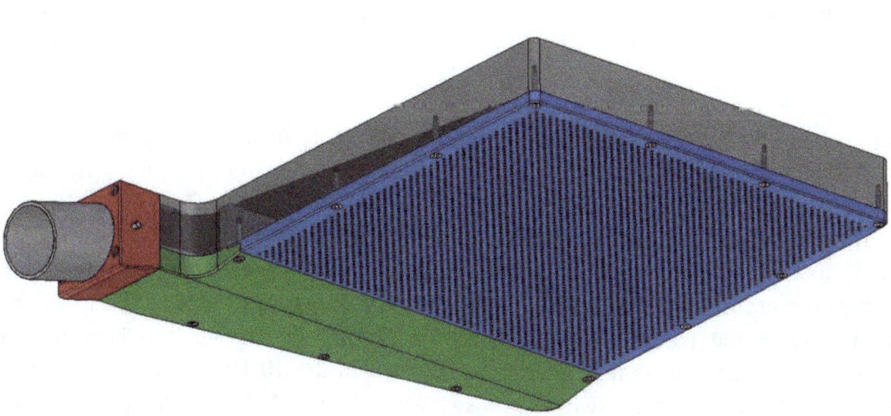

Fig. 3 The canopy air curtain underneath the roof of the shuttle car (drawing by Marshall University and J.H. Fletcher, Inc.)

Fig. 4 Photo showing the setup of the center-drive cab for testing with the manikin used for centering the CAC over the shuttle car operator

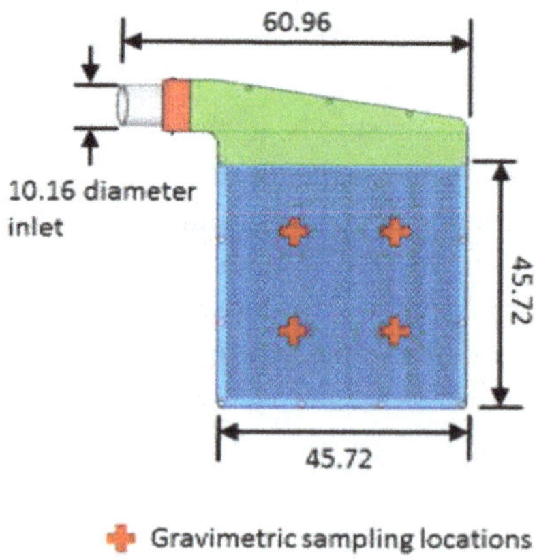

Fig. 5 The canopy air curtain showing gravimetric filter locations (measurements on figure are in centimeters)

1.3 Test procedure

Three ventilation air velocities of 4.3 m/s (850 fpm), 2.0 m/s (400 fpm), and 0.61 m/s (120 fpm) were tested in the return section where the CAC test stand was located. The 0.61 m/s (120 fpm) represented the air velocity encountered when being loaded by the continuous miner, and the 4.3 m/s (850 fpm) represented the max air velocity encountered when tramming to the feeder. The 2.0 m/s (400 fpm) ventilation air velocity was tested as an approximate midway point between the other two ventilation velocities. A dust feeder was used to obtain an upwind

Fig. 6 Photo showing the sampling locations, both the pDR-1000 and 4 gravimetric samplers, underneath the CAC

respirable dust concentration targeting approximately 6.0 mg/m^3. This concentration was selected because previous NIOSH studies have shown that this level of dust is encountered in the return of continuous miners using scrubbers (Colinet et al. 2013). The instantaneous sampler was used to monitor dust concentrations during testing.

Once the 6.0 mg/m^3 concentration was achieved, the CAC blower was turned on to supply airflow to the plenum underneath the cab roof. Sampling was then started and tests were conducted for 30 min. Initially, three trials for each test were conducted, resulting in a total of nine tests for each cab type—three tests at 0.61 m/s (120 fpm), three tests at 2.0 m/s (400 fpm), and three tests at 4.3 m/s (850 fpm). The data was analyzed using the estimation of the mean of a population using a single sample (Natrella 1963) to determine the need for additional trials, which were added as needed.

The effectiveness of the CAC was determined by comparing the respirable dust concentrations from the four gravimetric filters underneath the CAC with the respirable dust concentrations from the two gravimetric filters upstream from the test stand. The following equation was used for calculating the respirable dust control:

$$\%\,reduction = \left[1 - \left(\frac{Average\ canopy\ gravs}{Average\ upstream\ gravs}\right)\right] \times 100\% \tag{1}$$

where *% reduction* = respirable dust reduction in percentage. *Average canopy gravs* = average of the dust concentrations (mg/m^3) from the four gravimetric filters underneath the canopy. *Average upstream gravs* = average of the dust concentrations (mg/m^3) from the two gravimetric filters upstream of the test canopy.

During all testing, the blower velocity measurements taken in the 10.16 cm (4 in.) PVC pipe varied from 17.45 to 25.53 m/s (3436–5025 fpm). These measurements were only taken before and after the trial was completed. The average of the before and after measurements was used to calculate the velocity during the trial. Converting the velocities to air quantities showed that the blower provided anywhere from 0.14 to 0.21 m^3/s (299–439 cfm) to the CAC. Correlation coefficients were calculated comparing the airflow quantities to the dust reductions. The results of the correlation calculations, − 0.38 for end drive cars and -0.53 for center drive cars, never showed a strong correlation (correlation coefficients > |0.75|) of airflow quantity provided by the blower to dust reductions provided by the plenum. The dry temperature ranged from 16.7 to 23.9 °C (62–75 °F), with the relative humidity ranging from 23.4 to 67.4%. The barometric pressure ranged from 991 to 1029 mbar.

1.4 Results

In each case, for both center-drive and end-drive cabs, the column labeled "Count" represents the number of trials tested. Initially, three trials were tested. Additional trials were added based upon the equation (Natrella 1963):

$$n = \frac{\sigma^2 z^2_{\alpha=0.05}}{d^2} \qquad (2)$$

where n = sample size required [number of trials required], σ = standard deviation of number of trials, $z_{\alpha=0.05}$ = standard normal distribution value = 1.64 at 90%[1] confidence for two-tailed test, d = allowable error.

In reviewing all trials, the number of trials performed was sufficient with the allowable error for the resulting dust reductions set at ± 5%.

The dust control results from the testing are shown in Tables 2 and 3 along with associated statistics, such as standard deviation and upper and lower 95% confidence intervals. It can be seen that the canopy air curtain on the center-drive cab has a dust control efficiency of approximately 74% in 0.61 m/s (120 fpm) ventilation airflow. In 4.3 m/s (850 fpm) ventilation airflow, the dust reduction was very low at 16%, demonstrating that the dust control efficiency in high airflow velocities was reduced.

For the end-drive cab, it can be seen that the canopy air curtain has a dust control efficiency of up to approximately 83% in 0.61 m/g (120 fpm) ventilation airflow. In 4.3 m/s (850 fpm) ventilation airflow, the dust reduction was reduced to 6%, demonstrating that its dust control efficiency in high airflow velocities will also be reduced for end-drive cabs.

Past research has shown that as the ventilation velocity increases, the dust control efficiency of the CAC decreases (Engel et al. 1987). The results of this testing proved no different. However, the contract requires the ability to reduce the shuttle car operator's respirable coal dust exposure by at least 60% and poor performance during tramming—where relative velocities are highest—could result in efficiencies below this level.

1.5 Improvement of canopy air curtain dust control efficiencies

Because the contract focuses on shuttle cars with center-drive cabs, subsequent testing to improve the dust control efficiency was completed on the canopy air curtain installed on center-drive cabs. During this testing personal dust monitors (PDM) were used. The PDM allows downloading of the dust concentration data, which provides for quicker turnaround of results compared to weighing gravimetric filters. Therefore, PDMs were used in place of all gravimetric samplers, while the pDRs were still used to provide instantaneous dust concentration readings for monitoring dust inside the test facility.

Reviewing the previous reports on canopy air curtain development from Marshall University/J.H. Fletcher & Co., Inc., the cross-sections resulting from computational fluid dynamic (CFD) simulations were studied for potential improvements. Of interest were the CFD model results of the canopy air curtain with approximately 4.2 m/s (830 fpm) lateral ventilation flow (Salem et al. 2016). Figure 7 presents the results of the CFD analysis showing results in side, plan, and isometric views. The CAC plenum is centrally located atop the CFD modeled volume and is outlined in red. The airstream lines (blue to light green color) are shown emanating from the plenum with the streamflow immediately pushed downstream due to the high 4.2 m/s (830 cfm) ventilation flow. Normally, airstream flow from the plenum in low velocity ventilation air emanates straight from the plenum with no deflection. The alignment of these airstream lines due to the deflection caused by the high 4.2 m/s (830 cfm) ventilation airflow demonstrates that the airflow protection the canopy offers to the worker seems to shift downwind in high-velocity ventilation airflows. From the observation of the different views displayed in this figure, generated by CFD, the canopy air curtain was shifted approximately 22.86 cm (9 in.) forward of the shuttle car operator's position, and a front spoiler was added in an attempt to redirect ventilation airflow to improve its performance for dust control.

The front spoiler, consisting of a 12.7–15.2 cm (5–6 in.) wide piece of plywood that extended across the entire

[1] The confidence level was relaxed to 90% from the commonly used 95% level for this lab testing. While lab testing has much better control over influences on testing, there are still factors which cannot be controlled such as air properties, dust feed moisture content, airflow fluctuations, dust feed variations, etc. Additionally, laboratory availability can place time constraints limits on lab testing limiting the number of trials completed. A 90% confidence level is still a robust standard.

Table 2 Dust control reduction and associated statistics for center-drive shuttle car cabs at different ventilation airflows

Description	Vent airflow (m/s)	Average reduction (%)	Standard deviation	Count	Upper CI (95%)	Lower CI (95%)
Center drive	0.61	73.7	3.1	5	76.4	71.0
Center drive	2.03	42.6	2.7	3	45.6	39.6
Center drive	4.32	16.3	1.5	5	17.6	15.0

CI confidence interval

width of the center-drive cab, was added to the roof of the cab. Additionally, it was oriented at 64° from the horizontal and extended approximately 5.08 cm (2 in.) below the canopy plenum outlets as shown in Fig. 8. The first series of tests placed the canopy centrally in the cab underneath the cab roof with a front spoiler. The sampling locations were not moved with the CAC and were located over the operator position. However, for this first series of tests, two additional samplers were added in front of the existing sampling rows to include sample locations 25.4 cm (10 in.) directly underneath the CAC (Fig. 10).

The second series of tests used the same spoiler, but moved the canopy air curtain from the centered position to a location 22.9 cm (9 in.) forward from the original over-operator position (Fig. 9). Again, the sampling locations were not moved with the CAC. They were left in place, centered over the operator position. However, for this second series of tests, one additional row of samplers was added to include sample locations 25.4 cm (10 in.) underneath the CAC (Fig. 10). During testing two CPDM per row were used, totaling six CPDMs, replacing the gravimetric samplers. In all subsequent analysis for CAC

Table 3 Dust control reduction and associated statistics for end-drive shuttle car cabs at different ventilation airflows

Description	Vent airflow (m/s)	Average reduction (%)	Standard deviation	Count	Upper CI (95%)	Lower CI (95%)
End drive	0.61	82.8	1.4	5	84.0	81.5
End drive	2.03	38.6	6.5	5	44.3	32.8
End drive	4.32	6.2	3.4	6	8.9	3.5

CI confidence interval

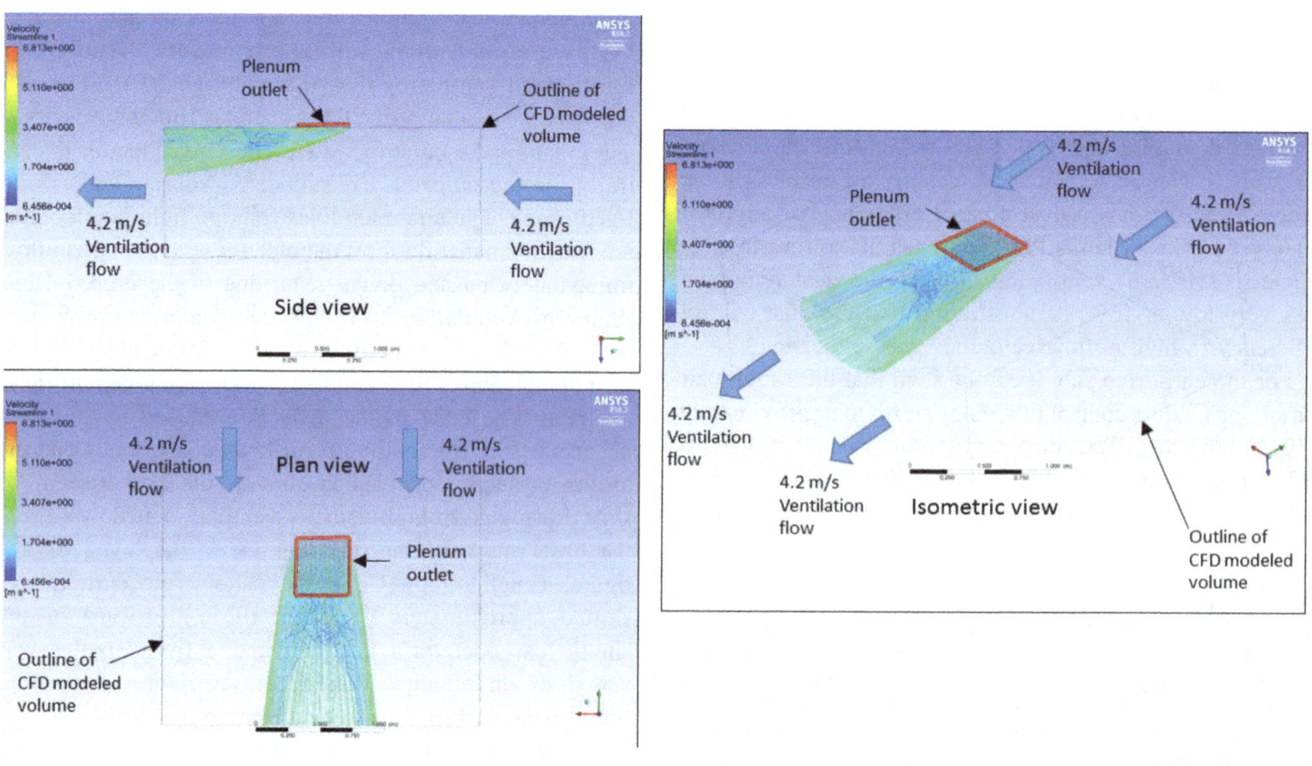

Fig. 7 CFD simulation results showing airflow streams from shuttle car CAC plenum in 4.2 m/s ventilation airflow (Marshall University)

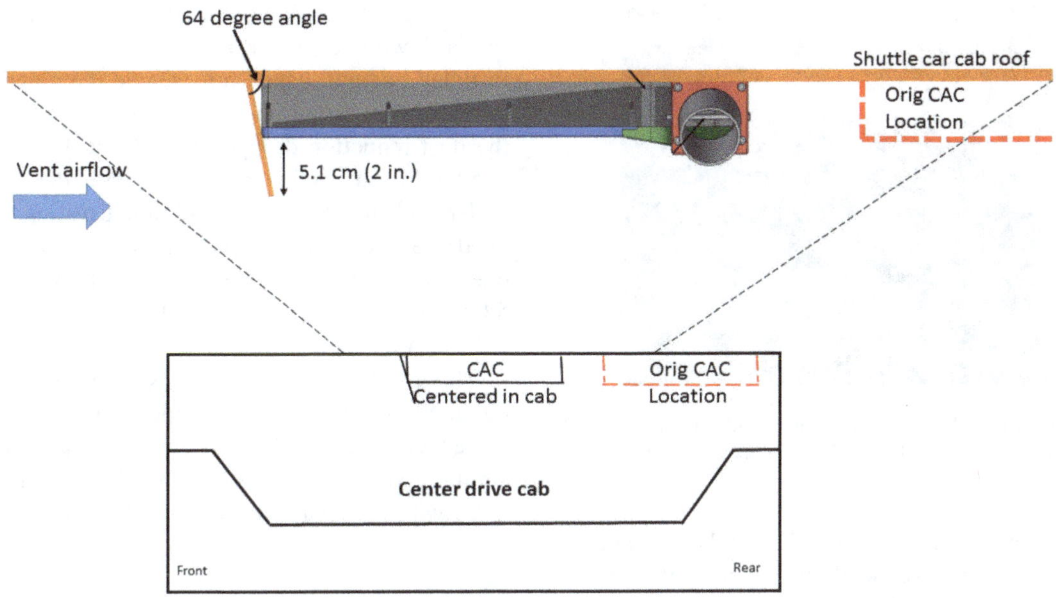

Fig. 8 Canopy air curtain underneath cab roof displaying spoiler to redirect ventilation airflow

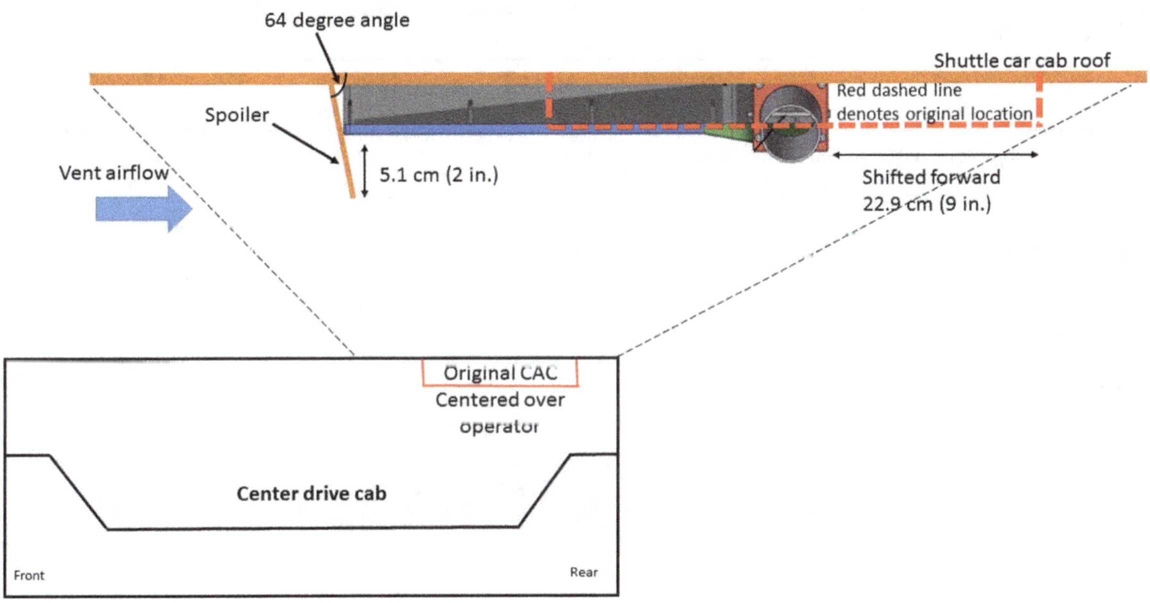

Fig. 9 Canopy air curtain underneath cab roof, displaying forward shift of canopy location 22.9 cm (9 in.) and spoiler to redirect ventilation airflow

respirable dust reductions, *Average Canopy Gravs* from Eq. (1) averaged the results of all six samplers underneath the CAC area of influence. Results of these tests can be seen in Table 4.

The canopy centered underneath the cab increased the average dust reduction to 36% (from 16%). Reviewing the CFD analysis of the CAC showed that moving the CAC only 22.86 cm (9 in.) forward might provide better results. Therefore, tests that centered the CAC underneath the cab were discontinued and tests that moved the CAC 22.86 cm (9 in.) forward were tested. Test results showed that the combination of the spoiler and moving the CAC 22.86 cm (9 in.) forward of the operator seemed to allow the CAC to perform better in the high ventilation airflows, having a 51% average reduction in 4.3 m/s (850 fpm). The CAC also performed better with 66% dust reduction at 2.0 m/s (400 fpm) ventilation airflow. At 0.61 m/s (120 fpm), the respirable dust reduction was 70%. The 51% dust reduction at 4.3 m/s (850 fpm) is a substantial increase in dust reduction from the canopy without any modifications to the

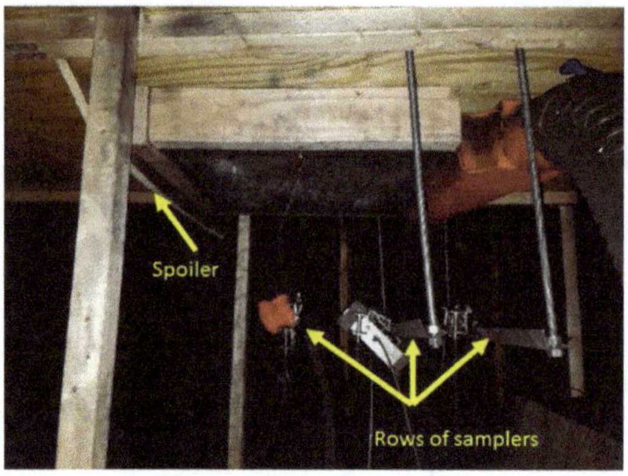

Fig. 10 Photo showing additional sampling row added to provide full sampling coverage underneath the CAC. Spoiler is shown on the front of the CAC

CAC itself. In reviewing all results, the number of samples was sufficient when setting the allowable error at ± 5% in dust reduction. It is assumed that these results will translate to equivalent results for the end-drive cabs.

A single test with the CAC moved 22.86 cm (9 in.) forward with a spoiler and an additional dummy CAC installed in front, which represented the planned design of two CACs installed underneath the cab roof, showed that the dust reduction dropped to 35% at 4.3 m/s (850 fpm), similar to the results in Table 4 with the CAC centered underneath the cab roof. Additional testing, which simulated the two canopies flush mounted onto the roof with one spoiler with a 10.16 cm (4-in.) drop instead of a 5.08 cm (2 in.) drop (Fig. 11), showed better results [a 55% reduction at 4.3 m/s (850 fpm)].

The results in Table 5 show that the canopies flush mounted onto the shuttle car roof will have better success in reducing respirable dust for the shuttle car operator; similar results, as shown in Table 4, were provided by the CAC with the spoiler moved 22.86 (9 in.) forward.

Table 4 Dust control reduction and associated statistics for a center-drive shuttle car at different ventilation airflows with different canopy configurations as shown in Figs. 8 and 9

Description	Vent airflow (m/s)	Average reduction (%)	Standard deviation	Count	Upper CI (95%)	Lower CI (95%)
Canopy with spoiler, canopy centered underneath cab	4.32	36.2	0.43	3	36.7	35.7
Canopy with spoiler, canopy moved 22.9 cm (9″) forward of operator	0.61	70.4	2.09	6	72.1	68.7
Canopy with spoiler, canopy moved 22.9 cm (9″) forward of operator	2.03	65.9	0.42	6	66.3	65.6
Canopy with spoiler, canopy moved 22.9 cm (9″) forward of operator	4.32	51.3	6.78	3	59.0	43.6

CI confidence interval

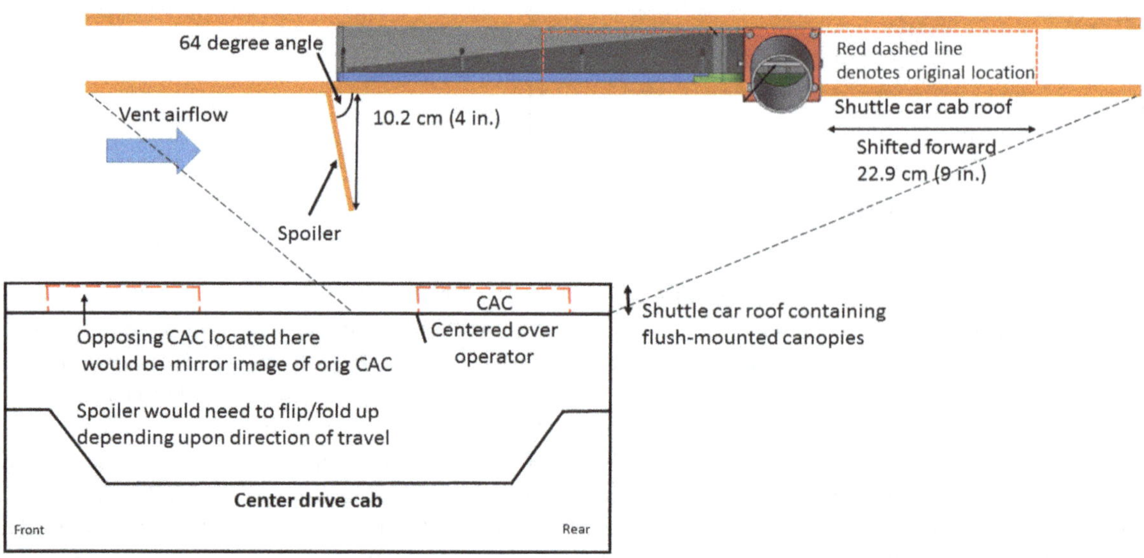

Fig. 11 Canopy air curtain built into cab roof, displaying forward shift of canopy location (9 in. forward) and spoiler to redirect ventilation airflow

Table 5 Dust control reduction and associated statistics for center-drive shuttle car for an improved canopy configuration shown in Fig. 11

Description	Vent airflow (m/s)	Average reduction (%)	Standard deviation	Count	Upper CI (95%)	Lower CI (95%)
Two canopies, 22.9 cm (9″) forward, mounted underneath roof with spoilers on both	4.32	26.6	NA	1	NA	NA
Two canopies, 22.9 cm (9″) forward, mounted underneath roof with spoiler on operator position	4.32	35.4	NA	1	NA	NA
Two canopies, 22.9 cm (9″) forward, flush-mounted underneath roof with spoiler on operator position	4.32	54.8	1.26	3	56.2	53.4

CI confidence interval

1.6 Conclusions

Laboratory tests conducted with the canopy air curtain as currently designed, and with its planned location on the shuttle car cab, has shown to be more than sufficient for dust reductions in airflow velocities up to 0.61 m/s (120 fpm), with reductions of 74% (center drive) and 83% (end drive). However, at 4.3 m/s (850 fpm) ventilation airflow, the reductions were very low at 16% (center drive) and 6% (end drive) and not sufficient to meet the contract requirements of 60% dust reduction during the entire operation of the shuttle car.

To improve dust reductions from the CAC, modifications to the canopy locations were tested. Installing a spoiler and moving the CAC 22.86 cm (9 in.) forward from the operator location seem to provide the best improvement in the performance of the CAC. Dust reductions of 70% at 0.61 m/s (120 fpm), 66% at 2.0 m/s (400 fpm), and 51% at 4.2 m/s (850 fpm) ventilation airflows were achieved. Although the 60% dust reduction threshold was not met at 4.2 m/s (850 fpm), a 51% reduction is close and is sufficient to proceed with field testing. In addition, the 4.2 m/s (850 fpm) air ventilation quantity threshold was originally based upon a mine ventilation velocity measured in an intake entry and the maximum shuttle car speed of 9.6 km/hr (6 mph) (Joy Global 2016). Recent studies completed by NIOSH show that the maximum air ventilation quantity threshold may not be equal to mine intake air plus the maximum shuttle car speed (Shahan and Reed 2018), and relative velocity may be significantly lower than 850 fpm in mines. In addition, the higher airflow a shuttle car encounters is generally associated with lower dust concentrations. In fact, the shuttle car operators' highest respirable dust exposure occurred when being loaded by the continuous miner in blowing face ventilation. Therefore, the modification of moving the CAC 22.9 cm (9 in.) forward and flush mounted onto the shuttle car cab roof should be sufficient to achieve the targeted 60% reduction.

These laboratory tests showed that the canopy air curtain for the shuttle car is successful at reducing respirable coal mine dust exposure at low ventilation velocities. Redesign of the CAC to shift it forward 22.9 cm (9 in.) with a spoiler and flush mount it onto the shuttle car roof should successfully protect the shuttle car operator from respirable coal mine dust as seen by the results from these tests. Other options to improve the dust control efficiency of the CAC could be evaluated, such as increasing the airflow to the CAC and/or enlarging the size of the CAC to have a single CAC over the entire canopy. But, these enhancements are probably not necessary for improving dust control efficiency. The laboratory testing on the shuttle car CAC demonstrates that it can successfully protect the shuttle car operator by reducing exposure to coal mine respirable dust. A field test is being planned to test the CAC at a mine site that uses blowing face ventilation.

References

Colinet JF, Reed WR, Potts JD (2013) Impact on respirable dust levels when operating a flooded-bed scrubber in 20-foot cuts. U.S. Department of Health and Human Services, Centers for Disease Control and Prevention, National Institute for Occupational Safety and Health, DHHS (NIOSH) Publication 2014-105, RI 9693, Pittsburgh, PA

Engel M, Johnson D, Raether T (1987) "Improved canopy air curtain systems". USBM Open File Report 25–88. US Bureau of Mines, Washington

Goodman JVR, Organiscak JA (2002) Evaluation of methods for controlling silica dust exposures of roof bolters. Trans Soc Min Metall Explor 312:133–137

Joy Global (2016) Haulage systems product overview. Komatsu America Corp, Rolling Meadows

Krisko WJ (1975) "Develop and test canopy air curtain devices". USBM Contract Report HO232067. US Dept. of the Interior, Bureau of Mines/Donaldson Company, Inc., Pittsburgh

Listak JM, Beck TW (2012) Development of a canopy air curtain to reduce roof bolters' dust exposure. Min Eng 64(7):72–79

Natrella MG (1963) Experimental statistics. National Bureau of Standards Handbook 91. U.S. Department of Commerce, National Bureau of Standards, Washington

Potts JD, Reed WR, Colinet JF (2011) Evaluation of face dust concentrations at mines using deep-cutting practices. NIOSH

Report of Investigations 9680. U.S. Dept. of Health and Human Services, Centers for Disease Control and Prevention, National Institute for Occupational Safety and Health, Office of Mine Safety and Health Research, Pittsburgh, PA

Reed WR, Joy GJ, Kendall B, Bailey A, Zheng Y (2017) Development of a roof bolter canopy air curtain for respirable dust control. Min Eng 69(1):33–39

Salem A, Begley R, Ross G (2016) "Progress Report #2". NIOSH Contract #:200–2015-63485. Marshall University Research Corp, Huntington

Shahan M, Reed WR (2018) Field investigation to measure shuttle car airflow velocities at a central Appalachian underground coal mine. 2019 SME Annual Meeting and Exhibit, Feb. 25-28, 2018, Minneapolis, MN. Pre-print 18-032

16

Estimation of carbon sequestration in reclaimed coalmine degraded land dominated by *Albizia lebbeck*, *Dalbergia sissoo* and *Bambusa arundinacea* plantation

Rimi Das[1] · **Subodh Kumar Maiti**[1]

Abstract Reclaimed mined lands provide an excellent opportunity to sequester carbon and combat global warming. Carbon sequestration on reclaimed sites depend on age of reclamation, composition of species, geomining conditions (soil characteristics) and prevailing climate. The aims of the present study were to calculate carbon (C)—stock of biomass of 4 years old plantation (dominated by *Albizia lebbeck*, *Dalbergia sissoo* and *Bambusa arundinacea*), understorey vegetation and litter, soil organic carbon in reclaimed minesoil (RMS) and compare with reference forest site. Allometric equation was used for the estimation of biomass C stock and found 13.0 Mg C ha^{-1} (*A lebbeck* 7.8 Mg C ha^{-1}, *D sissoo* 3.5 Mg C ha^{-1} and *B. arundincea* 1.2 Mg C ha^{-1}), while stock of understorey vegetation was 0.98 Mg C ha^{-1}. In RMS, C stock was 16.3 Mg C ha^{-1}, out of which inorganic C contributed 1.7 g kg^{-1} (8 % of total soil C), Coal C 8.7 g kg^{-1} (43 % of total soil C) and biogenic C 9.8 g kg^{-1} (49 % of total soil C). Total C stock in reclaimed site was calculated as 30.3 Mg C ha^{-1} (equivalent to 111 Mg CO_2 ha^{-1}). The study concluded that (i) coal C is responsible for overestimation of C stock in RMS (ii) Maximun C stock stored in aboveground biomass component and (iii) reclaimed mined lands will take approximately 17 years to reach the level of C stock of reference forest site in dry tropical climate.

Keywords Coal · Opencast mining · Reclaimed mine soils · Biomass carbon · Carbon sequestration · Carbon stock

1 Introduction

Carbon (C) sequestration is defined as "the process of increasing the C content of a C pool other than the atmosphere" (IPCC 2000). Terrestrial C sequestration is the process of (i) transforming or transferring atmospheric CO_2 through photosynthesis into biomass components such as trees, shrubs, vegetation, and soil organic matter (SOM), and (ii) incorporation of biomass into the soil as humus. This leads to the effective storage of atmospheric CO_2 in these biomass components until decomposition of biomass into the soil. Soil contains approximately 75 % of the terrestrial C pool, which is three times more than the amount stored in living plants and hence it plays a vital role in the global C cycling (Schlesinger and Bernhardt 2013).

In India, massive expansion of opencast mines is envisaged by adoption of latest technologies to meet the target and about 90 % of the coal is extracted by the opencast mining. During the process of surface coal mining, vegetation cover is completely stripped, and soils and overburden (OB) rocks are removed to reach the coal-deposit, which generates huge volume of heterogeneous mass, stored as OB dumps (external dumps) or used as backfill materials to fill open cast voids (internal dumping). These OB dumps, initially devoid of soil organic carbon (SOC) and generally reclaimed into a forest land through forestry reclamation approach (FRA) (Zipper et al. 2011; Skousen and Zipper 2014). Establishment of vegetation cover with high biomass production in reclaimed minesoils (RMS) reduces soil degradation and improves SOC with time. During technical reclamation, after dressing the

✉ Subodh Kumar Maiti
skmism1960@gmail.com

[1] Ecological Restoration Laboratory, Department of Environmental Science and Engineering (Centre of Mining Environment), Indian School of Mines, Dhanbad, Jharkhand 826 004, India

surface, topsoil is applied and slope is blanketing with coir mat, and seeded with grass-legume mixture as an initial colonizer to the site (Maiti and Maiti 2015; Xiao et al. 2014). In the top flat surface of the dump, fast growing tree species saplings are planted with addition of topsoil in plantation pits (Maiti and Ghose 2005; Maiti et al. 2007).

Studies across the globe on C sequestration through reclaimed coal mine soils have received considerable attention in recent past (IPCC 2000; Akala and Lal 2000, 2001; Ussiri and Lal 2005; Sperow 2006; Amichev et al. 2008; Shrestha and Lal 2010; Tripathi et al. 2014, 2016; Das and Maiti 2016). The CO_2 sequestration in reclaimed mine sites is calculated by estimating C-stock of biomass (aboveground and belowground), litter and reclaimed minesoils (Fig. 1). The aboveground biomass (AGB) C-stock consists of all living vegetation above the soil, inclusive of stems, stumps, branches, bark, seeds and foliage, while the belowground biomass (BGB) C-stock consists of the biomass contained within live roots. The greatest fraction of total biomass of forests is represented by trees, whereas the other carbon pools (understory, dead wood, litter and BGB) represent only a fraction of the total tree biomass. The understorey was estimated 3 % of AGB, dead wood 5 %–40 %, and fine litters only 5 % of that in the AGB (Brown 1997). Destructive sampling is the most accurate method to estimate C-stock of biomass, whereby vegetation is harvested, dried to a constant mass and the dry-to-wet biomass ratio established and correlated with biomass vis. DBH (Diameter at Breast Height). Two other approaches for estimating the tree biomass are: (a) regression equations or allometric equations which provide estimates of biomass per tree based on three main climatic zones, regardless of species; and (b) the conversion of wood volume to biomass density using biomass expansion factors (Brown et al. 1989; Brown 1997). Allometric equations play important role in accurate estimation of above and belowground biomass stock of tree species based on measured independent variables (like, DBH) (Brown et al. 1989). Allometric equations can be represented in many ways, but the most commonly represented by a linear equation (Dudley and Fownes 1992). However, they may also vary according to diameter classes (in particular for higher diameter classes and for smaller diameter class <10 cm). Hence the biomass estimation also varies according to various factors such as DBH of tree species, stand age of the species and topography. Cairns et al. (1997) formulated regression equations to estimate root biomass based on above-ground biomass carbon in tropical, temperate and boreal forests and found a mean root-to-shoot (RS) ratio of 0.26, ranging between 0.18 and 0.30 and stated that the most of the root biomass (heavy roots >2 mm diameter) is close to 20 percent of the total AGB. According to MacDicken (1997), the ratio of belowground to aboveground biomass in forests is about 0.2, depending on species.

In India, several authors have published biomass estimations through destructive sampling methods and developed allometric equations for a few tree species >10 cm DBH in reclaimed minesoils and forests (Singh and Singh 1995; Lodhiyal et al. 2002). Amichev et al. (2008) also derived allometric equations for the estimation of biomass in reclaimed minesoil of Midwestern and Appalachian coalfields of USA.

Litter accumulation is also an important parameter for consideration in the C sequestration study. Litterfall consists of dead plant material, such as leaves, bark, seeds, logs, or reproductive organs, and twigs that has fallen to the ground. Litterfall depends upon type of tree species, plant debris, the type of ecosystem (forest or grassland) and various factors. For example forest ecosystem accounts for 70 % of litterfall (leaf), but woody litter increase with forest age, however in grassland ecosystem, the annual litterfall is very low due to very little aboveground perennial tissue (Lonsdale 1988).

Mining and associated disturbances disrupt the SOC equilibrium relationship, causing severe losses of SOC due to unfavorable physico-chemical characteristics, leading to a C deficit and poor conditions for both plant and microbial growth and SOC. Revegetation and proper management practices help to regain the lost C; improve the soil quality and restore the SOC by reabsorbing it from the atmosphere (Lal and Bruce 1999). Several researchers in USA studied C accumulation in minesoils which were reclaimed to cropland, pasture or forest (Akala and Lal 2000, 2001; Amichev et al. 2008; Sperow 2006). Total soil C (TSC) content in reclaimed or biologically restored minesoils consist of three types of pools namely (i) soil inorganic C (SIC), (ii) geogenic C (coal-C or fossil C) and (iii) biogenic C (plant derived recent SOC) (Ussiri and Lal 2008). Sources of SIC are parent rock of overburden materials which consists of carbonates or calcite and dolomite in significant concentrations. "Geogenic C" in minesoils originates from the incorporation of coal particles or coal dust during overburden removal, coal mining and reclamation operations, while "biogenic C" originates from recent biological inputs. Microbial decomposition of biogenic C sources (litter) along with different pedogenic process leads to their mixing with coal C and also the humus fraction. The stable fractions of"biogenic C" such as the humus and mineralizable fractions are responsible for maintaining long term CO_2 sequestration as it controls the SOC pool. However, standard procedures for quantifying SOC concentrations in RMS cannot distinguish coal C from biogenic C due to presence of large amount of coal carbon. This leads to overestimation of C pools and

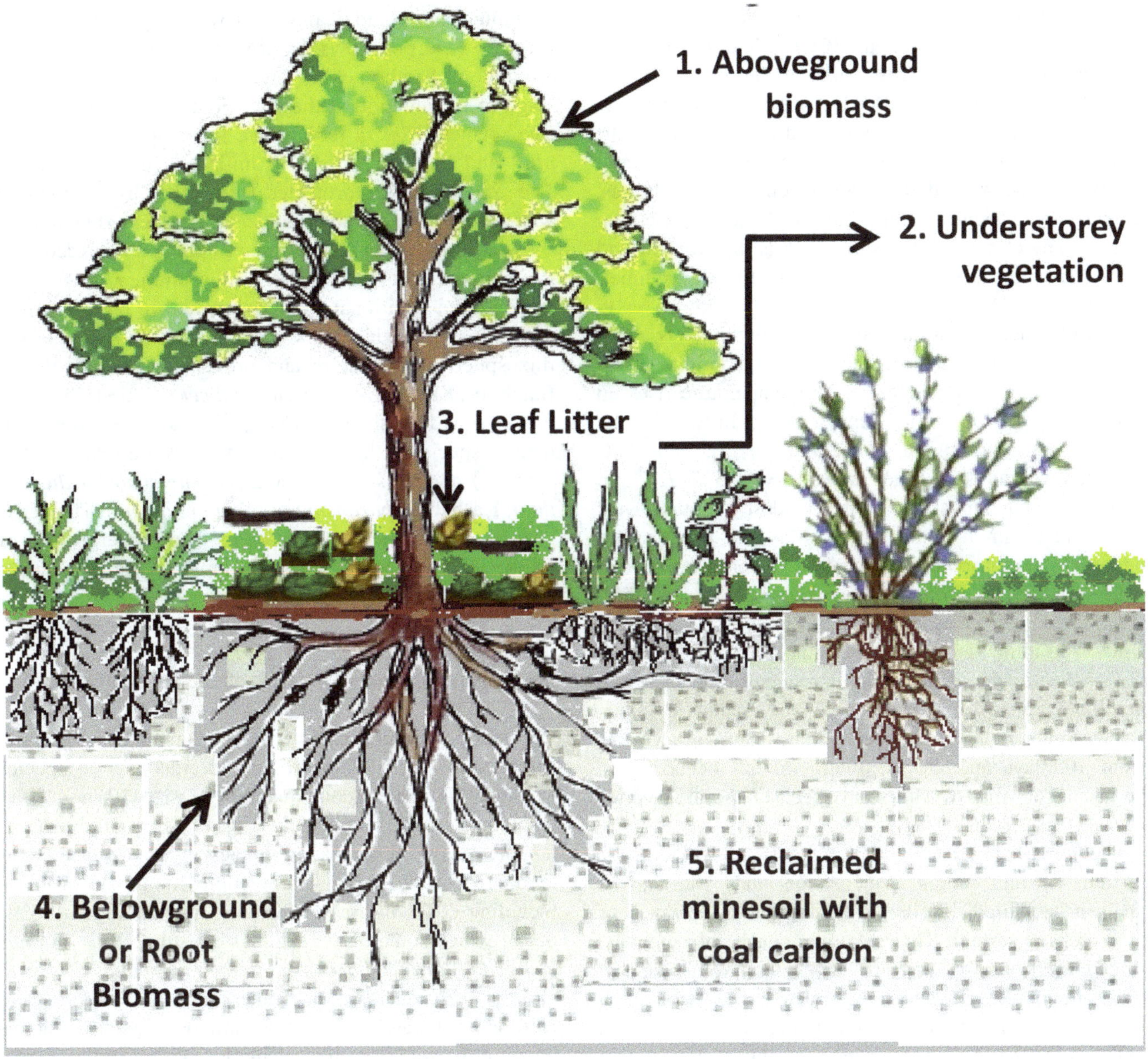

Fig. 1 Schematic diagram showing distrribution of carbon-stock stored in reclaimed coalmine site (*1* to *5*) and summation of all components gives CO seqquestration values (Million gram carbon per hectar, Mg C ha^{-1}). *1* Aboveground biomass, *2* and *3* Litterfall and understorey vegetation which includes shrubs, herbs, *4* Root biomass and *5* Soil organic carbon in reclaimed minesoils intermixed with coal carbon

sequestration rates, which must be corrected (Ussiri et al. 2014),

Scanty literature is available on C sequestration potential in RMS of India inspite of plentiful reports available worldwide. Direct tree harvest data is difficult to obtain and very less numbers of studies are available in the literature. In the present study, non harvest technique and allometric equations were used to estimate biomass and total carbon sequestered by different plant species in the reclaimed site. The aims of the present study are: (i) to estimate biomass C stock by using allometric equation; (ii) to estimate C stock of understorey vegetation, (iii) to study accumulation of C stock in mine soil and (iv) to study the distribution of sequestered C in different components of RMS (SOC or biogenic carbon; after removal of coal and inorganic carbon) and, (iv) to estimate rate of C sequestration and predict recovery of C stock in reclaimed mine soil to reference forest site.

2 Materials and methods

2.1 Location of study area

Carbon sequestration potential study was carried in a ecologically restored coalmine overburden dumps located in the Damoda colliery, Barora area, Jharia Coalfields (Bharat Coking Coal Limited, BCCL), Jharkhand, India. The Jharia coalfields falls between latitudes 23°39′–23°48′N and longitudes 86°11′–86°27′ E covering an area of 450 km^2. The site was located in the western part of Jharia coalfield which is characterized for the long history of land degradation, hostile climatic conditions, different tree species composition and slow succession process. The climate of the study area is dry tropical and characterized by hot and dry summer (March to June), rainy (July to October) and winter season (November to February). The average annual relative humidity is about 63 %. In summer months, the relative humidity (RH) varies between 32 % and 72 %. The temperature rises up to 42 to 46 °C on some summer days. In winter, the maximum and minimum temperature is 22 and 8 °C respectively, and sometimes drops down to 5–7 °C, in December and January, are the coldest months. Dust storms are common in dry season (May and June) before the onset of monsoon with increase in temperature and wind speed in the afternoon coupled with low humidity. The area receives annual rainfall of about 1140–1700 mm (average: 1306 mm), out of which 75 %–80 % of the annual rainfall occurs during the three months of June to September with smaller amounts during winter months. Number of rainy days in a year is about 100 days and the maximum rainfall occurring in the month of August. The wind speed of the area varies from 1.5 to 2.8 ms^{-1}.

The mining has been carried out by shovel–dumper combination and carbon sequestration study was carried out in a backfilled, lavelled OB dumps with presence of large rocks at surface. The rock types consist of sandstone (45 %), carbonaceous shale (20 %), intermixed shale and sandstone (16 %), Jhama (heat affected coal) with mica peridotite (3 %), subsoil (4 %) and coal (12 %) (Chandra 1992). The chemical composition of the rocks is very important because significant quantities of various plant nutrients are usually made available during the weathering process (Coppin and Bradshaw 1982).

The total study area was approximately 4 ha (40000 m^2) and the age of the entire restored site is 4 years. Location of study area is shown in Fig. 2. A 3-tier plantation consisting of grasses, shrubs and several tree species (native to the nearby forest) were raised before the onset on monsoon in the year 2011 by the BCCL in association with Forest Department to re-establish forest vegetation in the overburden dump. For raising tree species, saplings were planted in the plantation pits (30 cm × 30 cm × 30 cm) with the addition topsoil with weathered OB materials (soil: spoil; 1:4). Tree saplings used in the restoration purposes, are characterized by low in stature, water and nutrient demands, deciduous, and has economic and ecological values with high survival rate (Fig. 3). Figure 4 shows the arising of tender shoots from the node of bamboo culms after the propagation at initial stage of reclamation at the site. Figure 5 shows growing tree species (*Albizia lebbeck*), a shrub species (*Z. oenoplia*) and bamboo clumps at initial stage of plantation. The vegetation was protected from cattle by a boundary wall made of stones and boulders. A cemented signboard showing the type and no. of species planted is placed near the boundary wall (Fig. 6). During summer, trees were watered and additional soil was applied in the plantation pits of new growing trees or saplings to ameliorate the rhizospheric temperature and prevent them from dying. Bamboo, Kala siris (*Abizzia lebbeck*), Safed Siris (*A. procera*) and Shisham (*Dalbergia sissoo*) were the dominating species in the site along with Amla (*Phyllanthus emblica*) and grasses Dennanath *(Pennisetum pedicellatum)*, Kash (*Saccharam munja, S. spontaneum*). The dense growth of all these tree species and undergrowth of grasses and shrubs established complete 3-tier vegetation (Fig. 7)

2.2 Vegetation characteristics of the study area

The study area consists of 3-tier plantation of grasses, shrubs and several fast growing deciduous tree species (Table 1). Three random quadrats of size 10 m × 10 m (each quadrat of 100 m^2 area) (Fig. 8) were laid down and no. of trees of a particular species in each quadrat was recorded. Total 13 species were present in the quadrate including different grasses, bamboo and saplings of trees, out of which some of the saplings constituted *Phyllanthus emblica, Albizia lebbeck, D. sissoo, Terminalia arjuna, Lannea coromandelica*, and others. Bamboo clumps were recorded separately since the bamboo plantation dominated in the study site for calculation of density ha^{-1}. The girth of each tree species was measured at the height of 1.37 m above the ground level (i.e., DBH). The relative density of tree species with height >1.5 m and DBH >3 cm was calculated separately and the saplings height <1.5 m and DBH <3 cm were not considered for carbon sequestration study.

Two dominant tree species, *Albizia lebbeck* (58 %) and *Dalbergia sissoo* (33 %) together constitute 90 % of total plants population followed by *Phyllanthus emblica* (4 %), *Mitragyna parviflora* (2 %), *Syzizium cumini* (1 %) and *Ziziphus oenoplia* (1 %). Relative distribution of tree species in reclaimed site is shown in Fig. 9. Sixty one nos of bamboo clumps were observed in the three quadrats. Therefore the density of bamboo clumps was calculated as 2033 clumps ha^{-1}; whereas the tree density was 2500 trees ha^{-1}.

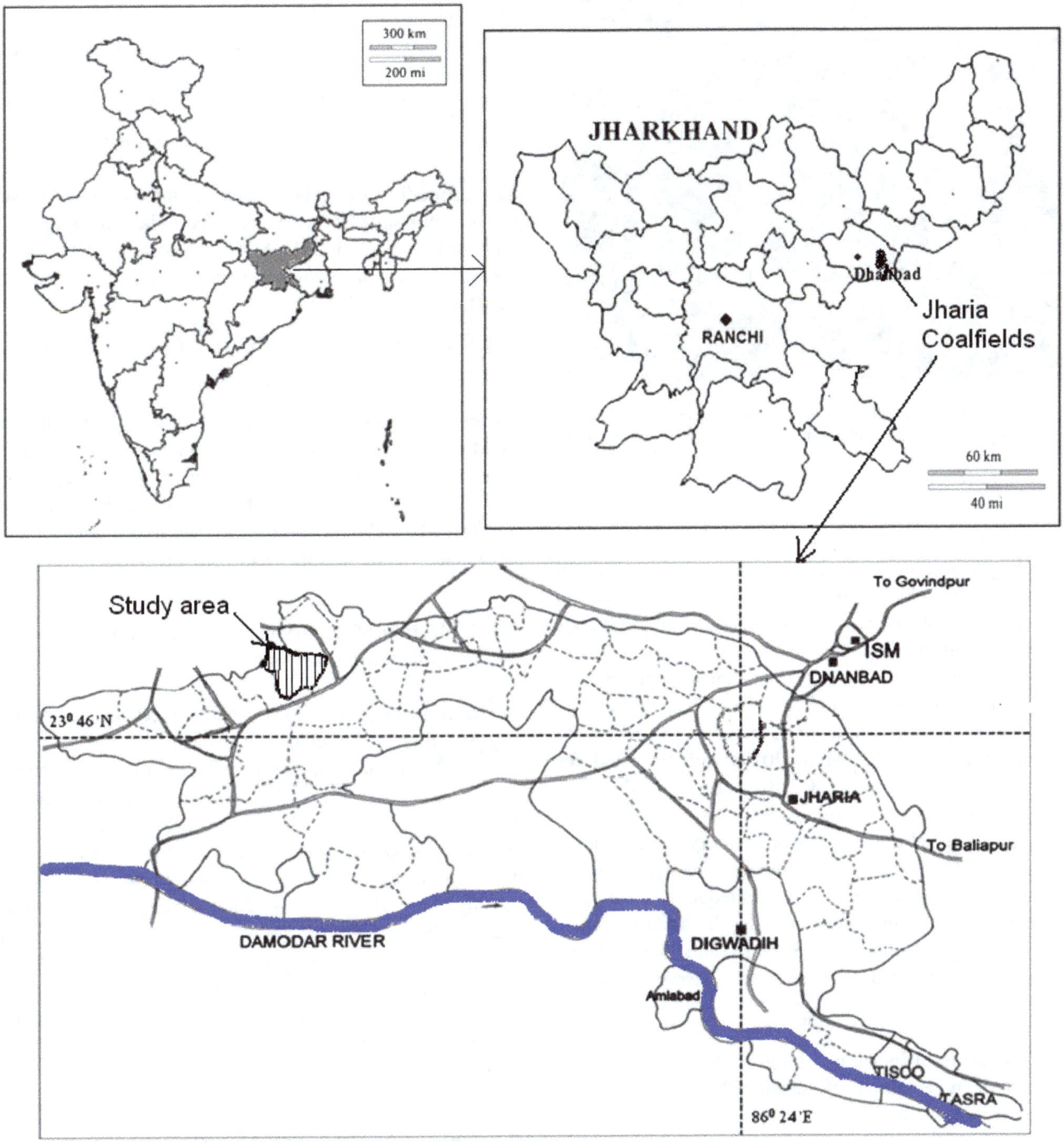

Fig. 2 Location map of study area

2.3 Reclaimed minesoil sampling and analysis

Sampling sites were selected underneath the canopy of different tree species *A. lebbeck, B. monosperma* and bamboo plantation. Total 5 replicates of minesoil samples were randomly collected by laying metal quadrat of 50 cm × 50 cm using the soil corer at 0–15 cm depth after the removal of litter layer (if any). Samples were properly packed in air tight sampling bags and brought carefully to laboratory for physical and chemical analysis.

Separate soil core were also collected with corer (15 cm height) for the measurement of bulk density at these sites. The soil samples were air dried for a week, crushed lightly by using mortar and pestle to remove the soil particles

Fig. 3 Initial stage of establishment of tree saplings of Safed Siris (*A. procera*) by pit plantation. Natural colonisation of Palash (*Butea monosperma*) on the site

sticking to the non soil particles, followed by sieving through a 2-mm sieve to separate the soil and non soil parts, reweighted to record the proportion of soil fraction (<2 mm size), labeled and kept in air tight sampling bags for further analysis.

The field moisture content of fresh soil was determined by gravimetric method and bulk density was determined by core method (dividing the oven dried mass of soil by the volume of the sample) (Maiti 2013) and corrected for gravel content. The paste pH and electrical conductivity (EC) was determined in soil: water ratio (1:1; w/v) suspension with a pH meter and Conductivity meter respectively (McLean 1982). Soil organic carbon was estimated by rapid dichromate oxidation technique (Nelson and Sommers 1996), available nitrogen by the alkaline potassium permanganate method (Subbiah and Asija 1956), available phosphorus by Bray's method (Bray and Kurtz 1966).

2.4 Estimation of inorganic, biogenic and coal carbon in reclaimed mine soil (RMS)

In reclaimed mine soil, the fraction of inorganic C, biogenic C (labile and stable) and coal C was determined by the methods suggested by Ussiri and Lal (2008), Chaudhuri et al. (2013) and Das and Maiti (2016). Total soil carbon (TSC) was determined in the CHNS elemental analyser (model: Euro EA) using CRM soil # 3 as a standard reference material after the sieving through the RMS samples <250 μ sieve size. The soil was pretreated with 1 M HCl, and loss of weight gives amount of inorganic C(IC).

$$\text{Inorganic C } (\%) = \text{TSC } (\%) - \text{TSC } (\%) \text{ after 1 N HCl treatment}$$

$$\text{Biogenic C } (\%) = \text{TSC } (\%) - (\text{Coal C } \% + \text{IC } \%)$$

2.4.1 Estimation of C stock in reclaimed minesoil

The estimation of carbon sequestration is based on soil organic carbon (SOC) (%), corrected bulk density (g cc^{-1}) and depth (m) of soil. The SOC sequestration potential for RMS in mine soil samples is estimated using the following equation (Lal et al. 1998):

$$\text{Mg C ha}^{-1} = [\%\text{C} \times \text{Corrected B}_d \times \text{d (m)} \times 10^4 \text{m}^2\text{ha}^{-1}]/100$$

where Mg C ha^{-1} is the mega grams C per hectare (1 Mg = 10^6), B_d (Mg m^{-3}) is the corrected soil bulk density (Mega gram per cubic meter), d soil depth (m), and % C is the biogenic carbon fraction in RMS as %.

Corrected soil bulk density was determined after removal of coarse fraction (>2 mm sieve size) from the sample.

2.5 Collection of understorey vegetation and litter and estimation of C stock

Understorey vegetation (grass, herbs, creepers, shrubs) and dry leaf litter accumulated underneath the canopy of different tree species (*D. sissoo, Albizzia spp.*) and bamboo (*B. arundinacea*) were collected within the metal quadrat 50 cm × 50 cm (Fig. 10a). In the RMS, understorey vegetation consists of dominant grass species *P. pedicellatum, C. ciliaris, C. setigerus,* etc. and shrubs (*Eupatorium, Lantata* etc.) (Figure 10b). Total 5 replicate of samples were collected from the study site. Samples were dried in an oven at 80 °C to constant weight to obtain the moisture free dry weight as followed by others (Karu et al. 2009; Singh et al. 2011) and expressed as Mg ha^{-1} C stock was calculated by multiplying a factor 0.4 assuming 40 % carbon present in the litter.

2.6 Estimation of biomass C stock

(a) *Calculation of above ground biomass* The DBH of individual tree species were divided according to their DBH classes (like, 3–5, 5–7, 7–9, 9–11, 11–13 cm). In the present study, non-destructive method of estimation of AGB was used. Mean DBH in each diameter class of individual tree species was used to calculate the AGB per tree (in kg) following regression equation, proposed for dry forest in India

Fig. 4 Tender shoots arising from the node of bamboo culms, which was planted during initial stage of reclamation

with rainfall >900 mm $year^{-1}$ (Brown et al. 1989; Brown 1997):

$$AGB = \exp\{-1.996 + 2.32 \times \ln(D)\}$$

where AGB is the aboveground biomass (kg) and D is the mean diameter at breast height (DBH) of each tree species in each diameter class (cm).

The average AGB multiplied to the tree density per hectare in the diameter classes to convert Mg ha^{-1}.

(b) *Calculation of below ground biomass* (BGB): In the present study 20 % of total above ground biomass was considered for the calculation of root biomass (MacDicken 1997). Several studies assumed 25 % of above ground biomass as the root biomass for hardwood species (IPCC 2006). Although there are limitations of using a constant factor for calculation of root biomass (Brown et al. 1989; Cairns et al. 1997), IPCC has proposed this method of estimating root biomass for reporting carbon stocks. Tree specific factors are available, but such factors vary with the age of the tree as well as with the soil characteristics (Brown et al. 1989). The use of constant factor may over/under estimate the total C stock (Cairns et al. 1997). Most of the research work revealed that AGB is strongly correlated with tree diameter (Brown et al. 1989). Also, it is accepted that simple model with only diameter as input is a good estimator of aboveground biomass.

Fig. 5 Growth of vegetation cover in the reclaimed site **a** *Albizia lebbeck* (Siris tree), **b** *Ziziphus oenoplia* (Wild jujube, a naturally colonizing shrub), **c** *Bambusa tudla* (Timber bamboo)

Fig. 6 Outside view of the reclaimed site showing different tree species; boundary wall and board indicating types and no. of species planted

Fig. 7 3-Tier vegetation established in the reclaimed site

$$\text{Below ground biomass}\left(\text{Mg ha}^{-1}\right) = \text{Average aboveground biomass}\left(\text{Mg ha}^{-1}\right) \times 0.2.$$

(c) *Estimation of total biomass* (Mg ha^{-1}) = AGB + BGB

(d) *Estimation of bamboo biomass*: In this study C stock of bamboo plantation was calculated by considering the work of Singh and Singh (1999).

(e) *Estimation of biomass C stock* The amount of carbon sequestered was calculated by reducing the biomass yield to its 50 % as per the guidelines of IPCC (2006). Hence biomass value was converted to

Table 1 Type of vegetation growing in Damoda reclaimed site

S no.	Scientific name (Family)	Common name	Remarks
1.	*Albizia lebbeck* Benth. (Fabaceae)	Silk flower (Siris)	Fast growing, medium-large tree, self-regenerated, common
2.	*Alstonia scholaris* (Linn.) R. Br. (Apocynaceae)	Devil tree (Chatim)	Small- medium size tree, poor growth
3.	*Azadirachta indica* A. Juss. (Meliaceae)	Margosa(Neem)	Large tree, native, poor growth
4.	*Bauhinia variegate* Linn. (Cesalpineae)	Bahunia	Medium size tree, not very common
5.	*Butea monosperma* L. (Fabaceae)	Flame of the forest (Palash)	Very common, grow naturally in reclaimed site
6.	*Dalbergia sissoo* Roxb. (Fabaceae)	Sissoo (Sisham)	Medium size tree, native, very common, luxuriant growth
7.	*Bambusa tulda* Roxb. (Poaceae)	Indian timber bamboo	Common in reclamation site
8.	*Bambusa arundinacea* (Retz.) Willd. (Poaceae)	Spiny bamboo	Common in reclamation site
9.	*Ficus infectoria* (Miq.) Domin. (Moraceae)	White Fig (Pakur)	Large tree, naturally colonise on reclaimed site
10.	*Lannea coromandelica* (Houtt.) Merr. (Anacardiaceae)	Indian ash tree (Doka)	Infrequent, naturally colonize
11.	*Matragyna perviflora* (Roxb.) Korth (Rubiaceae)	Kairn (Gulikadam)	Small- medium size tree, naturally colonize
12.	*Melia azedarach* L. (Meliaceae)	Persian lilac (Bakain)	Medium size tree, common in dump reclamation
13.	*Phyllanthus emblica* L. (Phyllanthaceae)	Indian gooseberry	Medium size tree, common, poor growth
14.	*Pongamia pinnata* Pierre. (Fabaceae)	Indian beech (Karanja)	Medium- large tree, common in dump reclamation
15.	*Syzygium cumini* (L.) Skeels (Myrtaceae)	Java plum	Medium size free tree, evergreen
16.	*Ziziphus oenoplia (L.)* Mill.	Wild jujube (Wild ber)	Evergreen shrub, spreading crown, self colonize

*Grass species Pennisetum pedicellatum (*Dennanath); *Saccharam munja, S. spontaneum* (Kans); *Cenchrus ciliaris (*Sui grass); *C. setigerus* (Dhaman grss); *Cynodon dactylon* (Doob)

Others Hemidesmus indicus (L.) R.Br. (common creeper, Asclepediaceae), *Tephrosia purpurea* (L.) Pers. (small shrubs, Fabaceae)

Fig. 8 Portion of quadrant sampling (shown by the rope) in the reclaimed site showing growth of *Albizzia lebbek* (kala siris), Bamboo and other tree species. Accumulation of litter at the floor consists of dry leaves and annual shrubs (*Eupatorium*)

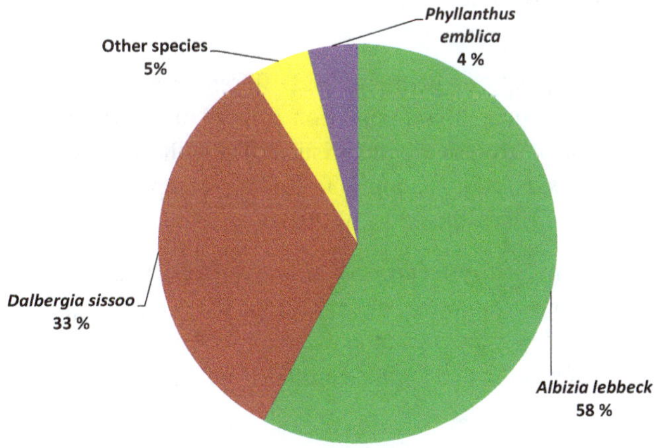

Fig. 9 Relative distribution of tree species in reclaimed site (others species - *Alstonia scholaris*, Mitragyna *parviflora, Syzygiumcumini* and *Ziziphus oenoplia*)

carbon stocks by multiplying a factor 0.5 and expressed as Mg C ha^{-1}.

2.7 Estimation of total C stock and CO_2 sequestration in reclaimed site

In RMS, there are three potential C sequestration pools, mainly: (i) above and below ground tree biomass, (ii) understorey vegetation and litter layer, and (iii) soil organic matter (SOM). Terrestrial plants can sequester carbon through photosynthesis which is stored in various above-ground (leaf, coarse woody material, branch, and logs) and belowground (live and dead roots, and SOM) biomass. This carbon is then transformed from live to dead organic matter

Fig. 10 **a** Close of view of quadrat sampling for litter collection (50 cm × 50 cm) taken within the sampling plot (10 m × 10 m). **b** Observed dried mulch of *Eupatorium* and other shrubs and grasses in the reclaimed site

by decomposition via microbes. Different tree species have different degree of influence on C pool in RMS due to variation in growth characteristics. Thus, the total C pool can be assessed by adding following components of an ecosystem (Shrestha and Lal 2010):

(a) Sum of Aboveground biomass (AGB) and belowground biomass (BGB) C-pool (CPool $_{(AGB+BGB)}$)
(b) Understorey and litter C stock (C Pool $_{UL}$)
(c) SOC at 0 to 15 cm depth (C Pool $_{soil\ 0\text{-}15\ cm}$)

$$\text{Eco C Pool} = \text{C Pool}_{(AGB+BGB)} + \text{C Pool}_{UL} + \text{C Pool}_{soil(0-15cm)}$$

where Eco C Pool is the total ecosystem C stock (Mg C ha^{-1}).

Total C stock is converted to CO_2 equivalent by multiplying a factor of 3.67 (by 44/12 ratio). Carbon sequestration rate (Mg C ha^{-1} $year^{-1}$) is calculated by dividing total C stock in RMS (Mg C ha^{-1}) with the age of the dump.

3 Results and discussion

3.1 Physico-chemical characteristics of reclaimed mine soil

Physico-chemical characteristics of reclaimed mine soil (RMS) significantly influence vegetation growth and accumulation of C stock. Generally mine soil is characterized with high stone content, lesser soil forming materials, low water holding capacity, low soil organic matter and humus, lower plant available nitrogen and phosphorous (Maiti 2013). Physico-chemical characteristics of reclaimed mine sites and reference forest site are given in Table 2. Soil fraction in RMS found 62 %, which was significantly lower than reference forest site (81 %), while non-soil or coarse fraction (>2 mm size) were found double than reference site. Higher coarse fraction in RMS attributed due to nature of mining, characteristics of rocks, measures taken during technical reclamation and amount of top soil application. However, it was also reported as high as 80 %–85 % due to geo-mining conditions in most of the reclaimed coalmine overburden dumps (Maiti and Saxena 1998). Excessive amount of coarse fragments limit the fine earth volume required for root proliferation, water-holding capacity, and long-term nutrient availability of plants (Mukhopadhyay et al. 2013; Maiti 2013). It also reduces with time in surface horizons due to weathering processes (Haering et al. 1993; Johnson and Skousen 1995). Corrected bulk density was found 1.12 g cc^{-1} (corrected by subtracting courser fraction for soil), which is lower than reference site. Higher values of bulk densities (1.67 g cc^{-1}) in reclaimed dumps were reported, which is caused due to compaction and lack of organic matter (Barnshiel and Hower 1997; Maiti 2007; Kumar et al. 2015). New mine soils had a slightly higher bulk density than the native soils due to the differences in mining method rather than pedogenic processes. The bulk density decreased with increase in age of reclamation, which is due to root system development, addition of biomass, and improvement in soil structure (Mukhopadhyay et al. 2014). Bi and Zhang (2014) also reported that soil bulk density of surface soil decreased and soil porosity and field capacity increased after afforestation in coal mining subsidence land. Akala and Lal (2001) also reported that growth and development of roots over time incorporates soil organic carbon and loosens up the soil, thus decreasing soil bulk density. The field moisture content in the RMS was lower (4.7 %) compared to forest site (7.7 %). Lower moisture content in RMS depends on the time of sampling, height of dump, stone content, soil texture, amount of organic carbon, and litter layer on dump surface (Mukhopadhyay and Maiti 2011). pH of the mine soil mainly depends on the

Table 2 Physsico-chemical characteristics of reclaimed minesoils (n = 5)

Parameter	Reclaimed mine soil	Reference forest site
Soil fraction (<2 mm size) %	61.90 ± 11.89[a] (46.78–74.75)	80.62 ± 5.42 (75.3–87.4)
Non-Soil fraction (>2 mm size) %	38.10 ± 6.75 (24.25–53.22)	19.40 ± 2.98 (15.6–28.2)
Corrected Bulk Density (g cc^{-1})	1.12 ± 0.05 (1.02–1.16)	1.34 ± 0.06 (1.23–1.40)
pH (1:1; w:v)	7.27 ± 0.49 (6.78–7.85)	5.92 ± 0.19 (5.52–6.32)
pH (1:2.5; w:v)	7.41 ± 0.54 (6.89–8.01)	6.25 ± 0.42 (5.95–6.40)
EC(1:1; w:v) (dS m^{-1})	0.182 ± 0.06 (0.10–0.26)	0.107 ± 0.017 (0.085–0.125)
Moisture content (%)	4.66 ± 1.04 (4.1-6.5)	7.29 ± 1.54 (6.1–10.4)
SOC (%)[b]	4.27 ± 0.91 (3.23–5.41)	2.02 ± 0.20 (1.75–2.15)
Av. N (mg kg^{-1})	94.50 ± 11.47 (84–110)	126.77 ± 11.02 (108.2–144.2)
Av-P (mg kg^{-1})	2.45 ± 0.69 (1.82–3.32)	6.85 ± 1.41 (4.5–8.84)

[a] Avg value ± SD (Min – Max)

[b] By Walkley–Black method (rapid dichromate digestion)

substrate and nature of the geological material of the overburden dump (the amount of acid-producing or acid-neutralizing material) (Mukhopadhyay and Maiti 2011). In the present study pH and electrical conductivity was found within the range and anticipate no detrimental effects on plant growth.

Variations in SOC may be mainly due to nature of mine soil, as well as decomposition of leaf litter of different tree species and humus to a large extent. The average SOC was found higher (4.27 %), due to accumulation of old, new and coal carbon fraction or mixing of recent carbon with coal carbon (Ussiri and Lal 2008). It has been reported that coalmine soil contains both recent (new) plant derived carbon fractions added through biological processes, detritus matter; and ancient (old) carbon fractions added by fossilized plant matter (humus), coal and inorganic carbonates (Maharaj et al. 2007). Plant available nitrogen was 94 mg kg^{-1}, which is lower than reference forest site, due to younger age of RMS. In the RMS, plant available P was 2.5 mg kg^{-1}, which also much lower than reference forest site. SOC, N and P have been reported as growth limiting factors on mined sites, but usually within the first 10-year after disturbance (Andrews et al. 2002). Increase in concentration of available N with increase in age of plantation has been observed by several researchers (Ahirwal and Maiti 2016; Mukhopadhyay and Maiti 2014; Maiti 2006, 2007). Low available N stock in the younger RMS suggests that N is limiting initially after reclamation when very little organic matter is present. Nitrogen accumulation is controlled by organic carbon input and N fixation, while phosphorus content is determined by the organic matter, pH of the soil substrate and weathering process (Andrews et al. 2002). Several chronosequence studies (Akala and Lal 2000; Maharaj et al. 2007; Shrestha and Lal 2010; Mukhopadhyay et al. 2013, 2014) have reported accumulation of C and N pools over time in RMS.

3.2 Tree biomass C stock and CO_2 sequestration

Tree species with DBH >3 cm found within the quadrat were classified according to DBH classes as applicable. *A. lebbeck,* the most dominant species was classified into 5 DBH classes ranging from 3 to 5 cm, 5 to 7 cm, 7 to 9 cm, 9 to 11 cm and 11 to 13 cm, with a total of 1433 nos of trees ha^{-1} and average DBH ranging from 4.08 to 12.42 cm (Table 3). Likewise, average DBH for other tree species were calculated. Out of 7 species, *A. lebbeck* contributed maximum AGB (13.04 Mg ha^{-1}) followed by *D. sissoo* (5.96 Mg ha^{-1}). The root biomass contributed by *A. lebbeck* was also highest (2.61 Mg ha^{-1}) than other tree

Table 3 Estimation of above ground biomass (AGB) and root biomass (RB) based on DBH (diameter at breast height) class for individual tree species in reclaimed site

Tree species name	DBH range (cm)	Avg DBH (cm)	No of trees ha^{-1}	AGB per tree (kg)	AGB of all trees (Mg ha^{-1})	Root biomass (RB) (Mg ha^{-1})[a]
Alzebia. lebbeck	3–5	4.08	600	3.54[b]	2.12[c]	0.42[d]
	5–7	5.88	533	8.27	4.41	0.88
	7–9	7.91	200	16.47	3.29	0.66
	9–11	9.43	67	24.76	1.65	0.33
	11–13	12.42	33	46.94	1.56	0.31
		Total	1433		13.04	2.61
Alstonia scholaris	3–5	3.18	33	2.00	0.07	0.01
		Total	33		0.07	0.013
Dalbergia sissoo	3–5	4.31	300	4.02	1.21	0.24
	5–7	5.80	467	8.02	3.74	0.75
	7–9	7.64	67	15.22	1.01	0.20
		Total	833		5.96	1.19
Mitragyna parviflora	3–5	3.82	33	3.05	0.10	0.02
		Total	33		0.10	0.02
Phyllanthus emblica	3–5	3.40	100	2.32	0.23	0.05
		Total	100		0.23	0.05
Syzygium cumini	3–5	4.78	33	5.11	0.17	0.03
		Total	33		0.17	0.03
Ziziphus oenoplia	3–5	3.82	33	3.05	0.10	0.02
		Total	33		0.10	0.02

[a] Root biomass (MacDicken 1997) = AGB (Mg ha^{-1}) × 0.2

[b] AGB (kg) = exp [−1.996 + 2.32 × ln DBH (cm)]; = exp [−1.996 +2.32 × ln (4.08)] = 3.54 kg

[c] 3.54 kg x 600 (no of trees ha^{-1}) × 10^{-3} = 2.12 (Mg ha^{-1})

[d] 2.12 (Mg ha^{-1}) × 0.2 = 0.42 (Mg ha^{-1})

species. It is noticed that AGB (kg $tree^{-1}$) increases exponentially at the higher DBH. Similar exponential relationship was noticed between biomass and DBH of tropical trees (Brown 1997).

Total tree biomass was estimated 23.6 Mg ha^{-1} (AGB 19.67 Mg ha^{-1} + RB 3.93 Mg ha^{-1}) (Table 4) and the biomass of bamboo clumps was estimated as 2.54 Mg ha^{-1}. Singh and Singh (1999) reported bamboo plantation developed on reclaimed mine spoil (Singrauli, India) accumulated substantial amount of biomass (46 Mg ha^{-1} for 3 year plantation and 74.7 Mg ha^{-1} for 5 year plantation). Several other studies in natural ecosystem of bamboo forests and plantations in India recorded 0.8 to 24 Mg ha^{-1} aboveground biomass (Rao and Ramakrishnan 1989; Tripathi and Singh 1996). In the present study, total biomass (tree and bamboo) was estimated 25.97 Mg ha^{-1}. Studies by Singh and Singh (1991) reported higher aboveground biomass in the native dry tropical deciduous forest, in the range between 42 and 78 Mg ha^{-1}. Lodhiyal et al. (2002) estimated the AGB of Tarai Shisham forest, India (5–15 years age) in the range of 41–103 Mg ha^{-1} and total biomass in the range of 59–136 Mg ha^{-1}.

Biomass C stock in reclaimed mine sites depends on tree species, climatic conditions and nature of substrate. A comparative C stock data under different climatic conditions, age of vegetation and forest type is presented in Table 5. In the present study, total biomass C stock was estimated as 13 Mg C ha^{-1}(equivalent to CO_2 sequestered of 47.96 Mg CO_2 ha^{-1}), which is much lower than reference forest site. In Indian conditions (19 years old RMS of Singrauli, India), Tripathi et al. (2014) reported higher C stock (41.73 Mg C ha^{-1}) with an accumulation rate of 2.20 Mg C ha^{-1} $year^{-1}$. Average C accumulation rate in aboveground biomass for reclaimed pine and hardwood forests ranged from 1.4 Mg C ha^{-1} $year^{-1}$ in Indiana to 3.3 Mg C ha^{-1} $year^{-1}$ in West Virginia (Amichev et al. 2008). Terakunpisut et al. (2007) compared the C stock of the aboveground biomass in different forest ecosystems of Thong Pha Phum National Forest, Thailand based on DBH $\geq$ 4.5 cm using allometric equation and reported that tropical rain forest had higher C stock than dry evergreen

Table 4 Distribution of carbon stock in total above ground biomass (AGB), root biomass (RB) of trees and bamboo in reclaimed area

Tree species name	Nos ha^{-1}	AGB of tree species ($Mg\ ha^{-1}$)	RB of tree species ($Mg\ ha^{-1}$)	Biomass C-stock ($Mg\ C\ ha^{-1}$)
Albizia spp	1433	13.04	2.61	7.82
Dalbergia sissoo	833	5.96	1.19	3.58
Other tree species[a]	234	0.67	0.13	0.40
B. arundinecea[b] (Clumps)	2033	2.36	0.18	1.27
Total		22.03	4.11	13.07

[a] *Alstonia scholaris, Mitragyna parviflora, Phyllanthus emblica, Syzygium cumini* and *Ziziphus oenoplia*

[b] Singh and Singh (1999) studied the biomass C of bamboo in the reclaimed mine sites of Singrauli, India, and reported that after 4 years, nos of clumps were 22475, which contributes above ground biomass of 26.1 Mg ha^{-1} (or 13.05 Mg C $ha^{-1)}$ and root biomass contributes 7.7 % of total biomass. So, in the present study, above ground bamboo biomass of 2033 nos clumps ha^{-1},is calculated as 2.36 Mg ha^{-1}. CO_2 sequestered = 13.07 t ha^{-1} × 3.67 (factor to convert C to CO_2) = 47.96 $MgCO_2$ ha^{-1}. Therefore, rate of CO_2 sequestration = [47.96 Mg ha^{-1}/ 4 years) = 12 Mg CO_2 ha^{-1} $year^{-1}$

and mixed deciduous forest in order of 137.7 ± 48, 70.3 ± 7.4 and 48 ± 16.7 Mg C ha^{-1}, respectively.

3.3 C stock in understorey vegetation and litter and CO_2 sequestration

Litter accumulation underneath of different trees and bamboo plantation were estimated in this study, which includes dry leaves, grasses, herbs and shrubs. In the present study average litter accumulation was 2.45 Mg ha^{-1}, and total C stock estimated as 0.98 Mg C ha [1] (0.3–1.86 Mg C $ha^{-1)}$, which is lower than reference forest site (Table 6). Dutta and Agarwal (2003) estimated overall litterfall rate in the range of 1.2–3.6 Mg ha^{-1} $year^{-1}$ in reclaimed coalmine site (Singraulli, India), whereas range 1-5.9 Mg ha^{-1} $year^{-1}$ was reported in established bamboo ecosystem of the same site (Tripathi and Singh 1996; Singh and Singh1999). Litterfall rate in Indian shisham forest (5–15 yrs age) was estimated in similar range of 2.2–5.1 Mg ha^{-1} $year^{-1}$ (Lodhiyal et al. 2002).

In the present study, carbon stored in the understorey vegetation and litter components was 0.98 Mg C ha^{-1}, which is lower than other reported values. (Karu et al. 2009; Pragasan and Parthasarathy 2005). C stock in litter component of reclaimed minesoil of Midwestern and Appalachian coalfields (USA) comprised of pine, hardwood and mixed stand was estimated 21.2, 14 and 6 Mg ha^{-1} respectively (Amichev et al. 2008). Other studies by Shrestha and Lal (2010) also estimated higher C stock than the present study in litter component of reclaimed minesoil of Ohio (USA) in forest ecosystem (3.96 Mg Cha^{-1}). In Indian mining conditions (revegetated minespoils of 19 years age, Singrauli, India) C accumulation by litter component was estimated 2.80 Mg C ha^{-1} @ 0.15 Mg C ha^{-1} $year^{-1}$ (Tripathi et al. 2014), which is higher than the present study. Studies reported that, with increase age of vegetation C stock in the understorey vegetation and litter fall also increases (Amichev et al. 2008; Singh et al. 2011).

3.4 C stock in reclaimed mine soil and CO_2 sequestration

C-stock in reclaimed mine soil (RMS) was estimated 16.33 MgC ha^{-1} and CO_2 sequestered was 59.93 Mg CO_2 ha^{-1}, which is lower than reference forest site (Table 7). In Indian reclaimed coal mine sites (Jayant project, Singrauli) accumulation of higher C-stock (22.9 Mg C ha^{-1}) was reported with an average accumulation rate of 1.20 Mg C ha^{-1} $year^{-1}$ (Tripathi et al. 2014). C stocks of reclaimed forest and grassland coal mine soils (0-30 cm) of Ohio, USA were reported in higher range of 37–45 Mg C ha^{-1} and 47–79 Mg C ha^{-1} over 21 and 25 year periods respectively (Akala and Lal 2000, 2001). Rate of accumulation of C stock was observed 0.1 to 3.1 Mg C ha^{-1} $year^{-1}$ and 0.7–4 Mg C ha^{-1} $year^{-1}$ in grass and forest RMS ecosystem respectively (Shrestha and Lal 2006; Akala and Lal 2001). Post and Kwon (2000) also reported a higher average C accumulation of 0.34 Mg C ha^{-1} $year^{-1}$ (at the top 30 cm soil in tropical to temperate forest) than permanent grasslands (0.74 Mg ha^{-1} $year^{-1}$ tropical dry lands to cool temperate ecoregion) which is lower than the RMS.

3.5 Total soil carbon fractionation of RMS

The total soil carbon content is considered as a key component of minesoil quality and its storage in reclaimed soil is recognized promising measure for mitigating global climate change through carbon sequestration (Lal 2004, 2005). Distribution of Total carbon (%), Inorganic

Table 5 C-stock in Total biomass, Above ground biomass (AGB) and accumulation rate under forest and reclaimed coal mine spoils in India and outside countries

Forest type/vegetation and location	Age (year)	Total biomass C (Mg C ha^{-1})	AGB C stock (Mg C ha^{-1})	Total biomass C accumulation rate (Mg C ha^{-1}yr^{-1})	Reference
Reclaimed minesoil, Northern Europe (*Pinus sylvestris*)	14	6.76	5.86	0.48	Karu et al. (2009)
	21	28.93	25.65	1.37	
	36	101.6	90.01	2.82	
Reforested post-mining sites, Poland (*Pinus sylvestris*)	17	24.41	22.13	1.43	Pietrzykowski and Daniels (2014)
	30	54.74	48.17	1.82	
	Con (30 yr)	71.10	64.50	2.37	
Reforested coal mine sites of Appalachian Coal Basin, USA	5	2.49	0.27	0.50	Avera et al. (2015)
	11	55.43	53.80	5.04	
	21	54.10	48.86	2.58	
	30	66.91	64.27	2.23	
	Unmined	136.09	125.71	–	
D. sissoo forests, Tarai forest, India	5	29.3	20.9	5.86	Lodhiyal et al. (2002)
	10	53	39.3	5.30	
	15	68	50.65	4.53	
D. sissoo forests, Bhabar forest, India	5	17.55	–	3.51	Lodhiyal and. Lodhiyal (2003)
	10	30.15	–	6.03	
	15	44.75	–	2.98	
Mixed plantation, Terai, India[a]	5	5.88	4.385	1.176	Singh et al. (2011)
D. sissoo,		0.325	0.285	0.065	
A. lebbeck		1.51	1.30	0.302	
A. catechu					
D. sissoo, Tarai, India	10	43.39	35.77	2.73	Kanime et al. (2013)
Tropical moist deciduous forests, *Techtona grandis* Tarai, India	5	15.8	13.9	3.16#	Jha (2015)
	11	35.4	31.2	3.21	
	18	39.0	34.2	2.17	
	24	61.5	54.3	2.65	
	30	73.2	63.8	2.44	
Reclaimed coal mine spoil, Singrauli, India *B. arundinacea*	3	23.45	21.65	7.81	Singh and Singh (1999)
	5	37.35	34.7	7.47	
Reclaimed coal mine spoil, Singrauli, India	2	1.74	1.52	0.87	Tripathi et al. (2016)
	4	5.45	4.75	1.36	
(Mixed plantation)	10	15.64	13.68	1.56	
	14	25.96	23.04	1.85	
	16	31.61	27.89	1.97	
	19	44.52	39.25	2.34	
Reclaimed coal mine spoil, Singrauli, India	3	10.95	7.8	3.65	Singh et al. (2011)
	6	33.9	25.55	5.65	
A. lebbek	3	6.55	3.2	2.18	
A. procera	6	20.1	13.65	3.35	
Reclaimed coalmine, *A. lebbeck, D. sissoo,*	4	13	11.01	3.25	*Present study*
B. arundinacea	Reference forest[b]	66.1	52.88	–	

[a] Total tree density—1332 trees ha^{-1}; (*D. sissoo*—770 ha^{-1}; *A. catechu*—376 ha^{-1} and *A. lebbeck*—176 ha^{-1}). [b] Natural forest, close proximity to the reclaimed site

Table 6 Carbon stock of understory vegetation and litter in 4 years old reclaimed dump

Forest type/vegetation, location	Components	Age of reclamation (years)	C- stock (Mg C ha^{-1})[a]	Reference
Reclaimed minesoil, Appalachian, USA	Pine stands	–	21.2	Amichev et al. (2008)
	Hardwood stands		13.9	
	Mixed stands		6.1	
Reclaimed minesoil, Northern Europe (*Pinus sylvestris*)	Herb + Shrub	14	6.76	Karu et al. (2009)
		21	0.62	
		36	7.8	
Reclaimed minesoil of Ohio, USA	Forest	25	3.96	Shrestha and Lal (2010)
Tropical dry evergreen forests, Coromandel coast, south India	Total fine litter		5.3–5.4	Pragasan and Parthasarathy (2005)
	Standing crop		1.64–1.94	
Terai Forest, India	Understorey	5	1.62	Singh et al. (2011)
	Floor litter		0.608	
Revegetated mine spoils, India		19	2.80 (0.15)[b]	Tripathi et al. (2014)
Temperate forests, Kashmir, India	Understorey	Natural forest-	0.08–1.18	Dar and Sundarapandian (2015)
	Detritus		2.01 -8.06	
Reclaimed coalmine, site, Jharia coalfields	Understory vegetation and litter	4	0.98 (0.288–1.856)	Present study
	Reference forest site	–	1.65 (0.842–2.103)	

Undestorey vegetation includes herbs, shrubs and grasses

[a] Total biomass is multiplied by 0.4 to calculate C-stock

[b] Values within brackets are C sequestration rate of understorey vegetation and litter

carbon (%), Coal carbon (%) and Biogenic carbon (%) of RMS samples collected under the canopy of different vegetation cover is shown in (Fig. 11). Total soil carbon (TSC) was found in the range of 1.72 %–2.31 % and the average value was 2.01 %. Carbon fractionation is important since intermixing of biogenic C with coal C leads to overestimation of C sequestration values (Ussiri and Lal 2005; Ussiri et al. 2014). In the present study, inorganic C was estimated as 0.17 %, which is 8 % of TSC. Stahl et al. (2009) reported IC concentration from 1.82 to 8.47 g C kg^{-1} in RMS (17 %–60 % of TSC) of Wyoming (USA). Ussiri and Lal (2008) reported coal C contributes 47.4 % of TSC (0–10 cm) in the RMS of Ohio. The quantity and quality of biogenic C have strong influences on other essential soil characteristics and is a major determinant of soil productivity and long-term stability of the reclaimed minesoil (Ussiri et al. 2014). Biogenic C was estimated in the present study in the range of 0.84 %–1.08 %, whereas coal C was estimated 0.63 %–1.03 %.

3.6 Coal C distribution in reclaimed minesoils

Depending on mining and reclamation techniques, the amount of coal C can be substantial (Table 8). In the present study, coal C contributed 43 % of total organic carbon (TOC) in RMS. In Indian conditions, average contribution of coal C to TOC in RMS ranged between 24 % and 68 % (Das and Maiti 2016). Coal contamination contributed to the irregular depth distribution of SOC in RMS (Thurman and Sencindiver 1986). In Lusatia mining district (Germany), lignite-C accounted for 25 % to 99 % of the total SOC in rehabilitated and reforested RMS (Rumpel et al. 1998, 2000). Coal C accounted for up to 80 % of the OC in the A- horizon of cultivated Mollisol contaminated by brown coal emissions from a briquette factory (Schmidt et al. 1996). Chabbi et al. (2006) reported lignite C contribution of 20 %–80 % of total SOC in RMS and sediment of Lusatia (Germany). Generally, the concentration of coal C increases with increase in soil depth, and nearly 100 % of OC in the C- horizon (mine spoil) can be of geogenic origin (Ussiri et al. 2014).

Age since reclamation had little effect on the relative abundance of Coal C in RMS probably due to its biochemical recalcitrance. For example, Rumpel et al. (2003) showed that lignite accounted for 80 %–93 % of the total C in the top 0–5 cm mineral horizon of the age-chronosequence series of minesoils ranging from 11 to 32 year old at Lusatian mining district in Germany. Rumpel et al. (1998) observed that after 36 year of restoration, more than half of the OC in 0–5 cm soil layer under red oak (*Querus*

Table 7 Soil carbon stock and rate of accumulation in afforested reclaimed coal mine soils and natural forest in India and abroad

Land use	Vegetation/land-use system	Age (year)	Soil depth (cm)	C stock (MgC ha^{-1})	Rate of C accumulation (MgC ha^{-1} $year^{-1}$)	Reference
Reclaimed minesoil Ohio, USA	Pasture	25	0–30	36.7	1.468	Akala and Lal (2000)
	Forest	25	0–30	37.1	1.484	
Reclaimed minesoil, Ohio, USA	Pasture (with top soil)	0–25	0–15	9.2–55.4	3.1–0.5	Akala and Lal (2001)
			15–30	7.8–37.8	1.9–0.4	
	Forest (with top soil)	0–21	0–15	14–48.4	2.3–07	
			15–30	8.4–14.5	0.4–0.3	
	Pasture (No top soil)	30–40	0–15	60–67	–	
			15-30	34–40		
	Forest (No top soil)	30–50	0–15	50-70	–	
			15–30	27-47		
Reclaimed mine soil, Central Appalachian, USA	Mixed hardwood forest	40	0–10	–	4.0	Burger (2004)
Reclaimed minesoil, Ohio USA	Hardwood forest	–	–	81	–	Jacinthe et al. (2004)
	Grassland			71		
Reclaimed minesoil Appalachian, USA	Pine stands	–	–	11	–	Amichev et al. (2008)
	Hardwood stands			13		
	Mixed stands			17.7		
Reclaimed minesoil, northern Europe (Estonia)	Forest (Scots pine)	14	O- and A-horizon (0.2.8)	0.1 & 0.2	0.33 & 0.51	Karu et al. (2009)
		21		4.6 & 0.4		
		36		12.3 & 19.0		
Reclaimed mine soils, Poland	Scots pine (*Pinus sylvestris*)	12–30	–	0.73[a]	1.45	Pietrzykowski and Krzaklewski (2010)
				2.17[b]		
				5.26[c]		
Reclaimed minesoil of Ohio, USA	Forest	25	0–15	38	1.52	Shrestha and Lal (2010)
	Pasture	25	0–15	35	1.4	
Reforested coal mine sites, Appalachian, USA	Mixed plantation[1]	5	–	7.02	–	Avera et al. (2015)
		11		13.52		
		21		21.35		
		30		16.62		
		Unmined		42.85		
Central Himalayan Tarai, India	*D. sissoo* monoculture plantation	10	0–30	36.3	–	Kanime et al. (2013)
Temperate forests Kashmir, India	Broad leaved species and conifers	Natural forest	0–30	39.1–91.4	–	Dar and Sundarapandian (2015)
Revegetated coal mines, Singrauli, India	Afforest site	19	0–30	22.9	1.20	Tripathi et al. (2014)
Reclaimed coalmine, site, Jharia coalfields	Afforest site	4	0–15	16.33 ± 1.04 (14.96–17.25)	4.08	Present study
	Forest (reference)	–	0–15	35.5 ± 5.8 (30.4–42.5)		

[a] Poorest sandy soils

[b] Sandy-clayish soils

[c] Carboniferous substrate soils

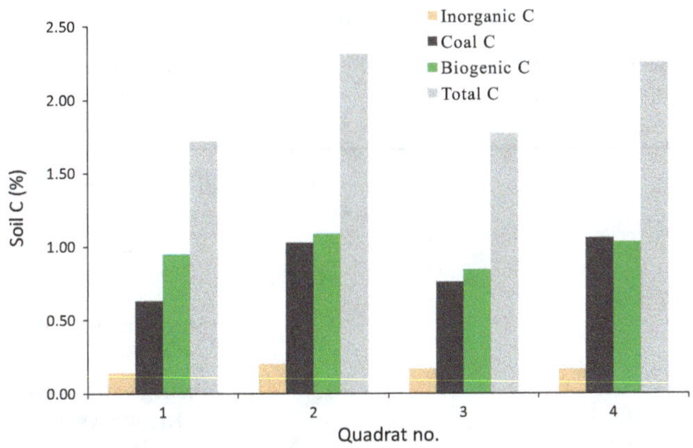

Fig. 11 Distribution of total carbon (%), inorganic carbon (%), coal carbon (%) and biogenic carbon (%) of different reclaimed mine soil samples

spp) was derived from decomposing plant material. Ussiri and Lal (2008) determined the coal C content of RMS restored with topsoil application using radiocarbon analysis and chemi-thermal method, observed that coal C accounted for 2 % to 40 % of the total OC in the top 30 cm soil layer, while in the subsoil (below 30 cm depth) coal C accounted up to 92 % of the total organic C. Overall, the reviewed data suggests that the distribution of coal C in coal-contaminated RMS can be highly variable; therefore, intensive sampling is necessary in order to obtain adequate measure of SOC in restored ecosystems (Ussiri et al. 2014).

3.7 Total C stock in reclaimed site and CO_2 sequestration

Total C stock of the study area was calculated 30.3 Mg C ha^{-1}.equivalent to 111 Mg CO_2 sequestered ha^{-1}. Total C stock in biomass, understorey vegetation and litter and mine soil is given in Fig. 12. In reference forest site, total C stock was estimated as 103 Mg C ha^{-1}, which is higher than RMS. More than 98 % of C-stock contributed by biomass and soil organic carbon in reference forest site, which is also similar to the RMS. Differences in C stock between RMS and reference forest site attributed to the biomass C stock. Tripathi et al. (2014) reported total C sequestration potential of 69.21 Mg C ha^{-1} in 19 years old RMS of Singrauli, India, (equivalent to 253.96 Mg ha^{-1} capture of atmospheric CO_2 and C sequestration rate of 3.64 Mg ha^{-1} $year^{-1)}$ which indicates that mine spoil can act as a significant sink for atmospheric CO_2 through revegetation. Amichev et al. (2008) estimated total C sequestration (all ecosystem components of RMS of Midwestern and Appalachian coalfields of USA) in pine, hardwood and mixed stands as 148, 117 and 130 Mg C ha^{-1} respectively. Another study by Shrestha and Lal (2010) reported total ecosystem C sequestration potential and average rate of RMS (Ohio USA, 25 years age) as 107 Mg ha^{-1} (5.1 Mg ha^{-1} $year^{-1}$) and 21 Mg ha^{-1} (1.0 Mg ha^{-1} $year^{-1}$) in forest and pasture ecosystems respectively.

3.8 Future projection of C stock

Climate, soil conditions and species composition have major influence on accretion of C stock in reclaimed mine soil. Similarly, the rate of C stock accretion for a particular site depends on site characteristics (texture, soil fraction, water holding capacity, bulk density, soil fertility), prevailing climatic conditions and types of species. Vogt et al. (1995) reported 40 %–62 % of total ecosystem C contained in living biomass, and SOM contained 33 %–50 % of the total. Rate of accumulation of total biomass C stock was found 3.25 Mg C ha^{-1} $year^{-1}$, which is close to the study of Shrestha and Lal (2010) and Singh et al. (2011) and it will take about 17 years to reach the level of C stock of reference forest site (Table 9). Most noticeable amount of tree C increased from 5-year stands (0.3 Mg C ha^{-1}) to 11-year stands (53.8 Mg C ha^{-1}) (Avera et al. 2015) and decreases gradually. The present rate of accumulation is

Table 8 Contribution of coal C (geogenic C) in total organic carbon (TOC) in the reclaimed coalmine soils

Post land use (age)	Age (yrs)	Depth (cm)	Coal C (g/kg)	Coal C (% of TOC)	Reference
Red Oak	30	–	37–56	25–96	Rumpel et al. (1998, 2000)
Old Pine	15–36	(O-horizon)	8–71	79–91	Rumpel et al. (2003)
Grassland	28	–	0–24.4	0–92	Ussiri and Lal (2008)
Reclaimed coal mines	0–22	0–6	0.6–1.76	2.5–14	Chaudhuri et al. (2013)
Reforested coalmine site	10	0–10	35.97	47.4	Das and Maiti (2016)
Unreclaimed (Reference site)	10	10–20	14.4	24.2	
		0–10	36.60	68.5	
		10–20	19.15	55.9	
Reclaimed coal mines	4	0–10	8.7	43	Present study

Table 9 Prediction recovery of C stock (biomass and minesoil) in reclaimed minesoil to reference forest site

Study site	Total biomass C accumulation rate (Mg C ha^{-1} $year^{-1}$)	Predicted year	SOC accumulation rate in reclaimed mine soil (Mg C ha^{-1} $year^{-1}$)	Predicted year	Reference
Reclaimed coal mine, Singrauli, India	1.658 (0.87–2.34)	32	1.20	16	Tripathi et al. (2014; 2016)
Reforested coal mine, USA	2.5875 (0.50–5.04)	20.5	1.05	18.3	Avera et al. (2015)
Reclaimed minesoil (forest), USA	3.365	15.8	1.52	12.6	Shrestha and Lal (2010)
Reclaimed mine soil, Poland	–	–	1.45	13.2	Pietrzykowski and Krzaklewski (2010)
Reclaimed coal mine, Singrauli, India	3.7075 (2.18–5.65)	14.32	–	–	Singh et al. (2004)
Reclaimed minesoil (forest), Ohio USA	–	–	1.484	13	Akala and Lal (2000)
Reclaimed coal mine, Jharia coalfields, India	3.25	16.4	4.08[a] 1.305[b]	5[a] 15	Present study

[a] Rate of SOC accumulation initially calculated is very high, but with time, SOC accumulation going to be decreased and reported in the range of 1.305 (1.05–1.52) Mg C ha^{-1} $year^{-1}$

[b] Assumed average rate

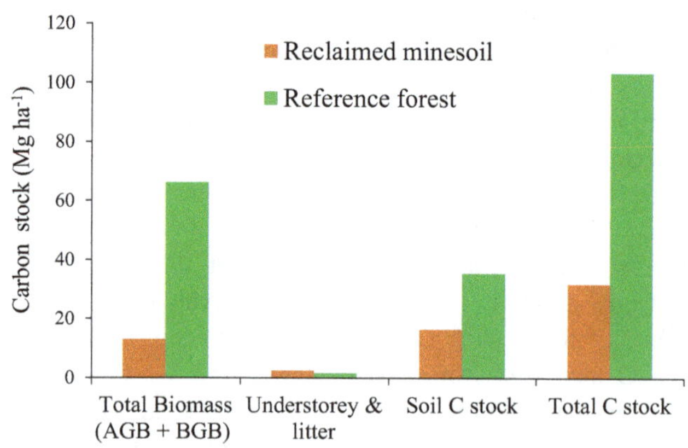

Fig. 12 Comparison of distribution of C stock (Mg ha-1) in various components of reclaimed coalmine site and reference forest site

much higher than the reported values of 1.658 Mg C ha^{-1} $year^{-1}$ (Singrauli, Tripathi et al. 2014, 2016), which is due to the younger age of site, however, it is expected that, rate of accumulation will decrease with age. Based on rate of SOC accumulation rate, C stock of RMS will reach close to the level of reference forest site is predicted as 15 years.

Alka and Lal (2001) stated that SOC increased rapidly over the first 20 years following reclamation and then accumulated at lower rates between 20 and 30 years. Based on 0–25 year reclaimed sites chronosequence data, Akla and Lal (2000, 2001) predicted that, it may take 100–150 years for SOC stock to reach an equilibrium state in RMS. The accumulation of C stock will also increase in biomass and minesoil components, with the gradual increase in age of reclamation and this accumulation of C stock is expected to increase highest in the biomass component. Avera et al. (2015), while studying the chronosequence reforested surface-mined lands in the Central Appalachian Coal Basin (USA) reported that, total ecosystem C increased from 9.5 Mg C ha^{-1} in the 5-year stands to 83.5 Mg C ha^{-1} in the 30-year stands and after 30 years, reforested surface-mined lands contains 47 % of the total C of the unmined ecosystem C (179 Mg C ha^{-1}). Aboveground tree C was the greatest proportion of ecosystem C, accounting for an average of 73 % of the total ecosystem C. SOC content tripled from the 5-year to the 11-year stands, with the most rapid SOC sequestration rate of 0.7 ± 0.1 Mg SOC ha^{-1} $year^{-1}$ measured in the 11-year stands. Amichev et al. (2008) reported similar results of at least 75 % of total ecosystem C contributed by aboveground tree C on reforested mined land.

Reclamation by development of 3-tier canopy over is better alternative than simple plantation, because it leads to reinstatement of ecosystem in the degraded site. It not only stabilizes and minimizes pollution but also act as potential sink of CO_2. Thus, reclaimed coalmine sites can be a viable alternative to sequester carbon both in soil and vegetation and can be useful for combating climate change and global warming issues. The widespread problem of land degradation and soil erosion is growing at an exponential rate due to opencast mining activities in agriculture based countries like India, where resources are limited. Concurrently, reclamation of those sites through forestry development approach is a common practice. Interpretation of carbon sequestration will be worthwhile since they can be used to create carbon credits which can be generate significant source of income to fulfill the demands of ever increasing pressure of population.

4 Conclusions

Reclamation process is an integral part of mining operation aimed to stabilize mine degraded sites which resulted carbon sequestration. C stock in reclaimed mine soil is dependent on the nature and type of soil and vegetation; age and geo-climatic conditions. Out of total C stock was 30.3 Mg C ha^{-1}, 43 % is stored in biomass and 53 % as soil organic matter. Coal C contributes a significant amount of organic carbon in reclaimed mine soils, and for better assessment of C stock, fractionation study is recommended. Limited field studies data is available on accretion of C stock in reclaimed minesoil in Indian conditions (only Singrauli and Jharia coalfields) along with very little work of USA and European conditions. Based on the present study, the C stock will tentatively reach similar to the reference forest site in 17 years, as the rate of carbon accumulation decreases with age. As species composition, climate conditions and nature of mine soils is site specific, this study will a guidelines for a specific geo-climatic conditions.

Compliance with ethical standards

Conflict of interest The authors declare no conflict of interest in any concern or context in the present study.

References

Ahirwal J, Maiti SK (2016) Assessment of soil properties of different land uses generated due to surface coal mining activities in tropical Sal (*Shorea robusta*) forest, India. Catena 140:155–163

Akala VA, Lal R (2000) Potential of mine land reclamation for soil organic carbon sequestration in Ohio. Land Degrad Dev 11(3):289–297

Akala VA, Lal R (2001) Soil organic carbon pools and sequestration rates in reclaimed minesoils in Ohio. J Environ Qual 30(6):2098–2104

Amichev BY, Burger JA, Rodrigue JA (2008) Carbon sequestration by forests and soils on mined land in the Midwestern and Appalachian coalfields of the US. Forest Ecol Manag 256(11):1949–1959

Andrews SS, Karlen DL, Mitchell JP (2002) A comparison of soil quality indexing methods for vegetable production systems in Northern California. Agric Ecosyst Environ 90:25–45

Avera BN, Strahm BD, Burger JA, Zipper CE (2015) Development of ecosystem structure and function on reforested surface-mined lands in the Central Appalachian Coal Basin of the United States. New For 46(5–6):683–702

Barnshiel RL, Hower JM (1997) Coal surface mine reclamation in the eastern United States: the revegetation of disturbed lands to hayland/pasture or cropland. Adv Agron 61:233–275

Bi Y, Zhang Y (2014) Role of the different planting age of seabuckthorn forests to soil amelioration in coal mining subsidence land. Int J Coal Sci Technol 1(2):192–197

Bray R, Kurtz LT (1966) Determination of total, organic and available forms of phosphorus in soil. Soil Sci 59:39–45

Brown S (1997) Estimating biomass and biomass change of tropical forests: a primer, vol 134). Food and Agriculture Organization (FAO) of the United Nations, Rome

Brown S, Gillespie AJ, Lugo AE (1989) Biomass estimation methods for tropical forests with applications to forest inventory data. For Sci 35(4):881–902

Burger JA (2004) Restoring forests on mined land in the Appalachians: results and outcomes of a 20-year research program. In: Barnhisel RI (ed) Proceedings of a joint conference of American Society of Mining and Reclamation. 21st annual national conference, 25th West Virginia surface mine drainage task force symposium Morgantown, WV, p 260

Cairns MA, Brown S, Helmer EH, Baumgardner GA (1997) Root biomass allocation in the world's upland forests. Oecologia 111:1–11

Chabbi A, Rumpel C, Grootes PM, Mariotti A, Huttl RF (2006) Isotopic tracers for theanalysis of vegetation-derived organic matter in lignite-containing soils and sediments along a transect ranging from a forest soil to submerged lake sediment. Org Geochem 37:740–753

Chandra D (1992) Jharia coalfield. Geological Society of India, Bangalore, pp 1–11

Chaudhuri S, McDonald LM, Pena-Yewtukhiw EM, Skousen J, Roy M (2013) Chemically stabilized soil organic carbon fractions in a reclaimed minesoil chronosequence: implications for soil carbon sequestration. Environ Earth Sci 70(4):1689–1698

Coppin NJ, Bradshaw AD (1982) Quarry reclamation. Mining Journal Books, London, pp 8–17

Dar JA, Sundarapandian S (2015) Variation of biomass and carbon pools with forest type in temperate forests of Kashmir Himalaya, India. Environ Monit Asses 187(2):1–7

Das R, Maiti SK (2016) Importance of Carbon Fractionation for the estimation of Carbon sequestration in Reclaimed Coalmine soils–a case study from Jharia Coalfields, Jharkhand, India. Ecol Eng 90:135–140

Dudley NS, Fownes JH (1992) Preliminary biomass equations for eight species of fast-growing tropical trees. J Trop For Sci 5(1):68–73

Dutta RK, Agrawal M (2003) Restoration of opencast coal mine spoil by planting exotic tree species: a case study in dry tropical region. Ecol Eng 21(2):143–151

Haering C, Daniels WL, Roberts JA (1993) Changes in mine soil properties resulting from overburden weathering. J Environ Qual 22:194–200

IPCC (2000) Special report on land use, land use change, and forestry, summary for policy makers. Intergovernmental Panel on Climate Change, Cambridge University Press, Cambridge

IPCC (2006) IPCC Guidelines for national greenhouse gas inventories. Intergovernmental Panel on Climate Change, NGGIP Publications, IGES, Japan

Jacinthe PA, Lal R, Ebinger M (2004) Carbon sequestration in reclaimed mined lands. In: Proceedings of the second annual carbon sequestration conference, 5–8 May, Alexandria, VA

Jha KK (2015) Carbon storage and sequestration rate assessment and allometric model development in young teak plantations of tropical moist deciduous forest, India. J For Res 26(3):589–604

Johnson CD, Skousen JG (1995) Minesoil properties of 15 abandoned mine landsites in West Virginia. J Environ Qual 24:635–643

Kanime N, Kaushal R, Tewari SK, Raverkar KP, Chaturvedi S, Chaturvedi OP (2013) Biomass production and carbon sequestration in different tree-based systems of Central Himalayan Tarai region. For Trees Livelihoods 22(1):38–50

Karu H, Szava-Kovats R, Pensa M, Kull O (2009) Carbon sequestration in a chronosequence of Scots pine stands in a reclaimed opencast oil shale mine. Can J For Res 39(8):1507–1517

Kumar S, Maiti SK, Chaudhuri S (2015) Soil development in 2-21 years old coalmine reclaimed spoil with trees: a case study from Sonepur Bazari opencast project, Raniganj Coalfield, India. Ecol Eng 84:311–324

Lal R (2004) Soil carbon sequestration impacts on global climate change and food security. Science 204:1623–1627

Lal R (2005) World crop residues production and implication of its use as a biofuel. Environ Int 31:575–586

Lal R, Bruce JP (1999) The potential of world cropland soils to sequester C and mitigate the greenhouse effect. Environ Sci Policy 2:177–185

Lal R, Kimble JM, Follett RF (1998) Land use and soil C pools in terrestrial ecosystems. In: Lal R, Kimble JM, Follett RF (eds) Management of carbon sequestration in soil. CRC Lewis Publisher, Boca Raton, pp 1–10

Lodhiyal N, Lodhiyal LS (2003) Biomass and net primary productivity of Bhabar Shisham forests in central Himalaya, India. For Ecol Manag 76(1):217–235

Lodhiyal N, Lodhiyal LS, Pangtey YPS (2002) Structure and function of Shisham forests in Central Himalaya, India: dry matter dynamics. Ann Bot 89(1):41–54

Lonsdale WM (1988) Predicting the amount of litterfall in forests of the world. Ann Bot 61(3):319–324

MacDicken KG (1997) A guide to monitoring carbon storage in forestry and agroforestry projects. Winrock Int Institute for Agricultural Development, Arlington, p 87

Maharaj S, Barton CD, Karathanasis TA, Rowe HD, Rimmer SM (2007) Distinguishing" new" from" old" organic carbon in reclaimed coal mine sites using thermogravimetry: II. Field validation. Soil Sci 172(4):302–312

Maiti SK (2006) Properties of minesoil and its affects on bioaccumulation of metals in tree species: a case study from a large opencast coalmining project. Int J Min Reclamat Environ 20(2):96–110

Maiti SK (2007) Bioreclamation of coalmine overburden dumps—with special emphasis on micronutrients and heavy metals accumulation in tree species. Environ Monit Assess 125:111–122

Maiti SK (2013) Ecorestoration of the coalmine degraded lands. Springer, New York

Maiti SK, Ghose MK (2005) Ecorestoration of acidic coalmine overburden dumps- an Indian case studies. Land Contam Reclamat 13(4):361–369

Maiti SK, Maiti D (2015) Ecological restoration of waste dumps by topsoil blanketing, coir-mating and seeding with grass-legume mixture. Ecol Eng 77:74–84

Maiti SK, Saxena NC (1998) Biological reclamation of coalmine spoils without topsoil: an amendment study with domestic raw sewage and grass-legumes mixture. Int J Surf Min Reclamat Environ 12:87–90

Maiti SK, Shee C, Ghose MK (2007) Selection of plant species for the reclamation of mine degraded land in the Indian context. Land Contam Reclamat 15(1):55–66

McLean EO (1982) Soil pH and lime requirement. In: Page AL et al (eds) Methods of Soil Analysis, Part 2, 2nd edn. ASA and SSSA, Madison, pp 199–224

Mukhopadhyay S, Maiti SK (2011) Trace metal accumulation and natural mycorrhizal colonisation in an afforested coalmine overburden dump- a case study from India. Int J Min Reclamat Environ 25(2):187–207

Mukhopadhyay S, Maiti SK (2014) Soil CO_2 flux in grassland, afforested land and reclaimed coalmine overburden dumps: a case study. Land Degrad Dev 25(3):216–227

Mukhopadhyay S, Maiti SK, Masto RE (2013) Use of reclaimed mine soil index (RMSI) for screening of tree species for reclamation of coal mine degraded land. Ecol Eng 57:133–142

Mukhopadhyay S, Maiti SK, Masto RE (2014) Development of mine soil quality index (MSQI) for evaluation of reclamation success: a chronosequence study. Ecol Eng 71:10–20

Nelson DW, Sommers LE (1996) Total carbon, organic carbon, and organic matter. Methods of soil analysis, Part 3(3):961–1010

Pietrzykowski M, Daniels WL (2014) Estimation of carbon sequestration by pine (*Pinus sylvestris* L.) ecosystems developed on reforested post-mining sites in Poland on differing mine soil substrates. Ecol Eng 73:209–218

Pietrzykowski M, Krzaklewski W (2010) Potential for carbon sequestration in reclaimed mine soil on reforested surface mining areas in Poland. Nat Sci 2(9):1015–1021

Post WM, Kwon KC (2000) Soil carbon sequestration and landuse change: processes and potential. Glob Change Biol 6(3):317–328

Pragasan LA, Parthasarathy N (2005) Litter production in tropical dry evergreen forests of south India in relation to season, plant life-forms and physiognomic groups. Curr Sci 88(8):125–126

Rao KS, Ramakrishnan PS (1989) Role of bamboos in nutrient conservation during secondary succession following slash and burn agriculture (Jhum) in North-East India. J Appl Ecol 26:625–633

Rumpel C, Knicker H, Kogel-Knabner I, Skjemstad JO, Huttl RF (1998) Types and chemical composition of organic matter in reforested lignite-rich mine soils. Geoderma 86:123–142

Rumpel C, Skjemstad JO, Knicker H, Kogel-Knabner I, Huttl RF (2000) Techniques for the differentiation of carbon types present in lignite-rich mine soils. Org Geochem 31:543–551

Rumpel C, Balesdent J, Grootes P, Weber E, Kogel-Knabner I (2003) Quantification of lignite- and vegetation-derived soil carbon using ^{14}C activity measurements in a forested chronosequence. Geoderma 112:155–166

Schlesinger WH, Bernhardt ES (2013) Biogeochemistry: An analysis of global change. Academic Press Inc., San Diego

Schmidt MWI, Knicker H, Hatcher PG, KogelKnabner I (1996) Impact of brown coal dust on the organic matter in particle-size fractions of a Mollisol. Org Geochem 25:29–39

Shrestha RK, Lal R (2006) Ecosystem carbon budgeting and soil carbon sequestration in reclaimed mine soil. Environ Int 32:781–796

Shrestha RK, Lal R (2010) Carbon and nitrogen pools in reclaimed land under forest and pasture ecosystems in Ohio, USA. Geoderma 157(3):196–205

Singh AN, Raghubanshi AS, Singh JS (2004) Comparative performance and restoration potential of two Albizia species planted on mine spoil in a dry tropical region, India. Ecol Eng. 22(2):123–140

Singh L, Singh JS (1991) Species structure, dry matter dynamics and carbon flux of a dry tropical forest in India. Ann Bot 68:263–273

Singh S, Singh JS (1995) Microbial biomass association with water-stable aggregates in forest, savanna and cropland soils of a seasonally dry tropical region, India. Soil Biol Biochem 27:1027–1033

Singh AN, Singh JS (1999) Biomass net primary production and impact of bamboo plantation on soil redevelopment in a dry tropical region. For Ecol Manag 119(1):195–207

Singh V, Tewari A, Kushwaha SP, Dadhwal VK (2011) Formulating allometric equations for estimating biomass and carbon stock in small diameter trees. For Ecol Manag 261(11):1945–1949

Skousen J, Zipper CE (2014) Post-mining policies and practices in the Eastern USA coal region. Int J Coal Sci Technol 1(2):135–151

Sperow M (2006) Carbon sequestration potential in reclaimed mine sites in seven East- Central states. J Environ Qual 35(4):1428–1438

Stahl PD, Wick AF, Ganjegunte G, Norton U, Ingram LJ (2009) Redevelopment of soil carbon pools on reclaimed surface mine lands. In: Barnhisel RI (ed) Revitalizing the environment: proven solutions and innovative approaches. ASMR, Lexington, pp 1348–1370

Subbiah BV, Asija GL (1956) A rapid procedure for the determination of available nitrogen in soils. Curr Sci 25:259–260

Terakunpisut J, Gajaseni N, Ruankawe N (2007) Carbon sequestration potential in aboveground biomass of Thong Pha Phum National Forest, Thailand. Appl Ecol Environ Res 5(2):93–102

Thurman NC, Sencindiver JC (1986) Properties, classification and interpretation of minesoils at two sites in Western Virginia. Soil Sci Soc Am J 50:181–185

Tripathi SK, Singh KP (1996) Culm recruitment, dry matter dynamics and carbon flux in recently harvested and mature bamboo savannas in the Indian dry tropics. Ecol Res 11:149–164

Tripathi N, Singh RS, Nathanail CP (2014) Mine spoil acts as a sink of carbon dioxide in Indian dry tropical environment. Sci Total Environ 468:1162–1171

Tripathi N, Singh RS, Hills CD (2016) Soil carbon development in rejuvenated Indian coal mine spoil. Ecol Eng 90:482–490

Ussiri DAN, Lal R (2005) Carbon sequestration in reclaimed minesoils. Cr Rev Plant Sci 24:151–165

Ussiri DAN, Lal R (2008) Method for determining coal carbon in the reclaimed minesoils contaminated with coal. Soil Sci Soc Am J 72(1):231–237

Ussiri DAN, Jacinthe PA, Lal R (2014) Methods for determination of coal carbon in reclaimed minesoils: a review. Geoderma 214:155–167

Vogt KA, Vogt DJ, Brown S, Tilley JP, Edmonds RL, Silver WL, Siccama T (1995) Dynamics of forest floor and soil organic matter accumulation boreal, temperate, and tropical forests. In: Lal R, Kimble J, Levine E, Stewart BA (eds) Soil management and greenhouse effect. CRC Lewis Publishers, Boca Raton, pp 159–175

Xiao W, Hu Z, Fu Y (2014) Zoning of land reclamation in coal mining area and new progresses for the part 10 years. Int J Coal Sci Technol 1(2):177–183

Zipper CE, Burger JA, Skousen JG, Angel PN, Barton CD, Davis V, Franklin JA (2011) Restoring forests and associated ecosystem services on Appalachian coal surface mines. Environ Manag 47:751–765

17

Support failure of a high-stress soft-rock roadway in deep coal mine and the equalized yielding support technology

Lihui Sun[1,2] · Haoyuan Wu[1,2] · Bensheng Yang[1,2] · Qiangqiang Li[1,2]

Abstract There are many soft-rock roadway coal mines in China. The surrounding rocks of the high-stress soft-rock roadways in deep mine are especially difficult to be supported using the traditional supporting way. In this study, the south wing rail roadway on the second level of Yunjialing coal mine in China was used as an example to analyze the deformation and failure characteristics and influencing factors of roadway. On this basis, this study proposed the equalized yielding support idea which employs the yielding rings to realize the pressure equalization on the bolts and cables in the section. To achieve this purpose, the first bolt-mesh-cable equalizing pressure yielding support was integrated with the second grouting reinforcement. The results proved that the yield rings of the bolts and cables on the spandrel of the arched roadway firstly developed yielding deformation; then the deformation extended to the vault of the roadway; the bolts and cables achieved a yielding extreme value of 15 and 18 tonnes, respectively. The roadway surrounding rock tended to be stable at the 26th day after the maintenance. The equalizing pressure yielding supporting technology plays a moderate pressure-releasing and actively controlling role on the surrounding rocks in the soft-rock roadway with large deformation.

Keywords Deep-well high-stress · Soft-rock roadway · Equalized yielding support · Grouting reinforcement

1 Introduction

Soft rock mines widely distribute in China (Niu et al. 2011) and extend to the depth at a speed of 8 m/a. The coal and rock masses in the deep mine are characterized by "three highs and one disturbance". The large deformation and failure of surrounding rocks and the floor heave of the roadway etc. seriously affect the safe production of the coal mine (He et al. 2005, 2006; He 2014). Therefore, the supporting problem of such roadway restricts the development of coal mine (Bai et al. 2008). Owning to differences of the geological structure, mining depth, and surrounding rock properties, a variety of roadway-controlling technologies have been proposed. However, these technologies fail to be promoted and applied widely. The high-stress soft-rock roadway controlling technology in the deep mine has been the hot research focus. Lots of existing studies have been made to improve the surrounding rock strength, the supporting material strength, and the coupling of surrounding rock with support and got fruitful achievements (Bai et al. 2007; Kang et al. 2010). However, the equalized yielding support technology for the high-stress soft-rock roadway has been rarely reported (Cheng 2004; Shen and Li 2005). In this study, the south wing rail roadway on the second level in Yunjialing coal mine in China was used as an example to analyze the characteristics, deformation rule, and deformation characteristics of the surrounding rocks. On this basis, it proposed the equalized yielding support technology for the roadway with high stress and soft rocks. Field industrial experiment suggested the equalized yielding support technology could avoid the uncontrollable deformation of surrounding rocks

✉ Lihui Sun
sunlihui78@163.com

[1] College of Resource, Hebei University of Engineering, Handan 056038, China

[2] Collaborative Innovation Center of the Comprehensive Development and Utilization of Coal Resource, Handan 056038, China

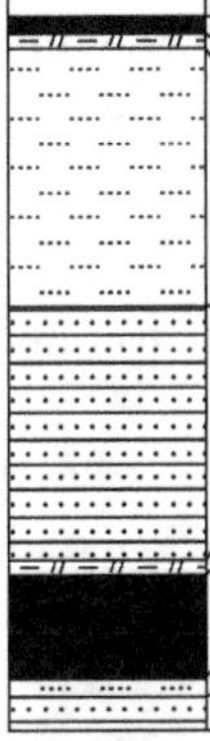

Columnar	Layer thicknessr (m)	Description
	0.5	#1coal
	0.35	Carbonaceous mudstone
	7.6	Siltstone
	7.5	Fine sandstone
	0.35	Carbonaceous mudstone
	3	#2 coal
	0.6	Siltstone
	1	Fine sandstone

Fig. 1 Histogram of the surrounding rocks

and applied uniform stresses on the internal support in the section. The surrounding rock controlling effect of the support technology was significant.

2 Basic conditions of test roadway

The second level in Yunjialing coal mine refer to −150 to −450 m in depth respectively. The south wing rail roadway is 650–750 m deep and 850 m long. The strata in this roadway basically include carbonaceous mudstone, sand-mudstone, and fine sandstone with developed joints. Since these strata are vulnerable to bulking deformation, the roadway exhibits a poor overall stability. Figure 1 is a histogram of the surrounding rocks on the two levels.

The roadway in this study was originally supported using bolt-beam-mesh-cable. The bolt used the screw-thread steel bolt spaced by 800 mm × 700 mm and connected by #14 steel ladder beam; the specification of top cable was 17.8 mm × 8000 mm, while the side cable was 17.8 mm × 4500 mm; the tray of the bolt was 10 mm × 100 mm × 100 mm, while that of the cable was 10 mm × 300 mm × 300 mm; the steel bar mesh was welded using 8 steel. During the construction, the bolt and cable were applied with a pretension force of 80 and 100 kN, respectively. I-steel was used to strengthen the support on local crushing area. The roadway was completed on September 2008 and experienced two maintenances subsequently. However, the roadway is still in large deformation, which is beyond the normal use.

3 Deformation and failure characteristics of roadway with high stress and soft rocks

3.1 Support failure

The roadway was completed on December 2008. However, most of roadway was repaired for twice 1 year later.

Fig. 2 Failure of roadway surrounding rocks. **a** Failure of the roadway. **b** Failure of the wall

Through field observation and geological condition analysis, the deformation and failure law of the roadway was concluded as follows:

(1) The surrounding rocks show fierce deformation, poor stability, and high ground stress. Shortly after the excavation and the two maintenances, the arch angle of the roadway deformed greatly and even the surrounding rocks tend to be instable. Figure 2 illustrates the roadway failure effect.
(2) The surrounding rocks are rheological. The mine pressure observation after the two maintenances suggest that the surrounding rocks from the surface to the depth of 8 m moved significantly. Moreover, after the roadway maintenances, the roadway movements spent 80 days to turn stable. The roof and floor of the roadway had a deformation of 560 mm on average, while the two sides converged by 890 mm on average. The deformation of the two sides exceeded the deformation of roof and floor in this roadway.

3.2 The causes for roadway deformation

(1) The complex geological conditions. The south wing rail roadway passes through several strata, including mudstone, san-mudstone, fine sandstone, and coal

seam. The ground stresses of two holes were measured by using the stress relieving method in the substation of the mining area at two levels. The maximum principal stresses are 19 and 18.52 MPa respectively; while the minimum principal stresses are 9.12 and 8.63 MPa; the intermediate principal stresses are 16.25 and 15.67 MPa, respectively. Both the maximum and minimum principal stresses are horizontal principal stresses. The maximum principal stresses are at NE85.9 and NE87.1°. The intermediate principal stresses are vertical principal stresses in downward directions. So the roadway is in near south-north direction, with the maximum horizontal principal stress of 19 MPa in the near east–west direction. The axial direction of the roadway is approximately perpendicular to the maximum horizontal principal stress direction. This situation is unfavorable for the roadway maintenance.

(2) The hydrophilic expansion of the surrounding rocks. The mineral compositions in the rocks, such as kaolinite and montmorillonite, were in high contents. In case of water accumulating in the roadway during roadway repairment, the deformation of corresponding position would be intensified and expanded. These components contain amounts of expansive clay minerals, which have poor water stability and are liable to weathering and deliquescence.

(3) The anchorage structure fails to form the active bearing as soon as possible. The key for the active control of the soft roadway support is to improve the residual strength of loose rocks to give full play to the active bearing capacity of the surrounding rocks. Mine pressure observation shows that the cable can achieve a pretension of 100 kN, while the bolt has a pretension force of 30–40 kN generally. Therefore, the shallow anchoring structure formed by bolt and surrounding rocks present insufficient active bearing capacity. Moreover, the cable support is generally lagged behind of heading end by certain distance. The anchoring structure formed by the bolt, cable, the surrounding rocks fail to control intense surrounding rock deformation due to the insufficient overall active bearing capacity at the initial stage of the excavation. Therefore, serious deformation of the surrounding rocks appears and calls for repeated maintenances.

(4) The overall uneven stressing of bolt and cable. The stressing of bolt and cable on the spandrel of the roadway is as high as 85 % of breaking force. While the stresses on the bolt and cable on other positions are merely 50 % of the breaking force.

Figure 3 illustrates the stress distribution of traditional bolt-mesh supporting and equalized yielding support, as well as the stress states of bolts and cables. It can be found that there was little difference in the stress distribution of roadway surrounding rock by using these two supporting schemes, except that the uneven high-stress zone formed in the deep roadway floor. However, there were great difference being found in the stress states of bolts and cables. The stress of the bolts in the whole section of the roadway with traditional bolt-mesh supporting was uneven, where the stress of the vault of roadway was great, and that of the two sides of roadway and floor bolt was small. In terms of the stress state of cables, it reflected that the stress of cables at the shallow section was great, while that of the cables at the deep section was small. On the contrary, the stress of bolts and cables in the roadway with equalized yielding support was relatively even. It suggested that the stress of bolts and cables in the section was more even and reasonable after yielding to reduce the fractures of local bolts and cables.

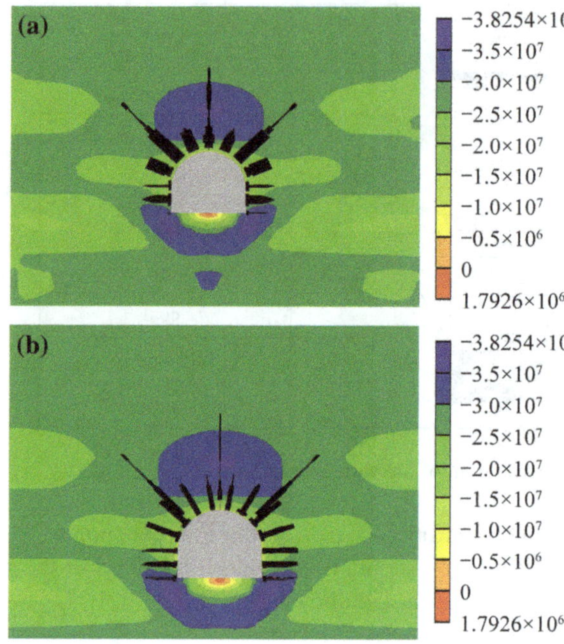

Fig. 3 The contrast of traditional bolt-mesh supporting and equalized yielding support. **a** Traditional bolt support of roadway. **b** Qualized yielding support of roadway

4 Equalized yielding pressure support technology for roadway with high stress and soft rocks

4.1 Equalized yielding support theory

The support of the soft-rock roadway in the deep mine are likely to be stressed unevenly. The bolt-mesh-cable coupling can eliminate the nonuniform stressing condition of

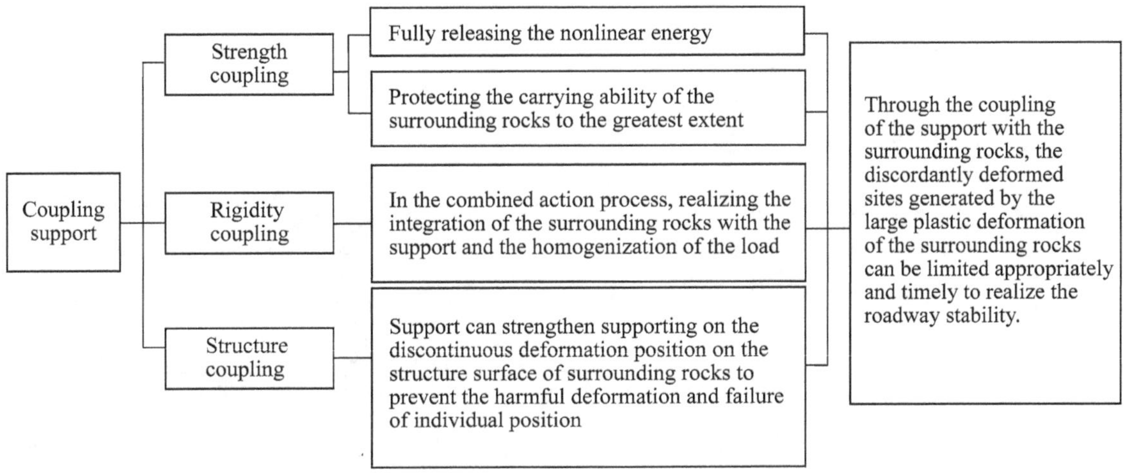

Fig. 4 Coupling support theory

the support structure through the coupling of the bolt mesh and surrounding rocks. The coupling way coordinates the deformation of the surrounding rocks and realizes the integration of the supporting and the uniform distribution of the load. Finally, the stability of the surrounding rocks can be achieved. The support of the soft-rock roadway in the deep mine is basically characterized the coupling of the strength and rigidity of the surrounding rocks with the support, as shown in Fig. 4 (Ren 2010; Lu et al. 2011).

The mechanism of equalized yielding support is as follows: due to the redistribution of stress, the stress concentrated at part of roadway after the excavation and support of roadway. Meanwhile, the surrounding rocks deformed sharply so that the stress of the bolts and cables at this part increased. Once the stress reached the yielding limit of bolts and cables, they yielded through the deformation of yield rings. At the same time, the high stress of surrounding rock gradually transferred to the neighboring bolts and cables. When the bolts and cables reached the yielding limit, they yielded immediately so that the high stress of surrounding rock passed on to the bolts and cables around. The process cycled in this way until the deformation of roadway surrounding rock stabilized. Thereby, the stress of bolts and cables was even, as seen in Fig. 5.

4.2 Equalized yielding support plan

Basing on the characteristics and causes of the deformation and failure of the roadway, the support technology in this study aims to control the local violent deformation, improve the overall bearing ability of the support structure, and realize the equalized stress on the support. The bolt with high strength and cable bearing high pretension force were used to bear the pressure on the anchoring structure. Moreover, yielding ring was employed to realize the uniform stress on the support components. The secondary grouting in the surrounding rocks can further improve the self-bearing ability of the surrounding rocks and thus to keep the stability of the roadway constantly. Figures 6 and 7 show the section of the supporting way, and the yielding ring respectively.

(1) First support.

The bolt with ultra-high strength of 20 mm × 2800 mm, lengthening anchorage length of 1400 mm, pretension force of 80 kN, and tensile strength of 250 kN was used. The protective metal mesh was in specification of 6.5 mm × 100 mm × 80 mm. The bolts were connected using the reinforced bar ladder beam of 14 mm. The tray was a square high-strength plate in size of 200 mm × 200 mm × 10 mm and strength of 280 kN above. The bolts were spaced by 800 mm × 800 mm.

The roof is supported by the nest bolt of 18.9 mm × 8000 mm; the two sides were supported using nest cables of 18.9 mm × 4500 mm, with anchorage of 2.2 m long and pretension force of 100 kN; the tray for the anchor rope was the spheroidal stray of 300 mm × 300 mm × 12 mm with high-strength. The cables were spaced by 1810 mm × 1600 mm.

The pressure yielding pipe was prepared by the thread steel with high strength. The anchor yielding ring is 20 mm long; the pressure yielding pipe of the anchor cable is 30 mm in length. And the yielding values of the anchor bolt and the anchor cable are designed to be 15 and 18 tonnes, respectively. If yielding ring is used in coal roadway, the length of the yielding ring needs to be properly increased according to the measured value of surrounding rock deformation in the field roadway so as to adapt to the deformation.

(2) Second reinforcing-grouting in the surrounding rocks.

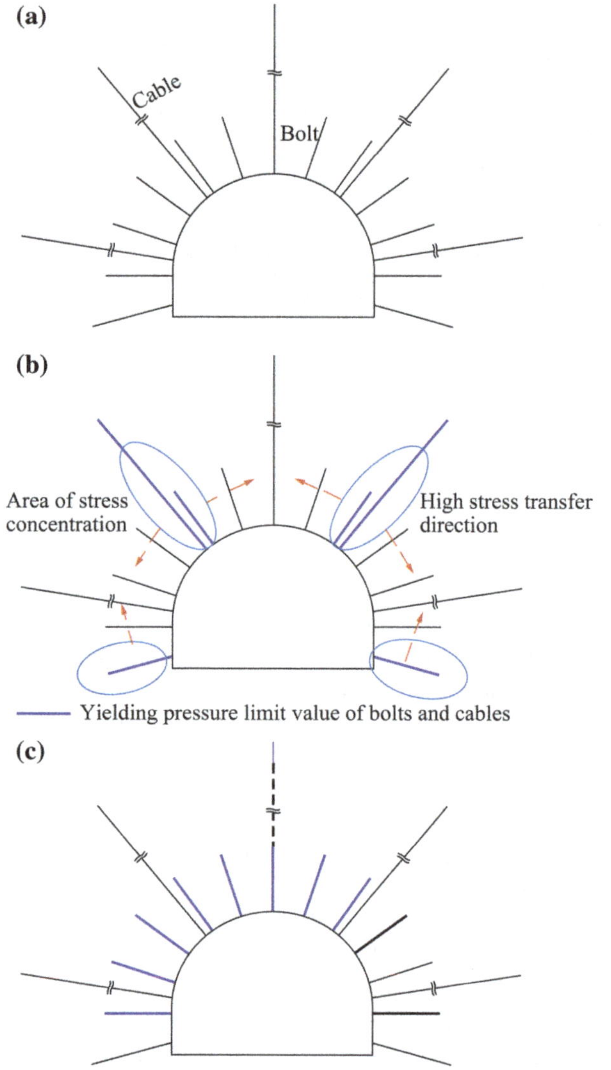

Fig. 5 Evolution model of qualized yielding support of roadway. **a** Initial stage of bolt support of roadway. **b** Yielding pressure stage of local roadway. **c** Equalized pressure of bolts and cables of whole roadway

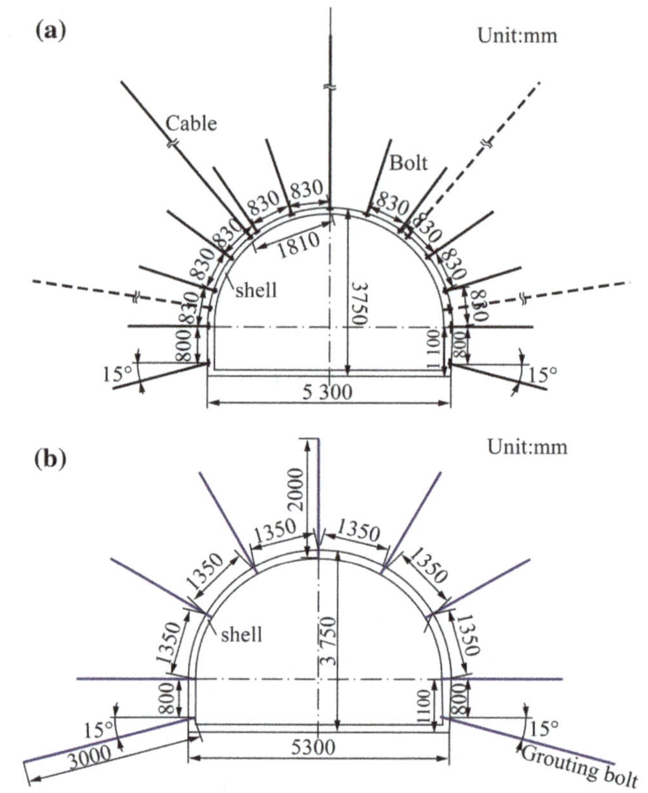

Fig. 6 Support parameter of roadway. **a** The first support. **b** The second reinforcing-grouting

The grouting anchor rode was 2000 mm long. The bolt for base angle grouting was 3000 mm long and spaced by 1350 mm × 1600 mm. The grouting in the surrounding rocks was implemented after the first support. #525 portland cement and ACZ-I cement grout additive were utilized as grouting material. The amount of additive is 8 % of that of cement; water cement ratio is 2:1; the grouting pressure is controlled at 3–5 MPa. The grouting amount is determined by considering three perspectives: the grouting requires to be ceased in the cases when grouting pressure reaches 3–5 MPa, or the cement more than two bags is grouted. The grouting also has to be stopped when cement leakage occurs in the rear grouting pipe or the zone around the grouting hole in the grouting process.

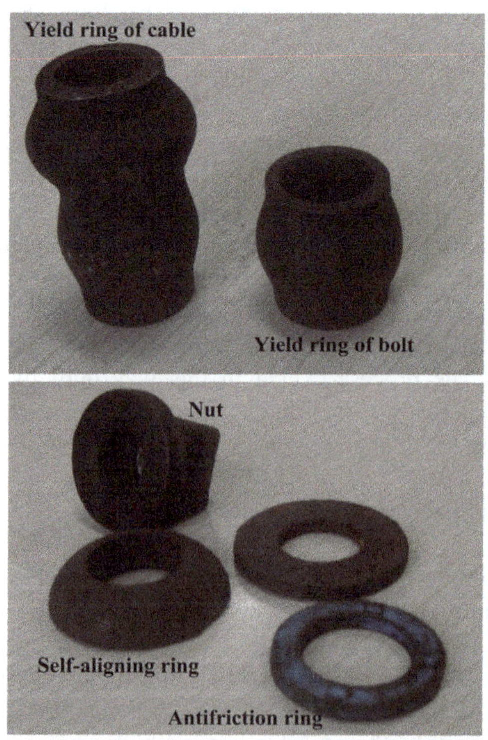

Fig. 7 Yelding ring and auxiliary components of bolts and cables

4.3 Numerical simulation on the equalized yielding support

According to the geological conditions of the south rail roadway on second level, a numerical model of 50 m × 50 m × 50 m (length × width × height) was established using FLAC3D numerical simulation software. The sides and bottom of the model were set with limited displacement boundary. The upper surface of the model was free boundary and was applied with the vertical load of 19 MPa. The bolts and cables were simulated using cable unit. Section 4.2 shows the support parameters. The simulation steps include excavation, support, and grouting. Given the role that grouting played in improving the strength of surrounding rock, so simulation grouting was realized by improving the cohesion force and friction angle of surrounding rock.

As seen in Fig. 8, in the case of using equalized yielding support plan, the roof and floor are shown a cumulative deformation of 6 cm, while the two sides show a cumulative deformation of 10 cm. The large deformation of the roof and bottom are mainly observed in vault and the shallow surrounding rocks within 1.5 m of the roadway. By reinforcing, roadway floor show smaller deformation and deformation influencing area. The large deformation on the two sides of the roadway is concentrated on the stringcourse of the roadway. Moreover, the influences of horizontal deformation in the surrounding rocks are larger than that of the roof and floor.

As shown in Fig. 9, the horizontal stress concentration of the surrounding rocks in the roadway mainly distributes

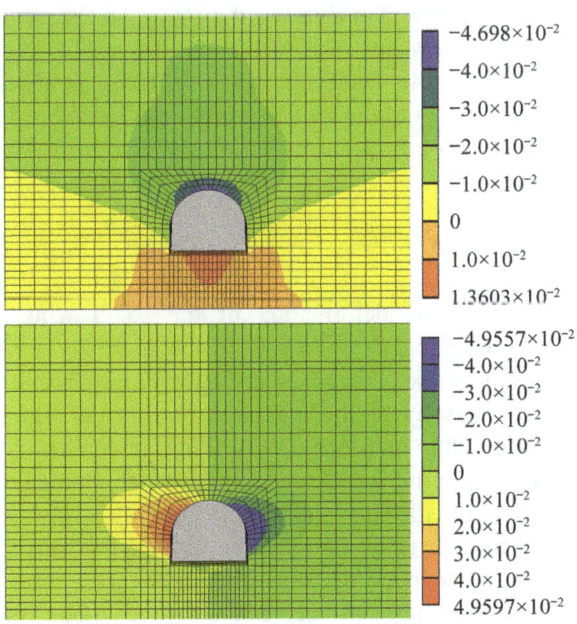

Fig. 8 Deformation of qualized yielding support of roadway

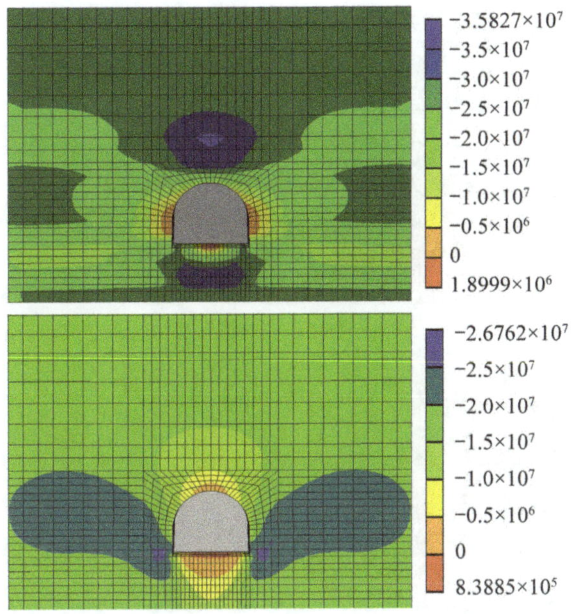

Fig. 9 Stress distribution of qualized yielding support of roadway

in the deep surrounding rocks of the roof and floor of the roadway. The maximum horizontal stress is 4 m away from the roof. The stress in the roadway maximizes at 35.82 MPa, with a stress concentration coefficient of 1.36. The vertical stress concentration area of the surrounding rocks of the roadway is mainly in the slope angel on the middle-to-low part of the two sides, with a maximum stress of 26.76 MPa and the stress concentration coefficient of 1.53. Especially, the concentration of the vertical stress on the two slope angles of the roadway requires to be given priority in the making controlling measures. In case of the deformation and failure, grouting needs to be applied in time. In addition, the bolt and cable are under the stress of 136 and 204 kN, respectively, which are both in the stressing ranges of them. The grouting reinforces the strength of the surrounding rocks in the roadway. The shallow surrounding rocks are in high strength and are conductive to maintain the stability of the roadway. Therefore, the equalized yielding support plan proposed is proved to be effective.

Figure 10 shows the deformation of roadway by traditional bolt-mesh supporting. It is obvious that the roadway surrounding rock deformed greatly owing to the unreasonable support parameters and without reinforcing the surrounding rocks by grouting. The subsidence of the roadway roof was up to 60.67 cm and the floor heave was 18.12 cm. In addition, the maximum deformation occurred at the waist line of the two sides of roadway with accumulative deformation of 83.68 cm. It can be known that the deformation of the roadway with traditional bolt-mesh supporting was greater than that of the roadway by equalized yielding support.

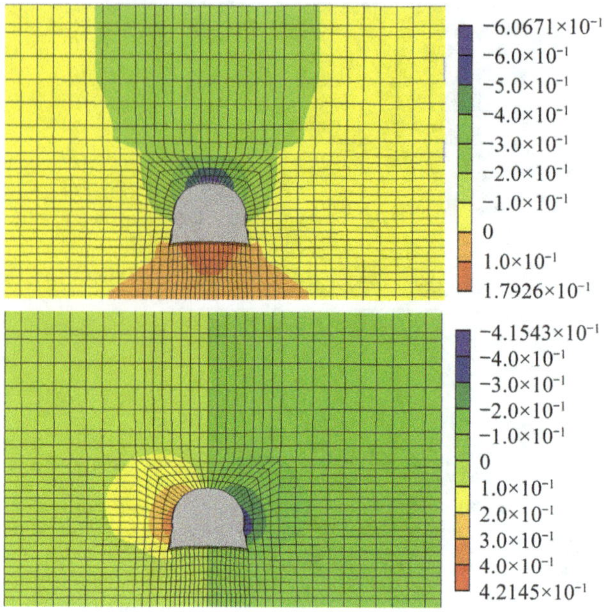

Fig. 10 Deformation of traditional bolt support of roadway

5 Field industrial test and analysis on test results

The exampled roadway was repaired by using equalized yielding support plan. The test roadway was 200 m long. In field, the roadway section was firstly repaired so as to meet the requirements of the designed section. Then, the first support was applied. Five days later, the secondary grouting reinforcement was conducted.

5.1 Displacement on the roadway surface

The displacement of the surrounding rocks after maintenance is shown in Fig. 11. It can be seen that the deformation of the surrounding rocks attenuates rapidly and tends to be stable 26 days later. After the roadway deformation tends to be stable, the two sides, and roof and floor move close by 161 and 42 mm respectively. This technology controls the deformation of the surrounding rocks in the deep roadway with high stress and soft rocks and ensures the stability of the surrounding rocks in the roadway.

5.2 Stressing analysis on the anchor rod and anchor rope

Figure 12 shows the stress variations of the bolt and cable. Before the secondary grouting reinforcement, the stresses on the bolt and cable increase rapidly. The stresses on the supports in the section are sequenced in a decreasing order: spandrel > vault > two sides. The yielding ring fails to shift the pressure. After grouting, the stress on the support

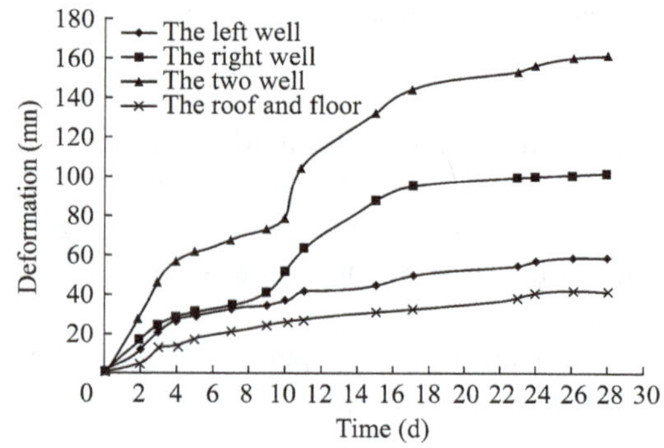

Fig. 11 Deformation of roadway

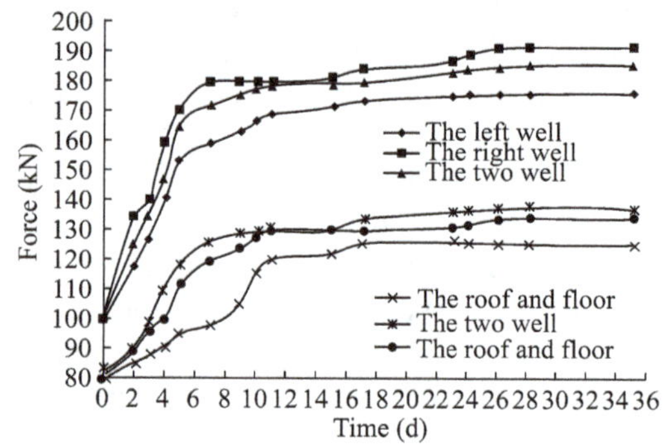

Fig. 12 Force of bolts and cables

grows slowly. Three days later after the grouting, the cable on the spandrel shifts the stress to the vault. The bolt on the vault functions on the 6th day after the grouting. The cable on the spandrel begins to convert the stress on the 5th day after the grouting to the vault. The cable on the vault functions on the 6th day after the grouting. Thirty days after the roadway maintenance, the stress on the support basically is constant. The stress on the cable lies in 176–191 kN, while that on the bolt is in 125–137 kN. The stresses on the bolt and cable on the both sides are both lower than yielding values.

6 Conclusions

(1) The serious deformation of the south wing rail roadway mainly results from the high stress environment, the perpendicular roadway axis direction of the roadway to the maximum horizontal principal stress, the high-content swelling mineral(such as kaolinite) in the rocks, and the poor self-stabilization ability of the surrounding rocks etc.

(2) According to the pressure law of the roadway, this study integrated the equalized yielding support with the grouting reinforcement and used a bolt-mesh-cable + anchor grouting support technology.

(3) The roof and both sides of the roadway deformed by 42 mm and 161 mm respectively; in the section, the bolt and cable were uniformly stressed in the normal working state. The deformation of the surrounding rocks in the roadway could be controlled effectively.

Acknowledgments Supported by the National Natural Science Foundation of China (51174070).

References

Bai JB, Wang XY, Yao Z (2007) Study of coupling support in soft rock roadway under high stress. J China Univ Min Technol 36(4):421–425

Bai JB, Wang XY, Jia MK, Hou CJ (2008) Theory and application of supporting in deep soft roadways.Chinese. J Geotech Eng 30(5):632–634

Cheng JY (2004) Application on supporting technology of release equilibrium pressure in high stressing and deformation roadway. Coal Eng 5:79–80

He MC (2014) Progress and challenges of soft rock engineering in depth. J China Coal Soc 39(8):1409–1417

He MC, Xie HP, Peng SP, Jiang YD (2005) Study on rock mechanics in deep mining engineering. Chin J Rock Mech Eng 24(16):2803–2814

He YN, Han LJ, Shao P, Jiang BS (2006) Some problems of rock mechanics for roadways stability in depth. Chin J Rock Mech Eng 25(3):288–295

Kang HP, Wu YZ, Li JB (2010) Analysis on mechanical performances and supporting function of combination components for rock bolting. J China Coal Soc 35(7):1057–1065

Lu YL, Wang LG, Zhang B (2011) An experimental study of a yielding support for roadways constructed in deep broken soft rock under high stress. Min Sci Technol (China) 11:839–844

Niu SJ, Jing HW, Zhang ZY, Yang SQ (2011) Study on control technology of surrounding rocks in deep soft roadway and its application. J China Coal Soc 36(6):914–919

Ren XL (2010) Application on supporting technology of release equilibrium pressure in Yun Jia-ling coal mine. In:The youth technology innovation proceedings of the fifth session of the national coal industry production line, pp 77–81

Shen RS, Li DS (2005) The study of tunnel yield and shotcrete technology. China Min Mag 14(9):63–65

18

Experimental forward and reverse in situ combustion gasification of lignite with production of hydrogen-rich syngas

Yong Cui · Jie Liang · Zhangqing Wang · Xiaochun Zhang · Chenzi Fan · Xuan Wang

Abstract This research focused on the feasibility of applying the forward and reverse combustion approach to the in situ gasification of lignite with the production of hydrogen-rich syngas (H_2 and CO). The so-called forward combustion gasification (FCG) and reverse combustion gasification (RCG) approach in which oxygen and steam are simultaneously fed to the simulated system of underground coal gasification (UCG) was studied. A simulated system of UCG was designed and established. The underground conditions of the coal seam and strata were simulated in the system. The combustion gasification of lignite has been carried out experimentally for almost 6.5 days. The average effective content (H_2 + CO) of syngas during the FCG phase was 62.31 % and the maximum content was 70.92 %. For the RCG phase the corresponding figures are 61.33 % and 67.91 %. Thus, the feasibility of using RCG way for UCG has been demonstrated. The temperature profiles have been provided by using of 85 thermocouples during the model experiment, which portrayed the several nephograms of thermal data in the gasifier were of significance for the prospective gasification processes.

Keywords Underground coal gasification · Reverse combustion gasification · Forward combustion gasification · Hydrogen-rich · Syngas

1 Introduction

In essence the underground coal gasification (UCG) process is the in situ conversion of coal into syngas using two boreholes drilled from the surface, one for the injection of gasification agents and the other for the egress of the syngas. Accompanied by coal combustion to provide heat, the gasification agents react with the coal in a series of combustion and gasification reactions to form H_2, CO, CH_4, CO_2 and other minor constituents (Friedmann et al. 2009; Evgeny and Arvind 2009). UCG technology differs from conventional coal gasification in surface reactors, in that it is an invisible process, so it is difficult to control the reaction process and consistently produce hydrogen-rich syngas in the underground situation (Friedmann et al. 2009). While they were engaged in field trials in the 1980s, several researchers began to establish simulation systems of UCG processes so as to study the actual phenomena and reactions in the UCG cavity (Skafa 1960; Singh et al. 1980; Hurloff 1983; Park and Edgar 1987; Prabu and Jayanti 2011).

At present, there are three configurations of experimental UCG units (Liu et al. 2011; Stanczyk et al. 2011; Krzysztof et al. 2012). The UCG model reactor was established in the shape of a cylinder with the external contour size 7.4 m (length) and 3.5 m (diameter) by researchers in the State Key Laboratory of Coal-Based Low Carbon Energy ENN Group Co., Ltd. Specifically, there were four windows to enable observations of coal combustion to be available, using closed circuit industrial television. A total of 96 thermocouples were located in the coal seam and strata to measure temperature gradients during the trial (Stanczyk et al. 2011). The ex situ reactor (Liu et al. 2011) was simulated in the gasification of lignite and hard coal with oxygen and enriched air (air + oxygen)

Y. Cui · J. Liang (✉) · Z. Wang · X. Zhang · C. Fan · X. Wang
School of Chemical and Environmental Engineering, China University of Mining & Technology (Beijing), Beijing 100083, China
e-mail: ucgrc@sohu.com

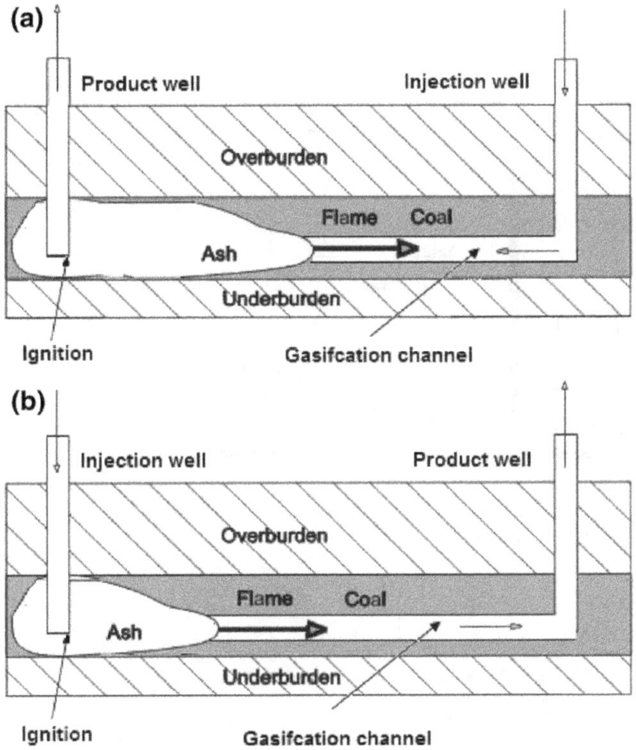

Fig. 1 Diagrammatic sketches of reverse combustion gasification (**a**) and forward combustion gasification (**b**)

in the 50 and 30 h experiments, respectively. Only the oxygen gasification experiment was relatively successful regarding product gas quality. Although the two-stage gasification approach in which oxygen and steam were separately fed to the reactor was studied in both of these experimental UCG units, hydrogen rich syngas should be obtained (Liu et al. 2011; Krzysztof et al. 2012), but the practical operation of the UCG process is difficult as regards the measurement of the temperature in the reaction zone and the control of the switching between oxygen and steam stages according to the change in temperature. It is concluded that the two-stage gasification approach is not suitable for application to UCG. The UCG system in the China University of Mining and Technology was investigated in a model experiment of underground coal gasification and is the subject of this paper.

Several different approaches have been applied to UCG technology (Yang and Liu 2003; Yang et al. 2003; Yang 2004; Khadse et al. 2007; Yang et al. 2008; Shugin et al. 2009), including reverse combustion gasification (RCG) and forward combustion gasification (FCG). In RCG, the gasification agents are injected into one borehole and coal is ignited in the other borehole so that the flame propagates towards the opposite direction of gas flow in the channel as shown in Fig. 1a. In the FCG process coal is ignited in the injection borehole, and the flame propagates along the direction of gas flow in the channel as depicted in Fig. 1b.

Generally forward and reverse combustion approaches are used in shaftless type UCG process. In the practical operations, reverse combustion was widely applied to link the injection and production wells in UCG (Blinderman et al. 2007), but was seldom used to gasify coal. Forward combustion has been usually applied to gasify the coal seam with appropriate gasification agents (Skafa 1960; Blinderman et al. 2007). When the flame gradually moved to the product borehole in the late stage of FCG, the huge cavity should be formed in the coal seam due to coal combustion gasification and overburden roof spalling. It made coal gasification efficiency reduce and the quality of production gas decrease. The high intensity forward gasification process would be difficult to be continued. The reasons are as follows (Yang et al. 2008): (1) The dry distillation zone becomes shorter and shorter in the later forward gasification. (2) The reaction activity of coal significantly decays after coal seams underwent dry distillation during forward gasification. (3) The reaction rates of coal combustion and gasification fell for descending concentration of gasification agents absorbed on coal surfaces of the cavity wall. To make coal seams further gasify and enhance syngas quality, the injection borehole and product borehole should be exchanged in practical operation. Then coal seams will be continued to gasify by shifting the direction of injection gas. New gasification conditions formed again. If the gasification flame moved to the direction of injection gas flow, therefore the residual coal seams around former product borehole would not be gasified. It is the waste for coal resource. In order to keep on gasifying residual dried and distilled zone, and solve the problem of the late stage of FCG, the reverse combustion gasification could be applied in UCG process. To research the feasibility and characteristics of the RCG approach, gasification experiments were performed in a system of UCG to simulate the FCG and RCG processes.

2 Materials and methods

Large-scale lignite and rock blocks were sampled at the same mine from the region and placed in the simulated UCG gasifier. The oxygen, enriched oxygen and steam as gasification agents were continuously fed into the gasifier in the model experiments. Detailed conditions used for the simulated underground gasification are described below.

2.1 Simulation of the inclined lignite seams and strata in simulated UCG units

The diagram of the simulated system for UCG for lignite is shown in Fig. 2. It consisted of gas flow inlets with valves, flow meters and regulators, gas flow temperature and

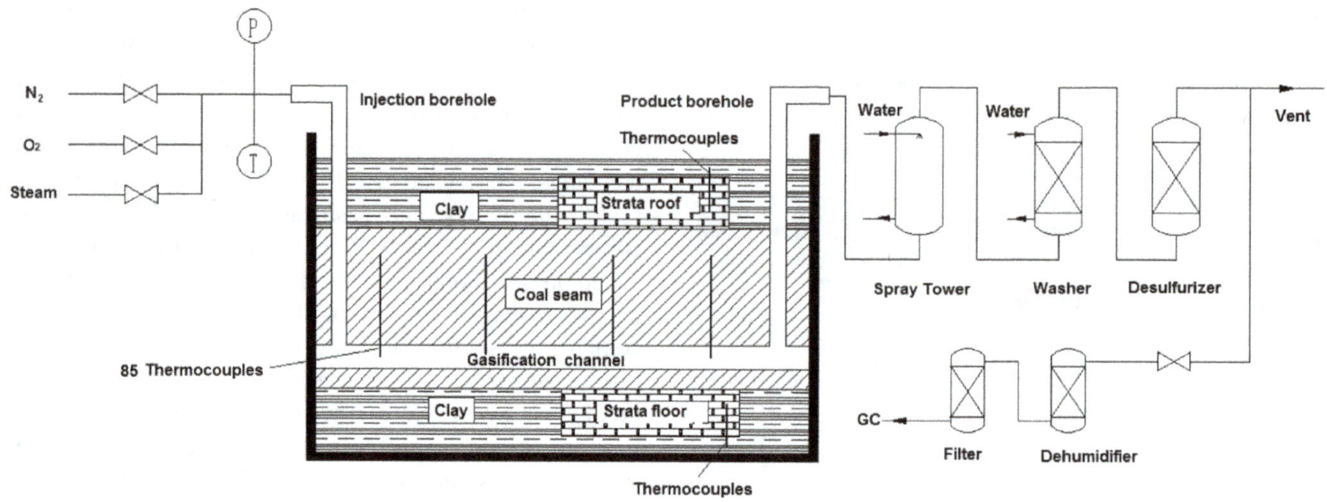

Fig. 2 Diagram of the units used for experimental lignite gasification in the UCG simulation

pressure instrumentation, the simulated UCG gasifier, spray tower, gas washer, desulfurizer, dehumidifier, gas filter system, and a gas chromatograph.

A pilot-scale simulated underground gasifier was constructed in the rectangular shape with the internal dimensions of 4.45 m (length) × 1.17 m (width) × 1.57 m (height). External walls of the gasifier were composed of fire retardant layers, thermal insulation layers, stainless steel sealed layers and reinforced concrete anti-pressure layers in order from inside to outside. There were some holes for measurement and observation structured on the gasifier body, hereinto, 19 holes were used to measure experimental temperatures, six holes were used to monitor the gasifiers pressures, and four boreholes used to observe the internal conditions of gasifier by cameras installed in them. Otherwise, four holes were utilized as injection gas inlets or production gas outlets.

Gasification agents like oxygen or oxygen-steam mixture gas were fed into the pilot system by supplying gas equipment. Oxygen agent was supplied from ten steel cylinders of parallel connection under 0.2 MPa pressure. Water was changed into saturated aqueous vapor under pressure of 0.7 MPa in the electric steam generator with its rated evaporation capacity of 80 kg/h. In view of the pilot system safety, valve and steel cylinder of nitrogen were installed in the supplying system to extinguish a fire in the gasifier in case of emergency.

The product gas was transported into a suit of pilot-scale purification system by a 0.10 m diameter pipe. The purification system was mainly composed of spray tower, gas washer and desulfurizer. The product gas suffered from purification process and was combusted at flare stack. Partial the syngas flow was led to the online gas chromatograph for component analysis. Before the sample gas entered into chromatographic columns it need be further purified by the dehumidifier and gas filter made of quartz.

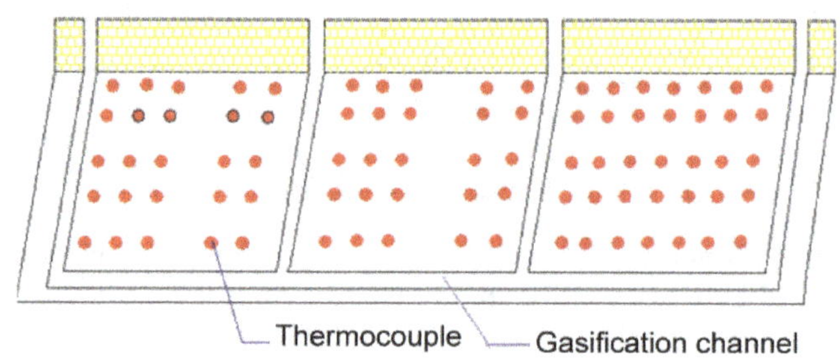

Fig. 3 Diagram of the 85 thermocouple horizontally located in coal seam

The schematic view of the simulated lignite seam and strata is shown in Fig. 2. The gasification channel was excavated in the bottom of the coal seam with a length of 4 m and a diameter of 0.08 m. The strata including roof and floor were established in the construction of the gasifier whose design is as follows: Grit stone and shale were laid in the strata floor, and two grit stones and shales were laid in the strata roof one by one. Before the lignite seams were constituted, the floor comprised rock and clay with a 17° inclination. The dimensions of the lignite block were 0.4 m × 0.5 m × 0.5 m. The roof comprised a layer with a thickness 0.5 m that was filled with rock blocks and clay above the coal seams. The expanded perlites with 0.1 m thickness were used as an insulating layer above the roof.

During the experiments, several parameters need to be monitored. The temperature data were collected by means of Ni–Cr/Ni–Si (K type) armored thermocouples uniformly placed in the seam seen in Fig. 3. This provides crucial information about coal combustion/gasification and explains the experimental results. The mass flow of the product syngas was measured at the outlet with a flow meter and its composition was continuously monitored by

Table 1 Proximate and ultimate analysis of lignite

Proximate analysis w_{ad} (%)			Ultimate analysis w_{daf} (%)				
M	*A*	*V*	C	H	O	N	S
32.5	16.32	46.24	74.46	4.80	18.34	1.41	0.98

the two-TCD online gas chromatograph (Shimadzu GC-2014) every 12 min, and the sampling interval was 3 min.

2.2 Materials

The coal seam comprised of blocks of lignite and strata and was simulated by blocks of rocks supplied from the coal mine. The proximate and ultimate analysis of the lignite samples acquired according to appropriate standards are presented in Table 1.

3 Experimental procedure

3.1 Cold-state test

Cold tests should be performed to check that all systems are leak-free. All valves (both inlet and outlet) were open. The air was blasted in from the injection inlet by means of a blower, and the product gas was released from the outlet. The air flow at both inlet and outlet was recorded to calculate gasifier leakage. Additionally, the pressure at the inlet and outlet and in the gasifier was monitored. Gasifier leaks could be detected using soap-water. When the blast volume was in the range of 2.5–10 m^3/h, and the gasifier leakage was <3 %, the test data can be recorded. In the cold-test the leak rate was 2.8 %. When the static pressure of the gasifier reached 25 kPa, pressure release took 35 min. This means that the velocity of pressure release was 0.714 kPa/min which satisfied ignition requirements.

3.2 Ignition

At first, an electric igniter was placed inside the gasification channel from the injection borehole. Pure oxygen was blown into the coal seam, and the electricity turned on. It could be seen that the coal seam was ignited because the temperature in the gasification channel exceeded 600 °C (Stanczyk et al. 2011). Subsequently, the ignition process was terminated. Next, the electric igniter was removed and pure oxygen was passed to the gasifier. The gasification process began after a 2 h ignition period.

The experiments were divided into three phases. In phase I, 98 % oxygen was fed to the reactor to heat up the coal seam and to provide sufficient heat energy before feeding oxygen and steam. In phase II, oxygen and steam were continuously supplied to the reaction zone, and the FCG experiment was performed. The primary purpose of phase II was to explore the optimal ratio of steam to oxygen and the flow rate required to obtain high-quality syngas in the FCG process. This was achieved by directly analyzing the product gas composition under different operational conditions and various temperature profiles in the coal seam and strata. In phase III, 93 % oxygen and steam were reversely supplied to the gasifier for improved syngas quality in the last stage of the UCG process. The characteristics and feasibility of RCG were largely investigated based on the RCG technique in phase III.

4 Results and discussion

Model experiments in the UCG gasifier were performed to investigate lignite gasification with oxygen or oxygen-steam in the forward and reverse combustion processes.

4.1 Oxygen gasification (phase I)

Pure oxygen was supplied to the gasifier and phase I was initiated after ignition. In the next 26 h only pure oxygen was fed to the gasifier with a flow rate of 3.6–5 m^3/h to heat the coal seam and to accumulate adequate heat energy for subsequent lignite gasification.

The composition of the product gas, the average gas composition as well as the heating value of the product gas in phase I of lignite seam gasification are presented in Fig. 4; Table 2.

The quality of the product syngas was relatively high although steam was not fed in this phase (2–28.5 h). The average heating value of the syngas was 9.13 MJ/m^3. The average value of effective syngas composition (CO and H_2) reached about 58 %, and H_2/CO was equal to 1.58 since lignite contained a relatively high moisture content of >30 %. At high temperatures, the moisture changed into steam and reacted with the coal to generate hydrogen and

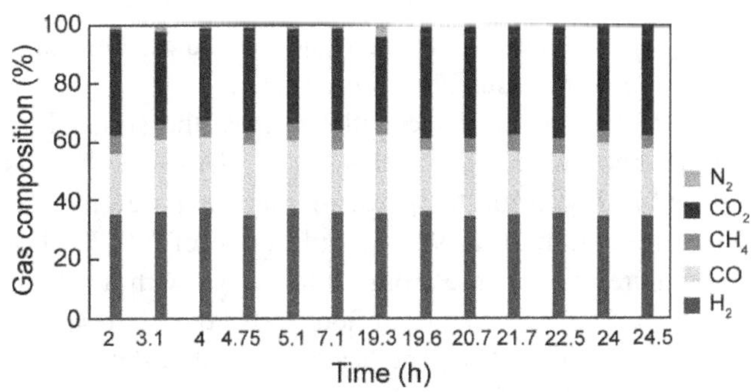

Fig. 4 The change of product syngas composition in phase I

Table 2 Average syngas composition and heating values of product syngas in phase I, II (FCG) and III (RCG)

Gasification phase	Time (h)	H_2 (%)	CO (%)	CH_4 (%)	CO_2 (%)	Heating value (MJ/m^3)
Pure oxygen (98 %)	2–28	35.5	22.5	4.91	35.3	9.13
FCG Phase	28.5–117	39.9	22.9	3.49	32.7	9.20
RCG phase	121–152	39.7	21.7	1.71	36.2	8.30

CO. Though the major purpose of this step was to preheat the coal seam, gasifying high moisture lignite could produce hydrogen-rich syngas in phase I. The coal seam temperature and composition of syngas were crucial factors to decide the time when the switch to pure oxygen should take place. The temperature profiles are presented in Fig. 12. The temperature of the coal seam changed from 300 to 1,000 °C after 15 h (Fig. 13a, b). The temperature in the combustion zone was relatively high, attributable to the release of a great deal of heat and volatiles by reaction of lignite with pure oxygen. The combustion of volatiles significantly heated the coal. The process is referred to as the main overall reaction (Yang 2004; Perkins and Sahajwalla 2005; Yang et al. 2007).

$$\text{Pyrolysis for coal:} \rightarrow \text{C} + \text{CO} + \text{CO}_2 + \text{H}_2 + \text{CH}_4 + 44.7\ \text{MJ/k mol}$$

$$\text{Carbon oxidation: C} + \text{O}_2 \rightarrow \text{CO}_2 - 394\ \text{MJ/k mol}$$

$$\text{Volatiles combustion: 2CO} + \text{O}_2 \rightarrow 2\text{CO}_2 - 571\ \text{MJ/k mol}$$

$$\text{H}_2 + 1/2\text{O}_2 \rightarrow \text{H}_2\text{O} - 242\ \text{MJ/k mol}$$

$$\text{CH}_4 + 2\text{O}_2 \rightarrow \text{CO}_2 + 2\text{H}_2\text{O} - 890\ \text{MJ/k mol}$$

There was sufficient thermal energy for moisture in the lignite (30 %) to evaporate at high temperature. The steam decomposition reaction with char took place in the gasification channel. The higher content of hydrogen is attributable to the steam gasification reaction, favored at temperatures above 800 °C (Yang and Liu 2003). Moreover, the composition of the syngas from 5 to 28 h of the phase I state remained constant and the temperature profiles in the combustion zone were stable.

The temperatures of the outlet syngas, the strata floor and the roof reflect the status of the UCG gasifier as depicted in Fig. 5. In this phase, the temperature of outlet syngas rose from 20 to 94 °C, with a rate of 2 °C/h. The temperature of roof rose from 20 to 76 °C, with a rate of 2.15 °C/h. The temperature of floor rose from 20 to 40 °C, with a rate of 0.76 °C/h. The heating rate of outlet syngas, strata floor and roof was relatively slow as the main purpose of this phase was to preheat the coal seam and

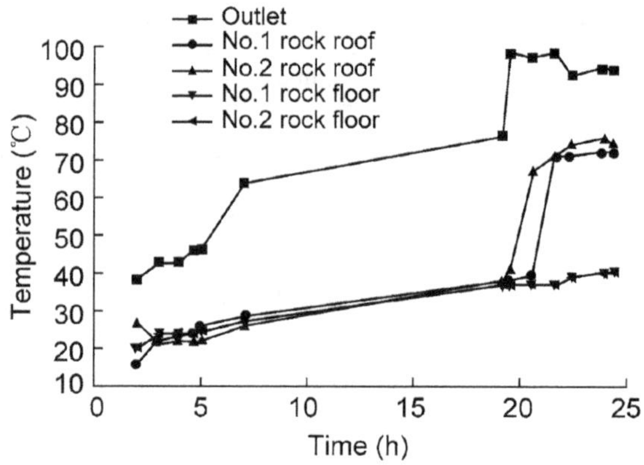

Fig. 5 Change in temperature of outlet syngas and of strata floor and roof in phase I

accumulated thermal energy, in preparation for latter phase (Stanczyk et al. 2011; Krzysztof et al. 2012).

The oxygen gasification process lasted for 26 h in which 387 m^3 syngas were produced. An average carbon consumption rate was 0.336 kg/m^3 carbon and 299 kg coal was gasified. The average gas production rate in the phase I was 14.9 m^3/h.

4.2 Oxygen and steam FCG (phase II)

Following phase I, steam and pure oxygen were injected into the gasifier to increase (H_2 + CO) content in syngas to about 60 %. Experiments during phase II were carried out between about 28 and 117 h of the test. This phase was divided into two stages. One involved pure oxygen and steam gasification for 52.5 h, and the other was oxygen-enriched air and steam gasification for 36.5 h. The pure oxygen supply rate was in the range 4–9 m^3/h. The rate of steam supplied should be adjusted according to the oxygen supply rate for keeping the (H_2 + CO) content in the syngas of about 60 %, in which H_2 accounted for 35 %–45 % and CO for 20 %–30 %. The supply rate of the stream was increased as the hydrogen content decreased, whereas the supply rate of the stream was reduced as CO content decreased to <20 %. The main aim of the experiment in phase II was to establish the optimal volume ratio of steam to oxygen. The optimal range was found to be in the range 1.5:1–2:1 (Yang et al. 2008).

In the gasification process, temperature was usually a decisive factor, because this not only affected the composition of the product gas but was also helpful in determining the progress of the gasification reaction. It is important that a high temperature in the combustion zone be maintained in the actual UCG process by adjusting the flow rate of gasification agents and the H_2O/O_2 ratio (Perkins and Sahajwalla 2005; Prabu and Jayanti 2011). Gasification theory indicates that steam should be

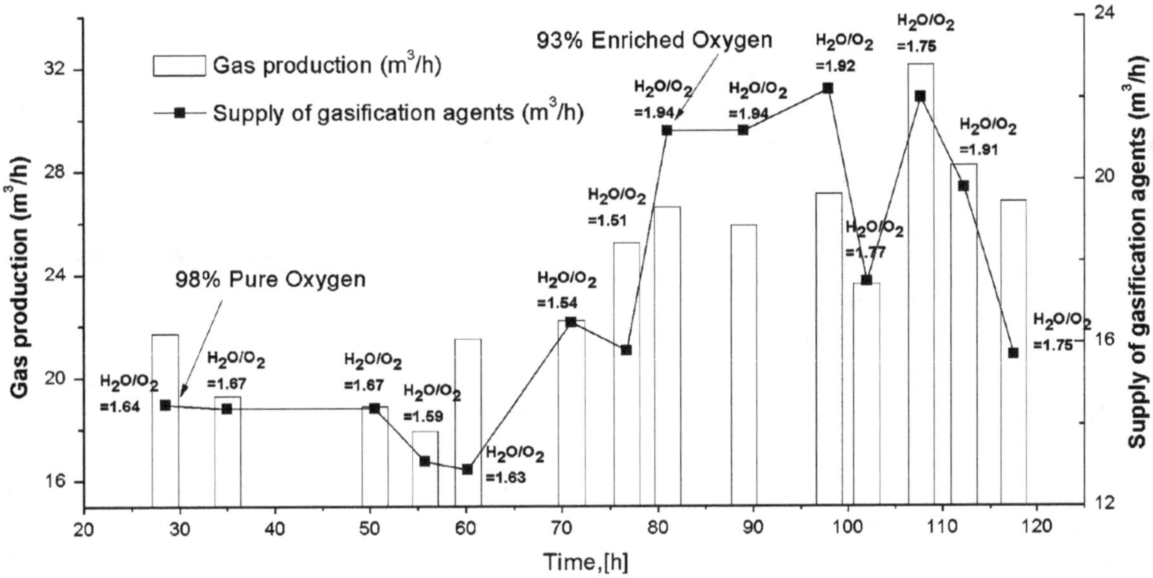

Fig. 6 Rates of gas production as a function of gasification agents supply rates and total supply of gas during phase II

decomposed with a high reaction rate to ensure a hydrogen-rich product gas while the gasification temperature exceeds 1,000 °C (Yang et al. 2007; Hossein et al. 2010). The production rates of CO and H_2 content were determined by two main reactions (Alejandro and Fanor 1998):

$$\text{Steam gasification reaction: } C + H_2O\ (g) \rightarrow CO + H_2 + 131.1\ \text{MJ/kmol}$$

$$\text{Water} - \text{gas} - \text{shift reaction: } CO + H_2O\ (g) \rightarrow CO_2 + H_2 - 41\ \text{MJ/kmol}$$

The key factor in producing hydrogen-rich product gas is the steam gasification reactions, but steam decomposition is a highly endothermic reaction. This led to a significant decrease in the rate of water decomposition and chars gasification when the temperature dropped below 700 °C Yang et al. 2007; Hossein et al. 2010). The water–gas-shift reaction made a significant impact on the composition of the syngas. The ($CO + H_2$) content of the syngas obviously increased at higher temperatures. The melting point of lignite ash was about 1,300 and, for this reason, the supply ratio of oxygen to steam was adjusted to maintain a gasifier temperature between 1,000 and 1,300 °C in the gasifier. The rates of gas production as a function of gasification agents supply rates and composition are presented in Fig. 6.

The average ratio of steam and oxygen in the first stage was about 1.60, but it was about 1.85 in the second stage to keep temperatures in the range between 1,100 and 1,200 °C. During phase II, the temperature in the gasifier was often observed to ensure suitable conditions for water decomposition. The most favorable gasification conditions to produce syngas in the gasifier were reached between 50 and 120 h from the beginning of the test, as shown in Fig. 7, since relatively high levels of thermal energy had been accumulated and the gasification zone was well developed (Krzysztof et al. 2012).

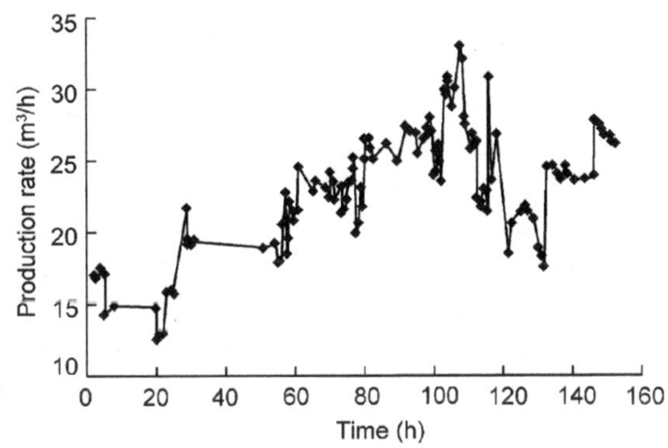

Fig. 7 Changes in gas production rate during the gasification experiment

When the rate of product gas reached 25.1 m^3/h and this status had been maintained for 2 h, the pure oxygen agent was changed into 93 % enriched oxygen, of interest for economic reasons. In the later stage of phase II, the rate of supply of oxygen and steam steadily increased. The supply rate of oxygen increased to maintain high temperature of about 1,000 °C in the UCG gasifier. The supply rate of steam was increased to promote water decomposition into hydrogen-rich combustible gas under high temperature conditions. The average product gas rate steadily increased to 26.9 m^3/h, and the effective content of the syngas is shown in Fig. 8. The results showed that the gasifier worked well, and hydrogen-rich syngas was produced in

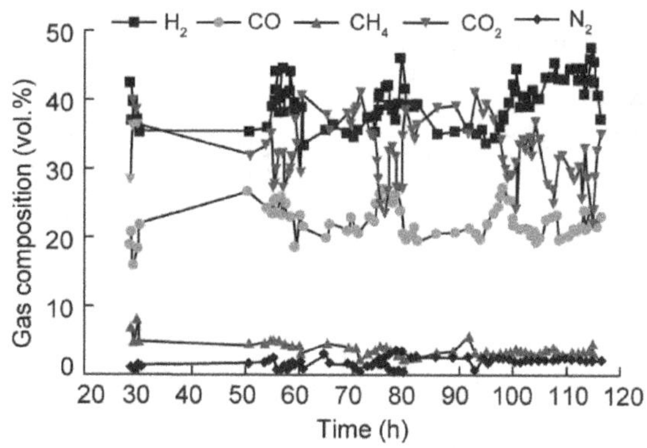

Fig. 8 The composition of product gas during phase II

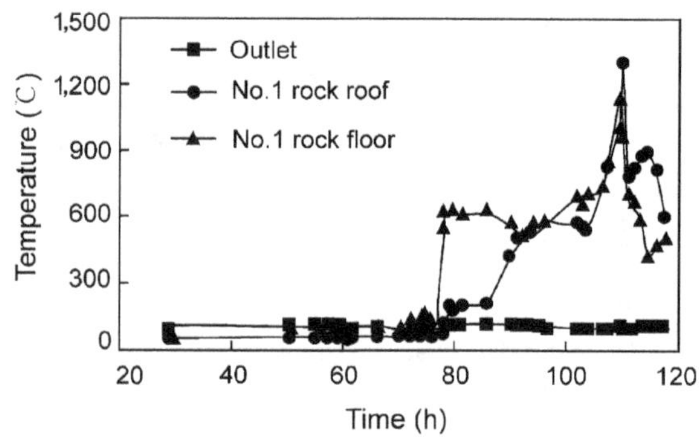

Fig. 9 Change in the temperature of the outlet syngas, strata floor and strata roof in phase II

the later stage of phase II. Changes in gas production rate during the complete gasification experiment are presented in Fig. 7. The temperature profiles in phase II were presented in Fig. 13c, d and e. The high temperature field gradually extended from the injection borehole to the middle of the gasifier. The maximum temperature reached 1,300 °C and the oxidation/gasification reaction zone moved to the center. These experimental results showed that the lignite was gasified to produce a hydrogen-rich syngas through the use of 93 % enriched-oxygen gasification. Thus, the gasification experiment may be carried out in the future with a lower concentration of oxygen in the oxygen-enriched air.

At the beginning of phase II, the strata temperatures were low (Fig. 9) since vertical propagation of the reaction zone took place relatively slowly (Krzysztof et al. 2012). Conduction was a less significant contributor to heat transfer. This is attributed to the low thermal conductivity of the lignite seam and strata (Yang and Liu 2003). In the latter stage of phase II, strata temperatures rapidly increased because the flame front moved to the gasifier center. Nevertheless, the roof temperatures rose more slowly than the floor temperatures in the period 75–91 h since the distance of the roof from the flame front was longer than that of the floor. The heating rate for the roof was lower than that for the floor, brought about by conduction when the coal was combusted and gasified in the middle of the channel. The highest temperature of the roof was at 1,300 °C while that of the floor was 1,135 °C. This phenomenon shows that it is important to maintain heat transport by means of radiation because the peak temperature is very high (Perkins and Sahajwalla 2007), when the top of the gasification cavity of coal seam reached the strata roof. Moreover, the strata roof is located at the top of the combustion flame. The result showed that the reaction zone of oxidation/gasification had gradually moved to the gasifier center.

The oxygen and steam FCG process lasted for 91 h in which 2,170 m^3 syngas were produced, 0.316 kg/m^3 carbon was consumed and 1,580 kg coal was gasified. The average gas production rate in this gasification period was 23.7 m^3/h.

The average hydrogen content obtained in this period was less than 45 %, which demonstrated that appropriate thermal conditions for the steam decomposition as well as for the intensification of the pyrolysis phenomena were not achieved, because the UCG gasifier was limited in the accumulation of thermal energy and the gasified lignite exhibited a low effective carbon content.

4.3 Oxygen and steam RCG (phase III)

The quality of syngas declined in the later stages of phase II because a large cavity was formed along the gasification channel. The oxygen concentration in the surface of the incandescent coal was so low that the coal combustion reaction rate declined. In addition, the reactivity of the coal declined because the residual coal was partially pyrolyzed and was converted into char, which possessed a lower reactivity owing to the larger pore diameter (Alejandro and Fanor 1998; Yang et al. 2008). The high intensity forward gasification process would be difficult to be continued. Then, the RCG approach was proposed to solve the problem in the late stage of UCG process. And a new UCG technique of the reverse combustion gasification combined with forward combustion gasification creatively formed. The objective of reverse combustion gasification is optimized to the UCG process further. At the moment the injection and product boreholes were switched, oxidants were injected into the product borehole and product gas was removed from the injection borehole, but the flame then moved to the product borehole. At the beginning of phase III, the steam was blown into the two boreholes to remove the product gas. The 93 % oxygen was injected from the product borehole to improve the gasifier temperature, and its initial supply rate was 5 m^3/h. After 0.5 h, the temperature of the gasifier center returned to above 1,000 °C, and the steam was fed continuously until the total process was terminated.

The flame front propagated faster with oxygen than with air (Skafa 1960). The global combustion process can be expressed as:

$$C + (1 - a/2)O_2 \rightarrow aCO + (1 - a)CO_2$$

The mass concentration of oxygen and the heat release were crucial factors to maintain RCG. The relationship between the initial mass concentration of oxygen $(\rho O_2)_0$ and heat release Q_h (T_b,...) can be expressed by (Blinderman et al. 2007; Blinderman and Klimenko 2007; Dmitry et al. 2010):

$$Q_h(T_{b,\ldots}) = \overline{\lambda}\left(\frac{\partial T}{\partial x}\right)_{-r} = \sqrt{(\rho_{O_2})_0 \overline{A}\theta\overline{\lambda}_r\varphi(T_b)},$$

where subscript " $-r$" denotes the value just upstream of the reaction zone, $\overline{\lambda}_r$ is the diffusion coefficient and $\varphi(T_b)$ is indicated by:

$$\varphi(T_b) = \int_{-\infty}^{T_b} K(T)\mathrm{d}T = \frac{K_d}{Z}\ln\left(1 + \frac{K_r(T_b)}{K_d}\right)$$

$$Z \equiv \frac{E_a/R}{T_b^2}$$

$$\theta = \frac{2H(\mu_s^-)}{\alpha}$$

$H(\mu_s^-)$ is the enthalpy of combustion reaction. Since coal was incandescent in the later stages of gasification, the coal had been preheated prior to combustion. The flame can readily propagate upstream and a higher temperature in the combustion zone leads to more intense reactions and an increase in the burning rate. Then, a greater mass flow of oxygen is needed to sustain the higher temperature flame front. If the flame temperature is low, the flame cannot move at an appreciable speed upstream since this would induce additional heat losses. The flame at very low temperatures or low flow rates of oxidants would be practically non-existent. As the flow of oxidants increases, the flame temperature rises (Blinderman et al. 2007; Blinderman and Klimenko 2007; Dmitry et al. 2010). To ascertain the appropriate enriched oxygen flow, the oxygen supply rate can be calculated using the above-mentioned equations. Some parameters in the calculations were selected as follows; $A = 1.06e4\ \text{min}^{-1}$, $Ea = 12{,}331.5$ J/mol, $T_b = 1{,}000$ K, $a = 0.4$; and other parameters were derived from the literature (Gort 1995; Hans et al. 2008; Blinderman et al. 2007; Blinderman and Klimenko 2007; Dmitry et al. 2010). According to our calculations, the initial supply rate of enriched oxygen was 5 m^3/h. When the temperature of the gasifier center exceeds 1,000 °C, the oxygen mass flow

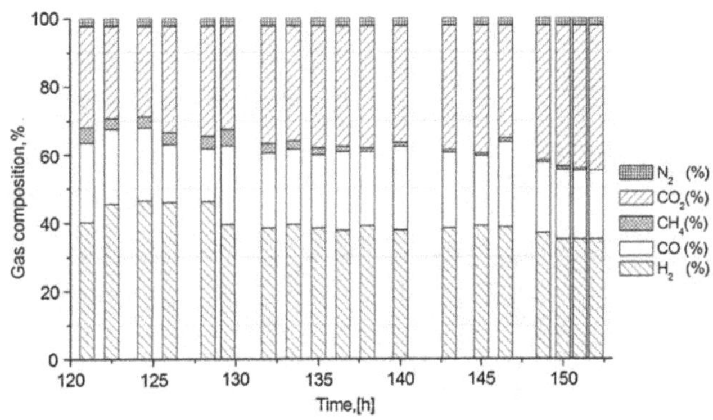

Fig. 10 The composition of product gas in the phase III

gradually increased to sustain coal/char combustion as shown in Fig. 10. The rate of production of syngas increased accordingly. The highest temperature in the UCG gasifier was 1,100 °C as seen in Fig. 13f. Simultaneously, the ratio of oxygen to steam in the supply gas was adjusted to maintain the appropriate temperature in the UCG gasifier for coal gasification. The experimental results showed that the RCG approach performed well and it was feasible as a UCG process.

The ratios of steam and oxygen in the supply gas should be adjusted to produce high-quality syngas. The composition of the syngas, the average gas composition and the heating value obtained in phase III are shown in Fig. 10; Table 2. The average gas production rates as a function of supply rates of the gasification agents and its composition are given in Fig. 11. The average ratio of steam to oxygen was 1.68, which was lower than the value in the latter stages of phase II. These results agreed with those reported by Skafa (1960). In practical UCG operations, a pear-shaped cavity should be formed in the gasification channel by forward combustion, while a predominantly tube-like cavity could be made by the reverse combustion approach (Skafa 1960; Sateesh et al. 2010). This resulted in a much lower consumption of coal for the reverse combustion process. The concentration of CH_4 gradually decreased from about 3.9 % at the beginning of the RCG to zero at the end of phase III, which demonstrates that the gasification process would finish and that H_2 or H_2O would not react with coal/char to produce CH_4 because of the low temperatures involved (Gregg and Edgar 1978; Kreinin and Shifrin 1993). The gasification time of UCG was extended by more than 30 % using the RCG approach and the gasification rate of coal seams can be improved.

At the end of phase II, the temperature of the strata roof is lower than 690 °C, but it rapidly increased in the reverse combustion technique to reach higher temperatures as depicted in Fig. 12. The highest temperature reached was 1,300 °C. Then reverse combustion gasification takes

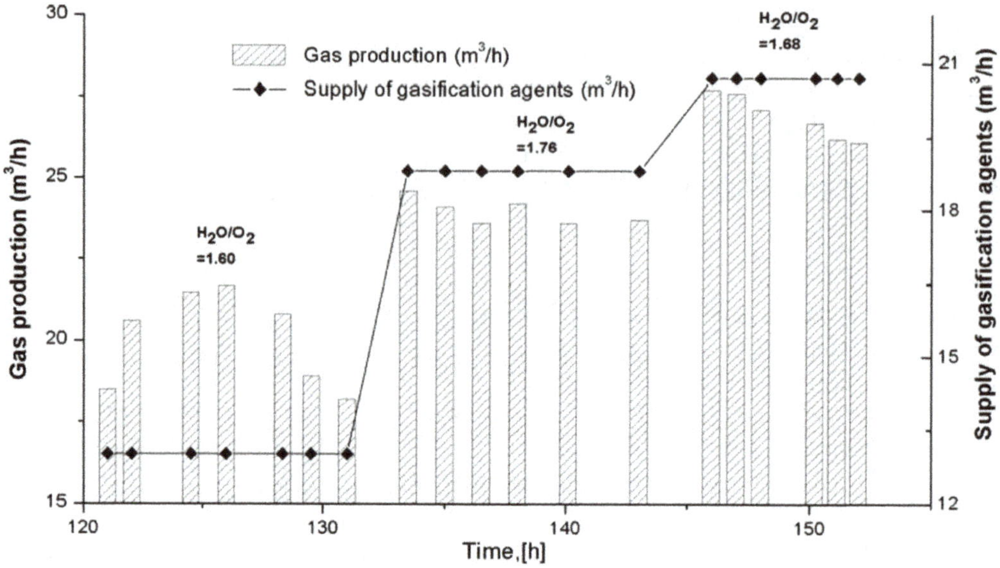

Fig. 11 Rates of gas production as a function of gasification agents rates and composition in the phase III

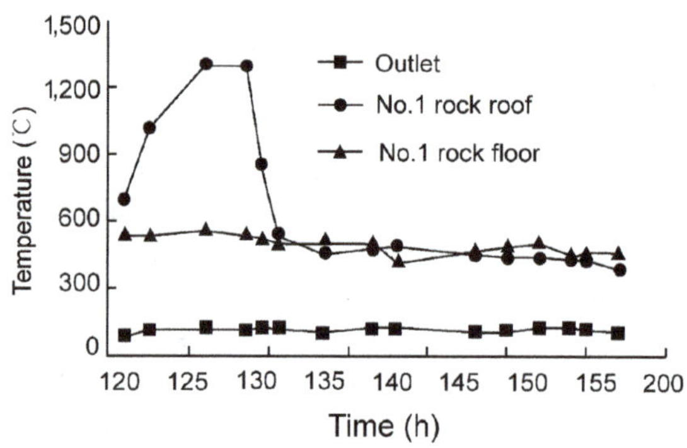

Fig. 12 Changes in temperatures of outlet syngas, strata floor and roof in phase III

place. After this state had been maintained for a few hours, the temperature in the strata roof and in the strata floor rapidly reduced. This phenomenon indicated that the reverse combustion flame rapidly moved from the gasifier center and proved that the reverse combustion rate is much higher than that of forward combustion (Blinderman et al. 2007). The temperature of outlet was stable at 100 °C, which showed that the structure of the UCG gasifier could be maintained.

The oxygen and steam RCG process lasted for 33 h, produced 770 m^3 syngas, consumed an average of 0.319 kg/m^3 carbon and gasified 565 kg coal. The average gas production rate in this gasification period was 23.3 m^3/h.

4.4 Nephograms of horizontal UCG temperature profiles

The 2D nephogram of temperature profiles was used to investigate the temperature changes of the UCG process.

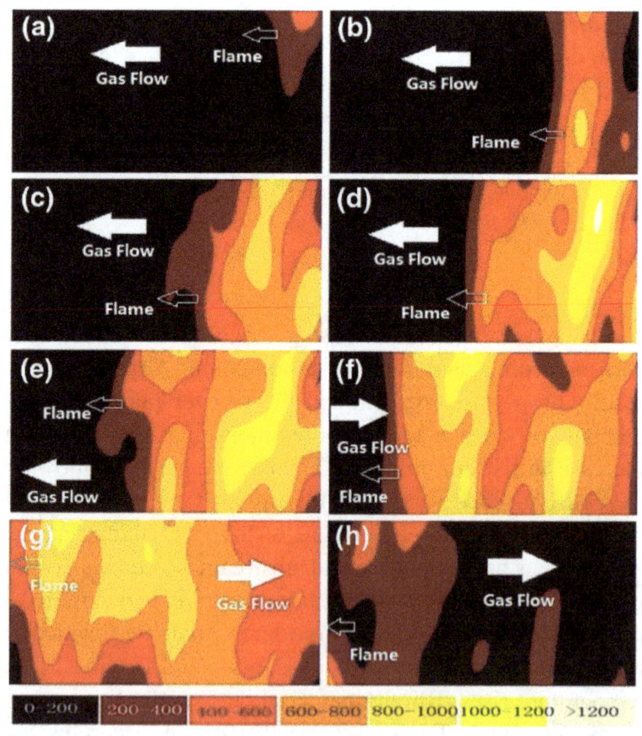

Fig. 13 Nephogram of temperature profiles from the commencement of the process. **a** 5 h **b** 20 h **c** 40 h **d** 70 h **e** 90 h **f** 120 h **g** 135 h **h** 140 h

This method used Tecplot 360 CFD Visualization software to solve the conservation equations according to real temperature data from the 85 thermocouples in the coal seam, and horizontal temperature profile nephograms of the coal seam as a function of time are presented in Fig. 13. The computational domain on the level of the gasification channel was meshed using approximately 720 quadrilateral control volumes. Grid refinement studies showed that the solutions were grid independent (Perkins and Sahajwalla 2007).

The gasification flame propagated in the same direction during the whole experiments though the gasification gas agents flew in different orientation seen from Fig. 13. The results proved the experiment of RCG technology was successfully carried out. It was proved the feasibility of using RCG technique to gasify lignite with hydrogen-rich syngas whether FCG or RCG was applied to gasify lignite in UCG gasifier, the highest temperature was observed in those areas of the coal seam that were in close proximity to the oxidation/steam gasification center, i.e. at the flame head. The zone of oxidation/gasification reflected not only the gasification channel but also the level of the universal direction of coal seam due to the high permeability of lignite. Finally, the whole gasification flame surface gradually proceeded along the gasification channel and moved the product borehole. Since there was a 17° inclination in the coal seam of the experimental system, the mass and heat should mainly transfer to the top left corner of inclination. Thus, the coal in this zone was essentially gasified.

5 Conclusions

(1) The model experiments proved the feasibility of producing hydrogen-rich and heating value syngas by applying the FCG and RCG techniques in underground coal gasification. In particular, the reverse combustion approach was used to successfully gasify lignite. The gasification time of UCG was extended by more than 30 % using the RCG approach and the gasification rate of coal seams can be improved.

(2) An oxygen and steam continuous gasification approach, in which the oxygen and steam are fed simultaneously, can steadily maintain the composition of the syngas produced during the complete UCG process. The average hydrogen concentration was 39.6 % with a maximum of 47.4 %, and the average CO concentration was 22.8 % with a maximum of 27.4 % during the FCG phase. The average hydrogen concentration was 39.7 % with a maximum of 46.4 %, and the average CO concentration was 21.7 % with a maximum of 25.9 % during the RCG phase. Though the main role of the oxygen was to heat the coal seam, the average H_2 and CO concentrations became 35.5 % and 22.5 %, respectively. The optimal steam: oxygen volume ratio was the critical parameter for maintaining the high temperature in the gasifier and preventing lignite ash melting. The optimal range was found to be 1.5:1–2:1 by analysis of experimental data. The average ratio value of steam and oxygen in FCG was slightly higher than the required for RCG.

(3) The theory of reverse combustion was briefly introduced in this work. According to the calculation, the supply rate of enriched oxygen was ascertained at 5 m^3/h, which was in agreement with experimental observations.

(4) From the gas composition measurements and the strata temperature profiles, it was found that an appropriately stable operation of the gasifier was achieved after about 50 h from the start of the process. The abundant heat was largely accumulated in the gasifier. It led to a favorable composition of the syngas produced in the UCG process and resulted in a heating value of the gas of 9.20 MJ/m^3.

(5) To maintain appropriate conditions for obtaining hydrogen-rich syngas, the temperatures of the strata need to be monitored continuously in the various gasification phases. Temperature changes in the strata floor and the roof reflect the operating status of a simulated UCG gasifier. The reaction zones of oxidation/gasification could be deduced by analysis of the strata temperature and especially that of the roof.

Acknowledgments The work has been supported by the National High Technology Research and Development Program 863 of China (2011AA050106) and the Fundamental Research Funds for the Central Universities (2012YH01).

References

Alejandro M, Fanor M (1998) Reactivity of coal gasification with steam and CO_2. Fuel 77(15):1831–1839

Blinderman M, Klimenko A (2007) Theory of reverse combustion linking. Combust Flame 150:232–245

Blinderman M, Saulov D, Klimenko A (2007) Forward and reverse combustion linking in underground coal gasification. Energy 33:446–454

Dmitry N, Ovid A, Klimenko A (2010) Flame propagation in a gasification channel. Energy 35:1264–1273

Evgeny S, Arvind V (2009) Underground coal gasification: a brief review of current status. Ind Eng Chem Res 48(17):7865–7875

Friedmann SJ, Ravi U, Fung-Ming K (2009) Prospects for underground coal gasification in carbon-constrained world. Energy Procedia 2(274): 4551–4557

Gort R, 1995. On the propagation of a reaction front in a packed bed. Ph.D. thesis, Twente University, Enschede, The Netherlands

Gregg DW, Edgar TF (1978) Underground coal gasification. AIChE J 24:753–781

Hans V, Jeroen V, Rob B, Philip D (2008) Reverse combustion: kinetically controlled and mass transfer controlled conversion front structures. Combust Flame 153:417–433

Hossein N, Mohammad K, Zhangxin C, Jalal A (2010) Simulation study of underground coal gasification in Alberta reservoirs: geological Structure and process modeling. Energy Fuels 24:3540–3550

Hurloff G (1983) Underground coal gasification cavity growth model. J Energy 7:410–415

Khadse A, Qayyumi M, Mahajani S, Aghalayam P (2007) Underground coal gasification: a new clean coal utilization technique for India. Energy 32:2061–2071

Kreinin E, Shifrin E (1993) Mathematical model of coal combustion and gasification in a passage of an underground gas generator. Combust Explos Shock Waves 29:148–154

Krzysztof S, Krzysztof K, Marian W, Jerzy S, Adam S, Jan R, Andrzej K (2012) Experimental simulation of hard coal underground gasification for hydrogen production. Fuel 91(1):40–50

Liu HT, Chen F, Pan X, Yao K, Liu SQ (2011) Method of oxygen-enriched two-stage underground coal gasification. Min Sci Techno (China) 21:191–196

Park K, Edgar T (1987) Modeling of early cavity growth for underground coal gasification. Ind Eng Chem Res 26:237–246

Perkins G, Sahajwalla V (2005) A mathematical model for the chemical reaction of a semi-infinite block of coal in underground coal gasification. Energy Fuels 19(4):1679–1692

Perkins G, Sahajwalla V (2007) Modelling of heat and mass transport phenomena and chemical reaction in underground coal gasification. Chem Eng Res Des 85:243–329

Prabu V, Jayanti S (2011) Simulation of cavity formation in underground coal gasification using bore hole combustion experiments. Energy 36:5854–5864

Sateesh D, Ramesh N, Sanjay M, Anuradda G (2010) Laboratory studies on combustion cavity growth in lignite coal blocks in the context of underground coal gasification. Energy 35:2374–2386

Shugin L, Yuanyuaon W, Ke Z, Ning Y (2009) Enhanced-hydrogen gas production through underground gasification of lignite. Min Sci Tech 19:384–394

Singh S, Weil S, Babu S (1980) Thermodynamic analysis of coal gasification processes. Energy 5:905–914

Skafa P (1960) Underground coal gasification. Gosgortechizdat, Moscow

Stanczyk K, Howaniec N, Smolinski A, Swiadrowski J, Kapusta K, Wiatowski M, Grabowski J, Rogut J (2011) Gasification of lignite and hard coal with air and oxygen enriched air in a pilot scale ex situ reactor for underground gasification. Fuel 90(12):1953–1962

Yang L (2004) Study on the model experiment and numerical simulation for underground coal gasification. Fuel 83:573–584

Yang L, Liu S (2003) Numerical simulation on heat and mass transfer in the process of underground coal gasification. Numer Heat Transf Part A 44:537–557

Yang L, Liang J, Yu L (2003) Clean coal technology–study on the pilot project experiment of underground coal gasification. Energy 28:1445–1460

Yang L, Liu S, Yu L, Liang J (2007) Experimental study of shaftless underground gasification in thin high angle coal seams. Energy Fuels 21:2390–2397

Yang L, Zhang X, Liu S, Yu L, Zhang W (2008) Field test of large-scale hydrogen manufacturing from underground coal gasification (UCG). Int J Hydrog Energy 33:1275–1285

19

Electrochemical performance and stability of Sr-doped $LaMnO_3$-infiltrated yttria stabilized zirconia oxygen electrode for reversible solid oxide fuel cells

Hui Fan · Minfang Han

Abstract Porous Sr-doped lanthanum manganite–yttria stabilized zirconia (LSM–YSZ) oxygen electrode is prepared by an infiltration process for a reversible solid oxide fuel cell (RSOFC). X-ray diffraction and SEM analysis display that perovskite phase LSM submicro particles are evenly distributed in the porous YSZ matrix. Polarization curves and electrochemical impedance spectra are conducted for the RSOFC at 800 and 850 °C under both SOFC and SOEC modes. At 850 °C, the single cell has the maximum power density of ~726 mW/cm^2 under SOFC mode, and electrolysis voltage of 1.35 V at 1 A/cm^2 under SOEC mode. Fuel cell/water electrolysis cycle shows the cell has good performance stability during 6 cycles, which exhibits the LSM–YSZ oxygen electrode has high electrochemical performance and good stability. The results suggest that network-like LSM–YSZ electrode made by infiltration process could be a promising oxygen electrode for high temperature RSOFCs.

Keywords Reversible solid oxide fuel cell · Solid oxide electrolysis cell · Infiltration · Strontium-doped lanthanum manganite

1 Introduction

A reversible solid oxide fuel cell (RSOFC) is an electrochemical device which can be operated at high temperature under both power generation mode and electrolysis mode (Gopalan et al. 2006; Tao et al. 2009; Jung et al. 2012). In power generation mode, the RSOFCs produce electricity directly from electrochemically oxidizing fuel. While, in electrolysis mode, power is applied to the cell and the RSOFC acts as a solid oxide electrolysis cell (SOEC). RSOFCs can be integrated with renewable production of electricity and hydrogen when power generation and steam electrolysis are coupled in a system (Chen et al. 2010; Rao et al. 2012). With the RSOFC, fuel such as hydrogen can be used to produce electricity by an SOFC, and the electricity can then be used to produce hydrogen from water electrolysis.

Studies have shown that the state-of-the-art SOFC materials could be directly employed in an SOEC, with Ni/YSZ (8 mol% Y_2O_3 stabilized ZrO_2) cermet hydrogen electrode, YSZ oxygen-ion conducting electrolyte, and perovskite LSM (Sr-doped $LaMnO_3$) and/or LSM–YSZ oxygen electrode (Kim-Lohsoontorn et al. 2010; Chen et al. 2012a). Delamination at the oxygen electrode/electrolyte interface was reported to be one of the most important problems causing the performance degradation and cell failure for an SOEC (Keane et al. 2012; Kim et al. 2013). Virkar (2010) reported that the high oxygen pressure developed inside the electrolyte near the air electrode could lead to the delamination of the oxygen electrode. Chen et al. (2010) systematically studied the polarization behavior and delamination mechanism of LSM oxygen electrode under SOEC operation conditions. They characterized the oxygen electrode and electrolyte interface by SEM and the results indicated that the delamination of LSM oxygen electrode was caused by the disintegration of

H. Fan · M. Han (✉)
Union Research Center of Fuel Cell, School of Chemical & Environment Engineering, China University of Mining & Technology (Beijing), Beijing 100083, China
e-mail: hanminfang@sina.com

the LSM particles at the oxygen electrode and electrolyte interfaces (Chen and Jiang 2011). Knibbe et al. (2009) studied the degradation behavior of SOEC under high current densities and suggested that the degradation was ascribed to the nucleation and growth of oxygen clusters in the YSZ grain boundaries near the oxygen electrode. Keane et al. (2012) fabricated symmetric air/LSM//YSZ//LSM/air cells and measured electrical performance degradation with time at various applied voltage ranging from 0 to 0.8 V with respect to OCV. After test, a developed weak oxygen electrode–electrolyte interface resulting from the delamination between the oxygen electrode and electrolyte was observed. Microstructural analysis of the oxygen electrode–electrolyte interface indicated the formation of lanthanum zirconate and morphological changes at the interface.

In studies on SOFC, by depositing active ionic/electronic-conductive nano-particles on porous electrode surface, the number of oxygen reduction reaction sites could be largely increased. It thereby enhanced the active triple phase boundary (TPB) zone, lowered the area specific resistance (ASR), and improved the electrochemical performance of SOFCs (Jiang 2012; Fu et al. 2012; Kiebach et al. 2013). LSM particles infiltrated into pre-sintered porous YSZ network could create a high density of active sites for the oxygen reduction reaction (Sholklapper et al. 2006). The Ni–YSZ/YSZ/LSM–YSZ cell fabricated by infiltration process achieved an open circuit voltage at ~1.1 V and maximum power density at ~0.27 W/cm^2 at 923 K. In addition, the infiltrated cell had a polarization resistance of 2.9 Ω·cm^2, which was strikingly smaller than 110 Ω·cm^2 of the cell prepared by conventional method (Sholklapper et al. 2006). Yang et al. (2010) prepared porous LSM–YSZ oxygen electrode by an infiltration method for high-temperature water electrolysis. After electrolyzing at 800 °C and a current density of 0.33 A/cm^2 for 50 h with 50 vol% AH (absolute humidity, the vol% of humidity in the total gas volume), there was no deterioration of the electrochemical performance for the LSM-infiltrated cell and as well as the delamination of oxygen electrode from electrolyte. ASR of the LSM-infiltrated cell was substantially lower than that of the cell with LSM–YSZ oxygen electrode made by a screen printing method. Chen et al. (2012b) also indicated that LSM-infiltrated LSM–YSZ composite oxygen electrode showed high electrocatalytic activity and good stability under SOEC operation. By infiltrating porous YSZ matrix with LSM nitrate solution, the electrode heat-treated at 900 °C exhibited a low electrode polarization resistance of 0.21 Ω·cm^2 at 800 °C.

In this paper, the electrochemical performance of a H_2 electrode-supported Ni–YSZ/YSZ/LSM–YSZ RSOFC with LSM infiltrated oxygen electrode was investigated under both fuel cell and electrolysis modes. The fuel cell/water electrolysis cycle performance of the cell at 0.4 A/cm^2 was carried out as well for evaluating the cell stability.

2 Experimental

RSOFCs, which consisted of Ni–YSZ hydrogen electrode supports, thin YSZ electrolytes, and infiltrated LSM–YSZ oxygen electrodes, were fabricated by tape casting and infiltration technique. For fabricating the Ni–YSZ electrode, powders of NiO, YSZ (8 mol% yttria stabilized zirconia, with a median particle size of 0.114 μm) and graphite (Furunda Zirconium Material Co., Ltd., China) pore former were mixed and ball-milled in a weight ratio of 50:50:10 with appropriate amounts of ethanol-butanone solvent, caster oil dispersant, dibutyl phthalate (DBP) plasticizer, and polyvinyl butyral (PVB) binder. A two-stage milling process (Liu et al. 2010) was adopted. Firstly, NiO and YSZ powders were dispersed in mixed solvent and dispersant via ball-milling for 15–20 h. Secondly, a homogenous mixture including remaining solvent, plasticizer, and binder formed by thermal dissolution was added, and the resultant slurry was then ball milled for another 24 h before adding the last constituent of graphite. For YSZ electrolytes, the same way of tape casting as in the case of NiO–YSZ substrates was used excepted for no addition of NiO and graphite powders. Porous YSZ layers were prepared with higher ratio of pore formers. Through previous study on different weight ratios of graphite to YSZ, 50 wt% was thought as an optimal proportion. A tape casting machine (DR-150, made in Japan) was used for separately casting the NiO–YSZ layer, the electrolyte, and the porous YSZ electrode. These tapes were dried in air at room temperature for 12 h. One sheet of NiO–YSZ support (~300 μm), YSZ electrolyte layer (20–30 μm), and porous YSZ layer (~70 μm) were stacked together under a vacuum condition and laminated at 20 MPa for 10 min using a thermal isostatic press (30T Shanxi, China) to form a NiO–YSZ supported tri-layer structure. The resulting tri-layer tape was punched to discs, and then co-sintered at 1,300 °C for 10 h in order to densify the electrolyte layer, during which graphite was burned out, leaving a well-formed porous YSZ network. The microstructural characteristics of the fired tri-layer tapes were performed using a scanning electron microscope, and the estimated porosity of the porous YSZ layer was 40 %–50 %.

A solution of LSM ($(La_{0.8}Sr_{0.2})_{0.95}MnO_3$) precursor was prepared by dissolving stoichiometric amounts of $La(NO_3)_3 \cdot 6H_2O$, $Sr(NO_3)_2$, $Mn(NO_3)_2$ (50 % w/w aqueous solution) into deionized water (Chen et al. 2012b). The solution was infiltrated into pre-treated porous YSZ layer and fired at 450 °C for 1 h to decompose the nitrate. In

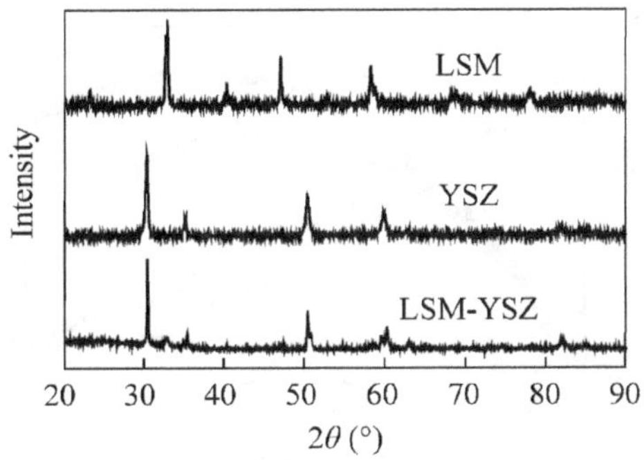

Fig. 1 X-ray diffraction patterns of LSM–YSZ oxygen electrode prepared by LSM infiltration method after sintering at 900 °C

order to ensure sufficient infiltration of LSM into the porous layer, repeated infiltration was used, followed by firing at 450 °C for 1 h after each infiltration. The infiltrated electrode was eventually fired at 900 °C for 2 h to reach a weight fraction of ~40 %. Ag paste (Beijing, China) was used as the current collector on electrode surfaces.

The RSOFCs were mounted on an alumina tube (Yu et al. 2005), sealed using ceramic paste as sealant (Aramco-552, USA) to separate the gas environment of the two electrodes, and tested in a tube furnace at temperatures from 800 to 850 °C. The area of electrolyte and hydrogen electrode was about 1.5 cm^2, the external area of oxygen electrode was 0.2 cm^2. For SOFC testing, 80 sccm flow rate of humidified hydrogen (3 vol% H_2O) was fed to the hydrogen electrode, while the oxygen electrode was exposed to ambient air. For SOEC operation, 50 % H_2O, 25 % H_2 and 25 % Ar were introduced into the Ni-based hydrogen electrode. Similarly, the oxygen electrode was left open to air. Polarization curves and electrochemical impedance spectra (EIS) under open circuit were conducted using an IM6 Electrochemical Workstation (ZAHNER, Germany). For EIS test, the frequency range was from 100 mHz to 100 kHz, and the AC amplitude was 20 mV. Microstructure of the LSM-infiltrated RSOFC was examined using scanning electron microscopy (SEM, JEOL, JSM 6700F). In addition, the phase formation of LSM was examined using a high power X-ray diffractometer (XRD, PANalytical X′Pert PRO, Netherlands) with CuKα radiation.

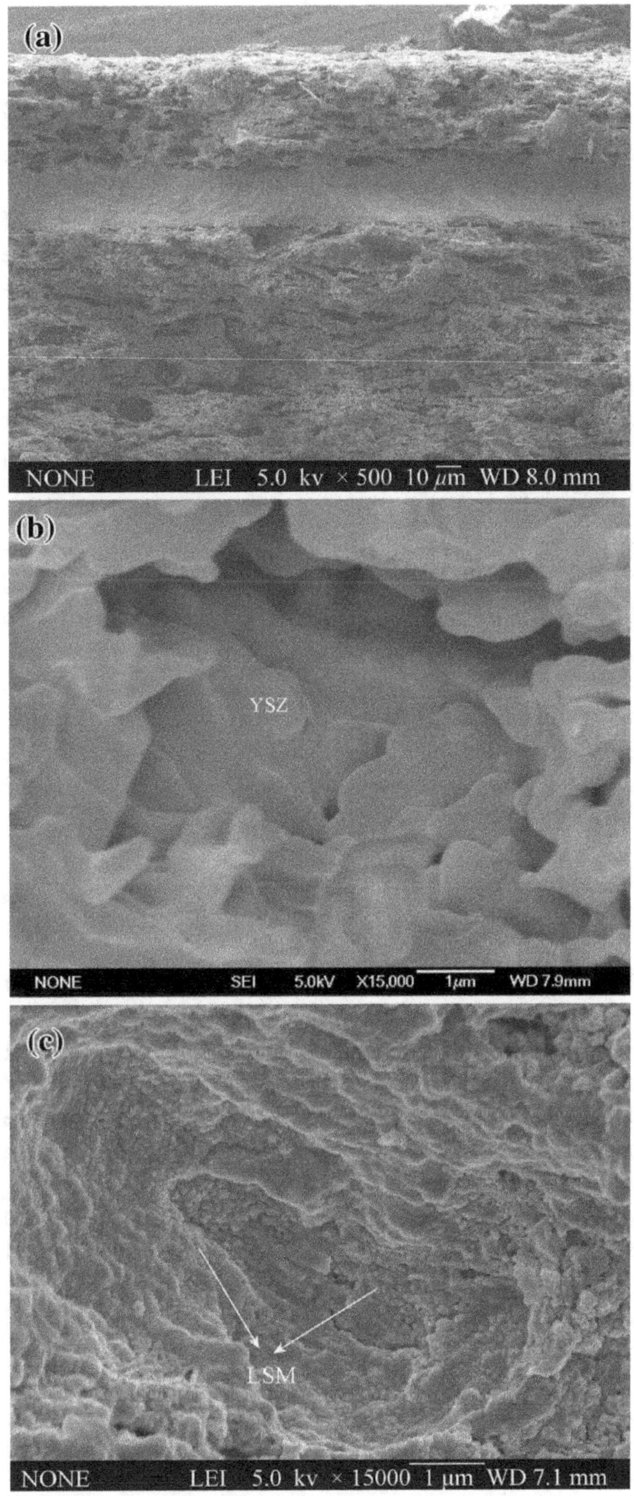

Fig. 2 Typical SEM images of cross-sectional view of **a** LSM-infiltrated NiO–YSZ/YSZ/LSM–YSZ single cell, **b** porous YSZ scaffold before infiltration and **c** LSM infiltrated LSM–YSZ oxygen electrode

3 Results and discussion

3.1 Physico-chemical characterization of RSOFCs

Figure 1 displays the XRD patterns of the LSM–YSZ oxygen electrode fabricated by infiltrating and sintered at 900 °C. In Fig. 1, the XRD patterns of pure perovskite LSM and YSZ are also shown as a reference. It is indicated that the infiltrated LSM–YSZ oxygen electrode has a mixture of perovskite LSM and YSZ phases, without observable impurity phase. The LSM–YSZ oxygen electrode is therefore of high phase-purity.

A cross-sectional SEM of NiO–YSZ/YSZ/LSM–YSZ RSOFC is given in Fig. 2a. A dense YSZ electrolyte and porous electrode layers were observed. It is indicated that the co-sintering at 1,300 °C for 10 h forms dense YSZ electrolyte and porous YSZ electrode backbone. The thickness of cell components is ~300, ~20, and ~60 μm for the hydrogen electrode, the electrolyte, and the oxygen electrode, respectively. Both the highly porous hydrogen and oxygen electrodes are in good contact with the dense electrolyte. Figure 2b and c are high-magnification SEM pictures focusing on the oxygen electrode before and after infiltration. In Fig. 2b, small grains of YSZ layer are formed at sintering temperature of 1,300 °C, and porosity for the porous YSZ layer is estimated at 40 %–50 % (Liu et al. 2010). Figure 2c reveals the microstructural feature of the LSM-infiltrated oxygen electrode. After infiltration and sintering, the submicro-sized LSM particles are uniformly and sufficiently covered on the pore walls of the YSZ network, forming a fairly densely packed microstructure to allow for sufficient electronic connectivity (Jiang et al. 2010). The small particle size of LSM is favorable for the catalytic activity of electrochemical reaction as well as the formation of more LSM/YSZ/gas TPB sites.

3.2 Initial electrochemical characterization under SOFC mode

Figure 3a shows the voltage and power density vs current density (V–I curve and P–I curve) at 850 and 800 °C for the LSM-infiltrated RSOFC. The open circuit voltage of the LSM-infiltrated cell is 1.065 V at 850 °C and 1.076 V at 800 °C, implying the YSZ electrolyte layer is fully dense after sintering at 1,300 °C and there is no leakage between the cell and the zirconia tube. The maximum power densities of the cell are ~726 and ~569 mW/cm^2 at 850 and 800 °C, respectively. In Fig. 3b, the ohmic impedance of the cell, determined from the high-frequency intercept on the real axis, was ~0.24 Ω·cm^2 at 850 °C and ~0.26 Ω·cm^2 at 800 °C. In addition, the polarization impedance of the cell, determined from the differences between high- and low-frequency intercepts on the impedance spectra, was 0.21 and 0.33 Ω·cm^2 at 850 and 800 °C, respectively. The high performance shown in infiltrated electrode is most likely due to the unique nanostructure fabricated at relatively low temperature (Sholklapper et al. 2006; Liu et al. 2010).

3.3 Electrochemical performance under SOEC mode

Figure 4a shows a high water electrolysis performance under SOEC mode at 850 and 800 °C for the LSM-infiltrated RSOFC, the input stream of 50 % H_2O, 25 % H_2 and 25 % Ar to hydrogen electrode and the air to oxygen electrode, respectively. The flow rate of inlet mixture gas was set as 150 sccm (standard cubic centimeters per minute). The electrochemical characteristic of LSM-infiltrated cell was recorded as a function of the operation temperature. The OCV of the cell was 0.906 V at 850 °C and 0.923 V at 800 °C. Both were near the theoretical OCV value calculated from Nernst formula (Liang et al. 2009; Laguna-Bercero et al. 2011). At 850 °C, the cell voltage for water electrolysis mode is only 1.35 V at 1 A/cm^2, and at 800 °C, the voltage is 1.47 V at 1 A/cm^2. At 850 and 800 °C, the electrochemical impedance spectra of LSM-infiltrated RSOFC under SOEC mode are measured at OCV, as shown in Fig. 4b. It can be seen that the electrode polarization losses decrease obviously with the increased operating temperature from 0.266 Ω·cm^2 at 800 °C to 0.158 Ω·cm^2 at 850 °C. The ohmic resistance shows no obvious change, from 0.272 Ω·cm^2 at 850 °C to 0.273 Ω·cm^2 at 800 °C. Yang et al. (2010) have reported that LSM-infiltrated cell has ohmic and polarization impedance smaller than those of the cell with LSM–YSZ electrode prepared by a mechanically mixing process and firing. The deposition of submicro-sized LSM on the porous YSZ network will effectively increase the number of the TPB sites for oxygen ion oxidation reaction. The LSM-infiltrated

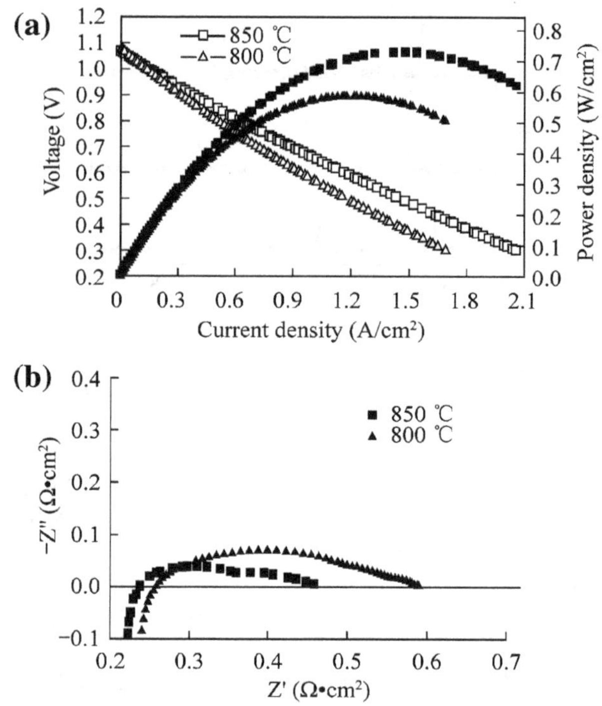

Fig. 3 **a** Voltage and power density vs current density, and **b** impedance spectrum at open circuit at 850 and 800 °C for LSM-infiltrated Ni–YSZ/YSZ/LSM–YSZ cell under SOFC mode operated with a 97 % H_2 and 3 % H_2O mixture at 80 sccm in the hydrogen electrode and ambient air in the oxygen electrode

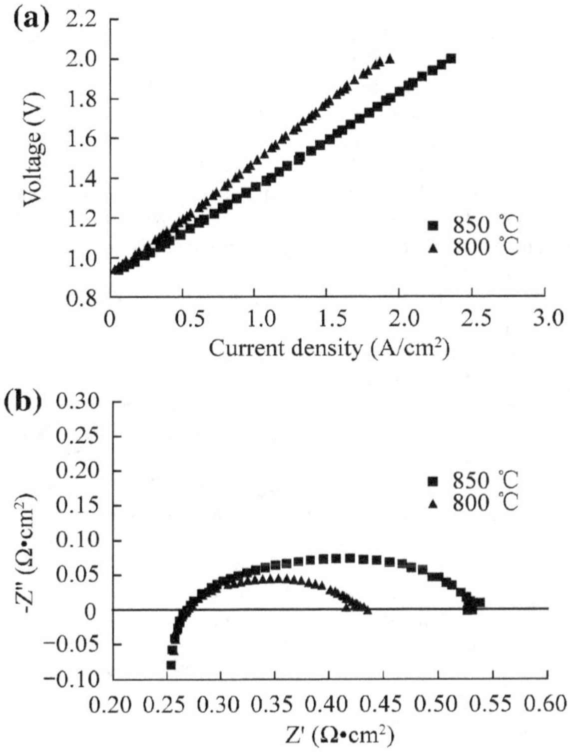

Fig. 4 **a** Voltage vs current density, and **b** impedance spectrum at open circuit at 850 and 800 °C for LSM-infiltrated Ni–YSZ/YSZ/LSM–YSZ cell under SOEC mode operated with a 50 % H_2O, 25 % H_2 and 25 % Ar gas mixture at 150 sccm in the hydrogen electrode and ambient air in the oxygen electrode

electrode consequently has much lower polarization loss and higher performance.

3.4 Fuel cell/water electrolysis cycle

Charge/discharge cycle performance (1 h of water electrolysis and 1 h of fuel cell mode for each) of the LSM-infiltrated RSOFC at 800 °C is shown in Fig. 5. The cell has a slight performance loss during 6 cycles, with the discharge voltage decreased from 0.827 to 0.806 V and the charge voltage increased from 1.122 to 1.189 V at current density of 400 mA/cm^2, respectively. The larger voltage variation on charge than that on discharge is likely resulted from larger diffusion resistance of water molecules than that of hydrogen gas molecules for inlet gas mixture. However, the cell shows a small degradation rate of about 5 % or below after charge/discharge cycles. Thus, the LSM-infiltrated oxygen electrode can substantially avoid performance degradation and delamination from electrolyte, with good performance stability (Chen et al. 2012a).

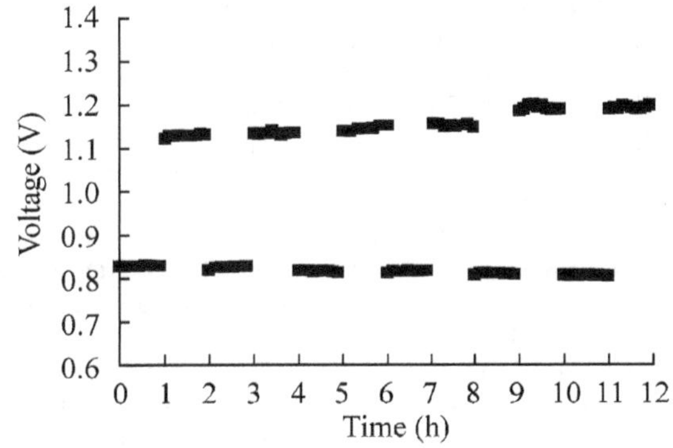

Fig. 5 Fuel cell/water electrolysis cycle of LSM-infiltrated Ni–YSZ/YSZ/LSM–YSZ cell at 800 °C and 400 mA/cm^2

4 Conclusions

In this study, porous LSM–YSZ oxygen electrode was prepared by infiltration process to fabricate Ni–YSZ/YSZ/LSM–YSZ RSOFC. X-ray diffraction and SEM analysis showed that perovskite phase LSM submicron particles were evenly distributed in the porous YSZ matrix. The electrochemical performance of the RSOFC was investigated in both fuel cell and water electrolysis modes at 800 and 850 °C. By infiltrating submicron-sized LSM into porous YSZ, the RSOFC cell performance has been significantly improved (Sholklapper et al. 2006). Under SOFC mode, the cell has a maximum power density of ~726 mW/cm^2, ohmic impedance of ~0.24 Ω·cm^2 and polarization impedance of 0.21 Ω·cm^2 at 850 °C. Under SOEC mode, the electrolysis voltage is 1.35 V at current density of 1 A/cm^2. Ohmic resistance is 0.272 Ω·cm^2 and polarization resistance is 0.158 Ω·cm^2 at 850 °C. Fuel cell/water electrolysis cycle performance shows that the cell has slight performance degradation during 6 cycles. Thereby, the LSM-infiltrated oxygen electrode has high electrochemical performance and good stability. It is concluded that the LSM–YSZ electrode made by infiltration process could be a promising oxygen electrode for high temperature RSOFC.

Acknowledgment This project was sponsored by financial supports from the Major State Basic Research Development Program of China (973 Program, No. 2012CB215406).

References

Chen KF, Jiang SP (2011) Failure mechanism of (La, Sr)MnO_3 oxygen electrodes of solid oxide electrolysis cells. Int J Hydrogen Energy 36:10541–10549

Chen GB, Zhang HM, Zhong HX, Ma HP (2010) Gas diffusion layer with titanium carbide for a unitized regenerative fuel cell. Electrochim Acta 55:8801–8807

Chen KF, Ai N, Jiang SP (2012a) Performance and stability of (La, Sr)MnO_3–Y_2O_3–ZrO_2 composite oxygen electrodes under solid oxide electrolysis cell operation conditions. Int J Hydrogen Energy 37:10517–10525

Chen KF, Ai N, Jiang SP (2012b) Reasons for the high stability of nano-structured (La, Sr)MnO_3 infiltrated Y_2O_3–ZrO_2 composite oxygen electrodes of solid oxide electrolysis cells. Electrochem Commun 19:119–122

Fu YP, Ouyang J, Li CH, Hu SH (2012) Characterization of nanosized $Ce_{0.8}Sm_{0.2}O_{1.9}$-infiltrated $Sm_{0.5}Sr_{0.5}Co_{0.8}Cu_{0.2}O_{3-\delta}$ cathodes for solid oxide fuel cells. Int J Hydrogen Energy 37:19027–19035

Gopalan S, Ye GS, Pal UB (2006) Regenerative, coal-based solid oxide fuel cell-electrolyzers. J Power Sources 162:74–80

Jiang SP (2012) Nanoscale and nano structured electrodes of solid oxide fuel cells by infiltration—advances and challenges. Int J Hydrogen Energy 37:449–470

Jiang ZY, Lei ZW, Ding B, Xia CR, Zhao F, Chen FL (2010) Electrochemical characteristics of solid oxide fuel cell cathodes prepared by infiltrating (La, Sr)MnO_3 nanoparticles into yttria-stabilized bismuth oxide backbones. Int J Hydrogen Energy 35:8322–8330

Jung GB, Chen JY, Lin CY, Sun SY (2012) Fabrication of hydrogen electrode supported cell for utilized regenerative solid oxide fuel cell application. Int J Hydrogen Energy 37:15801–15807

Keane M, Mahopatra MK, Verma A, Singh P (2012) LSM–YSZ interactions and anode delamination in solid oxide electrolysis cells. Int J Hydrogen Energy 37:16776–16785

Kiebach R, Knöfel C, Bozza F, Klemensø T, Chatzichristodoulou C (2013) Infiltration of ionic-, electronic- and mixed-conducting nano particles into LSM–YSZ cathodes—a comparative study of performance enhancement and stability at different temperatures. J Power Sources 228:170–177

Kim J, Ji HI, Dasari HP, Shin D, Song H, Lee JH, Kim BK, Je HJ, Lee HW, Yoon KJ (2013) Degradation mechanism of electrolyte and air electrode in solid oxide electrolysis cells operating at high polarization. Int J Hydrogen Energy 38:1225–1235

Kim-Lohsoontorn P, Brett DJL, Laosiripojana N (2010) Performance of solid oxide electrolysis cells based on composite $La_{0.8}Sr_{0.2}$-$MnO_{3-\delta}$-yttria stabilized zirconia and $Ba_{0.5}Sr_{0.5}Co_{0.8}Fe_{0.2}O_{3-\delta}$ oxygen electrodes. Int J Hydrogen Energy 35:3958–3966

Knibbe R, Drennan J, Love JG (2009) Effect of alumina additions in YSZ on the microstructure and degradation of the LSM–YSZ interface. Solid State Ion 180:984–989

Laguna-Bercero MA, Kilner JA, Skinner SJ (2011) Development of oxygen electrodes for reversible solid oxide fuel cells with scandia stabilized zirconia electrolytes. Solid State Ionics 192:501–504

Liang MD, Yu B, Wen MF, Chen J, Xu JM, Zhai YC (2009) Preparation of LSM–YSZ composite powder for anode of solid oxide electrolysis cell and its activation mechanism. J Power Sources 190:341–345

Liu Z, Zheng ZW, Han MF, Liu ML (2010) High performance solid oxide fuel cells based on tri-layer yttria-stabilized zirconia by low temperature sintering process. J Power Sources 195:7230–7233

Rao YY, Wang ZQ, Zhong W, Peng RR, Lu YL (2012) Novel Ni–$Ba_{1+x}Zr_{0.3}Ce_{0.5}Y_{0.2}O_{3-\delta}$ hydrogen electrodes as effective reduction barriers for reversible solid oxide cells based on doped ceria electrolyte thin film. J Power Sources 199:142–145

Sholklapper TZ, Lu C, Jacobson CP, Visco SJ, Jonghe LCD (2006) LSM-infiltrated solid oxide fuel cell cathodes. Electrochem Solid-State Lett 9:A376–A378

Tao Y, Nishino H, Ashidate S, Kokubo H, Watanabe M, Uchida H (2009) Polarization properties of $La_{0.6}Sr_{0.4}Co_{0.2}Fe_{0.8}O_3$-based double layer-type oxygen electrodes for reversible SOFCs. Electrochim Acta 54:3309–3315

Virkar AV (2010) Mechanism of oxygen electrode delamination in solid oxide electrolyzer cells. Int J Hydrogen Energy 35:9527–9543

Yang CH, Jin C, Coffin A, Chen FL (2010) Characterization of infiltrated $(La_{0.75}Sr_{0.25})_{0.95}MnO_3$ as oxygen electrode for solid oxide electrolysis cells. Int J Hydrogen Energy 35:5187–5193

Yu HC, Zhao F, Virkar AV, Fung KZ (2005) Electrochemical characterization and performance evaluation of intermediate temperature solid oxide fuel cell with $La_{0.75}Sr_{0.25}CuO_{2.5-\delta}$ cathode. J Power Sources 152:22–26

20

Composition and mode of occurrence of minerals in Late Permian coals from Zhenxiong County, northeastern Yunnan, China

Tian Li · Shi-Feng Dai · Jian-Hua Zou · Sen Zhang · He-Hing Tian · Li-Xin Zhao

Abstract Minerals in the Late Permian coals from the Niuchang-Yigu mining area, Zhenxiong County, northeastern Yunnan, China, were investigated using optical microscopy and low temperature ashing plus X-ray diffraction (LTA + XRD). The results showed that minerals in the coal LTAs are mainly quartz, kaolinite, chamosite, mixed-layer illite/smectite (I/S), pyrite, and calcite, with trace amounts of marcasite, dolomite, and bassanite. The authigenic quartz generally occurs in collodetrinite or as a filling in cleats or cell cavities. This silica was mainly derived from aqueous solutions produced by the weathering of basaltic rocks in the Kangdian Upland and from hydrothermal fluids. The presence of β-quartz paramorph grains in collodetrinite probably indicates that these grains were detrital and came from a volcanic ash. Clay minerals are generally embedded in collodetrinite and occur as cell-fillings. Pyrite occurs as framboidal, anhedral, and euhedral grains and a cell-filling. The coals are high in pyrite and the high pyrite content probably results from seawater invading during the stage of peat accumulation. Calcite generally occurs as vein-fillings, indicating an epigenetic origin.

Keywords Late Permian coals · Minerals in coal · Zhenxiong of Yunnan

1 Introduction

Understanding the abundance and occurrence modes of minerals in coal is significant both academically for developing conjectures concerning coal-forming processes and pragmatically for optimizing coal utilization (Ren 1996; Ward 2002; Dai et al. 2012c).

There are large coal resources in southwestern China in Yunnan Province, resources second in abundance only to those in neighboring Guizhou Province (Fig. 1). The coal-bearing strata in Yunnan are Carboniferous, Permian, Triassic, and Neogene in age. The major sediment source for the Eastern Yunnan region, which was an important coal accumulation area in the Late Permian, was the Kangdian Upland (China Coal Geology Bureau 1996). Previous work on the origin and mode of occurrence of mineral matter in the Late Permian coals from eastern Yunnan includes that of Chen et al. (1992), Dai and Chou (2007), Dai et al. (2008, 2014), and Wang et al. (2012, 2013). Tonsteins derived from pyroclastics in the Late Permian coal seams from eastern Yunnan have also been studied (Zhou 1999; Zhou and Ren 1983; Zhou et al. 1982, 1990; Dai et al. 2014). In this paper, we present new data on the abundances, origin, and mode of occurrence of minerals in coal from two cores, the "NYa" and "NYb" holes drilled through the Longtan and Changxing Formations in the Niuchang-Yigu mining area, Zhenxiong County, northeastern Yunnan.

T. Li (✉) · J. Zou · S. Zhang · H. Tian
Chongqing Key Laboratory of Exogenic Mineralization and Mine Environment, Chongqing Institute of Geology and Mineral Resources, Chongqing 400042, China
e-mail: lt_litian@163.com

T. Li · S. Dai · L. Zhao
State Key Laboratory of Coal Resources and Safe Mining, China University of Mining and Technology, Beijing 100083, China

T. Li · J. Zou · S. Zhang · H. Tian
Chongqing Research Center of State Key Laboratory of Coal Resources and Safe Mining, Chongqing 400042, China

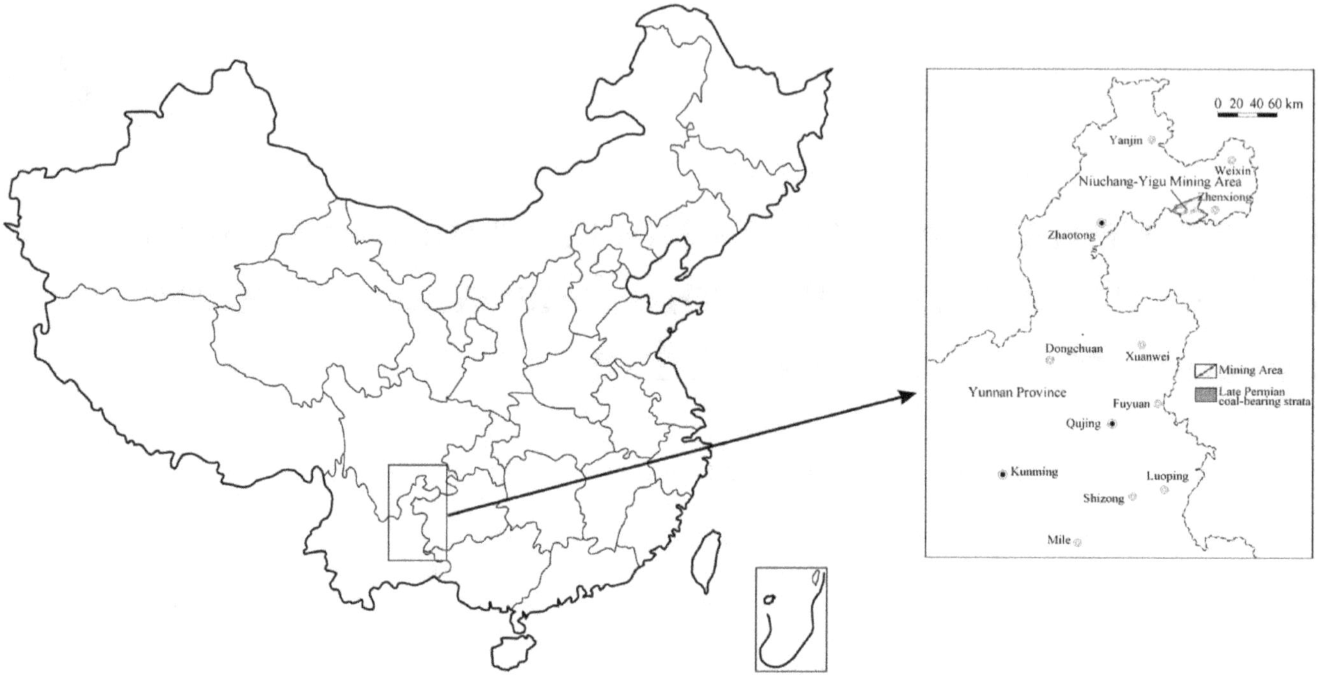

Fig. 1 Location of Niuchang-Yigu mining area, Zhenxiong County, northeastern Yunnan

2 Geological setting

The Niuchang-Yigu mining area is located in Zhenxiong County, northeastern Yunnan Province (Fig. 1). The Longtan and Changxing Formations are the major coal-bearing sequences of Late Permian age in this area (Fig. 2.; China Coal Geology Bureau 1996). The Longtan Formation mainly consists of fine-grained sandstone, siltstone, mudstone, sandy mudstone, carbonaceous mudstone and coal seams and has a thickness of > 230 m in borehole NYa and 202 m in hole NYb (Fig. 2). The Longtan Formation has a disconformable contact with the underlying Late Permian Xuanwuyan Formation and a conformable contact with the overlying Changxing Formation.

3 Sampling and analytical procedures

The coal samples were collected from two drill cores (the NYa and NYb holes) (Fig. 2). Four coal samples, NYa 2, NYa 3, NYa 4, and NYa 5, were collected from the NYa drill hole. Seven samples (NYb C3, NYb C5a, NYb C5b, NYb C6, NYb C7a, NYb C7b, and NYb C8a) were collected from NYb. The NYb C3 coal is from the Changxing Formation and the other coals are in the Longtan Formation (Fig. 2).

The proximate analyses, which determine moisture, volatile matter, and ash yield, were conducted according to Chinese Standards GB/T 212-2008 (2008). The total sulfur content was determined following Chinese Standards GB/T 214-1996 (1996). The content of C, H, and N was determined with a Vario MACRO elemental analyzer.

Polished grain mounts were prepared following Chinese Standards GB/T 474-2008 (2008) for microscopic analysis. Mean random reflectance of vitrinite was determined using a Craic QDI 302^{TM} spectrophotometer attached to a Leica DM-4500P microscope at a magnification of 500×. The standard reference gadolinium gallium garnet (manufactured by Klein & Becker) with a calculated standard reflectance of 1.722 % at $\lambda = 546$ nm in oil immersion was used for calibrating the vitrinite reflectance spectra.

Maceral components were identified using white-light reflectance microscopy in oil. The maceral classification and terminology are based on Taylor et al. (1998) and the ICCP System 1994 (International Committee for Coal and Organic Petrology (ICCP 1998, 2001).

Mineralogy was determined by optical microscopy and X-ray powder diffraction (XRD). Low-temperature ashing (LTA) of coal was performed using an EMITECH K1050X plasma asher prior to XRD analysis. XRD analysis was performed with a Rigaku D/max 2500pc powder diffractometer using Ni-filtered Cu-Kα radiation and a scintillation detector. The XRD pattern was recorded in a 2θ range of 2.6–70°, with a step size of 0.01°. X-ray diffractograms of the LTAs were subjected to quantitative mineralogical analysis using Siroquant^{TM} commercial interpretation software developed by Taylor (1991) based on the principles for diffractogram profiling set out by Rietveld (1969).

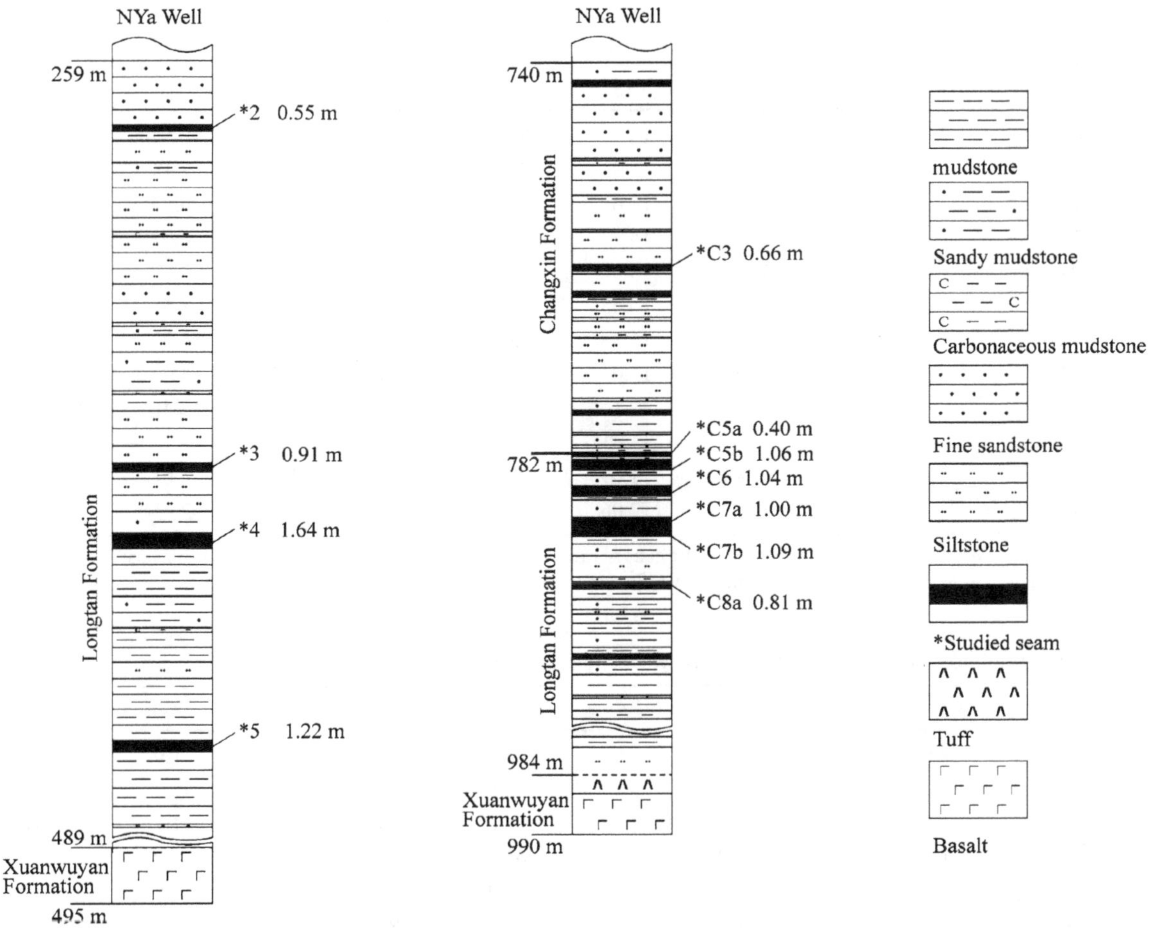

Fig. 2 Sedimentary sequences in the Niuchang-Yigu mining area

Further details describing the use of this technique for coal-related materials are given by Ward et al. (1999, 2001) and Ruan and Ward (2002).

4 Results and discussion

4.1 Coal chemistry

Proximate and ultimate analyses results, total sulfur, the gross calorific values, and random vitrinite reflectance data for the Niuchang-Yigu coals are shown in Table 1.

The weighted averages of moisture, volatile matter, gross calorific value, and random vitrinite reflectance are 0.86 %, 12.06 %, 22.46 MJ/kg, and 2.25 %, respectively. The ash yield from Niuchang-Yigu coals is high, ranging from 23.53 % to 48.63 %, with a weighted average of 34.92 %. The total sulfur content of the coals ranges from 0.07 % to 6.27 %, with an average of 2.45 %. According to Chinese Standard GB/T 15224.1-2010 (2010), the NYa 4, NYb C7a, and NYb C7b samples are medium-ash coals (20.01 % $< A_d \leq$ 30.00 %), whereas the NYa 2 and NYa 5 samples are high-ash coals (40.01 % $< A_d \leq$ 50.00 %). The other samples are medium–high ash coals (30.01 % $< A_d \leq$ 40.00 %). According to Chinese Standard GB/T 15224.2-2010 (2010), the special-low-, low-, medium-, medium–high-, and high-sulfur coals have $\leq$ 0.50 %, 0.51 %–1.00 %, 1.01 %–2.00 %, 2.01 %–3.00 %, and > 3.00 % total sulfur contents, respectively. According to this standard, more than half of the samples analyzed are medium–high- and high-sulfur coals.

4.2 Maceral composition

Maceral compositions are presented in Table 2. The weighted average of the vitrinite and inertinite contents in Niuchang-Yigu coals are 71.3 and 28.7 %, respectively.

The vitrinite is dominated by collodetrinite (Fig. 3a), collotelinite (Fig. 3a), and vitrodetrinite, with traces of telinite and corpogelinite (Fig. 3b). Corpogelinite was observed in only two samples, NYb C3 and NYb C5a; the abundances in those samples being 1.1 % and 0.4 %, respectively. The inertinite is mainly composed of fusinite (Fig. 3a), inertodetrinite (Fig. 3a), and semifusinite (Fig. 3d), along with a trace amount of macrinite (Fig. 3c).

Table 1 Proximate and ultimate analyses (%), total sulfur (%), gross calorific values (MJ/kg), and random vitrinite reflectance (%) in the Niuchang-Yigu coals

Samples	Thickness (m)	M_{ad}	A_d	V_{daf}	C_{daf}	H_{daf}	N_{daf}	$S_{t,d}$	$Q_{gr,d}$	$R_{o,ran}$
NYa 2	0.55	0.76	48.63	13.71	87.71	3.93	1.55	2.53	17.57	2.19
NYa 3	0.91	0.89	30.95	11.01	88.00	3.49	1.39	3.14	24.23	2.34
NYa 4	1.64	1.01	25.76	10.67	90.25	3.61	1.61	0.54	26.19	2.48
NYa 5	1.22	0.84	47.91	14.77	86.20	4.83	1.35	0.07	17.60	2.23
NYb C3	0.66	0.84	38.45	13.02	86.81	4.15	1.69	3.62	21.24	2.34
NYb C5a	0.40	0.77	39.37	10.45	88.39	3.69	1.54	1.58	20.58	2.48
NYb C5b	1.06	0.93	34.99	15.38	83.64	3.71	1.32	6.27	21.89	2.23
NYb C6	1.04	0.75	31.14	11.03	85.68	3.61	1.29	4.94	23.35	2.35
NYb C7a	1.00	0.84	28.92	10.14	87.30	3.71	1.43	2.45	24.55	2.39
NYb C7b	1.09	1.01	23.53	11.41	89.22	4.19	2.23	1.15	26.60	2.39
NYb C8a	0.81	0.77	34.49	11.09	90.35	3.55	1.73	0.63	23.23	2.46
WA		0.86	34.92	12.06	87.60	3.86	1.56	2.45	22.46	2.35

M moisture, *A* ash yield, *V* volatile matter, *C* carbon, *H* hydrogen, *N* nitrogen, S_t total sulfur, *ad* air-dry basis, *d* dry basis, *daf* dry and ash-free basis, $Q_{gr,d}$ gross calorific value, on dry basis, $R_{o,ran}$ random reflectance of vitrinite, *WA* weighted average

Table 2 Maceral compositions in the Niuchang-Yigu coals (vol. %, on mineral-free basis)

Samples	Cd	Ct	Cg	T	Vd	V	F	Sf	Mac	Id	I
NYa 2	43.5	2.5	bdl	1.9	18.0	65.8	12.4	9.3	1.9	10.6	34.2
NYa 3	54.6	9.2	bdl	5.6	5.1	74.5	11.2	3.1	2.0	9.2	25.5
NYa 4	45.1	14.5	bdl	6.7	3.3	69.6	11.4	10.3	1.7	7.0	30.4
NYa 5	49.1	1.8	bdl	0.9	13.4	65.2	17.0	3.6	0.9	13.4	34.8
NYb C3	55.3	10.6	1.1	1.8	4.4	73.3	11.4	4.4	4.8	6.2	26.7
NYb C5a	45.9	10.1	0.4	1.1	3.0	60.4	28.4	3.7	3.7	3.7	39.6
NYb C5b	62.3	6.2	bdl	bdl	6.8	75.3	16.7	3.1	0.6	4.3	24.7
NYb C6	69.3	5.1	bdl	bdl	bdl	74.5	17.5	0.7	0.7	6.6	25.5
NYb C7a	54.5	10.1	bdl	2.2	3.9	70.8	20.8	3.9	1.7	2.8	29.2
NYb C7b	50.6	19.7	bdl	2.5	2.5	75.3	10.9	5.4	2.9	5.4	24.7
NYb C8a	65.6	5.4	bdl	bdl	9.1	80.1	8.6	0.5	2.2	8.6	19.9
WA	54.2	8.7	0.3	2.1	6.3	71.3	15.1	4.4	2.1	7.1	28.7

Cd collodetrinite, *Ct* collotelinite, *Cg* corpogelinite, *T* telinite, *Vd* vitrodetrinite, *V* total vitrinite, *F* fusinite, *Sf* semifusinite, *Mac* macrinite, *Id* inertodetrinite, *I* total inertinite, *WA* weighted average, *bdl* below detection limit

4.3 Minerals in low temperature ash of Niuchang-Yigu coal

The minerals in LTAs of Niuchang-Yigu coal are mainly quartz, kaolinite, chamosite, mixed-layer illite/smectite (I/S), pyrite and calcite, with trace amounts of marcasite, dolomite and bassanite (Table 3).

4.3.1 Quartz

Quartz is a common mineral in most coals (Ren 1996; Tang and Huang 2004). In general, quartz is mainly of terrigenous detrital origin (Ren 1996) but authigenic quartz is quite common in the Late Permian coals from eastern Yunnan (Ren 1996; Dai and Chou 2007; Dai et al. 2008, 2014).

The quartz content of LTAs of Niuchang-Yigu coals is between 33.2 % and 67.6 %, with a weighted average of 48.8 %, an average close to that of the Late Permian coals from the Taoshuping area (43.8 %; Wang et al. 2012). Quartz in the Niuchang-Yigu coals has three main modes of occurrence, (1) embedded in collodetrinite with a variety of shapes (Fig. 4a), (2) as cell- and cleat-fillings (Fig. 4b, d), and (3) coexisting with pyrite and clay minerals (Fig. 4f). These modes of occurrence suggest an authigenic origin. The authigenic quartz in eastern Yunnan coals was derived from silica-containing solutions produced by the weathering of basaltic rocks in the Kangdian Upland (Ren

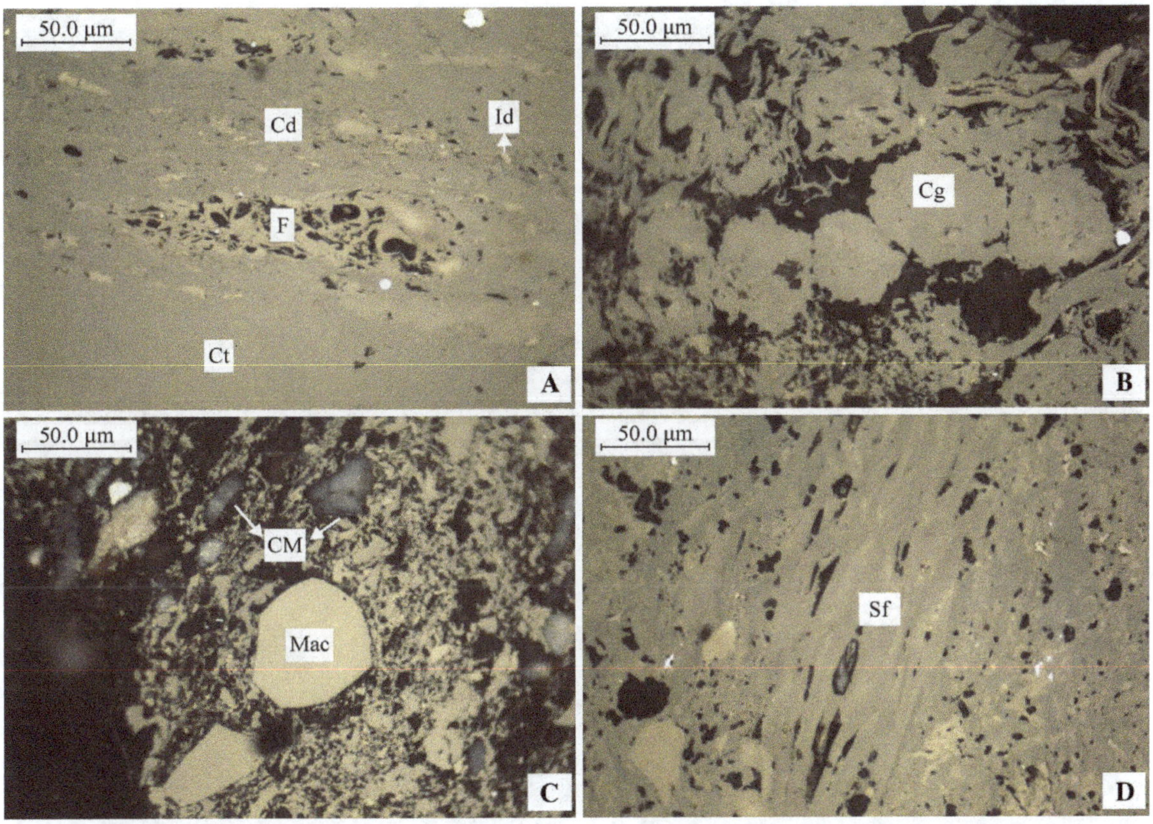

Fig. 3 Photomicrographs of macerals in the Niuchang-Yigu coals. **a** *Cd* collodetrinite, *Ct* collotelinite, *F* fusinite, and *Id* inertodetrinite in sample NYb C7b coal; reflected light, oil immersion. **b** *Cg* corpogelinite in sample NYb C3 coal; reflected light, oil immersion. **c** *Mac* macrinite and detrital *CM* carbonate minerals in sample NYb C7b coal; reflected light, oil immersion. **d** *Sf* semifusinite in sample NYb C3 coal; reflected light, oil immersion

Table 3 Low temperature ash yield (%) and mineral compositions of LTAs of Niuchang-Yigu coal by XRD analysis (%; on organic matter-free basis)

Samples	LTA yield	Quartz	Kaolinite	Chamosite	I/S	Pyrite	Calcite	Marcasite	Dolomite	Bassanite
NYa 2	52.18	67.6	bdl	4.4	16.5	7.7	3.6	bdl	0.2	bdl
NYa 3	33.79	60.8	12.5	2.8	6.9	11.1	1.2	4.7	bdl	bdl
NYa 4	28.10	37.2	29.8	7.4	19.4	2.1	4.2	bdl	bdl	bdl
NYa 5	50.81	38.0	38.7	9.0	14.0	bdl	0.3	bdl	bdl	bdl
NYb C3	42.10	58.5	bdl	7.9	11.6	15.5	3.4	1.5	bdl	1.6
NYb C5a	43.19	57.8	7.7	7.3	bdl	7.2	14.3	bdl	5.7	bdl
NYb C5b	42.40	33.2	10.9	3.1	16.6	25.5	3.6	bdl	0.8	6.2
NYb C6	37.78	49.3	7.1	1.8	bdl	17.5	14.6	3.8	0.9	5.1
NYb C7a	33.23	40.4	22.6	9.0	bdl	7.4	10.5	3.8	0.5	5.8
NYb C7b	26.26	35.5	44.7	8.8	bdl	3.9	5.9	bdl	0.3	1.0
NYb C8a	36.23	58.8	25.5	10.4	bdl	2.0	3.3	bdl	bdl	bdl
WA	38.73	48.8	18.2	6.5	7.8	9.1	5.9	1.3	0.8	1.8

WA weighted average, *bdl* below detection limit

1996) and then precipitated from hydrothermal fluids (Dai and Chou 2007; Dai et al. 2008). Most of the detrital quartz found in the coals also originated from the weathering of basaltic rocks in the Kangdian Upland (Fig. 4c).

Some quartz grains distributed in the collodetrinite have a well-developed β-quartz crystal paramorph. This crystal form probably indicates a volcanic origin for these grains (Fig. 4c). Several tonstein layers deposited in the middle-

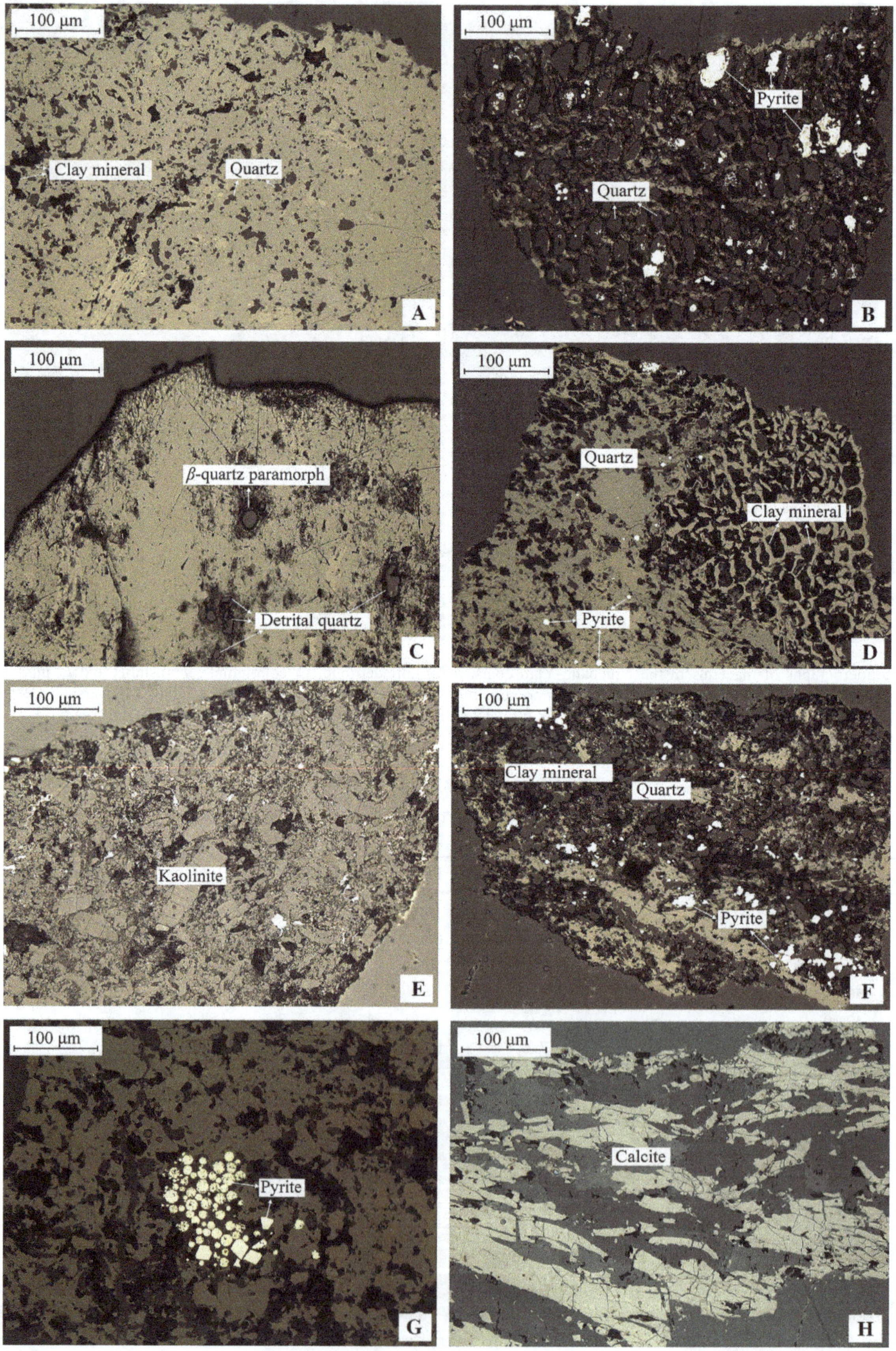
100 μm
Clay mineral
Quartz
A
100 μm
Pyrite
Quartz
B
100 μm
β-quartz paramorph
Detrital quartz
C
100 μm
Quartz
Clay mineral
Pyrite
D
100 μm
Kaolinite
E
100 μm
Clay mineral
Quartz
Pyrite
F
100 μm
Pyrite
G
100 μm
Calcite
H

◀**Fig. 4** Photomicrographs of minerals in the Niuchang-Yigu coals. **a** quartz grains with different shapes and clay minerals embedded in collodetrinite in sample NYb C7b coal; reflected light. **b** quartz and pyrite as cell-fillings in sample NYa 3 coal; reflected light. **c** Well-developed crystal of β-quartz and detrital quartz in collodetrinite in sample NYa 5 coal; reflected light. **d** cleat-filling quartz, cell-filling clay minerals, and isolated framboidal pyrite in sample NYb C5a coal; reflected light. **e** Vermicular kaolinite in sample NYb C7b coal; reflected light. **f** quartz along with pyrite and clay minerals in sample NYa 3 coal; reflected light. **g** irregular massive lumps of clay minerals and framboidal and euhedral pyrite in collodetrinite in sample NYb C5a coal; reflected light. **h** calcite occurs as vein-fillings in sample NYb C5a coal; plane-polarized and reflected light

late to Late Permian coal seams in this area (Zhou et al. 2000; Dai et al. 2014), demonstrate that there had been frequent volcanic eruptions during the Late Permian.

4.3.2 Clay minerals

Clay minerals in the LTAs of Niuchang-Yigu coal include kaolinite, mixed-layer I/S, and chamosite. The clay mineral content ranges from 8.9 % to 61.7 %, with a weighted average of 32.5 % (Table 3). In the coals studied, the clays occur in three different modes. They occur as irregular massive lumps in collodetrinite (Fig. 4a), as cell-fillings (Fig. 4d), and as clots coexisting with quartz and pyrite (Fig. 4f). Vermicular kaolinite was also found in the coal (Fig. 4e).

Kaolinite is the dominant clay mineral in most Niuchang-Yigu samples except for samples NYa 2 and NYb C3. The kaolinite content ranges from below detection limit (bdl) to 38.7 % and from bdl to 44.7 % in the LTAs of NYa and NYb cores, respectively. In general, the kaolinite content is roughly vertically zoned increasing from top to bottom in both holes.

Chlorite, with the two common forms being chamosite and clinochlore, is generally uncommon in coal (Bouška et al. 2000; Tang and Huang 2004) but it is present in our samples. Chamosite has the same diffraction peaks as clinochlore in XRD patterns but it can be identified by its weaker odd-order peaks (the 001 and 003 peaks) and stronger even-order peaks (the 002 and 004 peaks) in X-ray diffractograms (Dai and Chou 2007; Moore and Reynolds 1989). Figure 5 shows that the chlorite in this study is chamosite, which is actually common in the Late Permian coals from eastern Yunnan (Dai and Chou 2007; Dai et al. 2008, 2014; Wang et al. 2012, 2013).

Chamosite exists in each coal seam examined for this study with the content ranging from 1.8 % to 10.4 % (average of 6.5 %) and reaching its highest values in the lower coals in the both holes (sample NYa 5 at 9.0 % and NYb C8a at 10.4 %). The mode of occurrence of the chamosite, coexisting with kaolinite as cell-fillings, indicates that it is the product of Fe- and Mg-rich solutions that probably originated from volcanic activity (Dai and Chou 2007; Dai et al. 2008).

The mixed-layer I/S is an intermediate product in the smectite to illite transformation process. It is commonly observed in the coal LTAs of the NYa hole, and its content ranges from 6.9 % to 19.4 %. However, it appears only in two samples in hole NYb, (NYb C3 and NYb C5b). Mixed-layer I/S is also common in other Late Permian coals from eastern Yunnan (Dai and Chou 2007; Dai et al. 2008; Wang et al. 2012), and is the dominant clay mineral in Changxing coal (Wang et al. 2013). There it generally occurs as thin beds, massive lenses, and fracture- or cell-fillings (Wang 2011).

4.3.3 Pyrite, marcasite and bassanite

The content of pyrite in the LTAs ranges from bdl to 25.5 % (mean 9.1 %), lower than that of quartz and kaolinite (Table 3). It is, however, relatively high in the LTAs

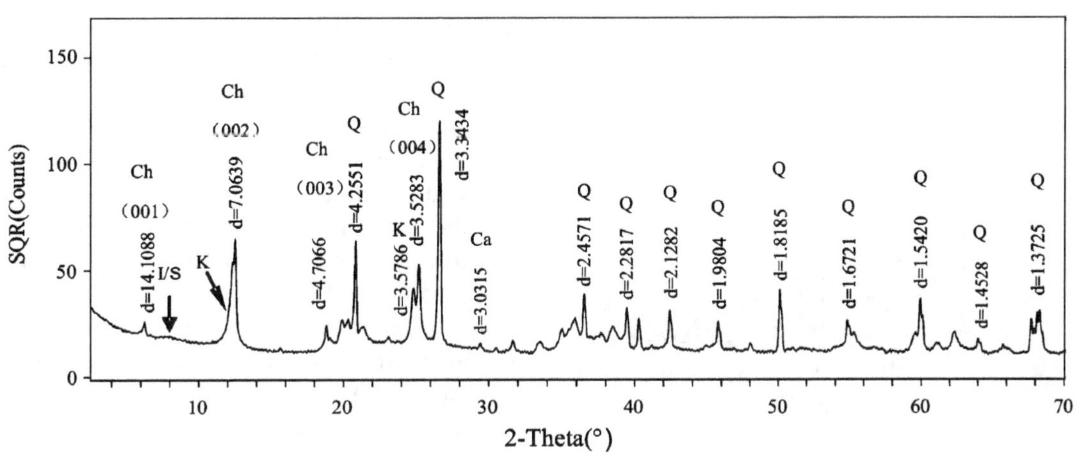

Fig. 5 XRD pattern of LTA of sample NYa 5 coal. *Ch* chamosite, *I/S* illite-smectite, *K* kaolinite, *Q* quartz, *Ca* calcite

of high-sulfur coals including the NYa 3, NYb C3, NYb C5a, and NYb C6 samples (11.1 %, 15.5 %, 25.5 %, and 17.5 %, respectively). The pyrite-total sulfur correlation coefficient (r) is 0.99—essentially all of the sulfur in the coal is contained in pyrite.

In our samples pyrite occurs in framboidal, anhedral and euhedral forms (Fig. 4d, f, g), and also as cell-fillings (Fig. 4b). Dai and Chou (2007) also found pyritized algae in the coals from this area. The high pyrite content is probably due to the invasion of seawater during early diagenesis (Dai and Chou 2007; Ren 1996).

Although having the same chemical compositions, marcasite and pyrite can be differentiated by both optical microscopy and XRD (Ward 2002; Fig. 6). Marcasite is present in at least four LTAs (NYa 3, NYb C3, NYb C6, and NYb C7a identified by XRD analysis), with the content in those four samples being 4.7 %, 1.5 %, 3.8 %, and 3.8 %, respectively. Marcasite commonly exists in the coal LTAs which have high content of pyrite as well. However, sample NYb C5b has the highest pyrite content, but the marcasite content is below detection limit in that sample.

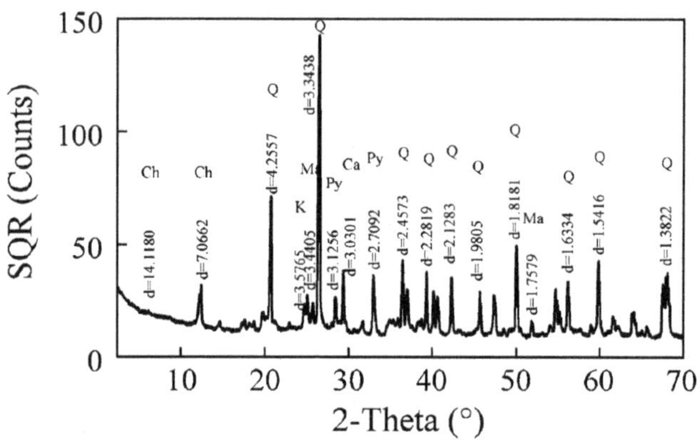

Fig. 6 XRD pattern of LTA of the sample NYa 3 coal. *Ch* chamosite, *K* kaolinite, *Q* quartz, *Ma* marcasite, *Py* pyrite, *Ca* calcite

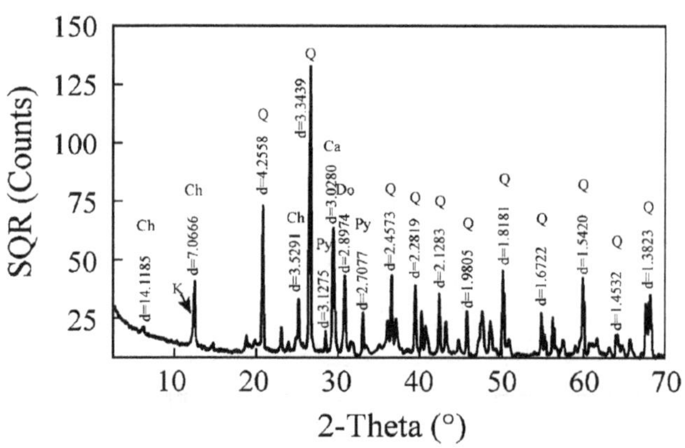

Fig. 7 XRD pattern of LTA of the sample NYb C5a coal. *Ch* chamosite, *K* kaolinite, *Q* quartz, *Py* pyrite, *Ca* calcite, *Do* dolomite

The content of bassanite in the LTAs is from bdl to 6.2 % (mean 1.3 %; Table 3). It has also been identified in LTAs of some other Chinese coal such as the Wulantuga and Guanbanwusu coals in Inner Mongolia and the Fusui coals in Guangxi (Dai et al. 2012a, 2012b, 2013). It is commonly thought that the bassanite in the coal LTAs was formed in the plasma-ashing process (Ward et al. 2001; Ward 2002).

4.3.4 Calcite and dolomite

The carbonate minerals in the LTAs are calcite and dolomite, with a weighted average content of 5.6 % and 0.8 %, respectively (Table 3). The calcite content ranges from 3.3 % to 14.6 % in the LTAs of NYb core, much higher than that in the NYa core (0.3 %–4.2 %). Although the dolomite content in most of the LTAs is low, sample NYb C5a has a high content of 5.7 % (Fig. 7, Table 3). The calcite mainly occurs as vein-fillings, (Fig. 4h), indicating an epigenetic origin. Detrital calcite is only rarely observed in coal as it can be easily decomposed in the acid environment of a peat mire (Kortenski 1992; Bouška et al. 2000). However, detrital carbonate grains do occur in some of our samples in the collodetrinite (Fig. 3c) and are present in other Late Permian coals from eastern Yunnan (Wang 2011).

5 Conclusions

Most coals analyzed for this study have high total sulfur contents and high ash yields. Vitrinite is the dominant maceral group in all the coals, followed by inertinite. The minerals in LTAs of Niuchang-Yigu coals are mainly quartz, kaolinite, chamosite, mixed-layer I/S, pyrite, and calcite, with trace amounts of marcasite, dolomite, and bassanite. Quartz is mainly embedded in collodetrinite and occurs as cleat- and cell-fillings. The silica was primarily transported by aqueous solutions originated from the weathering of basaltic rocks in the Kangdian Upland and by hydrothermal fluids. The presence of β-quartz paramorph grains, distributed in collodetrinite, probably indicates that quartz from volcanic ash was deposited during peat accumulation. Clay minerals generally occur in collodetrinite and as fill in coal-forming plant cells. The chamosite was the product of Fe- and Mg-rich solutions that were probably generated by volcanic activity. In most cases, pyrite occurs as framboidal, anhedral, and euhedral particles in collodetrinite but it also is present as cell-fillings. The high pyrite content is probably the result of seawater invasion during early diagenesis. A small proportion of the carbonate minerals present occur as detrital grains in collodetrinite but most of the calcite occurs as vein-fillings and no doubt has an epigenetic hydrothermal origin.

Acknowledgments This research was supported by the National Key Basic Research and Development Program (No. 2014CB238902), National Natural Science Foundation of China (Nos. 41272182 and 40930420), and the Program for Changjiang Scholars and Innovative Research Team in University.

References

Bouška V, Pešek J, Sykorova I (2000) Probable modes of occurrence of chemical elements in coal. Acta Mont. Serie B Fuel Carbon Miner Process Praha 10:53–90

Chen SQ, Chen JH, Zhao SJ (1992) Occurrence character of pyrite in coal of the upper part of the Late Permian coal measures in eastern Ynnan and western Guizhou. Bull Yichang Inst Geol Miner Resour (CAGS) 18:135–145 (in both Chinese and English)

China Coal Geology Bureau (1996) Sedimentary environments and coal accumulation of Late Permian coal formation in Western Guizhou, Southern Sichuan and Eastern Yunnan, China. Chongqing University Press, Chongqing (in Chinese with English abstract)

Chinese Standard GB/T 15224.1-2010 (2010) Classification for quality of coal. Part 1: Ash (in Chinese)

Chinese Standard GB/T 15224.2-2010 (2010) Classification for quality of coal. Part 2: Sulfur content (in Chinese)

Chinese Standard GB/T 212-2008 (2008) Proximate analysis of coal (in Chinese)

Chinese Standards GB/T 214-1996 (1996) Determination of total sulfur in coal (in Chinese)

Chinese Standards GB/T 474-2008 (2008) Method for preparation of coal sample (in Chinese)

Dai SF, Chou CL (2007) Occurrence and origin of minerals in a chamosite-bearing coal of Late Permian age, Zhaotong, Yunnan, China. Am Miner 92:1253–1261

Dai SF, Tian LW, Chou CL, Zhou YP, Zhang MQ, Zhao L, Wang JM, Yang Z, Cao HZ, Ren DY (2008) Mineralogical and compositional characteristics of Late Permian coals from an area of high lung cancer rate in Xuan Wei, Yunnan, China: occurrence and origin of quartz and chamosite. Int J Coal Geol 76:318–327

Dai SF, Wang XB, Seredin VV, Hower JC, Ward CR, O'keefe MKJ, Huang WH, Li T, Li X, Liu HD, Xue WF, Zhao LX (2012a) Petrology, mineralogy, and geochemistry of the Ge-rich coal from the Wulantuga Ge ore deposit, Inner Mongolia, China: new data and genetic implications. Int J Coal Geol 90–91:72–99

Dai SF, Jiang YF, Ward CR, Gu LD, Seredin VV, Liu HD, Zhou D, Wang XB, Sun YZ, Zou JH, Ren DY (2012b) Mineralogical and geochemical compositions of the coal in the Guanbanwusu Mine, Inner Mongolia, China: further evidence for the existence of an Al (Ga and REE) ore deposit in the Jungar coalfield. Int J Coal Geol 98:10–40

Dai SF, Ren DY, Chou CL, Finkelman RB, Seredin VV, Zhou YP (2012c) Geochemistry of trace elements in Chinese coals: a review of abundances, genetic types, impacts on human health, and industrial utilization. Int J Coal Geol 94:3–21

Dai SF, Zhang WG, Ward CR, Seredin VV, Hower JC, Li X, Song WJ, Wang XB, Kang H, Zheng LC, Wang PP, Zhou D (2013) Mineralogical and geochemical anomalies of late Permian coals from the Fusui Coalfield, Guangxi Province, southern China: influences of terrigenous materials and hydrothermal fluids. Int J Coal Geol 105:60–84

Dai SF, Li T, Seredin VV, Ward CR, Hower JC, Zhou YP, Zhang MQ, Song XL, Song WJ, Zhao CL (2014) Origin of minerals and elements in the Late Permian coals, tonsteins, and host rocks of the Xinde Mine, Xuanwei, eastern Yunnan, China. Int J Coal Geol 121:53–78

International Committee for Coal and Organic Petrology (ICCP) (1998) The new vitrinite classification (ICCP System 1994). Fuel 77:349–358

International Committee for Coal and Organic Petrology (ICCP) (2001) The new inertinite classification (ICCP System 1994). Fuel 80:459–471

Kortenski J (1992) Carbonate minerals in Bulgarian coals with different degrees of coalification. Int J Coal Geol 20:225–242

Moore DM, Reynolds RC (1989) X-ray diffraction and the identification and analysis of clay minerals. Oxford University Press, New York

Ren DY (1996) Mineral matters in coal. In: Han DX (ed) Coal Petrology of China. Publishing House of China University of Mining and Technology, Xuzhou, pp 67–77 (in Chinese with English abstract)

Rietveld HM (1969) A profile refinement method for nuclear and magnetic structures. J Appl Crystallogr 2:65–71

Ruan CD, Ward CR (2002) Quantitative X-ray powder diffraction analysis of clay minerals in Australian coals using Rietveld methods. Appl Clay Sci 21:227–240

Tang XY, Huang WH (2004) Trace elements in Chinese coals. The Commercial Press, Beijing (in Chinese)

Taylor JC (1991) Computer programs for standardless quantitative analysis of minerals using the full powder diffraction profile. Powder Diffr 6:2–9

Taylor GH, Teichmüller M, Davis A, Diessel CFK, Littke R, Robert P (1998) Organic Petrology. Gebrüder Borntraeger, Berlin 704

Wang XB (2011) Enrichment and differentiation mechanism of mineral matter in the Late Permian coals from Eastern Yunnan Province. Doctor's Thesis, China University of Mining and Technology, Beijing (in Chinese with English abstract)

Wang XB, Dai SF, Chou CL, Zhang MQ, Wang JM, Song XL, Wang W, Jiang YF, Zhou YP, Ren DY (2012) Mineralogy and geochemistry of Late Permian coals from the Taoshuping Mine, Yunnan Province, China: evidences for the sources of minerals. Int J Coal Geol 96–97:49–59

Wang XB, Zhang MQ, Zhang WG, Wang JM, Zhou YP, Song XL, Li T, Li X, Liu HD, Zhao LX (2013) Occurrence and origins of minerals in mixed-layer illite/smectite-rich coals of the Late Permian age from the Changxing Mine, eastern Yunnan, China. Int J Coal Geol 102:26–34

Ward CR (2002) Analysis and significance of mineral matter in coal seams. Int J Coal Geol 50:135–168

Ward CR, Spears DA, Booth CA, Staton I, Gurba LW (1999) Mineral matter and trace elements in coals of the Gunnedah Basin, New South Wales, Australia. Int J Coal Geol 40:281–308

Ward CR, Taylor JC, Matulis CE, Dale LS (2001) Quantification of mineral matter in the Argonne Premium coals using interactive Rietveld-based X-ray diffraction. Int J Coal Geol 46:67–82

Zhou YP (1999) The synsedimentary alkalinity-volcanic ash derived tonsteins of early Longtan age in South-western China. Coal geol Explor 27:5–9 (in Chinese with English abstract)

Zhou YP, Ren YL (1983) Genesis and distribution of tonsteins in late-Permian coalfields in Southwestern China. J China Coal Soc 8:76–88 (in Chinese with English abstract)

Zhou YP, Ren YL, Bohor BF (1982) Origin and distribution of tonsteins in Late Permian coal seams of Southwestern China. Int J Coal Geol 2:49–77

21

Simulation of coal char gasification using O_2/CO_2

Haibin Li · Yu Yu · Minfang Han · Ze Lei

Abstract The authors proposed an integrated gasification fuel cell zero-emission system. The coal char gasification is discussed using high temperature and concentration of CO_2 produced by solid oxide fuel cells and oxy-fuel combustion. The gasification is simulated by Aspen plus based on Gibbs free energy minimization method. Gasification model of pulverized coal char is computed and analyzed. Effects of gas flow rate, pressure, preheating temperature, heat losses on syngas composition, reaction temperature, lower heating value and carbon conversion are studied. Results and parameters are determined as following. The optimum O_2 flow rate is 20 kg/h. The reaction temperature decreases from 1645 to 1329 °C when the CO_2 flow rate increases from 0 to 5 kg/h, the CO_2 flow rate should be operated reasonably; lower heating value reduces and reaction temperature increases as the pressure increases; compared to the CO_2 preheating, O_2 preheating has greater influence on reaction temperature and lower heating value.

Keywords Coal char · Gasification · Aspen plus

1 Introduction

In recent years, global demand for fossil energy has been growing rapidly with the economic and social development of the world. Long-term use of fossil energy has led to serious environmental problems. Especially the green house effect is getting worse because of CO_2 emission, which has attracted worldwide concerns (Gleick et al. 2010; Mao 2010; Wei et al. 2011). Chinese government has proposed that by the year 2020, CO_2 emission of unit GDP could be reduced by 40 % to 45 % compared with the emission in 2005. As a result, CO_2 emission reduction and development of low carbon science and technology have become the most urgent scientific issues to solve for the world (Xie H 2010; Xie et al. 2012).

Coal production in China was 3.66 billion tons in 2012, increased by 4 % than in 2011. Coal utilization is the main source of CO_2 emission. The percentage of CO_2 emission produced by coal utilization is as high as 70 % (Huang 2012; Sun and Gao 2013). Clean and effective utilization of coal is the key to achieve sustainable development of energy for the world, especially for China. Coal gasification is the core technology in fuel cells, coal chemical synthesis, IGCC, and coal gasification-based poly-generation. It is the most efficient way to achieve clean and effective utilization of coal.

CO_2 can be used as gasification agent to gasify coal char. Then high purity CO can be used for chemical synthesis and Solid Oxide Fuel Cells (SOFCs), which can realize clean and effective utilization of coal (Peng and Han 2009).

Coal gasification in CO_2 rich gas atmosphere is recognized as one of the most promising technology for controlling CO_2 emission in pulverized coal-fired power plants. This technology has been reviewed (Antonio and Mara 2004). Large amount of syngas, mainly CO, can be produced by this technology. Syngas production from coal gasification under O_2/CO_2 atmosphere has been simulated using CFD software Fluent. A 3D geometry simulation

H. Li · Y. Yu · M. Han (✉) · Z. Lei
School of Chemical and Environmental Engineering, China University of Mining and Technology, Beijing 100083, China
e-mail: hanminfang@sina.com

model for gasification was established. The results show that the gas temperature decreases and the gross heating value of syngas increases with the increasing of CO_2 concentration; besides, effect of particle size on coal gasification is significant (Alam et al. 2012). Gasification of aquatic biomass under O_2/CO_2 atmosphere has been carried out. Effects of O_2, CO_2 concentrations, feeding rate and $[H_2O]/[C]$ ratio on O_2/CO_2 gasification behavior have been reported (Toshiaki et al. 2009; Toshiaki et al. 2013).

The syngas produced by coal gasification can be used for SOFCs. Then high temperature and concentration of CO_2 is produced after electrochemical reaction. The CO_2 could be used for gasification of char. For this closed cycle system, vast heat of exhaust could be used efficiently and energy needed in the gasification could be reduced. Also the system can realize CO_2 emission reduction and even zero emission. The specific principle is shown in Fig. 1. O_2 can be separated from air by oxygen transportation Membrane (OTM), which is prepared by perovskite powders and operated at high temperature from 800 to 1000 °C. Then high temperature O_2 is produced.

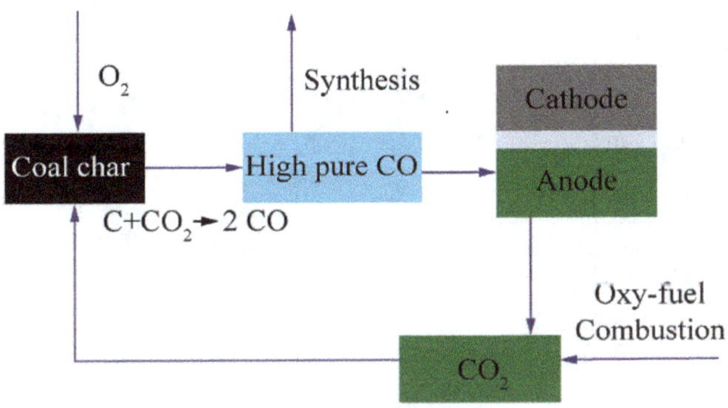

Fig. 1 Closed cycle system of CO_2 gasification of coal char

The process flowsheet simulation program Aspen plus is used in this paper. The simulation is based on entrained-flow gasifier. The physical property databases and unit module in Aspen plus have also been used in the simulation model. Coal char gasification model under O_2/CO_2 has been established, which can provide theory bases for the determination of process conditions.

2 Establishment of simulation model

As shown in Fig. 2, the simulation model consists of three unit modules, five material streams and two heat streams. They are the unit module of RYIELD, RGIBBS, SSPLIT; material streams of NATCOKE (coal char), CO_2, O_2, SYNGAS and ash; heat streams of QLOSE (heat loss of the gasifier) and QDECOMP (Heat of char cracking). The reaction blocks used are the RYIELD, RGIBBS and SSPLIT. Nonconventional material of coal char can crack to single element molecule in the RYIELD reactor and the cracking heat can be led into the RGIBBS reactor.

3 Results and discussion

3.1 Effect of O_2 flow rate on the gasification

The computation in the RGIBBS reactor is based on Gibbs free energy minimization method. SSPLIT module is used to simulate the separation of syngas and ash. The model is established under the assumptions that the gasifier runs

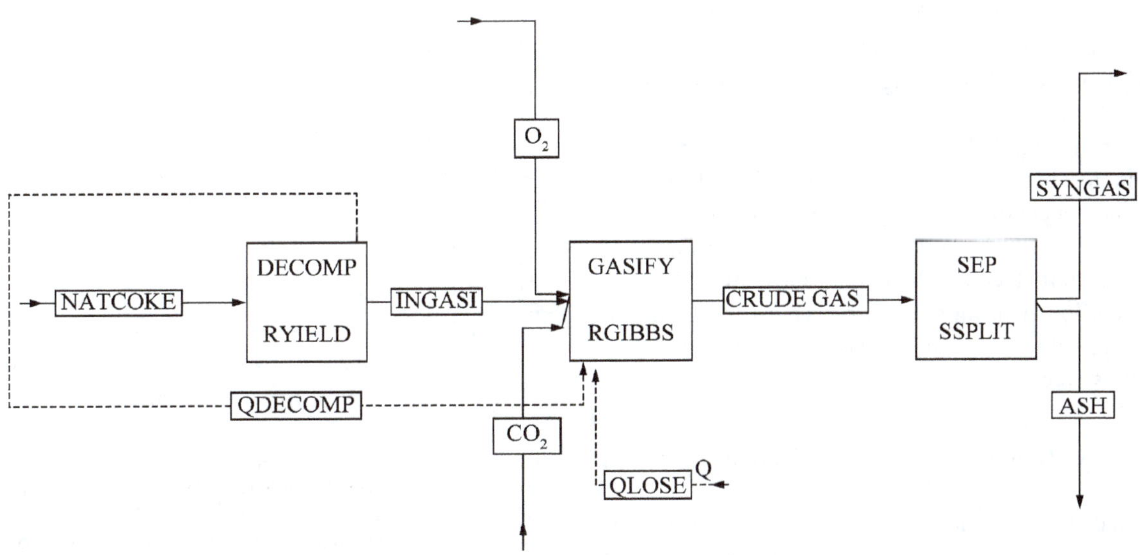

Fig. 2 Diagram of Aspen plus model

Table 1 Properties of coal char

Proximate analysis (%)				Ultimate analysis (%) (d)					Sulfur analysis/ % (d)			Heating value (MJ/kg)
M_{ad}	A_d	V_d	FC_d	C_d	H_d	O_d	N_d	S_d	S_p	S_s	S_o	
0.81	16.28	9.12	74.6	77.96	1.66	2.69	0.92	0.48	0.2	0.12	0.16	26.81

(ad-air dried base, *d*-dried base)

stably, no operating parameters change, the gasification agent and coal char particle can mix completely instantly, the elements hydrogen, oxygen, nitrogen and sulfur convert to gas phase totally except element carbon. There is no pressure drop in the gasifier, ash in the coal char is inert material that does not take part in gasification reaction, temperature distribution in coal particle is uniform, all the gas-state reactions are fast and attain equilibrium (Wang et al. 2004; Zhou et al. 2010).

Coal char used in this paper is from Xuzhou, China (Lin and Zhao 2012). The proximate analysis, ultimate analysis and sulfur analysis are shown in Table 1. O_2 and CO_2 are chosen as gasification agents. The handling capacity of coal char is 21.59 kg/h (The mass flow rates are set according to simulation requirement). Gasification pressure is 3 MPa when it keeps steady. Mass ratio of O_2, CO_2 to coal char is the key factor to the quality of syngas.

Other parameters keep steadily. The coal char and CO_2 mass flow rate are 21.59 and 2.16 kg/h respectively. The relationships among syngas composition, reaction temperature, lower heating value of syngas and O_2 content are shown in Fig. 3. The reaction temperature increases rapidly with the rising of O_2 content because of the increased combustion reaction. At the start of gasification there is not enough O_2, so imperfect combustion of char occupies a predominant position. As a result, molar ratio of CO increases gradually at first. Then the molar fraction of CO decreases because the complete combustion reaction increases as a result of more O_2 being injected into the gasifier. Combustion reactions of H_2 and CO increase with the rising of O_2 content, which leads to higher H_2O and CO_2 content.

As shown in Fig. 3 the concentration of CO increases with the increasing of O_2 flow rate, and then decreases. Although the concentration of H_2 decreases gradually, it is much lower than CO's. As a result, the concentration of CO is the leading factor, which promotes lower heating value and the lower heating value of syngas increasing to the maximum value and then decreasing with increasing of O_2 mass flow rate.

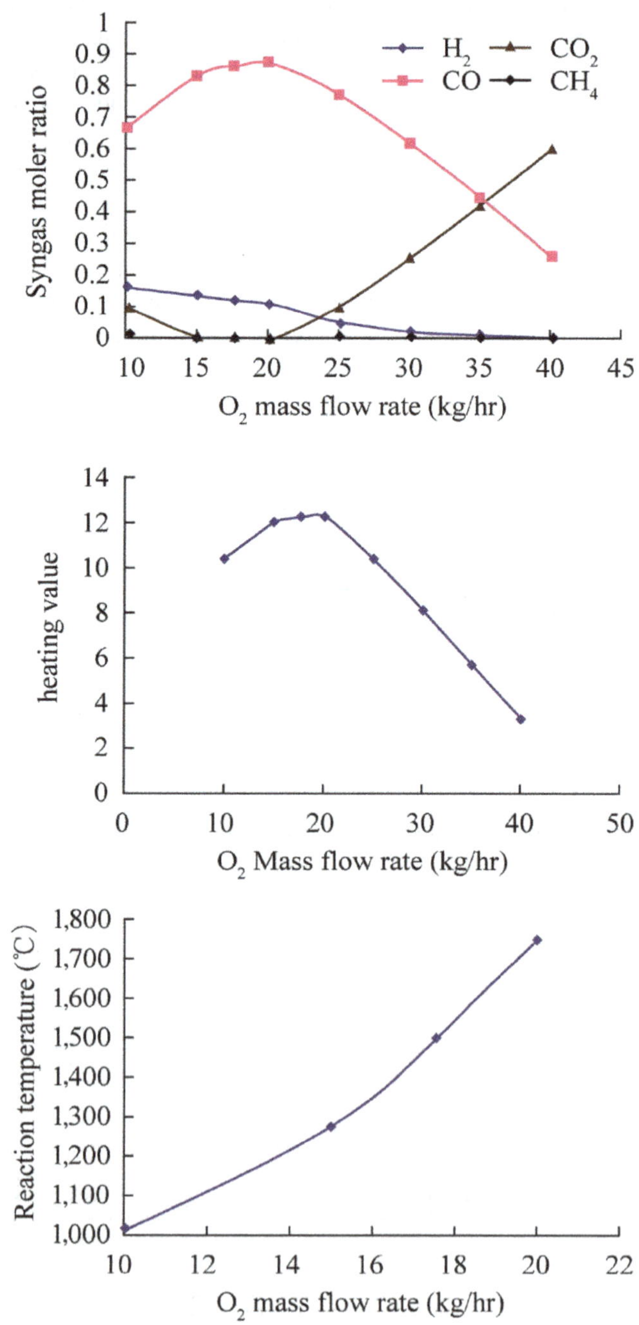

Fig. 3 Effects of O_2 content on syngas composition, reaction temperature and lower heating value of syngas (MJ/m^3). *Notes* The reaction temperature is much higher than 1,700 °Cwhen the O_2 mass flow rates are larger than 20 kg/h, which does not accord with the reality. As a result, the abscissa value in the last figure of Fig. 3 is no more than 20 kg/h

3.2 Effect of CO_2 flow rate on the gasification

As illustrated in Fig. 4 the reaction temperature decreases gradually with the increasing of CO_2 flow rate when the coal char and O_2 mass flow rates are 21.59 and 17.57 kg/h

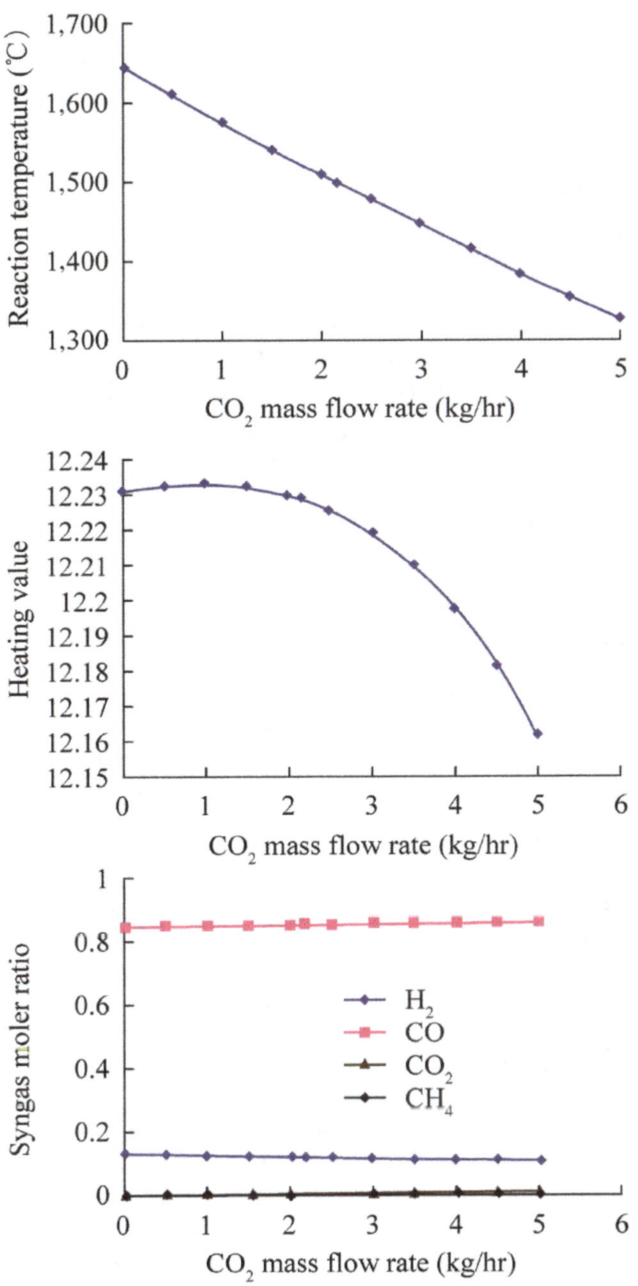

Fig. 4 Effect of CO_2 content on syngas composition, reaction temperature and lower heating value of syngas (MJ/m^3)

respectively. It decreases from 1645 °C to 1329 °C when CO_2 flow rate increases from 0 to 5 kg/h. Reaction temperature decreasing is caused by several factors listed below. First, diffusion rate of O_2 in the CO_2 atmosphere is slower with the increasing of CO_2 flow rate, which leads to slower combustion reaction. Second, the larger heat capacity of CO_2, increasing of CO_2 flow rate and keeping O_2 flow rate steady, leads to lower reaction temperature. Third, the Boudouard reaction rate is enhanced with the increasing of CO_2 flow rate and more heat is consumed, which leads to lower reaction temperature.

Reaction temperature decreasing is not favorable to Boudouard reaction. However, the increasing of CO_2 concentration promotes Boudouard reaction and its effect on Boudouard reaction is more dominant than the temperature. As a result, the molar ratio of CO increases with the rising of CO_2 flow rate. Reaction of coal char with H_2O is endothermic. Temperature decreasing is not favorable to the reaction. Temperature decreasing is good for water gas shift reaction and the production of CO_2 and H_2, but the concentrations of CO_2 and H_2 are very low. Therefore, the reaction of coal char with H_2O is more dominant in the production of H_2. As shown in Fig. 4, the concentration of H_2 decreases with the increasing of CO_2.

The reverse water gas shift reaction increases as a result of increasing of CO_2 and partly H_2 converts to H_2O. As a result, the lower heating value of syngas decreases slightly with the increasing of CO_2 mass flow rate as shown in the Fig. 4.

3.3 Effect of pressure on the gasification

The relationship of the pressure with gasification is shown in Fig. 5 when the coal char, O_2 and CO_2 mass flow rate are 21.59, 17.57 and 2.16 kg/h respectively. The reverse Boudouard reaction rate increases with the rising of pressure, which leads to slightly decreasing of CO molar ratio and increasing of heat and CO_2 content. The reaction temperature is high enough and the effect of pressure is not remarkable, but pressurizing can increase syngas production per unit time and production capacity. CO_2 content increasing is favorable to the reverse water gas shift reaction and the molar ratio of H_2 decreases. As a result, the lower heating value decreases slightly.

3.4 Effect of preheating temperature on the gasification

Effect of preheating temperature on the gasification with coal char, O_2 and CO_2 mass flow rates are 21.59, 17.57 and 2.16 kg/h is shown in Fig. 6. The preheating of O_2 and CO_2 can increase the reaction temperature and the lower heating value, but the effect of CO_2 preheating on reaction temperature and lower heating value is weaker than the O_2 preheating. The reaction temperature and the heating value increase by only 25 °C and 0.0063 MJ/M^3 respectively, when the preheating temperature of CO_2 increases from 300 to 1000 °C. However, the heat capacity of CO_2 is so high that it is unnecessary for high preheating temperature.

3.5 Effect of reaction temperature on the gasification

To obtain the effect of temperature on gasification, heat loss of gasifier is introduced in the gasification. The gasifier

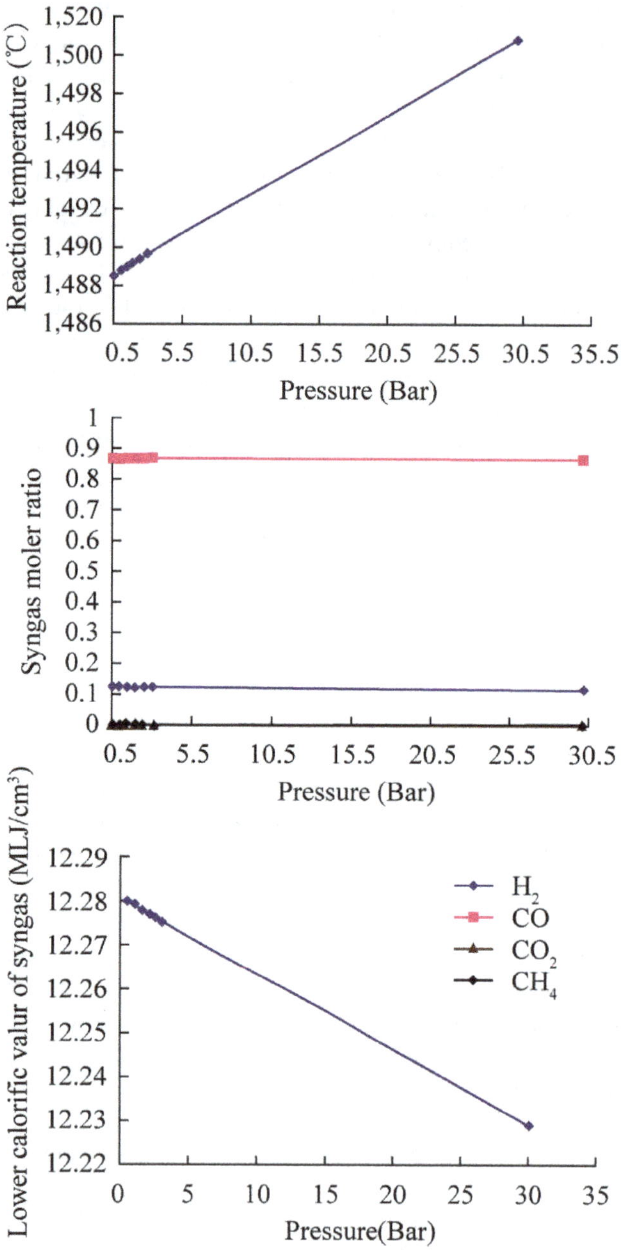

Fig. 5 Effect of pressure on syngas composition, reaction temperature and lower heating value of syngas (MJ/m^3)

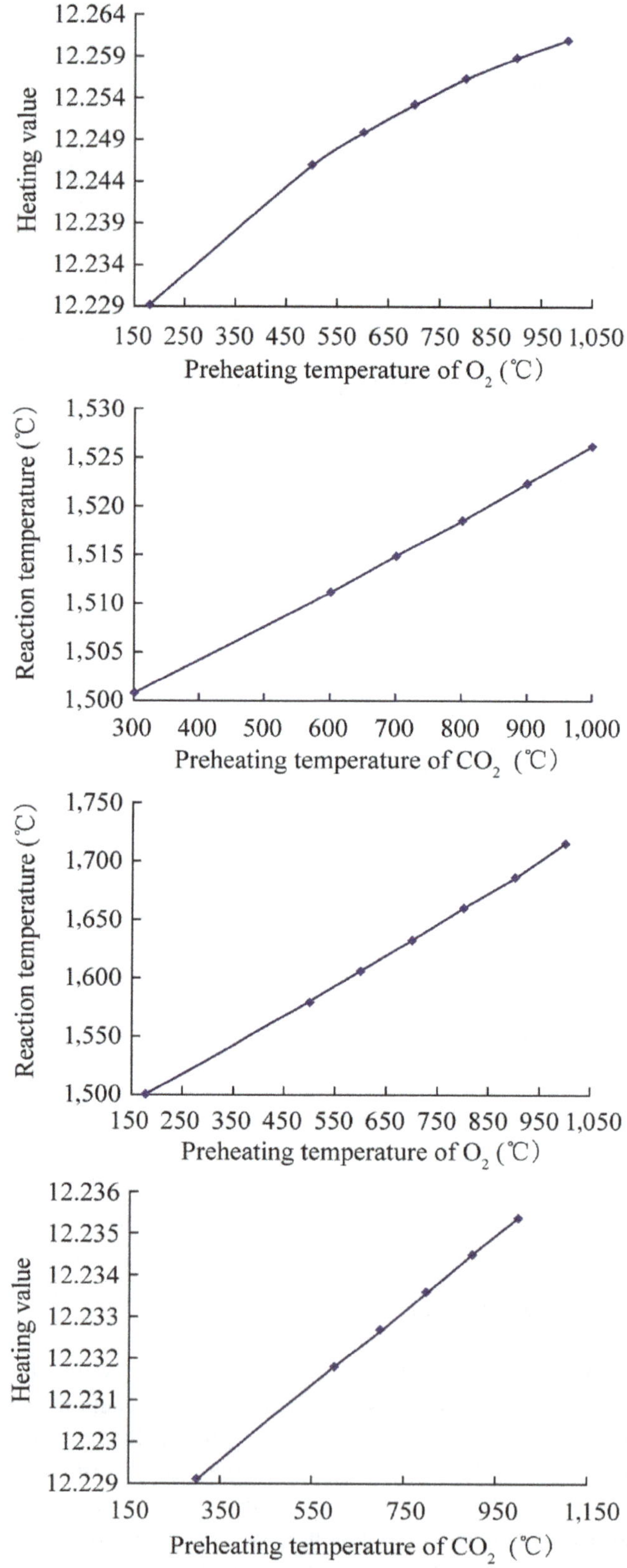

Fig. 6 Effect of preheating temperature on reaction temperature and lower heating value of syngas (MJ/m^3) *Notes* O_2 preheating temperature is much high because the O_2 is separated by OTM which is operated about 800 to 1000 °C

heat loss is the ratio of heat loss of the gasifier with heating value of coal. As coal char, O_2 and CO_2 mass flow rates are 21.59, 17.57 and 2.16 kg/h separately, the relationships between syngas composition, temperature, lower heating value and gasifier heat loss are illustrated in Fig. 7. Reaction temperature decreases as the heat loss increases from 0 % to 2 %, which leads to slower Boudouard reaction rate and lower molar fraction of CO. The H_2 content decreases because of the decreasing temperature and the reaction rate of char with H_2O.

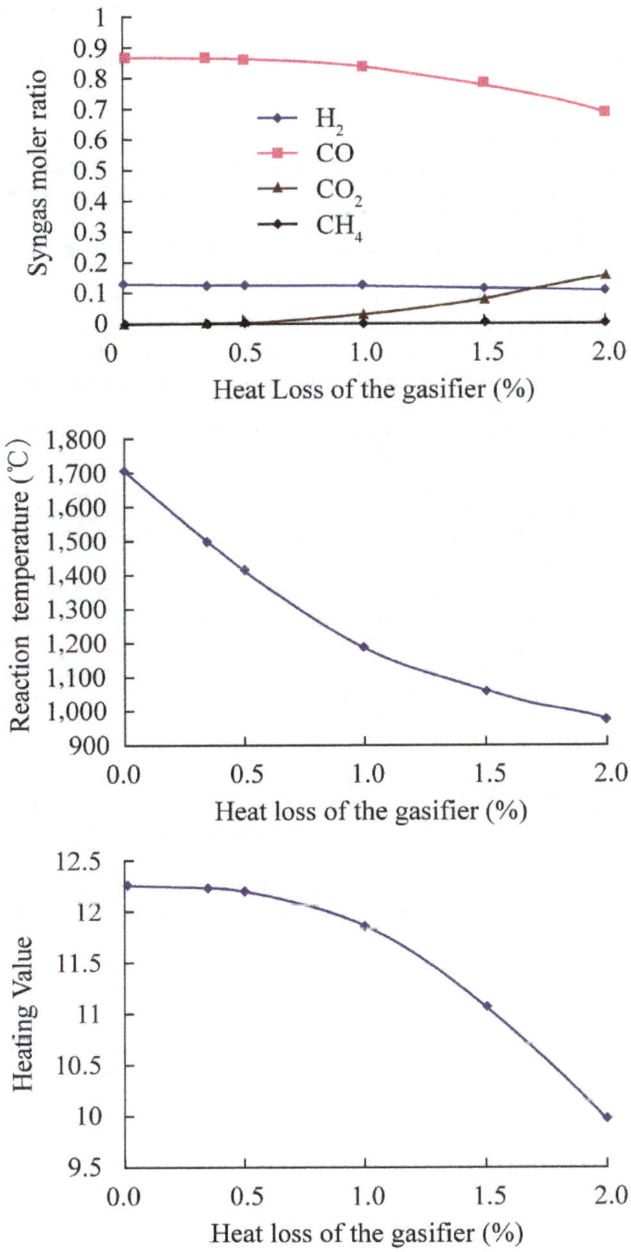

◀ **Fig. 7** Effect of heat losses of gasifier on syngas composition, reaction temperature and lower heating value of syngas (MJ/m^3)

3.6 Effects of different factors on the carbon conversion

Effects of O_2 and CO_2 mass flow rates, pressure, preheating temperature and heat loss on the carbon conversion were simulated. The results are shown in Table 2.

Carbon conversion increases with increasing of O_2 and CO_2 mass flow rates. Carbon combustion reactions are enhanced with the increasing of O_2 mass flow rates so carbon conversion increases greatly. We can get that the optimum O_2 mass flow rate is 20 kg/h for syngas heating value in Fig. 3. Increasing of CO_2 mass flow rates is good for Boudouard reaction and carbon conversion. Pressure, preheating and gasifier heat loss have also some effects on carbon conversion, but they are not apparent.

4 Conclusions

The coal char gasification using O_2/CO_2 based on IGFC was proposed. The thermodynamic analysis of the gasification under O_2/CO_2 atmosphere was studied using Aspen plus simulation method. The results are concluded as following.

(1) The molar ratio of CO increased gradually when the O_2 flow rate is lower than 20 kg/h. Then the molar fraction of CO decreases. H_2 molar ratio decreases gradually with the increasing of O_2 mass flow rate. The optimum lower heating value is obtained when the O_2 flow rate is 20 kg/h.

(2) The Mass flow rates of CO_2 has a significant effect on reaction temperature. The reaction temperature decreases from 1645 to 1329 °C when CO_2 flow rates increases from 0 to 5 kg/h because of the slower diffusion of O_2 in CO_2 atmosphere, larger heat

Table 2 Effect of different factors on carbon conversion

O_2 mass flow rates (kg/h)	10	15	17.5	20			
Carbon conversion	0.4343	0.7272	0.8537	0.9657			
CO_2 mass flow rates (kg/h)	0	0.5	1.5	2.16	3	4	5
Carbon conversion	0.8208	0.8286	0.8439	0.8538	0.8657	0.8791	0.8911
Pressure (bar)	1	2	3	4	5	30	
Carbon conversion	0.8581	0.8579	0.8578	0.8576	0.8574	0.8537	
O_2 preheating (°C)	100	200	400	600	700	800	900
Carbon conversion	0.8532	0.8538	0.8547	0.8555	0.8558	0.856	0.8563
CO_2 preheating (°C)	300	600	700	800	900	1000	
Carbon conversion	0.8537	0.8539	0.854	0.8541	0.8541	0.8542	
Heat loss of gasifier (%)	2.5	2	1.5	0.5	0		
Carbon conversion	0.8537	0.8545	0.8551	0.8561	0.8564		

capacity of CO_2 and promoting of Boudouard reaction. The syngas lower heating value decreases gradually with the increasing of CO_2 mass flow rates.

(3) The heating value is lower with the raising of pressure, but pressurize is favorable to reaction temperature. The molar ratios of CO and H_2 decreased slightly even under high pressure because of the high reaction temperature.

(4) Preheating of O_2 and CO_2 can both enhance reaction temperature and syngas heating value, but effect of CO_2 preheating is weaker than O_2 preheating.

(5) The carbon conversion increases with the increasing of O_2, CO_2 flow rates and O_2, CO_2 preheating temperature. Pressure is not favorable to the carbon conversion because it could inhibit the Boudouard reaction. The gasifier heat loss can reduce reaction temperature and then the carbon conversion decreases.

Acknowledgments We are grateful to the National Basic Research Program of China (No. 2012CB215404, 2012CB215406), the National Natural Science Foundation of China (No. 51261120378) for financial support of this work.

References

Alam MS, Wijayanta AT, Nakaso K, Fukai J (2012) Numerical investigation of syngas production from coal gasification under various CO_2/O_2 mixtures. Can J Comput Math Nat Sci Eng Med 3(4):88–97

Antonio C, Mara DJ (2004) Mild combustion. Prog Energy Combust Sci 30(4):329–366

Gleick PH, Adams RM, Amasino RM et al (2010) Climate change and the integrity of science. Science 328:689–690

Huang J (2012) Coal list and study on emission reduction policy. Doctor's Thesis, University of Fudan

Lin LS, Zhao CS (2012) Study of the simulation of natural coke-H_2O gasification reaction by using the software Aspen plus. J Eng Therm Energy Power 27(3):355–360

Mao ZQ (2010) Discussing about CO_2 emission and H_2 energy. Mod Phys 22(5):41–45

Peng SP, Han MF (2009) Development of coal/carbon based solid oxide fuel cell. Chin J Nat 31(4):187–192

Sun CS, Gao F (2013) Review of development of China's coal industry in 2012 and prediction of its development trend in 2013. China Coal 39(3):10–16

Toshiaki H, Hiroto S, Kinya S, Yusuke E (2009) Syngas production by woody biomass gasification with a CO_2/O_2 mixture. J Jpn Inst Energy 88:862–868

Toshiaki H, Shou H, Yusuke E (2013) Syngas production by CO_2/O_2 gasification of a aquatic biomass. Fuel Process Technol 116:9–15

Wang Y, Dai ZH, Yu GS, Yu ZH (2004) Simulation of entrained-flow bed coal gasifier by the method of Gibbs free energy minimization. Coal Convers 27(4):27–33

Wei W, Sun YH, Wen X, Sun NN (2011) Opportunities and challenges of carbon dioxide utilization as a resource. Chem Ind Eng Prog 30(1):216–224

Xie HP (2010) CO2 storage and climate change. Science and Technology Review 28(18)

Xie HP, Xie LZ, Wang YF, Zhu JY, Liang B, Ju Y (2012) CCU: a more feasible and economic strategy than CCS for reducing CO_2 emissions. J Sichuan Univ 44(4):1–5

Zhou JH, Chen XL, Guo Q, Wang YZ (2010) Process simulation of biomass and coal entrained flow co-gasification based on Aspen plus. Acta Energiae Solaris Sinica 31(9):1112–1116

PERMISSIONS

All chapters in this book were first published in IJCS&T, by Springer; hereby published with permission under the Creative Commons Attribution License or equivalent. Every chapter published in this book has been scrutinized by our experts. Their significance has been extensively debated. The topics covered herein carry significant findings which will fuel the growth of the discipline. They may even be implemented as practical applications or may be referred to as a beginning point for another development.

The contributors of this book come from diverse backgrounds, making this book a truly international effort. This book will bring forth new frontiers with its revolutionizing research information and detailed analysis of the nascent developments around the world.

We would like to thank all the contributing authors for lending their expertise to make the book truly unique. They have played a crucial role in the development of this book. Without their invaluable contributions this book wouldn't have been possible. They have made vital efforts to compile up to date information on the varied aspects of this subject to make this book a valuable addition to the collection of many professionals and students.

This book was conceptualized with the vision of imparting up-to-date information and advanced data in this field. To ensure the same, a matchless editorial board was set up. Every individual on the board went through rigorous rounds of assessment to prove their worth. After which they invested a large part of their time researching and compiling the most relevant data for our readers.

The editorial board has been involved in producing this book since its inception. They have spent rigorous hours researching and exploring the diverse topics which have resulted in the successful publishing of this book. They have passed on their knowledge of decades through this book. To expedite this challenging task, the publisher supported the team at every step. A small team of assistant editors was also appointed to further simplify the editing procedure and attain best results for the readers.

Apart from the editorial board, the designing team has also invested a significant amount of their time in understanding the subject and creating the most relevant covers. They scrutinized every image to scout for the most suitable representation of the subject and create an appropriate cover for the book.

The publishing team has been an ardent support to the editorial, designing and production team. Their endless efforts to recruit the best for this project, has resulted in the accomplishment of this book. They are a veteran in the field of academics and their pool of knowledge is as vast as their experience in printing. Their expertise and guidance has proved useful at every step. Their uncompromising quality standards have made this book an exceptional effort. Their encouragement from time to time has been an inspiration for everyone.

The publisher and the editorial board hope that this book will prove to be a valuable piece of knowledge for researchers, students, practitioners and scholars across the globe.

LIST OF CONTRIBUTORS

Paul T. Behum
Office of Surface Mining and Reclamation Enforcement, Alton, IL 62002, USA

Yoginder P. Chugh
National Academy of Inventors, Tampa, FL 33612, USA
Department of Mining and Mineral Engineering, Southern Illinois University Carbondale, Carbondale, IL 62901, USA

Liliana Lefticariu
Department of Geology, Southern Illinois University Carbondale, Carbondale, IL 62901, USA

Yi Luo
Department of Mining Engineering, West Virginia University, Morgantown, WV, USA

Deming Wang and Jianwei Cheng
College of Safety Engineering, China University of Mining Technology, Xuzhou, Jiangsu, China

Madhulika Dutta and Binoy K. Saikia
Polymer Petroleum and Coal Chemistry Group, Materials Sciences and Technology Division, CSIR-North East Institute of Science and Technology, Jorhat 785006, India

Madhulika Dutta and Durlov Saikia
School of Applied Sciences, Department of Chemistry, University of Science and Technology, Ribhoi 793101, Meghalaya, India

Madhulika Dutta
Department of Chemistry, CNB College, Bokakhat 785612, India

Puja Khare
Agrotechnology Division, CSIR- Central Institute of Medicinal and Aromatic Plants, Lucknow 226015, India

Sanchita Chakravarty
Analytical Chemistry Division, CSIR- National Metallurgical Laboratory, Jamshedpur 831007, India

Anna Krzyszowska Waitkus
Environmental Consulting, 80 Eagle Nest, Laramie, WY 82070, USA

Xiang Liu, Chaoqun Jiao and Aifen Yao
School of Electrical Engineering, Beijing Jiaotong University, Beijing 100044, China

Jiang Chang and Xiangxu Liu
School of Architecture and Design, China University of Mining and Technology, Xuzhou, China

Tinghao Hu
School of Mechanics and Civil Engineering, China University of Mining and Technology, Xuzhou, China

Xiaogeng Ren
School of Architecture, Southeast University, Nanjing, China

M. Onifade and B. Genc
The School of Mining Engineering, University of the Witwatersrand, Johannesburg 2050, South Africa

Huigui Li and Huamin Li
School of Energy Science and Engineering, Henan Polytechnic University, Jiaozuo 454000, Henan, People's Republic of China
Opening Project of Key Laboratory of Deep Mine Construction, Henan Polytechnic University, Jiaozuo 454000, Henan, People's Republic of China

Yu Liu and Yanming Zhu
Key Laboratory of Coal Methane Resource & Reservoir Formation Process of Ministry of Education, China University of Mining & Technology, Xuzhou 221008, Jiangsu, China
School of Resource and Earth Science, China University of Mining & Technology, Xuzhou 221116, Jiangsu, China

Ryszard Hejmanowski
AGH University of Science and Technology, Cracow, Poland

Yinli Bi, Yanxu Zhang and Hui Zou
State Key Laboratory of Coal Resources and Safe Mining, China University of Mining and Technology (Beijing), Beijing 100083, China

Amir Saffari, Farhang Sereshki and Mohammad Ataei
Faculty of Mining, Petroleum and Geophysics Engineering, Shahrood University of Technology, Daneshgah Blvd., Shahrood, Iran

Keramat Ghanbari
HSE Management of Eastern Alborz Coal Mines Company, Shahrood, Iran

Haifeng Hu and Xugang Lian
College of Mining Engineering, Taiyuan University of Technology, Taiyuan 030024, China

Linlin Cheng
Department of Surveying and Land Use, School of Geoscience and Surveying Engineering, China University of Mining and Technology (Beijing), D11 Xueyuan Road, Haidian District, Beijing 100083, China

Jeffrey G. Skousen
Division of Plant and Soil Sciences, West Virginia University, Morgantown, WV 26506-6108, USA

W. R. Reed, Y. Zheng and M. Yekich
Pittsburgh Mining Research Division, Centers for Disease Control and Prevention, National Institute for Occupational Safety and Health, 626 Cochrans Mill Rd., Pittsburgh, PA 15236, USA

G. Ross
J.H. Fletcher, Inc., Huntington, WV, USA

A. Salem
Marshall University, Huntington, WV, USA

Rimi Das and Subodh Kumar Maiti
Ecological Restoration Laboratory, Department of Environmental Science and Engineering (Centre of Mining Environment), Indian School of Mines, Dhanbad, Jharkhand 826 004, India

Lihui Sun, Haoyuan Wu, Bensheng Yang and Qiangqiang Li
College of Resource, Hebei University of Engineering, Handan 056038, China
Collaborative Innovation Center of the Comprehensive Development and Utilization of Coal Resource, Handan 056038, China

Yong Cui, Jie Liang, Zhangqing Wang, Xiaochun Zhang, Chenzi Fan and Xuan Wang
School of Chemical and Environmental Engineering, China
University of Mining & Technology (Beijing), Beijing 100083, China

Hui Fan and Minfang Han
Union Research Center of Fuel Cell, School of Chemical & Environment Engineering, China University of Mining & Technology (Beijing), Beijing 100083, China

Tian Li, Shi-Feng Dai, Jian-Hua Zou, Sen Zhang, He-Hing Tian and Li-Xin Zhao
Chongqing Key Laboratory of Exogenic Mineralization and Mine Environment, Chongqing Institute of Geology and Mineral Resources, Chongqing 400042, China

Haibin Li, Yu Yu, Minfang Han and Ze Lei
School of Chemical and Environmental Engineering, China
University of Mining and Technology, Beijing 100083, China

Xiaochun Li and Zhi-ming Fang
State Key Laboratory of Geomechanics and Geotechnical Engineering, Institute of Rock and Soil Mechanics, Chinese Academy of Sciences, Wuhan 430071, China

Xiaoming Ni
State Key Laboratory Cultivation Base for Gas Geology and Gas Control, Henan Polytechnic University, Jiaozuo 454000, China

Quanzhong Li
School of Energy Science and Engineering, Henan Polytechnic University, Jiaozuo 454000, China
State Key Laboratory Cultivation Base for Gas Geology and Gas Control, Henan Polytechnic University, Jiaozuo 454000, China

Wenxue Chen
School of Civil Engineering, Henan Polytechnic University, Jiaozuo 454000, China

Index

www.ingramcontent.com/pod-product-compliance
Lightning Source LLC
LaVergne TN
LVHW080906210926
848485LV00004B/605